JN296045

6カ国語
ばね用語事典

英語／日本語／中国語／インドネシア語／韓国語／タイ語

社団法人 日本ばね工業会　編

日本規格協会

On the Occasion of the Publication of Six Language Spring Dictionary

I would like to congratulate JSA and all the experts who contributed to the production of this Technical Dictionary on springs.

These components are vital elements for many industrial equipments and numerous appliances used by the end consumer, both increasingly subject to global trade. The technical characteristics of springs determine their durability and safety, issues which are in the forefront of current concerns. Quality requirements from manufacturers and buyers become therefore more precise and targeted. Having a compendium of terms and definitions is all the more necessary to support good technical and business relations.

Standardization begins with terminology, so this dictionary should certainly be a welcomed contribution to future regional cooperation and international work. International standardization is a bottom up approach and can only benefit from national initiatives and experiences.

By organizing the development and publication of this dictionary, opening them to regional input and making sure that its content draws from industrial expertise, JSA and its partner organizations have demonstrated once more Japan's ability to contribute and innovate in the area of standardization. No doubt that this work will be favorably received by the world community.

August 2004

Alan Bryden
ISO Secretary-General

発刊にあたって

　ばねに関する専門的な事典の編集・制作に貢献したすべての専門家及び日本規格協会に祝辞を述べたい．

　これらの構成部品は，多くの工業用機器及び最終消費者によって使用される多くの機器にとって不可欠な部品であり，その国際的貿易・取引は，増加の一途にある．ばねの技術的な特性は，機器の耐久性，安全性，及び現在における最前線の技術仕様を決定することになる．したがって，製造者や購入者からの品質要求事項はより明確で，目標的なものになる．ばね及びこれに関連する用語と定義を取り纏めた本事典は，技術的にも，ビジネス上での良い相互関係を支援するためにも是非とも必要なものである．

　標準化は，用語から始まる．したがって，本事典は将来の地域協力及び国際的な取組みに対して，多大な貢献が期待される．国際標準化はボトムアップのアプローチであり，国家の自発性・主導権及び経験によってのみ恩恵を受けることができる．

　産業界の専門的知見をもとに記述された本事典の刊行に至る日本規格協会及び協力団体の尽力は，標準化の分野における貢献と革新のために今一度，日本の実力を示したと言える．この業績は，世界で好意的に受け入れられるだろうことを信じて疑わない．

2004 年 8 月

国際標準化機構（ISO）　事務総長
アラン・ブライデン

序　文

　海外との商取引が盛んな今日，技術業務の国際化は成熟期にある．我が国でも，多数の企業が生産拠点を欧米やアジアにシフトして活発に活動している．産業技術の国際標準化の果たす役割は，その重要度を増している．

　このような状況下で，各種専門用語が品質の維持・向上，技術交流等に果たす役割は非常に大きく，ばね用語も例外ではない．各種技術文書を構成するに必須な専門用語に関するグローバルな視点が欠かせないものとなっているのである．

　そこで，機械要素の中でも特に重要な「ばね」に関する用語を収録した事典を発刊する運びとなった．

　本事典では，約2,000のばね用語を厳選し，収録している．見出し用語として英語，日本語（JIS），中国語，インドネシア語，韓国語，タイ語の6カ国語をアルファベット配列に掲載するとともに，英語及び日本語の定義を収録している．さらに巻末には，各国語相互の検索が可能となる各国語別の索引を収録し，利用の便を図っている．

　本事典の刊行が，ばね用語のさらなる国際標準化推進の契機となることを切に願うものである．

　本事典の編纂は，（社）日本ばね工業会，（社）自動車技術会，（社）日本自動車部品工業会，ばね技術研究会等，関係団体の賛意並びに専門的な知見を得るとともに，中国，韓国，インドネシア，タイ等，多くの関係者の参画により推進されたものである．関係各位の多大なご尽力に対し，敬意を表するとともに心から感謝申しあげる．

　2004年8月

財団法人日本規格協会
理事長　坂　倉　省　吾

Forward

Business transaction across national borders is active, and internationalization of technical activities is blossoming today. Many of Japanese companies have moved their production bases to overseas such as Asia, Europe and the United States. And international standardization is now playing an increasingly important role than ever.

Under these circumstances, technical terminology is extremely important in quality maintenance and improvement, technology exchange, and terms of springs are no exception in that. The global perspective on technical terms is indispensable.

Therefore, we decided to compile a dictionary of spring terms, which are the important elemental parts hold up the foundation of industrial technology.

This dictionary includes about 2,000 highly-selected terms of springs. The direction words are in six languages, English, Japanese (JIS), Chinese, Indonesian, Korean, and Thai, and arranged in alphabetical order. Definitions are written both in English and Japanese. It also has index pages for each language at the end of this book for your convenience.

I sincerely hope that this dictionary be a chance to promote international standardization on springs.

I am grateful to Japan Spring Manufacturers Association, Society of Automotive Engineers of Japan, Inc., Japan Auto Parts Industries Association, Japan Society for Spring Research, and other concerned bodies that provided us with highly-specialized knowledge.

I would like to acknowledge the many contributors from China, Korea, Indonesia, Thailand, and others involved in the compilation of this dictionary.

August 2004
Shogo Sakakura
Japanese Standards Association - President

発刊にあたって

　21世紀に入り，急速なグローバル化が進展し，『地球は小さくなった』と実感することが多くなった．「ばね」の分野においても，海外との輸出入取引や技術交流などがますます活発になり，なおさらこの感が深い．

　「ばね」は，自動車，鉄道車両，航空機，電子機器，家庭電器，産業機械など，あらゆる産業の基盤を支える重要な機械要素であることを考えれば，海外と円滑に，明確に，意思の疎通を図るために，ばね関連用語を整理・体系化することの必要性が一段と高まってきている．

　さて，このたび発刊の運びとなった「6カ国語ばね用語事典」は，最新のJIS用語をはじめ現場で使われる用語から2000語を精選し，その用語の定義を和文・英文で併記した．さらに一部の用語には説明図を添えるなど，画期的な編纂内容とした．用語としては定義なしで中国，韓国，タイ，インドネシア語を収録し，6カ国語事典となっている．

　本事典の企画は当工業会の標準化会議メンバーが担当したが，国際標準化の流れに沿って作業を進めるべく，編集委員に(財)日本規格協会，(社)自動車技術会，(社)日本自動車部品工業会，ばね技術研究会の他，中国，韓国，タイ，インドネシアなど多くの関係者に参画をお願いした．

　編集作業の基本方針として，(財)日本規格協会のご指導を仰ぎながら，
　① わが国の工業標準化の方針に基づいて，JISばね用語の見直しをすること
　② 将来のISO規格として採用可能な原案を開発して，制定を図ること
という二つの理念を常に念頭においた．

　一方，欧州ばね工業会連合会（ESF）や米国ばね工業会（SMI）との間で，国際標準化を進めるための土壌づくりや，東アジア地域における主要国の規格制定機関との連携強化も，視野にいれている．

　そういう意味において，本事典が，国際標準づくりの先駆者的な役割を果たしうるであろうと，自負している次第である．

　本書の翻訳作業のために多数の方々にご尽力いただいた．中国語については中華人民共和国の機械科学研究院の王徳成副院長並びに全国弾簧標準化委員会

の皆様，上海理工大学の陳康民学長ならびに陳抱雪教授，韓国語については大韓民国の大圓鋼業(株)，許丞鎬社長ならびに李元副社長，インドネシア語についてはインドネシアの PT. CHUUHATU INDONESIA 社，梶原勇介社長ならびに ARIFIN マネージャー，タイ語についてはタイの日本発条（泰国）有限公司，大森義憲社長ならびに SOMCHAI 工場長など，関係者の多大なご努力とご協力に感謝したい．

　最後に，本事典の編集・出版にあたって，全面的なご指導ご協力をいただいた(財)日本規格協会に対して，心からの感謝を申し上げたい．

　なお，本書に使われた図版は(財)日本規格協会(JSA)，英国規格協会(BSI)，及び米国ばね工業会(SMI)のご好意によって現行ばね規格から転載した．ここに感謝の意を表する．

2004 年 7 月

社団法人日本ばね工業会

会長　前 田 次 啓

In Issuing Technical Dictionary of Springs

The rapid progress of globalization in the 21st century makes us feel that "the globe has become smaller". In the field of springs, an accelerated increase in production transfer and technology exchange as well as export and import causes us to feel so much more.

In such a situation, the fact that springs are important machine elements supporting bases for automobiles, rolling stocks, airplanes, electronic equipment, electric appliances, and industrial machinery necessitates the systematization of spring terms in order for us to strive for smoother and clearer mutual communication with overseas clients.

6 Language Spring Dictionary is a JSMA's epoch-making compilation. It contains 2,000 entries which are terms carefully selected from updated Japanese Industrial Standard (JIS) and shop floor jargon. The definitions are given both in Japanese and English with an illustration attached to some of the entries. Terms in Chinese, Korean, Thai, and Indonesian are also provided without definitions, thereby this dictionary is regarded as a six-language spring dictionary.

Our Standardization Committee as a planner for the compilation determined to tackle the work in line with the ongoing international standardization activities, and requested people concerned from a wide range of fields to take part in the project. They are people from Japanese Standards Association, Society of Automotive Engineers of Japan, Inc., Japan Auto Parts Industries Association, and Japan Society for Spring Research in Japan, and people from the People's Republic of China, the Republic of Korea, the Kingdom of Thailand, and the Republic of Indonesia abroad. Under the guidance of Japanese Standards Association, we kept in mind two principles:

① Spring terms in JIS should be reviewed under the government's policy in industrial standardization.

② The draft which could be adoptable as ISO standards should be developed and established.

On the other hand, we take into consideration that the foundation of an international standardization body among ESF, SMI, and JSMA is essential, and a closer connection with standard organizations in major East Asian countries is important. In this regard, we feel proud that this publication may play a pioneering role in the development of international standards.

We take this opportunity of expressing our gratitude to many participants for their great efforts and cooperation in performing translation jobs: Professor De Cheng Wang at China Academy of Machinery & Technology and members from China Standardization Technical Committee; President Kang-Min Chen and Professor Baoxue Chen at University of Shanghai for Science and Technology; Mr. Hur Seung Ho, President and Mr. Lee Byung Hoon, Adviser of DAE WON KANG UP CO., LTD. (for Korean); Mr. Yusuke Kajiwara, President and Mr. Arifin, Factory Manager of PT. CHUHATSU INDONESIA; Mr. Yoshinori Omori, President and Mr. Somchai Pimpisaes, Factory Manager of NHK SPRING (THAILAND) CO., LTD..

Last but not least, we wish to express our sincere thankfulness to people from Japanese Standards Association for their unreserved help in the compilation and publication of this dictionary.

Acknowledgement:

The illustration in this dictionary are reprinted from published standards.

We thank JIS (Japanese Standards Association), BSI (British Standards Institution), and SMI (Spring Manufacturers Institute) for their courtesy.

July 2004

T. Maeda

Tsuguhiro Maeda
President
Japan Spring Manufacturers Association

前　言

在继承原日文、泰文、印度尼西亚文、中文和英文五国语言《弹簧用语辞典》的优点与经验的基础上，日本弹簧工业会组织有关国家的弹簧组织、专家与学者及时吸纳相关国家弹簧标准最新术语和弹簧工业的先进技术，完成了包含韩文在内的六国语言《弹簧用语辞典》的编写与出版工作。这是日本弹簧工业会及所有编者对亚洲乃至世界弹簧工业的新贡献，是亚洲弹簧制造技术的集中展示，更是参编国家现代文化的交融和最新弹簧技术的结晶。作为中国弹簧工作者的代表深表欣喜和欢迎。

21世纪的今天，经济全球化日益深化，全球制造业大分工的格局已经呈现，东南亚已成为全球的制造中心之一。考虑弹簧工业伴随制造业共同成长的特征，东南亚也必将成为全球弹簧产品的制造基地与应用中心。跨国越区的弹簧设计、制造和贸易离不开弹簧用语的交流与沟通。回顾过去，是东南亚经济发展的呼唤诞生了多国语言的《弹簧用语辞典》。展望未来，六国语言《弹簧用语辞典》将促进东南亚的经济走向新的繁荣。进而，具备区域标准用语特征的《弹簧用语辞典》也将为国际弹簧用语标准的诞生奠定重要的基础。

全国弹簧标准化技术委员会是中国国家标准化管理委员会授权归口管理弹簧产品标准化工作的技术组织，中国机械科学研究院承担着弹簧标委会的秘书处工作。40余年的标准化历程，使中国的弹簧工业已拥有22项国家标准、24项机械行业标准，形成了完备的弹簧标准体系。

承蒙日本弹簧工业会前田次啓会长的委托，全国弹簧标准化技术委员会组织了由主任委员、委员（姜膺　弹簧标委会秘书长、曹辉荣　无锡泽根弹簧有限公司副总经理）、专家学者（张英会　北京科技大学教授）组成的审校工作组，完成了《弹簧用语词典（六国语言）》汉语部分的标准化审校工作，使《弹簧用语词典（六国语言）》的汉语表述与现行标准的弹簧术语相一致，与弹簧工业的惯用理解相对应。由于《弹簧用语词典（六国语言）》是一部多国语巨著，相应汉语表述的不妥之处恳望随时指正。

全国弹簧标准化技术委员会主任委员
中国机械科学研究院副院长
王　德　成

前 言

　　由日本弹簧工业会主持和组织编写的《弹簧用语词典（六国语言）》是在原《弹簧用语词典（五国语言）》的基础上，修正了部分用词的表述，新增了近三百条弹簧专业用语后，汇编而成。很大程度地反映了现代弹簧工业的发展和弹簧技术的进步。语言种类方面，在原来的日文、泰文、印度尼西亚文、中文和英文五种语言上，加入了韩文表述，以求适应近年来东南亚诸国在弹簧工业发展方面的需求。

　　弹簧及其应用技术在现代工业的几乎所有行业都发挥着十分重要的作用。伴随着21世纪东南亚地区高速的经济发展和工业进步，国与国之间、区域与区域之间的经济合作和技术交流已经成为不可抗拒的趋势，有理由相信这种趋势将进一步发展壮大，形成潮流，推动社会的进步。无疑本词典的问世将为促进东南亚地区的弹簧工业进步和技术交流做出积极的贡献。

　　弹簧工业有着悠久的发展历史，它的每一个进步和突破都凝聚了包括材料工业、化学工业、冶金工业、机械制造工业乃至电子工业等方面的技术成果，这个特征和倾向在工业技术高度发展的21世纪必将更加突出。技术开拓和科技创新过去是、今后也仍然是弹簧工业保持蓬勃发展的关键。

　　上海理工大学是一所以工科为主，兼有管理、经济、文科和理科的多学科性大学，有着近百年的发展历史。在步入新世纪的今天，我们正致力于培养具有旺盛的创新意识、科技开拓能力强和道德品行高尚的一大批面向21世纪的青年人才。我们诚挚地希望我校培养的年轻人才将来在包括弹簧工业在内的各行各业发挥生力军作用，更多更好地为人类社会的不断进步做出贡献。

　　蒙日本弹簧工业会的信任，我校受邀参加本词典的编写工作是一次宝贵机遇，特向前田次启会长为首的日本弹簧工业会致以衷心的感谢。本词典的中文部分的用语修订和新增用词的翻译工作由我校陈抱雪教授执笔，恐有谬误之处，恳望读者指正。

上海理工大学 校长

陈 康 民 教授

2004 年 7 月

Kata Pengantar

Edisi perdana kamus 4 bahasa, yang memuat bahasa Inggris, Jepang, Cina dan Korea di terbitkan pada tahun 1995. Setelah itu kamus ini berkembang menjadi kamus 6 bahasa, dengan penambahan bahasa Thailand dan bahasa Indonesia.

Waktu penyusunan kamus tersebut bersamaan dengan krisis ekonomi bersejarah, yang melanda Asia hingga menyebabkan hilangnya percaya diri negara di Asia terhadap masa depan.

Edisi terbaru ini yang didasarkan pada kamus 6 bahasa sebelumnya, dilengkapi penjelasan dalam bahasa Inggris yang menjadikan kamus ini lebih padat dan lengkap. Kami yakin kamus ini sangat bermanfaat bagi orang-orang yang bergelut dalam dunia pegas.

Kamus ini diharapkan dapat membantu teknisi muda Indonesia yang sedang mempelajari teknologi pegas yang tumbuh dan berkembang di Jepang, Eropa dan Amerika. Semoga kamus ini dapat memberikan manfaat pada perkembangan teknologi pegas di Negara Republik Indonesia.

Kami sangat berbahagia ikut serta menyusun kamus dunia pegas dalam berbagai bahasa negara-negara Asia, negara-negara yang sedang mengembangkan teknologi dan ekonomi global.

Kami ucapkan banyak terima atas kerjasama yang tak ternilai kepada Badan Standarisasi National dan semua pihak yang ikut dalam penyusunan kamus ini.

July, 2004
P.T. CHUHATSU INDONESIA
Yoshitaka Taniguchi

머 리 말

오늘날 눈부시게 발전하는 산업기술은 기계요소 중의 하나인 스프링 분야도 예외가 아닙니다. 스프링부문에 있어서도 설계, 제조, 사용에 이르기까지 여러 부문에서 국제적인 협력이 가속화되고 있는 이 때 이번 6개국 스프링용어 사전을 발간하게 되어 매우 기쁘게 생각합니다.

이번 스프링 용어 사전은 일본 스프링공업회에서 1995년에 발간한 바 있는 영어, 일본어, 중국어, 한국어의 4개국어를 포함하여 이번에 다시 타이, 인도네시아어를 포함하는 6개국 스프링 용어사전으로 확대 · 편찬하게 된 것입니다. 따라서 스프링 생산업계는 물론, 기계공학, 금속공학 관계의 제현과 학생, 그리고 스프링을 사용하는 모든 분야에 유용하게 활용될 수 있다고 생각됩니다. 또한 세계화 시대에 상대국의 용어를 이해하는데 큰 도움을 줄 것으로 확신해마지 않습니다. 용어에 있어서는 한국공업규격에 규정된 스프링 용어를 기본으로 하여 기계 · 금속 용어사전을 활용하였고, 여기에 수록되지 않은 용어는 원어인 영어를 한국어로 표현한 용어, 오랫동안 스프링 제조 현장에서 사용되어 온 용어도 있습니다. 이번 한글용어 작업에는 또 기편찬된 바 있는 4개 국어 스프링 용어사전에서의 오류를 최대한 바로잡고자 노력하였음에도 불구하고, 아직도 미진한 부분이 있을 것으로 사료되며 독자들의 관심과 질책을 당부하는 바입니다.

아무튼 이 국제적인 스프링 용어사전이 동북아는 물론 동남아시아에 이르기까지 스프링 기술을 향상시키는데 큰 역할을 할 것으로 믿어 의심치 않으며, 앞으로 유럽까지 더욱 확대된 세계 스프링 용어사전이 되기를 기대해 봅니다. 이번 용어집의 한국어편은 지난 번 책자와 마찬가지로 저희 회사가 담당하게 되어 큰 영광과 함께 무거운 사명감으로 편찬작업에 임하였으며, 이러한 작업을 수행함에 있어서 한국표준협회의 유영상 회장님의 협조에 감사를 표합니다.

2004년 7월
대원강업주식회사
대표이사 허 승 호

คำนำ

สภาพเศรษฐกิจของประเทศไทยในปัจจุบัน ได้ฟื้นตัวจากภาวะสับสนอันเนื่องมาจากการลดค่าเงินบาท ในเดือนกรกฎาคม ปี 1997 (พ.ศ. 2540) เป็นอย่างมาก ส่งผลให้ยอดผลิตรถยนต์ในประเทศพุ่งขึ้นสู่ 1 ล้านคัน ในปี 2007 (พ.ศ. 2550)

อีกทั้งจากความชัดเจนของการที่ ผู้ผลิตรถยนต์ทั้งหมด ได้รุกหน้าเสริมสร้างความแข็งแกร่ง ผลักดันด้านการออกแบบ และพัฒนาผลิตภัณฑ์ เพื่อเป็นการรองรับการให้ประเทศไทยเป็นฐานการผลิตรถยนต์เพื่อการส่งออก ในช่วงเวลาเดียวกันนี้ การที่สมาคมผู้ผลิตสปริงแห่งประเทศญี่ปุ่น Japan Spring Manufacturers Association ได้พิจารณาจัดทำ "พจนานุกรมศัพท์ช่าง 6 ภาษา สำหรับการผลิตสปริง" ขึ้น ซึ่งครั้งนี้ได้เพิ่มภาษาเกาหลีขึ้นอีก 1 ภาษานั้น (จากฉบับเดิมคือ"พจนานุกรมศัพท์ช่าง 5 ภาษา สำหรับการผลิตสปริง") นับว่าเป็นสิ่งที่แสดงให้เห็นอย่างชัดเจน ถึงความคาดหวังที่จะพัฒนาภูมิภาคอาเซียนให้มีความแข็งแกร่งยิ่งขึ้น

"พจนานุกรมศัพท์ช่าง 6 ภาษา สำหรับการผลิตสปริง" ฉบับใหม่นี้ ยังถูกจัดทำขึ้นตามระบบมาตรฐานสากล สำหรับให้ผู้คนส่วนใหญ่ใช้กันได้อย่างกว้างขวางทั่วโลก สำหรับงานแปลพจนานุกรมศัพท์ช่างฯเป็นภาษาไทยนั้น บริษัท เอ็น เอช เค สปริง (ประเทศไทย) จำกัด รู้สึกเป็นเกียรติเป็นอย่างยิ่งที่ได้รับการพิจารณาให้เป็นผู้รวบรวมจัดทำ

สุดท้ายนี้หวังอย่างยิ่งว่า อย่างน้อยพจนานุกรมเล่มนี้จะเป็นประโยชน์ ต่อการพัฒนาภูมิภาคเอเชียอันรวมถึงประเทศไทยด้วย และขอขอบคุณสมาคมผู้ผลิตสปริงแห่งประเทศญี่ปุ่น รวมถึงสำนักงานมาตรฐานผลิตภัณฑ์อุตสาหกรรมประเทศไทยเป็นอย่างสูง ที่ให้โอกาสแก่บริษัทฯอีกครั้งหนึ่ง

Yoshinori Omori
ประธานบริษัท
บริษัท เอ็น เอช เค สปริง (ประเทศไทย) จำกัด
กรกฎาคม 2004

6 カ国語ばね用語事典　編集委員名簿

委員長	金澤	健二	中央大学
副委員長	相羽	繁生	株式会社東郷製作所
幹事	田部	隆幸	日本発条株式会社
	熊澤	信雅	中庸スプリング株式会社
委員	小河	雄二	株式会社東郷製作所
	甲斐	和憲	株式会社ホリキリ
	加藤	功	株式会社東郷製作所
	小島	克己	社団法人日本自動車部品工業会
	坂口	辰雄	東海バネ工業株式会社
	流石	一郎	日本発条株式会社
	島田	英男	社団法人自動車技術会（富士重工業株式会社）
	髙橋	悌郎	株式会社パイオラックス
	竹内	康晃	中央発条工業株式会社
	谷口	義孝	中央発條株式会社
	都築	章雄	株式会社東郷製作所
	中	靖彦	特殊発條興業株式会社
	早坂	善広	三菱製鋼株式会社
	牧野	正	株式会社タカホ製作所
	宮本	正己	東洋発條工業株式会社
	村松	達也	中央発條株式会社
	山下	俊明	サンコールエンジニアリング株式会社
	吉永	洋一	サンコール株式会社
	桑原	敏夫	社団法人日本ばね工業会
	栗原	義昭	社団法人日本ばね工業会
	瀬下	和正	社団法人日本ばね工業会
	加山	英男	財団法人日本規格協会
	黒田	元信	財団法人日本規格協会
	石川	健	財団法人日本規格協会
海外委員	王	德成	机械科协研究院／中国全国弹簧标准化技术委员会
	姜	膺	中机生产力促进中心／中国全国弹簧标准化技术委员会
	曹	辉荣	无锡泽根弹簧有限公司／中国全国弹簧标准化技术委员会
	张	英会	北京科技大学／中国全国弹簧标准化技术委员会
	陈	抱雪	上海理工大学
	梶原	勇介	Chuhatsu–indonesia
	Heru WIBOWO		中央発條株式会社
	李	秉勲	大圓鋼業株式会社
	金	基銓	大圓鋼業株式会社
	加藤	忠一	HHK SPRING (THAILAND) Co., Ltd
	SOMCHAI PIMPISAES		HHK SPRING (THAILAND) Co., Ltd
	PEANGRUDEE DHIR		HHK SPRING (THAILAND) Co., Ltd

6 Language Spring Dictionary Staff

Chief	Kenji Kanazawa	Chuo University
Deputy Chief	Shigeo Aiba	Togo Seisakusyo Corporation
Senior Staff	Takayuki Tabe	NHK Spring Co., Ltd.
	Nobumasa Kumazawa	Chuyo Spring Co., Ltd.
Staff	Yuji Ogawa	Togo Seisakusyo Corporation
	Kazunori Kai	Horikiri. Inc.
	Isao Kato	Togo Seisakusyo Corporation
	Katsumi Kojima	Japan Auto Industries Association
	Tatuo Sakaguchi	Tokaibane Mfg. Co., Ltd.
	Ichiro Sasuga	NHK Spring Co., Ltd.
	Hideo Shimada	Society of Automotive Engineers of Japan,INC
	Yasuo Takahashi	PIOLAX INC.
	Yasuaki Takeuchi	Chuo Spring Industry Co., Ltd.
	Yoshitaka Taniguchi	Chuo Spring Co., Ltd.
	Akio Tsuzuki	Togo Seisakusyo Corporation
	Yasuhiko Naka	Tokuhatsu Spring Industry Co., Ltd.
	Yoshihiro Hayasaka	Mitsubishi Steel MFG. Co., Ltd.
	Tadashi Makino	Takaho Mfg. Co., Ltd.
	Masami Miyamoto	Toyo Spring Industrial Co., Ltd.
	Tatsuya Muramatsu	Chuo Spring Co., Ltd.
	Toshiaki Yamashita	Suncall Engineering Corp.
	Yoichi Yoshinaga	Suncall Corporation
	Toshio Kuwabara	Japan Spring Manufacturers Association
	Yoshiaki Kurihara	Japan Spring Manufacturers Association
	Kazumasa Seshimo	Japan Spring Manufacturers Association
	Hideo Kayama	Japanese Standards Association
	Motonobu Kuroda	Japanese Standards Association
	Takeshi Ishikawa	Japanese Standards Association
Overseas Staff	Wang Decheng	China Academy of Machinery Science and Technology
	Jiang Ying	China Productivity Center for Machinery
	Cao Huirong	Wuxi Sawane Spring Co.,Ltd
	Zhang Yinghui	Universtiy of Science and Technology Beijing china Standardization Technical Committee of spring
	Chen Baoxue	Universtiy of Shanghai for Science and Technology
	Yuusuke Kajiwara	Chuhatsu − indonesia
	Heru WIBOW	Chuo Spring Co., Ltd.
	Byung-Hoon Lee	Dae Won Kang Up Co., Ltd.
	Ki-Jeon Kim	Dae Won Kang Up Co., Ltd.
	Tadakazu Kato	NHK Spring (Thailand) Co., Ltd.
	SOMCHAI PIMPISAES	NHK Spring (Thailand) Co., Ltd.
	PEANGRUDEE DHIR	NHK Spring (Thailand) Co., Ltd.

この事典の使い方

1. 語順：本事典においては，英語見出し語をボールド体で示し英語のアルファベット順に配列している．
2. 記載形式：以下による．

   英語見出し語
   　　英語定義文
   　　日本語見出し語（発音）
   　　　日本語定義文
   　　中国語見出し語（発音）
   　　インドネシア語見出し語（発音）
   　　韓国語見出し語（発音）
   　　タイ語見出し語（発音）

3. 索引：英語を除く各国語の索引を巻末に設けた．

How to Use the Dictionary

1. Alphabetization: The terms in this dictionary are alphabetized on a letter-by-letter basis. They appear in boldface.
2. Format: Entries are defined as follows;

   Term in English
   　　Definition in English
   　Term in Japanese (pronouncing)
   　　Definition in Japanese
   　Term in Chinese (pronouncing)
   　Term in Indonesian (pronouncing)
   　Term in Korean (pronouncing)
   　Term in Thai (pronouncing)

3. Index: Indexes are provided for each language except English.

A

abrasion
the removal of surface material from any solid through the frictional action of another solid, liquid or gas.
摩耗　[mamou]
固体の表面に他の固体，液体あるいはガスが摩擦作用を及ぼすことにより表面が削りとられること．
磨損　[mó sǔn]
pengikisan　[pungikisan]
마모　[mamo]
การสึกจากการเสียดสี　[gaan suk chaak gaan siead sii]

abrasion resistance
a degree of resistance against abrasion.
耐摩耗性　[tai mamousei]
摩耗に対する抵抗の度合い．
耐磨性　[nài mó xìng]
resistensi pengikisan　[resisutensi pungikisan]
내마모성　[nae-mamoseong]
ความต้านทานการสึกจากการเสียดสี　[kwaam taan taan gaan suk chaak gaan siead sii]

abrasive
a hard substance to be used in polishing or grinding.
研磨材　[kenmazai]
金属等の硬い物質を研磨したり磨くのに用いられる硬い物質．
耐磨材料　[nài mó cái liào]
pengampelas　[punganperasu]
연마재　[yeonma jae]
สารขัด　[saan khad]

abrasive blasting
the cleaning of metal surfaces by the use of abrasive entrained in a blast of air.
研磨材吹付　[kenmazai hukituke]
研磨材を含んだ空気を吹き付けて金属表面を清浄にすること．
噴砂　[pēn shā]
semburan pengampelas　[senburan punganperasu]
연마재 브라스팅　[yeonmajae beuraseuting]
การยิงขัดผิว　[gaan ying khad piw]

abrasive cone
an abrasive shaped into a solid cone to be rotated by an arbor for abrasive machining.
円すいといし　[ensui toisi]
研磨加工用に円すい形の固体に成形されたといしで，回転軸によって回転する．
磨料　[mó liào]
kerucut pengampelas　[kerukuto punganperasu]
원추형 숫돌　[wonchuhyeong sutdol]
แท่งขัดรูปโคน　[taaeng khad ruup cone]

abrasive disk
an abrasive shaped into a disk to be rotated by an arbor for abrasive machining.
研磨ディスク　[kenma djisuku]
研削加工用にディスク形状に成形されたといしで，回転軸によって回転する．
研磨盘　[yán mó pán]
piringan pengampelas　[piringan punganperasu]
연마판　[yeonmapan]
แผ่นขัด　[paa-en khad]

1

abrasive machining

grinding or shaping by an abrasive tool.

研磨加工　[kenma kakou]
といし工具を使った研削あるいは形状仕上加工．
研磨加工　[yán mó jiā gōng]
olahan pengampelas　[orahan punganperasu]
연마가공　[yeonmagagong]
กระบวนการขัด　[gra-buan gaan khad]

absorbed energy

energy consumed in breaking a test piece in the impact test.

吸収エネルギー　[kyuusyuu enerugii]
衝撃試験において，試験片を破断するのに要したエネルギー．
吸收能　[xī shōu néng]
energi serapan　[enerugi serapan]
흡수 에너지　[heubsu eneoji]
พลังงานที่ถูกดูดซับไว้　[pha-lang ngaan tii took dood sap wai]

absorbing spring

a spring to be used for absorbing shock.

緩衝用ばね　[kansyouyou bane]
衝撃を緩和するために用いるばねの総称．
缓冲弹簧　[huǎn chōng tán huáng]
pegas penyerap　[pugasu punyerapu]
완충용 스프링　[wanchungyong spring]
สปริงกันกระแทก　[sa-pring gan gra ta-aek]

accelerated life test

operation of a product above its maximum ratings to produce premature failure to estimate life under normal condition.

促進寿命テスト　[sokusin zyumyou tesuto]
製品を最大規格を越えて使用して早期故障を引き起こし，通常時の製品寿命を推定すること．
促进寿命试验　[cù jìn shòu mìng shì yàn]
uji percepatan umur　[uji perusepatan umuru]
가속 수명 시험　[gasok sumyeong siheom]
การทดสอบอายุการใช้งานที่อัตราสูงสุด　[gaan tod-sorb aa-yu gaan chai ngaan tii at-tra soong-sud]

acceptable quality level

the maximum percentage of defects that has been determined tolerable as a process average for the test of a product.

許容品質水準　[kyoyou hinsitu suizyun]
製品検査において平均的な工程として許容され得ると決められた欠陥割合の最大値．
容许质量水平　[róng xǔ zhì liàng shuǐ píng]
tingkat kualitas yang diterima　[tingukato kuaritasu yangu diterima]
허용 품질 수준　[heoyong pumjil sujun]
ระดับคุณภาพที่ยอมรับได้　[ra-dab kun-na-parb tii yorm-rab dai]

acceptance test

a test used to determine conformance of a product to design specifications to be conducted by a buyer.

受入テスト　[ukeire tesuto]
設計仕様に合っているか否かを決めるために購入者によって行われるテスト．
接收试验　[jiē shōu shì yàn]
uji penerimaan　[uji punerimaan]
수입 시험　[suip siheom]
การทดสอบเพื่อการยอมรับ　[gaan tod sorb pheua gaan yorm-rab]

acid brittleness

brittleness of metal caused by acid pickling.

酸ぜい性　[sanzeisei]

酸洗によって鉄鋼の粘り強さが著しく損なわれる現象.
氢脆性 [qīng cuì xìng]
kerapuhan asam [kerapuhan asan]
산취성 [sanchwi-seong]
ความเปราะจากกรด [kwaam pra-or chaak grod]

acid resistance
    a property to withstand corrossion in acid environment.
耐酸性 [taisansei]
酸による腐食作用に耐える性質.
耐酸 [nài suān]
ketahanan terhadap asam [ketahanan teruhadapu asan]
내산성 [naesan-seong]
การทนกรด [gaan ton grod]

acid zinc plating
    galvanization without using cyanide.
酸性亜鉛めっき [sansei aen mekki]
シアン化合物を使用しない亜鉛めっき.
酸性电解液镀锌 [suān xìng diàn jiě yè dù xīn]
pelapisan seng asam [perapisan sengu asan]
산성 아연 도금 [sanseong ayeon dogeum]
ชุบซิงค์ชนิดกรด [chub zinc cha-nid grod]

active corrosion protection
    a corrosion protection method utilizing external electricity.
活性防食 [kassei bousyoku]
外部電流を流して防食する方法.
活性防腐 [huó xìng fáng fǔ]
perlindungan korosi aktif [perurindungan korosi akutihu]
활성 방식 [hwalseong bangsik]
การป้องกันการกัดกร่อน [gaan pong-gan gaan gad gron]

active deflection
    deflection based on which to determine the spring characteristics.
有効たわみ [yuukou tawami]
ばね特性を計算する基礎になるたわみ.
作用挠度 [zuò yòng náo dù]
lendutan aktif [rendutan akutihu]
유효 변위 [yuhyo byeonwi]
การยุบตัวที่มีผล [gaan yub toa tii mii pon]

active length
    length of the spring stroke.
作動長さ [sadou nagasa]
ばねとして作動する有効な長さ.
工作长度（工作高度）[gōng zuò cháng dù (gōng zuò gāo dù)]
panjang aktif [panjangu akutihu]
작동 길이 [jakdong giri]
ระยะสโตรคของสปริง [ra-yastroke khong sa-pring]

active stress
    stress under external force.
作用応力 [sayou ouryoku]
外力が作用しているときに生じる応力.
作用应力 [zuò yòng yīng lì]
tegangan aktif [tegangan akutihu]
작용 응력 [jagyong eungyeok]
ความเค้นจากภายนอก [kwaam ken chaak paai nork]

actuate
    to put into motion or mechanical action.
作動させる [sadou saseru]
運動あるいは機械の作動を始動させること.
从动 [cóng dòng]
aktuasi [akutuasi]
작동시킴 [jakdongsikim]
กระตุ้น [gra-tun]

actuator
    an instrument which converts energy

into power to run a machine.
アクチュエータ　[akutyueeta]
機械の運動を発生させるために，エネルギーを力，又は変位に変換する装置．
促动器　[cù dòng qì]
aktuator　[akutuatoru]
액추에이터　[aekchueiteo]
ตัวกระตุ้น　[toa gra-tun]

adhesion test
　a test which judges adhesive strength of a coat.
付着性試験　[hutyakusei siken]
下地と塗膜などの付着性を判定する試験．碁盤目試験，折り曲げ試験などがある．
附着性試驗　[fù zhuó xìng shì yàn]
tes adhesi　[tesu adohesi]
접착력 시험　[jeobchakyeok siheom]
การทดสอบการยึดติดของผิวเคลือบ　[gaan tod-sorb gaan yud-tid khong piw-kluab]

adjusting washer
　a washer to adjust load or a gap in machine parts.
調整用座金　[tyouseiyou zagane]
荷重や隙間などを調整するために用いる薄板の座金．
调整垫圈　[tiáo zhěng diàn quān]
ring pipih pengatur　[ringu pipihu pungature]
조정 와셔　[jojeong washyeo]
แหวนปรับแต่ง　[wa-aen prab taaeng]

age hardening
　time dependent hardness increment in steel.
時効硬化　[zikou kouka]
急冷又は冷間加工した鉄鋼などが時効によって硬化する現象．
时效硬化　[shí xiào yìng huà]
pengerasan dengan penuaan
[pungerasan dengan punuaan]

시효 경화　[sihyo gyeonghwa]
การทำให้แข็งตัวตามระยะเวลา　[gaan tam hai kha-ang toa taam ra-ya we-laa]

aging
　a phenomenon where the mechanical properties (such as hardness) of metal vary as time go on.
時効　[zikou]
急冷又は冷間加工などの後，時間の経過に伴い鋼などの機械的性質が（硬さなど）変化する現象．
时效（老化）　[shí xiào (lǎo huà)]
penuaan　[punuaan]
시효　[sihyo]
การกลายสภาพด้านแมคคานิค　[gaan glaai sa-paab daan mechanic]

aging treatment
　a process to stimulate aging.
時効処理　[zikou syori]
焼入れ又は加工によって生じた不安定相を安定化するための熱処理．
时效处理　[shí xiào chǔ lǐ]
pengolahan penuaan　[pungorahan punuaan]
시효처리　[sihyo cheori]
การทำให้เกิดการกลายสภาพ　[gaan tam hai gerd gaan glaai sa-paab]

agitator
　an equipment to mix a volume of liquid for uniformity of ingredient or temperature.
攪拌器　[kakuhanki]
2種以上の液体を均一に混合したり，温度を均一にするためにかき回す装置．
搅拌器　[jiǎo bàn qì]
pengaduk　[pungaduku]
교반기　[gyobangi]
เครื่องผสม　[kru-ang pa-som]

air brake
    an energy conversion mechanism activated by air pressure to reterd or stop a vehicle.
    エアブレーキ　[ea bureeki]
    空気圧によって作動させるエネルギー変換装置で，車両を減速させたり停止させるのに用いられる．
    气制动　[qì zhì dòng]
    rem angin　[remu angin]
    에어 브레이크　[eeo beureikeu]
    เบรคลม　[brake lom]

air circulation furnace
    a furnace whose heating is provided by the convection of heated air.
    流気式炉　[ryuukisiki ro]
    加熱空気の循環により，主に対流により加熱を行う炉．
    循环气体炉　[xún huán qì tǐ lú]
    tanur sirkulasi udara　[tanuru sirukurasi udara]
    열풍 순환 로　[yeolpung sunhwan ro]
    เตาที่ใช้ลมร้อนหมุนเวียนภายใน　[ta-o tii chai lom rorn soong mun wein paai nai]

air cooling
    cooling in air.
    空冷　[kuurei]
    空気中で冷却すること．
    空气冷却　[kōng qì lěng què]
    pendinginan dengan udara [pundinginan dengan udara]
    공냉　[gongnaeng]
    การทำให้เย็นโดยใช้ลมเย็น　[gaan tam hai yen dooi lom yen]

air hardening
    hardening in air or suitable gas atmosphere for steel with property of self hardening.
    空気焼入れ　[kuuki yakiire]
    空気中で冷却する焼入れ．自硬性をもつ鋼を焼入れする場合に行われる．
    空气淬火　[kōng qì cuì huǒ]
    pengerasan dengan udara　[pungerasan dengan udara]
    공기 경화　[gonggi gyeonghwa]
    การปล่อยให้แข็งตัวเอง　[gaan ploi hai khaaeng toa eeng]

air pollution
    atmospheric contamination caused by industrial activity.
    大気汚染　[taiki osen]
    工業生産活動による大気の汚染．
    大气污染　[dà qì wū rǎn]
    polusi udara　[porusi udara]
    대기 오염　[daegioyeom]
    มลภาวะทางอากาศ　[mon-la pha-wa taang aagaad]

air spring
    spring using the elasticity of air, which is one of the fluid springs.
    空気ばね　[kuuki bane]
    流体ばねの一種で，空気の弾性を利用するばね．
    空气弹簧　[kōng qì tán huáng]
    pegas udara　[pugasu udara]
    에어 스프링　[eeo spring]
    สปริงรับลมอัด　[sa-pring rab lom ad]

air suspension system
    a system to support the car body and frame by means of a cushion of air to absorb road shock caused by passage of wheels over bumps.
    空気ばね懸架システム　[kuuki bane kenga sisutemu]
    車両が路面の隆起を通過する際のショックを吸収するための空気のクッション性を利用した懸架システム．
    空气弹簧悬架系统　[kōng qì tán huáng xuán jià xì tǒng]
    sistem suspensi udara　[sisutemu susupensi

air suspension system — alloy steel

エアスプリング懸架システム
udara]
에어 스프링 현가 시스템 [eeo seupeuring hyeonga siseutem]
ระบบซัสเพนชั่นในรถยนต์ที่ทำงานโดยลม [ra-bob suspension nai rot-yon tii tam ngaan dooi lom]

aligning device
a gadget to align the center of rotating mechanism.
心出し装置 [sindasi souti]
機械などの中心位置を正しく出すための装置.
定心装置 [dìng xīn zhuāng zhì]
alat pengatur tengah [alato pungaturu tengahu]
심 맞추기 장치 [sim matchugi jangchi]
ตัวปรับศูนย์การหมุน [toa prab soon gaan mun]

alkaline test
a test of galvanized metal to be dipped in 70~80 degree sodium hydroxide solution.
アルカリ試験 [arukari siken]
試験片を 70〜80°Cの水酸化ナトリウム溶液に浸せきして水素発生の開始から反応が終止するまでの時間によってめっきの性状を調べること.
碱性试验 [jiǎn xìng shì yàn]
tes alkali [tesu arukari]
알카리 시험 [alkari siheom]
การทดสอบอัลคาไลน์ [gaan tod-sorb alkaline]

alkaline zinc coating
galvanization with cyanide.
アルカリ性亜鉛めっき [arukarisei aen mekki]
シアン化亜鉛めっき浴による亜鉛めっき.
碱性电解液镀锌 [jiǎn xìng diàn jiě yè dù xīn]
penyepuhan seng basa [punyepuhan sengu basa]
염기성 아연도금 [yeomgiseong ayeon dogeum]
การชุบซิงค์ชนิดด่าง [gaan chub zinc cha-nid daang]

allowable stress
maximum level of stress to be allowed for the safety use.
許容応力 [kyoyou ouryoku]
ばねの各部に生じる応力が，これ以内であれば安全であるとする許され得る最大の値.
许用应力 [xǔ yòng yīng lì]
tegangan yang diboleh kan [tegangan yangu diborehukan]
허용 응력 [heoyong eungnyeok]
ความเค้นสูงสุดที่ยอมรับได้ [kwaam ken soong-sud tii yorm rab dai]

alloy composition
metal or non-metal element to be added to a matrix metal in order to obtain favorable characteristics.
合金成分 [goukin seibun]
ある特性を付与する目的で，金属に添加される金属，又は非金属元素.
合金成分 [hé jīn chéng fēn]
komposisi paduan [konposisi paduan]
합금 성분 [hapgeum seongbun]
ส่วนประกอบเป็นโลหะผสม [su-an pra-gorb pen loo-ha pa-som]

alloy steel
steels which contain one or more than two kinds of alloy element for the improved mechanical properties.
合金鋼 [goukinkou]
鋼の性質を改善向上させるため又は所定の性質を持たせるため合金元素を1種類又は2種類以上含有させた鋼.
合金钢 [hé jīn gāng]
baja paduan [baja paduan]
합금강 [hapgeum gang]

เหล็กโลหะผสม [lek loo-ha pa-som]
alloy steel spring
　　a spring made of alloy steel.
　　合金鋼ばね [goukinkou bane]
　　合金鋼を用いたばね.
　　合金钢弹簧 [hé jīn gāng tán huáng]
　　pegas baja paduan [pugasu baja paduan]
　　합금강 스프링 [hapgeum gang spring]
　　สปริงเหล็กโลหะผสม [sa-pring lek loo-ha pa-som]

alloying element
　　elements to be used in alloy steel.
　　合金元素 [goukin genso]
　　合金鋼に含有させる元素.
　　合金元素 [hé jīn yuán sù]
　　unsur paduan [unsuru paduan]
　　합금원소 [hapgeumwonso]
　　ธาตุในโลหะผสม [gaan pa-som loo-ha]

Almen strip
　　test plate for measuring the strength of shot peening (arc height). This plate is 19 mm wide, 76 mm long, and has the hardness of 46 to 50 HRC, including A, C and N types according to the thickness.
　　アルメンストリップ [arumen sutorippu]
　　アークハイトを測定するための試験板. 幅19 mm, 長さ76 mm 及び硬さ46〜50 HRC で, 厚さによってA, C 及びN型がある.
　　阿尔曼试片 [ā ěr màn shì piàn]
　　kepingan almen [kepingan arumen]
　　알멘 스트립 [almen seuteurip]
　　แถบอัลเมน [taaeb Almen]

Almen test
　　a test to measure the strength of shot peening using Almen strip.
　　アルメンテスト [arumen tesuto]
　　アルメンストリップにショットを投射しアークハイトを測定してショットピーニングの強さを測定すること.
　　喷丸试验 [pēn wán shì yàn]
　　tes almen [tesu arumen]
　　알멘 테스트 [almen teseuteu]
　　การทดสอบช็อตพีนนิ่งโดยแถบอัลเมน [gaan tod sorb shot-peening doi taaeb Almen]

alternating immersion corrosion test
　　a corrosion test in alternating environmental condition (wet in air and immersed).
　　交互浸せき腐食試験 [kougo sinseki husyoku siken]
　　乾湿繰返し条件下での腐食評価法.
　　交替浸渍腐蚀试验 [jiāo tì jìn zì fǔ shí shì yàn]
　　uji karat celup bolak balik [uji karato cherapu boraku bariku]
　　건습 반복침적 부식시험 [geonseup banbokchimjeok busiksiheom]
　　การทดสอบการกัดกร่อนในสภาวะต่างๆสลับกัน [gaan tod sorb gaan gad gron nai sa-pa-wa taang-taang sa-lab gan]

alternating load
　　load which varies periodically within maximum and minimum magnitude.
　　繰返し荷重 [kurikaesi kazyuu]
　　一定の最大値と最小値の間を単純かつ周期的に変動する荷重.
　　循环变负荷 [xún huán biàn fù hè]
　　beban bolak balik [beban boraku bariku]
　　반복 하중 [banbok hajung]
　　โหลดที่ระดับต่างๆ [load tii ra-dab taang-taang]

alternating stress
　　stress under alternating load.
　　繰返し応力 [kurikaesi ouryoku]
　　繰返し荷重を受けて周期的に変動する応力をいう.
　　循环变应力 [xún huán biàn yīng lì]
　　tegangan bolak balik [tegangan boraku bariku]

alternating stress amplitude
    half of the difference between maximum and minimum stress in algebraic calculation.
    繰返し応力振幅 [kurikaesi ouryoku sinpuku]
    最大と最小応力の代数差の1/2.
    循环应力幅 [xún huán yīng lì fú]
    amplitude tegangan bolak balik [anpurichude tegangan boraku bariku]
    반복 응력 진폭 [banbok eungnyeok jinpok]
    แอมปริจูดของความเค้นณโหลดที่ระดับต่างๆ [amplitude khong kwaamken na load tii ra-dab taang-taang]

alternative characteristic
    measurable characteristics of an object which substitute an unmeasurable property.
    代用特性値 [daiyou tokuseiti]
    要求される品質特性を直接測定することが困難なため，その代用として用いる他の品質特性.
    代用特性值 [dài yòng tè xìng zhí]
    karakteristik alernatif [karakuterisutiku arerunatihu]
    대체 특성치 [daeche teukseongchi]
    คุณลักษณะที่วัดได้และไม่ได้ [kun-na-lak-sa-na tii wad dai laae mai-dai]

alumina grit
    a grain of natural or artificial alumina (aluminum oxide).
    アルミナグリット [arumina guritto]
    天然又は人造のアルミナ（酸化アルミニウム）の粒.
    氧化铝磨料 [yǎng huà lǚ mó liào]
    grit aluminium [gurito aruminiumu]
    알루미나 그릿 [allumina geurit]
    ผงอลูมิเนียมออกไซด์ [phong Aluminium]

aluminum coating of spring
    alminum coating on steel springs through a specialized treatment.
    ばねのアルミニウム被覆 [bane no aruminiumu hihuku]
    特殊な方法で施す鋼製ばねへのアルミニウム皮膜.
    弹簧的铝镀层 [tán huáng de lǚ dù céng]
    pelapisan aluminium pada pegas [purapisan aruminiumu pada pugasu]
    스프링의 알루미늄 피복 [springui alluminyum pibok]
    การเคลือบสปริงด้วยอลูมิเนียม [gaan kluab sa-pring doui aluminium]

analog instrument
    an instrument whose measurement is displayed in analog scale.
    アナログ計器 [anarogu keiki]
    測定量を物理的な連続量で表示する方式の計器.
    模拟测量仪 [mó nǐ cè liáng yí]
    instrumen analog [insutorumen anarogu]
    아날로그 계기 [anallogeu gyegi]
    อุปกรณ์ชนิดอนาล็อก [up-pa-gorn cha-nid analog]

analysis of variance
    a technique to divide variance of experimental data into elements of causes.
    分散分析 [bunsan bunseki]
    実験データの分散を，特定の原則に割り付けた成分に分割する技法.
    离散分析 [lí sàn fēn xī]
    analisis varian [anarisisu barian]
    분산분석 [bunsanbunseok]
    การวิเคราะห์ค่าที่เปลี่ยนไป [gaan wi-kra-or kaa tii plian paai]

angle measuring device
 a device which measures an angle like a protractor. An optical deduction machine, or a circumference scale tester, etc.
 角度測定装置 [kakudo sokutei souti]
  角度を測定する装置で, 光学式割出器, 円周目盛り検査器などがある.
 角度測量仪 [jiǎo dù cè liǎng yí]
 alat pengukur sudut [arato pungukuru suduto]
 각도 측정 장치 [gakdo cheukjeong jangchi]
 อุปกรณ์วัดมุม [up-pa-gorn wad mum]

angle of friction
 an angle at which the object start sliding down on the slope.
 摩擦角 [masatukaku]
  斜面上に物体を乗せて, 傾斜の傾きを次第に大きくし, 物体が滑り落ち始めるときの角度.
 摩擦角 [mó cā jiǎo]
 sudut gesek [suduto geseku]
 마찰각 [machalgak]
 มุมของการเสียดทาน [mum khong gaan siead-taan]

angled roop
 see "inclined side hook".
 斜めフック [naname maruhukku]
 "inclined side hook" 参照.
 斜圆钩环 [xié yuán gōu huán]
 kait bulat miring [kaito burato miringu]
 경사 원형 고리 [gyeongsa wonhyeong gori]
 ลูปวงเอียง [loop wong ieang]

angular frequency
 a quantity of frequency multiplied by 2π.
 角振動数 [kakusindousuu]
  振動数の 2π 倍の量.
 圆频率 [yuán pín lǜ]
 frekuensi sudut [hurekuensi suduto]
 각진동수 [gakjindongsu]
 ความถี่เชิงมุม [kwaam tii che-ong mum]

angular velocity
 the velocity of angle shift.
 角速度 [kakusokudo]
  角変位の時間的変化の割合.
 角速度 [jiǎo sù dù]
 kecepatan sudut [kesepatan suduto]
 각속도 [gaksokdo]
 ความเร็วเชิงมุม [kwaan rew che-ong mum]

anionic electrodeposition coating
 a painting method where the metal is given anode function while depositing compound presents anion property.
 アニオン電着塗装 [anion dentyaku tosou]
  解離可能な水溶性塗料中で金属を陽極として直流電圧を印加し塗装する方法.
 阴极电镀涂附 [yīn jí diàn dù tú fù]
 pelapisan [purapisan]
 아니온 전착 도장 [anion jeonchak dojang]
 การเคลือบอิออนโดยใช้แผ่นอิเลคโทรด [gaan kluab i-on dooi chai paaen electrode]

anisotropy of spring steel strip
 a difference of physical property in steel strip depending on the direction.
 ばね板の異方性 [baneita no ihousei]
  ばね板の物理的性質が方向(材料の幅, 長さ方向)によって異なること.
 弹簧钢片的各向异性 [tán huáng gāng piàn de gè xiàng yì xìng]
 kepingan pegas yang anisotrop [kepingan pugasu yangu anisotoropu]
 스프링 판 이방성 [spring pan ibangseong]
 คุณสมบัติที่ต่างกันของแผ่นสปริงตามทิศทาง [kun-na-som-bat tii taang gan khong paaen sa-pring taam tid-taang]

annealing

operation to heat an object to appropriate temperature, keep the temperature and then cool it. Its purpose includes the elimination of residual stress, reduction of hardness, improvement of machinability, improvement of cold workability, adjustment of crystal structure and obtaining required mechanical, physical and other properties.
焼なまし [yakinamasi]
適当な温度に加熱し，その温度に保持した後，冷却する操作．その目的は，残留応力の除去，硬さの低下，被削性の向上，冷間加工性の改善，結晶組織の調整，所要の機械的，物理的又はその他の性質を得ることなどである．
退火 [tuì huǒ]
penguatan, anneal [punguatan annearu]
풀림 [pullim]
การอบอ่อน [gaan ob-orn]

annealing time

holding time at elevated temperature for annealing process.
焼なまし時間 [yakinamasi zikan]
焼なまし時に適当な温度に加熱し，その温度に保持する時間．
退火时间 [tuì huǒ shí jiān]
waktu penguatan/anneal [wakutu punguatan / annearu]
풀림 시간 [pullim sigan]
เวลาในการอบอ่อน [we-laa nai gaan ob-orn]

antifriction coat

coat to prevent abrasion.
耐摩耗性被覆 [taimamousei hihuku]
摩耗を防ぐために用いられるコーティング．
耐磨镀层 [nài mó dù céng]
lapisan anti gesek [rapisan anti geseku]

내마모성 피막 [naemamoseong pimak]
การเคลือบลดการเสียดสี [gaan klu-ab lod gaan siead sii]

antilock braking system

a sensor controlled braking system which prevents wheel lockup while allowing the brake to continue showing the wheel.
アンチロックブレーキ [anti rokku bureeki]
車輪のロックを防ぎつつ，ブレーキが車輪の回転を低くし続けることを可能にするセンサ制御のブレーキシステム．
防抱死系统 [fáng bào sǐ xì tǒng]
sistem rem anti kunci [sisuten ren anti kunchi]
앤티로크 브레이크 시스템 [aentirokeu beureikeu siseutem]
ระบบเบรคแบบแอนตี้ล๊อค [ra-bob brake baaep antilock]

antiroll bar

generic term for the springs mounted on the body for reducing its rolling when the centrifugal force is applied. Generally made of bars and formed into a letter "U" like shape, including solid type (solid stabilizer bar) and tubular type (tubular stabilizer bar).
スタビライザ [sutabiraiza]
車体に遠心力が作用した場合の車体の横揺れを少なくするために取り付けられているばねの総称．丸棒をほぼコの字状に成形したものが一般的で，中実のもの（中実スタビライザ）及び中空のもの（中空スタビライザ）がある．
稳定器 [wěn dìng qì]
stabiliser [sutabiriseru]
스테빌라이저 [seutebillaijeo]
สตาบิไลเซอร์ [stabilizer]

antiroll bar — arc height

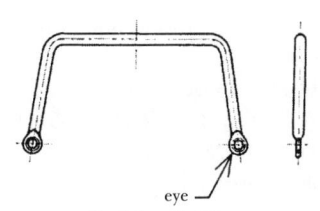

Fig A-1 antiroll bar

anti-slide band
band to secure the antiroll bar in position.
横ずれ防止バンド　[yokozure bousi bando]
スタビライザが横方向にずれるのを防止するために装着されるバンド又はバンド状の構造物.
防滑带　[fáng huá dài]
karet pencegah pergeseran　[kareto punchegahu perugeseran]
밀림방지 클램프　[millimbangji keullaempeu]
แผ่นกันเลื่อน　[paaen gan le-un]

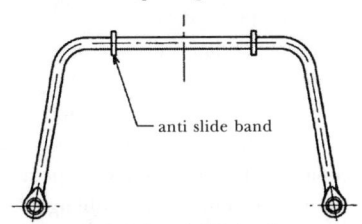

Fig A-2 anti-slide band

apparatus spring
a general term of springs used in apparatus and equipment.
機器用ばね　[kikiyou bane]
機器，装置に用いるばねの総称.
机器用弹簧　[jī qì yòng tán huáng]
pegas untuk peralatan　[pugasu untuku puraratan]
기기용 스프링　[gigiyong spring]
สปริงสำหรับเครื่องจักร　[sa-pring sam-rab klu-ang chaak]

arbor
a columnar metal parts whose tip is machined to a half-cut shape, hardened, and mirror finished to be used in a coiling machine for the center guide.
軸　[ziku]
コイリング機の心金に使われる円柱状の部品．先端が半割状に成形され，熱処理硬化され，鏡面仕上げを施されている．
刀杆　[dāo gǎn]
poros　[porosu]
아 - 버　[a-beo]
เพลาในเครื่องม้วน　[pla-o naai kruang muan]

arc compression spring
a compression coil spring in the shape of arc.
圧縮アークコイルばね　[assyuku aaku koiru bane]
圧縮円筒コイルばねを円弧状に成形したばね．
弧形螺旋弹簧　[hú xíng luó xuán tán huáng]
pegas lengkung kompresi　[pugasu rengukungu konpuresi]
압축 아 - 크 코일 스프링　[apchuk a-keu koil spring]
สปริงกดรูปอาร์ค　[sa-pring god roop arc]

arc height
magnitude of the curling of the test plate, indicating the strength of the shot peening.
アークハイト　[aaku haito]
ショットピーニングの強さを示す値で，試験板の反りの大きさ．
弧高　[hú gāo]
arc height　[aruchu heiguto]
아 - 크 하이트　[a-keu haiteu]
อาร์คไฮท์　[arc height]

11

A

arc welding
    a welding method utilizing arc heat. Both AC arc and DC arc are being used.
    アーク溶接　[aaku yousetu]
    アークの熱で行う溶接で，交流アーク溶接及び直流アーク溶接の2種類に大別される．
    电弧焊接　[diàn hú hàn jiē]
    las arc　[rasu aruchu]
    아크 용접　[akeu yongjeop]
    การเชื่อมแบบอาร์ค　[gaan shi-um baaep arc]

Archimedean spiral
    a spiral line whose distance from the center is proportional to an angle created by starting line and any line which goes through the center and any point on the locus.
    アルキメデスら旋　[arukimedesu rasen]
    中心からの距離が回転角に比例して大きくなる渦巻線．
    阿基米德螺旋　[ā jī mǐ dé luó xuán]
    spiral archimides　[supiraru aruchimidesu]
    아르키메데스 나선　[areukimedeseu naseon]
    เกลียวแบบอะคิมีเดียน　[gliaew baaep Archimedean]

artificial aging
    aging performed at elevated temperature.
    人工時効　[zinkou zikou]
    室温以上の適当な温度で加熱したときに起こる時効．
    人工时效　[rén gōng shí xiào]
    penuaan buatan　[punuaan buatan]
    인공 시효　[ingong sihyo]
    การแปรรูปสังเคราะห์　[gaan praae ruup sang-kraor]

aspect ratio of wire cross section
    ratio of the width to the thickness of a noncircular cross-section of the material.
    長短径比　[tyoutankeihi]
    異形断面コイルばねに用いる材料の横断面の幅と厚さとの比．
    异形截面钢丝的长短径比　[yì xíng jié miàn gāng sī de chǎng duǎn jìng bǐ]
    perbandingan panjang-pendek penampang　[purubandingan panjangu pundeku punamapangu]
    장단 경 비　[jangdan gyeong bi]
    อัตราส่วนความกว้างและความหนาของหน้าตัดเส้นลวด　[at-tra suan kwaam gwaang laae kwaam naa khong naa tad sen lu-ad]

assembly jig
    a jig devised to hold components together for easy assembling.
    組立治具　[kumitate zigu]
    組立を容易にするために構成部品を保持するように工夫された治具．
    组装夹具　[zǔ zhuāng jiā jù]
    jig perakitan　[jigu purakitan]
    조립지구　[jolibjigeu]
    จิ๊กสำหรับการประกอบ　[jig sam-rab pra-gorb]

assembly line
    a volume production arrangement whereby the work in progress is continuously transfered from one operation to next to complete assembling of the products.
    組立ライン　[kumitate rain]
    製品の組立を行うためにワークが連続的にある作業から次の作業へ移動するように配置した量産ライン．
    组装线　[zǔ zhuāng xiàn]
    lini merangkai　[rini merangukai]
    조립 라인　[jorip rain]
    ไลน์ประกอบ　[line pra-gorb]

assembly machine
    a machine that produces a configura-

assembly machine — ausforming

tion of equipment from discrete components.
組立機　[kumitateki]
別々の部品から装置の構成を送り出す機械．
组装机　[zǔ zhuāng jī]
mesin perakit　[mesin purakito]
조립기　[jolibgi]
เครื่องจักรสำหรับการประกอบ　[kru-ang chak sam-rab gaan pra-gorb]

asymmetric leaf spring
　asymmetrical leaf spring the center bolt or the center pin of which is not located at the center of the span. (see Fig A-3)
非対称ばね　[hitaisyou bane]
センタボルト又はセンタピンの位置が，スパンの中央にない非対称の重ね板ばね．(Fig A-3 参照)
非对称板簧　[fēi duì chèn bǎn huáng]
pegas daun asimetris　[pugasu daun asimetorisu]
비대칭 판 스프링　[bidaeching pan spring]
แหนบชนิดไม่สมมาตร　[naaep cha-nid mai som-maad]

atmospheric humidity
　the degree of steam amount contained in the atmosphere. Relative humidity and absolute humidity are being used.

大気湿度　[taiki situdo]
大気中に水蒸気の含まれている程度を表す語．絶対湿度と相対湿度との表し方がある．
大气湿度　[dà qì shī dù]
kelembaban atmosfer　[kerenbaban atomosuferu]
대기 습도　[daegi seupdo]
ความชื้นในบรรยากาศ　[kwaam shuun nai ban-yaa-gaad]

ausforming
　process where steel is heated to Ac3 transformation temperature or more, then cooled rapidly to the metas-table austenite range temperature. After providing plastic distortion process at this temperature, the steel shall be cooled rapidly.
オースホーミング　[oosuhoomingu]
変形加工を伴った熱処理の方法で，鋼をAc3変態点以上に加熱し，準安定オーステナイト範囲まで急冷し，ある程度大きな塑性変形加工を付与した後，急冷する処理．
形变热处理　[xíng biàn rè chǔ lǐ]
ausform　[ausuforumu]
오스포밍　[oseupoming]
ออสฟอร์มมิ่ง　[ausforming]

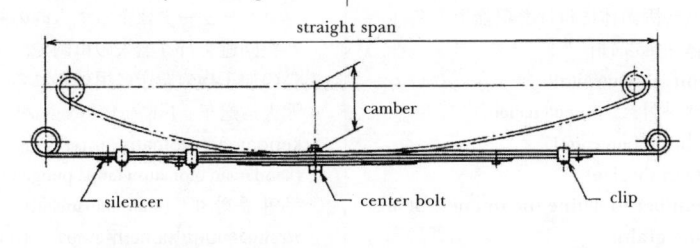

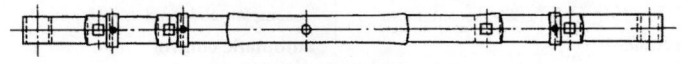

Fig A-3　asymmetric leaf spring
(straight span, camber, silencer, center bolt, clip)

austempering
heat treatment method utilizing constant temperature transformation. After cooling austenitized steel rapidly to the appropriate temperature at A1 or less, the temperature shall be kept until the transformation completes to the bainite transformation zone. This process is to provide desirable toughness to steel and to prevent thermal strain caused by the rapid cooling.
オーステンパ処理　[oosutenpa syori]
加熱してオーステナイト化した鋼を，A1点以下の適当な温度まで急冷し，この温度に保つことによってベイナイト変態域を横切って変態が進行する．この変化を恒温変態といい，この熱処理方法のことをいう．この目的は，鋼に良好なじん性を与えることと，急冷による熱ひずみを防止することにある．
等温淬火　[děng wēn cuì huǒ]
austemper　[ausutenperu]
오스템퍼링　[oseutaempeoring]
ออสเทมเพอริ่ง　[austempering]

austenite
the name of micro structure labeled to the solid solution of gamma iron.
オーステナイト　[oosutenaito]
γ鉄の固溶体に付けた組織上の名称．
奥氏体　[ào shì tǐ]
austenit　[ausutenito]
오스테나이트　[oseutenaiteu]
ออสเทไนท์　[austenite]

austenite grain size
a number to define the fineness of austenite grain.
オーステナイト結晶粒度　[oosutenaito kessyou ryuudo]
オーステナイト結晶粒の大きさ．粒度番号で表す．
奥氏体结晶　[ào shì tǐ jié jīng]
ukuran butiran austenit　[ukuran butiran ausutenito]
오스테나이트결정입도　[oseutenaiteu gyeoljeong ipdo]
ขนาดเกรนของออสเทไนท์　[kha-naad khong austenite]

austenitic stainless steel
stainless steel with austenite micro structure. It is generally non-magnetic.
オーステナイト系ステンレス鋼　[oosutenaitokei sutenresukou]
常温においてオーステナイト組織を示すステンレス鋼．熱処理によって硬化せず，一般に非磁性である．
奥氏体不锈钢　[ào shì tǐ bù xiù gāng]
baja tahan karat austenit　[baja tahan karato ausutenito]
오스테나이트계 스테인레스강　[oseutenaiteugye seuteilleseugang]
เหล็กสเตนเลสออสเทไนท์　[lek stainless austenite]

automated guided vehicle
a driverless computer-controlled vehicle equipped with guidance and collision-avoidance system to transport workpieces in the shop floor.
無人搬送車　[muzin hansousya]
コンピュータによって方向の制御と衝突回避を行う無人の車両で，工場内での加工物の搬送に用いられる．
无人运送车　[wú rén yùn sòng chē]
kendaraan otomatis tanpa pengemudi　[kendaraan otomatisu tanpa pungemudi]
무인 운반차　[muin unbancha]
รถขนส่งควบคุมโดยคอมพิวเตอร์　[rot khon-song kwuab-kum dooi computer]

automatic control
a control method to run the machine

automatically.
自動制御　[zidou seigyo]
機械・装置などを自動的に操作・調整すること．
自动控制　[zì dòng kòng zhì]
kontrol otomatis　[kontororu otomatisu]
자동 제어　[jadong jeeo]
การควบคุมแบบอัตโนมัติ　[gaan kuab-kum baap at-ta-no-mat]

automation
  the use of technology to ease human labor or extend the mental or physical capabilities of humans.
オートメーション　[ootomeisyon]
人間の労力を軽減したり，人間の知的あるいは身体的能力を発揮させるために技術力を用いること．
自动装置　[zì dòng zhuāng zhì]
automatis　[automatisu]
자동화　[jadonghwa]
การทำงานแบบอัตโนมัติ　[gaan tam ngaan dooi at-ta-no-mat]

automotive frame
  the basic structure of automotive vehicles supported by the suspension.
車台　[syadai]
懸架装置によって支えられている自動車の基本構造物．
台架装置　[tái jià zhuāng zhì]
bingkai kendaraan　[bingukai kendaraan]
차대　[chadae]
โครงรถยนต์　[kroong rot-yon]

automotive transmission
  a device for providing different gear ratios between the engine and drive wheels of an automotive vehicles.
自動車用変速機　[zidousyayou hensokuki]
自動車の駆動輪とエンジンの間で異なるギヤ比を与える装置．
汽车变速器　[qì chē biàn sù qì]

transmisi kendaraan　[toransumisi kendaraan]
자동차용 변속기　[jadongchayong byeonsokgi]
อุปกรณ์ปรับความเร็วสำหรับรถอัตโนมัติ　[uppa-gorn prab kwaam rew sam-rab rot at-ta-no-mat]

auxiliary leaf
  leaf to protect the main leaf against the load in reverse direction.
押さえばね板　[osae baneita]
親板を逆方向の荷重に対して保護するためのばね板．
压紧簧板　[yā jǐn huáng bǎn]
daun pegas penekan　[daun pugasu punekan]
누름판　[nureumpan]
แหนบช่วย　[naaep sho-ui]

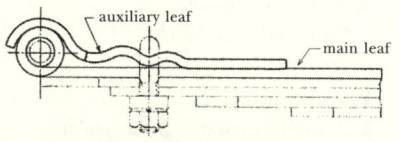

Fig A-4　auxiliary leaf

auxiliary spring
  spring which supports the load supplementary as the load increases in a progressive leaf spring. (see Fig A-5)
補助ばね　[hozyo bane]
プログレッシブ重ね板ばねにおいて，荷重の増加とともに補助的に働くばね．(Fig A-5 参照)
辅助弹簧　[fǔ zhù tán huáng]
pegas bantu　[pugasu bantu]
보조 스프링　[bojo spring]
สปริงช่วย　[sa-pring sho-ui]

axial deflection
  deflection toward axial direction.
軸方向たわみ　[zikuhoukou tawami]
物体の軸線方向にたわむ現象．

axial deflection — axle load

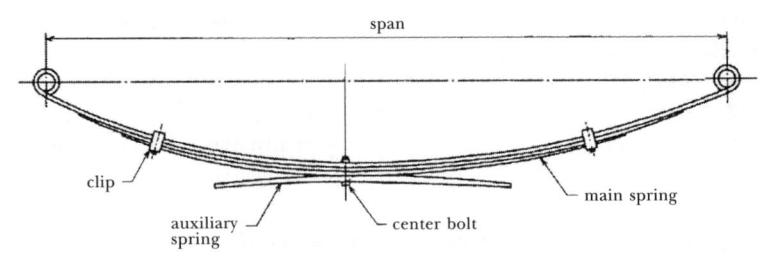

Fig A-5 auxiliary spring
(span, clip, auxiliary spring, center bolt, main spring)

軸向挠度　[zhóu xiàng náo dù]
defleksi sumbu(aksial)　[dehurekusi sunbu (akusiaru)]
축 방향 휨　[chuk banghyang hwim]
การยุบตัวในแนวแกน　[gaan yup to-a nai naao gaaen]

axial direction
　a line run through the center of body shaft.
軸方向　[zikuhoukou]
　物体の軸線又は軸面に平行方向のことをいう．
軸向　[zhóu xiàng]
arah sumbu(aksial)　[arahu sunbu (akusiaru)]
축 방향　[chuk banghyang]
ทิศทางสู่ศูนย์กลางของแกน　[tid taang suu soon-glaang khong gaaen]

axial load
　load generally applied in the direction of the coil axis.
軸荷重　[zikukazyuu]
　一般にコイル軸線方向に加わる荷重．
軸向负荷　[zhóu xiàng fù hè]
beban aksial　[beban akusiaru]
축 하중　[chuk hajung]
โหลดในแนวแกนของขด　[load nai nnaew gaaen khong khod]

axis line
　a line connecting the centers of coils of a coil spring. A line connecting the centers of leaves of a leaf spring in longitudinal direction. A line going through the center of the eye perpendicular to the longitudinal axis.
軸線　[zikusen]
　コイルばねの場合は，コイルばねのコイル中心を結んだ線，重ね板ばねの場合は，重ね板ばねの長手方向に中心を結んだ線又は目玉の円筒中心を結んだ線．
軸线　[zhóu xiàn]
garis sumbu　[garisu sunbu]
축선　[chukseon]
แนวแกน　[naaeo gaaen]

axle box spring
　a spring which supports axle boxes.
軸箱用ばね　[zikubakoyou bane]
　軸受け箱部に使用されているばね．
軸承座用弹簧　[zhóu chéng zuò yòng tán huáng]
pegas kotak gandar　[pugasu kotaku gandaru]
액슬 박스 스프링　[aekseul bakseu spring]
สปริงกล่องเพลา　[sa-pring glong pla-o]

axle load
　static load on wheel shaft.
軸荷重　[zikukazyuu]
　車軸に加わる静荷重で，軸線を通じて接地面に加わる各車軸あたりの荷重．
軸向负荷　[zhóu xiàng fù hè]
beban gandar　[buban gandaru]

차축 하중 [chachuk hajung]
โหลดของเพลา [load khong pla-o]

axle spring
spring used between the axle box and the bogie frame to absorb the vertical shock.
軸ばね [zikubane]
軸箱と台車枠との間に用いる上下方向の衝撃を緩和するばね.
轴弹簧 [zhóu tán huáng]
pegas poros [pugasu porosu]
액슬 스프링 [axle spring]
สปริงเพลา [sa-pring pla-o]

B

backlash
a gap between teeth for the smooth rotation of gear.
バックラッシュ [bakkurassyu]
歯車に滑らかな回転をさせるためにつけた歯と歯の間の遊び．
间隙 [jiān xì]
kelonggaran [kerongugaran]
백래시 [baengnaesi]
แบลคแลช [backlash]

back-up leaves
leaf other than the main leaf. (see Fig B-1)
子板 [koita]
親板以外のばね板．(Fig B-1 参照)
副弾簧 [fù tán huáng]
pegas daun pendukung [pugasu daun pundukungu]
자 판 [ja pan]
แหนบช่วย [naaep shu-oi]

bainite
microscopic structure of heat treated steel when quenched and held to a certain elevated temperature (150~500 °C).
ベイナイト [beinaito]
鋼を 150～500℃の熱浴に焼入れして恒温変態を起こさせたときにできる組織．
贝氏体 [bèi shì tǐ]
bainit [bainito]
베이나이트 [beinaiteu]
ไบไนท์ [bainite]

baking
heat treatment to eliminate the strain of the material or to eliminate hydrogen after plating.
ベーキング [beekingu]
素材のひずみ除去又はめっき後の水素除去を目的として行う熱処理．
低温干燥处理 [dī wēn gān zào chǔ lǐ]
membakar [menbakaru]
베이킹 [beiking]
การเบคกิ้ง [gaan baking]

balance spring
a spring to be used for a barance.
秤ばね [hakari bane]
秤用に用いるばね．
平衡弾簧 [píng héng tán huáng]
pegas pengimbang [pugasu punginbangu]
저울 스프링 [jeoul spring]
สปริงบาลานซ์ [spring balance]

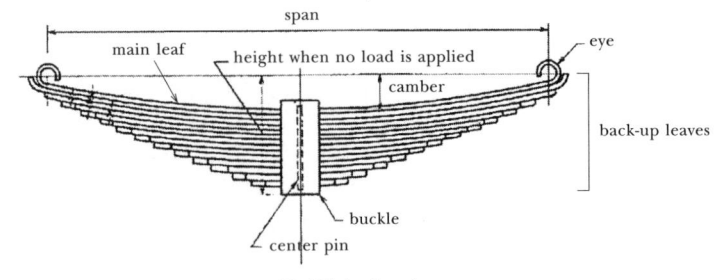

Fig B-1 back-up leaves

ball joint type
  eye of a bar stabilizer with a ball bushing.
  ボールジョイントタイプ [booru zyointo taipu]
  　バースタビライザ目玉部形状の一種で，ボールジョイントで固定する形式．
  万向节 [wàn xiàng jié]
  jenis sambungan bola [jenisu sanbungan bora]
  볼 조이트 형 [bol joiteu hyeong]
  หูสตาบิไลเซอร์ชนิดลูกบอล [hoo stabilizer cha-nid luuk-ball]

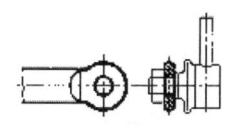

Fig B-2 ball joint type

banding press
  a press to place the band of the leaf spring in position.
  バンディングプレス [bandjingu puresu]
  　重ね板ばねの胴締めを焼きばめするプレス．
  卡箍压力机 [kǎ gū yā lì jī]
  kempa pengelompokan [kenpa pungeronpokan]
  밴딩 프레스 [baending peureseu]
  แบนดิ้งเพรส [banding press]

bar diameter
  バーの径 [baa no kei]
  杆径 [gǎn jìng]
  batang [batangu]
  봉의 지름 [bongui jireum]
  เส้นผ่าศูนย์กลางของบาร์ [sen-pa-soon-glang khong bar]

bar length
  バーの長さ [baa no nagasa]
  杆长 [gǎn cháng]

panjang batang [panjangu batangu]
봉의 길이 [bongui giri]
ความยาวของบาร์ [kwaam ya-o khong bar]

bar stabilizer
  stabilizer of which torsion section and the arm section are formed integrally by the continuous solid or tubular material.
  バースタビライザ [baa sutabiraiza]
  　トーション部とアーム部とを連続した中実材又は中空材で一体成形したスタビライザ．
  稳定杆 [wěn dìng gǎn]
  batang stabiliser [batangu sutabiriseru]
  바 - 스테비라이저 [ba-seutebiraijeo]
  สตาร์บิไลเซอร์ [stabilizer]

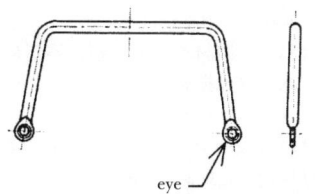

Fig B-3 bar stabilizer

barrel
  a flat cylindrical case for power spring where the spring touches to inner surface on released condition. The tips of the spring are fixed to the barrel.
  ぜんまいケース [zenmai keesu]
  　時計に使用されている動力用ぜんまいばねのケース．ばねは巻戻しの状態では，ケース内側に接触しており，端部はケースに固定されている．
  弹簧盒 [tán huáng hé]
  tong [tongu]
  스파이럴 스프링 케이스 [seupaireol spring keiseu]
  บาร์เรล [barrel]

barrel cam

a cylinder with a groove on its surface to perform cam function.

筒型カム　[tutugata kamu]
円筒上に溝を切ってカム機能を持たせたもの．

圓柱凸轮　[yuán zhù tū lún]

cam bulat　[chan burato]

원통형 캠　[wontonghyeong kaem]

บาร์เรลแคม　[barrel cam]

barrel polishing

process to remove burrs and scales of the spring in a rotating or vibrating container with polishing agents in it. This has also cleaning effect of the spring surface.

バレル研磨　[bareru kenma]
ばねを研磨材などと一緒に容器に入れて回転又は振動させることによって，ばり，スケールなどを除去する加工．ばねの表面を清浄にする効果もある．

滚动抛光　[gǔn dòng pāo guāng]

pengampelas bentuk tong
[punganperasu bentuku tongu]

배럴 연마　[baereol yeonma]

การเอาครีบและสเกลออก　[gaan aow kreep laae scale oorg]

barrel spring

a spring whose profile shows convex line like a beer barrel.

たる形ばね　[tarugata bane]
たるのような形をした（中央部が両端部の径より大きい）ばねの総称．

鼓形螺旋弾簧　[gǔ xíng luó xuán tán huáng]

pegas berbentuk tong　[pugasu berubentuku tongu]

볼록통형 스프링　[bolloktonghyeong spring]

บาร์เรลสปริง　[barrel sa-pring]

barrel tapered coil spring

barrel-shaped coil spring made of tapered round bar.

たる形テーパコイルばね　[tarugata teepa koiru bane]
たる形で，通常は，材料の直径が変化しているコイルばね．

中凸形锥形螺旋弾簧　[zhōng tū xíng zhuī xíng luó xuán tán huáng]

pegas tong bertirus　[pugasu tongu berutirusu]

볼록통 테이퍼 코일 스프링　[bolloktong teipeo koil spring]

สปริงขดรูปร่างบาร์เรลชนิดปลายเรียว　[sa-pring khod tii ruup-rang barrel cha-nid plaai rieaw]

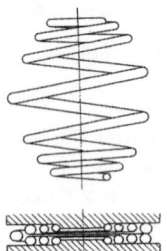

Fig B-4　barrel tapered coil spring

barrel-shaped spring

barrel-shaped coil spring made of the material with a constant diameter.

たる形コイルばね　[tarugata koiru bane]
たる形で，通常は，材料の直径が一定なコイルばね．

中凸形螺旋弾簧　[zhōng tū xíng luó xuán tán huáng]

pegas ulir bentuk tong　[pugasu uriru bentuku tongu]

볼록통형 코일 스프링　[bolloktonghyeong koil spring]

สปริงขดรูปร่างบาร์เรล　[sa-pring god ruup rang barrel]

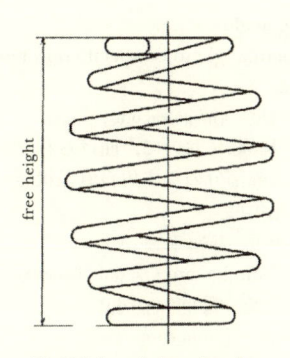

Fig B-5 barrel-shaped spring

batch process
　　a process which is carried out with discrete volume of materials in a non-continuous manner.
　　バッチ操作　[batti sousa]
　　不連続に一区分の量の材料に対して行われる操作．
　　批量生产　[pī liàng shēng chǎn]
　　proses batch　[purosesu batochi]
　　배치 프로세스　[baechi peuroseseu]
　　กระบวนการเดี่ยว　[gra-buan gaan dieaw]

bearing spring
　　a leaf spring for railway rolling stock.
　　担いばね　[ninai bane]
　　鉄道車両などの車体を支える重ね板ばね．
　　支撑板簧　[zhī chēng bǎn huáng]
　　pegas bantalan　[pugasu bantaran]
　　지지 겹판 스프링　[jiji gyeoppan spring]
　　สปริงสำหรับรางรถไฟ　[sa-pring san-rab raang rot-fai]

beehive compression spring
　　a compression coil spring whose profile shows conex line on one side and straight on the other side.
　　圧縮片絞りばね　[assyuku katasibori bane]
　　片側の先端部分だけが円すい形に成形された圧縮コイルばね．
　　蜂窝式压缩弹簧　[fēng wō shì yā suō tán huáng]
　　pegas tekan beehive　[pugasu tekan beehibe]
　　벌집형 코일 스프링　[beoljipyeong koil spring]
　　สปริงกดรูปร่างรังผึ้ง　[sa-pring god ruup rang rang pung]

Belleville spring washer
　　washers in the shape of coned disc springs when a load is applied, the washer tend to flatten with radial and circumferential strain. (patented J.F. Belleville)
　　ベルビルばね座金　[berubiru bane zagane]
　　円すい形円盤ばね形状のワッシャ．荷重が加わると径方向と円周方向のひずみを生じつつ平らになる（J.F. Belleville の特許に由来する）．
　　贝氏(碟形)弹簧垫圈　[bèi shì (dié xíng) tán huáng diàn quān]
　　pegas pipih belleville　[pugasu pipihu berurebirure]
　　접시 스프링형 와셔　[jeopsi springhyeong wasyeo]
　　แหวนสปริงเบลล์วิลล์　[waaen sa-pring Belleville]

bellows spring
　　a spring whose shape resembles a corrugated tube.
　　ベローズスプリング　[beroozu supuringu]
　　ばね用鋼板を蛇腹状に成形したばねで，エアサスペンションなどに用いられる．
　　波纹弹簧　[bō wén tán huáng]
　　pegas bellows　[pugasu berurousu]
　　벨로스 스프링　[belloseu spring]
　　สปริงหีบลม　[sa-pring heep lom]

belt polishing
　　a polishing method with abrasive cloth

belt polishing — bending fatigue strength

belt.
ベルト研磨　[beruto kenma]
研磨材を塗布したベルトによる研磨.
帯式研磨　[dài shì yán mó]
sabuk pengampelas　[sabuku punganperasu]
벨트 연마　[belteu yeonma]
การขัดเงาโดยสายพาน　[gaan khad nga-o dooi saai paan]

bench marking
a corporate activity to set a target of improvement based on the record of an excellent company in the field of the particular activity.
ベンチマーキング　[benti maakingu]
特定企業の優れた活動の状況を記録として残し,企業活動の一つの改善目標とする方法.
基准　[jī zhǔn]
benchmark　[benchimaruku]
벤치마-크　[benchima-keu]
เบนช์มาร์ค　[benchmark]

bend test
a test for bending capability of plate whereby a piece of plate is pressed against round bar. The evaluation is to check the cracks of outside surface under a certain bending angle and bending radius.
曲げ試験　[mage siken]
試験片を規定の内側半径で規定の曲げ角度になるまで曲げて,曲部の外側のき裂,そのほかの欠点を調べる試験.
弯曲试验　[wān qū shì yàn]
uji pembengkokkan　[uji punbengukokukan]
굽힘 시험　[gupim siheom]
การทดสอบการโค้งงอ　[gaan tod sorb gaan kong ngor]

bending angle
a bent angle to be specified in bending test.
曲げ角度　[mage kakudo]
曲げ試験において,曲げられた試験片の直線部分のなす角の直線状態からの変化角度.
弯曲角度　[wān qū jiǎo dù]
sudut bengkokan　[suduto bengukokan]
굽힘 각도　[gupim gakdo]
มุมโค้งงอ　[mum kong ngor]

bending and quenching machine
machine which quenches the heated leaf while cramping it to the curvature by the press.
成形焼入機　[seikei yakiireki]
加熱したばね板をプレスで挟んで成形したまま焼入れする機械.
成形淬火设备　[chéng xíng cuì huǒ shè bèi]
mesin penbengkok dan kejut　[mesin punbengukoku dan kejuto]
담금질 성형 기계　[damgeumjil seonghyeong gigye]
เครื่องขึ้นรูปและชุบแข็ง　[kruang khun ruup laae chub khaaeng]

bending cam
a cam to control bending process in forming machine.
曲げ用カム　[mageyou kamu]
曲げ加工をコントロールするカム.
弯曲凸轮　[wān qū tū lún]
nok(cam) pembengkok　[noku (chan) punbengukoku]
굽힘용 캠　[gupimnyong kaem]
แคมสำหรับขึ้นรูปโค้ง　[cam sam-rab khoon roop koong]

bending fatigue strength
maximum stress a material can endure for a given number of bending stress cycles without breaking.

曲げ疲れ強さ [mage tukare tuyosa]
　定められた回数の繰返し曲げに対する
　耐え得る最大応力．
弯曲疲劳强度 [wān qǔ pí láo qiáng dù]
kuat lelah pemengkokan [kuato rerahu pumengukokan]
굽힘 피로 강도 [gupim piro gangdo]
ความค้นโค้งอ่อ [kwaam ken koong ngor]

bending mandrel
　a round bar to make radius of plate in bending process.
曲げ心金 [mage singane]
　曲げ加工を行うときのワークを巻き付
　かせるために内側の心にする金具．
弯曲成形心棒 [wān qǔ chéng xíng xīn bàng]
mandrel(sumbu) pembengkok [mandoreru (sunbu) punbengukoku]
굽힘 가공용 맨드릴 [gupim gagongyong maendeuril]
แกนม้วนแบบโค้ง [gaaen muan baaep kong]

bending moment
　a product of force and distance.
曲げモーメント [mage moomento]
　力と距離の積．
弯矩 [wān jù]
momen lentur [momen renture]
굽힘 모멘트 [gupim momenteu]
โมเมนต์ของการโค้งอ่อ [moment khong gaan kong ngor]

bending radius
　the radius corresponding to the curvature of a bent specimen or part as measured at the inside surface of the bend.
曲げ半径 [mage hankei]
　曲げ加工を行ったとき曲げられた試験
　片又は部材の内側で測定した半径．
曲率半径 [qǔ lǜ bàn jìng]
jari-jari pembengkokan [jari-jari punbengukokan]
굽힘 반경 [gupim bangyeong]

รัศมีของการโค้ง [ras-sa-mii khong gaan kong]

bending strain
　strain defined by elongation devided by original length at outer surface when a plate is bent.
曲げひずみ [mage hizumi]
　板が曲げられるとき，外側表面の伸び
　を元の長さで割った値．
弯曲应变 [wān qǔ yīng biàn]
regang pembengkokan [regangu punbengukokan]
굽힘 변형률 [gupim byeonhyeongnyul]
ความเครียดของการโค้งอ่อ [kwaam kriead khong gaan kong ngor]

bending strength
　maximum stress whereby the test piece is broken under bending load.
曲げ強さ [mage tuyosa]
　曲げモーメントを加えて試験片が破壊
　するときの最大応力．
弯曲强度 [wān qǔ qiáng dù]
kuat pembengkokan [kuato punbengukokan]
굽힘 강도 [gupim gangdo]
ความค้นสูงสุดที่ทำให้หักขณะโค้งขึ้นรูป [kwaam-ken soong sud tii tam hai hak kha-na khoon ruup]

bending stress
　surface stress of flat bar when bending moment is applied.
曲げ応力 [mage ouryoku]
　曲げモーメントによって，梁の表面に
　生じる垂直応力．
弯曲应力 [wān qǔ yīng lì]
tegangan pembengkokan [tegangan punbengukokan]
굽힘 응력 [gupim eungnyeok]
ความค้นที่ผิวขณะโค้งขึ้นรูป [kwaam ken tii piw kha-na koong khoon ruup]

bending tool

tools for bending process. In coiling operation, a coiling pin and an arbor are taken for bending tools.

曲げ工具　[mage kougu]
　曲げ成形に用いる工具．コイリング機においては送りローラ，心金，コイリングピンが曲げ工具とみなされる．

弯曲工具　[wān qū gōng jù]

alat pembengkok　[arato punbengukoku]

굽힘 공구　[gupim gonggu]

อุปกรณ์ในการทำให้ค้องอ　[up-pa-gorn nai gaan tam hai kong ngor]

Berlin eye

eye aligned to the center line of the main leaf. This is advantageous in strength for horizontal and longitudinal loads caused by the sudden stop of the vehicles.

ベルリンアイ　[berurin ai]
　板ばねの目玉の名称で，その中心が親板のほぼ中心線上になるように丸めた目玉．車両が急停止した場合などに発生する水平前後方向の力に対し強度上有利である．

板簧吊耳　[bǎn huáng diào ěr]

gelang berlin　[gerangu berurin]

베를린 아이　[bereullin ai]

หูแหนบเบอร์ลิน　[huu naaep berlin]

Fig B-6 Berlin eye

bevel angle

the angle of chamfering.

面取角　[mentorikaku]
　面取りの角度．

倒角角度　[dǎo jiǎo jiǎo dù]

sudut tirus　[sudutu tirusu]

참퍼링 각도　[champeoring gakdo]

มุมบาก　[mum baak]

bias spring

generic term for coil springs and leaf springs for generating reciprocal actions repeatedly by applying external force.

バイアススプリング　[baiasu supuringu]
　コイルばね，板ばねなどを用いて外力を負荷して，2方向動作を繰り返して行わせるために用いるばねの総称．

偏置弹簧　[piān zhì tán huáng]

pegas miring　[pugasu miringu]

바이어스 스프링　[baieoseu spring]

สปริงไบแอส　[sa-pring bias]

billet

short thick bar of iron or steel before final rolling.

ビレット　[biretto]
　形鋼に圧延する前の角形断面の鋼材．

角钢　[jiǎo gāng]

bilet　[bireto]

빌렛　[billet]

ก้อนเหล็กก่อนรีดขึ้นรูป　[gorn lek gorn riid khoon ruup]

bimetal

two different metal sheet laminated together to detect temperature change.

バイメタル　[baimetaru]
　熱膨張率の違う2種の金属板を張り合わせたもの．熱膨張係数の違いによって生じる形状変化の性質を利用する．

双金属　[shuāng jīn shǔ]

bimetal　[bimetaru]

바이메탈　[baimetal]

ไบเมทัล　[bimetal]

bimetallic thermometer

a thermometer in which thin dissimilar metals bonded together in the shape of spiral is used to actuate a pointer utilizing the differential thermal ex-

pansion.
バイメタル温度計 [baimetaru ondokei]
異種金属が張り合わされてスパイラル状になったものが，熱膨張を利用して指針を動かす形式の温度計.
双金属温度計 [shuāng jīn shǔ wēn dù jì]
termometer bimetal [terumometeru bimetaru]
바이메탈식 온도계 [baimetalsik ondogye]
เทอร์โมมิเตอร์ชนิดไบเมทัล [thermometer cha-nid bimetal]

blanking
cutting of metal sheets into shapes by striking with a punch.
ブランク [buranku]
ポンチとダイスを用いて鋼板などをいろいろな形に打ち抜いたもの.
冲裁，(落料) [chōng cái, (luò liào)]
bentukan awal [bentukan awaru]
블랭킹 [beullaengking]
การขึ้นรูปด้วยการเจาะรู [gaan khoon roop do-ui gaan chor-ruu]

blanking force
a force to punch steel sheets in blanking process.
打抜力 [utinuki ryoku]
鋼板などをポンチとダイスを用いて様々な形に打ち抜くための力.
冲裁力 [chōng cái lì]
gaya bentukan awal [gaya bentukan awaru]
블랭킹 훠스 [beullaengking hwoseu]
แรงเจาะรู [raaeng chu-or ruu]

blanking tool
a set of punch and die.
打抜工具 [utinuki kougu]
ポンチとダイス一式.
冲裁工具 [chōng cái gōng jù]
alat untuk bentukan awal [arato untuku bentukan awaru]

블랭킹 툴 [beullaengking tul]
อุปกรณ์การเจาะรู [up-pa-gorn gaan chu-or ruu]

blasting
a cleaning process of steel products by blasting hard particles on its surface.
ブラスト加工 [burasuto kakou]
鉄鋼製品に研磨材を高速度で吹き付け，その表面を清浄化する加工法.
冲圧加工 [chōng yā jiā gōng]
semprotan [senpurotan]
블라스팅 [beullaseuting]
กระบวนการทำความสะอาดผิว [gra-buan gaan taam kwaam sa-ard piw]

blasting energy
energy to carry out blasting process.
投射エネルギー [tousya enerugii]
ブラスト加工を行うときの投射エネルギー.
冲击能 [chōng jī néng]
energi semprotan [enerugi senpurotan]
투사 에너지 [tusa eneoji]
พลังงานที่ใช้ในการทำความสะอาดผิว [pa-lang ngaan tii chai nai gaan taam kwaam sa-ard piw]

blister
state where a portion of the plating or coating layer is not sticked closely to the base metal or the base layer.
膨れ [hukure]
めっき層又は塗膜の一部が，素地又は下地層と密着しないで浮いている状態.
起泡 [qǐ pào]
lecet [recheto]
블리스터 (부풀음) [beulliseuteo (bupureum)]
การบวมของชั้นผิวชุบ [gaan buam khong shan piw shub]

blue brittleness
decrease in ductility of steel when heat-

ed to 200~300°C. In the heat range, bluish oxide film appears on the surface.
青熱ぜい性　[seinetu zeisei]
200〜300℃付近で伸び，絞りが減少してもろくなる性質．
蓝脆　[lán cuì]
kerapuhan biru　[kerapuhan biru]
청열 취성　[cheongyeol chwiseong]
ความเปราะจากการรมดำ　[kwaam pra-or chaak gaan rom dam]

blueing
　a heat trestment similar to a low-temperature annealing. Bluish oxide film appears on the surface.
ブルーイング　[buruuingu]
冷間成形ばねで，主に低温焼なましと同じ目的の処理をすること．この加熱によって表面は，黄色又は青色の酸化膜を生じる．
回火发蓝处理　[huí huǒ fā lán chǔ lǐ]
pembiruan　[punbiruan]
블루잉　[beulluing]
การรมดำ　[gaan rom dam]

bolster spring
　spring used between the bogie frame and the body.
枕ばね　[makura bane]
台車枠と車体との間に用いる重ね板ばね．
铁道车辆用悬架板弹簧　[tiě dào chē liàng yòng xuán jià bǎn tán huáng]
pegas pada bantalan　[pugasu pada bantaran]

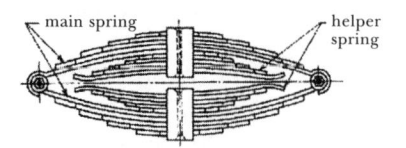

Fig B-7　bolster spring

볼스터 스프링　[bolseuteo spring]
สปริงหมอนรองรถไฟ　[sa-pring morn rorng]

bolt
ボルト　[boruto]
螺栓　[luó shuān]
baut　[bauto]
볼트　[bolteu]
โบลท์　[bolt]

both ends grinding
　process to grind the end face of the helical compression spring. When both ends of the spring are ground, it is called both-end grinding.
両端研削　[ryoutan kensaku]
主として，圧縮コイルばねの端面を研削する加工．ばねの両端を端面研削する場合を，両端研削という．
两端磨削　[liǎng duān mó xuē]
gerinda pada kedua ujung　[gerinda pada kedua ujungu]
양단 연삭　[yangdan yeonsak]
การเจียร์ปลายทั้ง 2　[gaan chia plaai tang song]

both ends slide contact type leaf spring
　a leaf spring with sliding contact on both ends.
両スライドタイプ重ね板ばね
[ryousuraido taipu kasaneita bane]
両端に目玉がなく，すべり支持構造の重ね板ばね．
两端滑块接触形片簧　[liǎng duān huá kuài jiē chù xíng piàn huáng]
pegas daun jenis slide　[pugasu daun jenisu suride]
양단 슬라이드형 겹판 스프링　[yangdan seullaideuhyeong gyeoppan spring]
แหนบสปริงชนิดปลาย 2 ข้างเลื่อนได้　[naaep sa-pring cha-nid plaai song khaang leun dai]

both ends slide contact type tapered leaf spring
 a taper leaf spring with sliding contact on both ends.
両スライドタイプテーパリーフスプリング　[ryousuraido taipu teepa riihu supuringu]
テーパリーフ形の重ね板ばねで両端に目玉がなく，すべり支持構造の重ね板ばね．
两端滑块接触形锥形片簧　[liǎng duān huá kuài jiē chù xíng zhuī xíng piàn huáng]
pegas daun yg kedua ujunnya bertiris [pugasu daun yugu kedua ujunnya berutirisu]
양단 슬라이드형 테이퍼 판 스프링 [yangdan seullaideuhyeong teipeo pan spring]
แหนบสปริงปลายเรียวชนิดเลื่อนได้ทั้ง 2 ข้าง [naaep sa-pring plaai riew cha-nid le-un dai tang song khaang]

bound stroke
 a stroke from normal load position to rebound position.
バウンドストローク　[baundo sutorooku]
懸架ばねのたわみの範囲を指す．常用荷重時からリバウンドストッパで止まるまでのストローク．
极限冲程　[jí xiàn chōng chéng]
stroke pantulan　[sutoroke panturan]
바운드 스트로크 [baundeu seuteurokeu]
ระยะสโตรคขึ้นลง　[ra-ya stroke khoon-long]

boundary sample
 a quality sample where numerical quality index is difficult to obtain.
限度見本　[gendo mihon]
良品又は不良品となる品質の限度を示した見本．
边界样品　[biān jiè yàng pǐn]
contoh batas　[chontohu batasu]
한도 견본 [hando gyeonbon]
ตัวอย่างแสดงขอบเขต　[toa-yaang sa-daaeng khorb khet]

bowing
 state where the axis of a coil spring is bent under no load or the phenomenon where it bends as the load increases. The bowing is caused by the initial distortion in manufacturing, the relative position of end seats, eccentric loading, and the inclination of end seats.
胴曲り　[doumagari]
コイルばねの軸線が，無荷重時に曲がっている状態又は荷重の増加とともに曲がる現象．製造時の初期変形，座の相対的な位置関係，偏心荷重，座の傾きなどが原因で発生する．
弹簧弓　[tán huáng gōng]
melengkung　[merengukungu]
평강의 직선도 [pyeonggang-ui jigsseondo]
การโค้งงอขณะยังไม่มีโหลด　[gaan koong ngor kha-na yung mai mii load]

braided wire
 a wire consisting of more than two thin wires stranded.
編み線　[amisen]
２本以上の細い線をより合わせた線．
绞线　[jiǎo xiàn]
kawat anyaman　[kawato anyaman]
브레이드 와이어 [beureideu waieo]
ลวดถัก　[luad tak]

brainstorming
 a procedure used to find a solution for a problem by collecting ideas from participant.
ブレインストーム　[bureinsutoomu]
参加者のアイデアを集めてある問題の解決策を発見する方法．
会诊　[huì zhěn]
kuras otak　[kurasu otaku]
브레인스토밍 [beureinseutoming]
การระดมสมอง　[gaan ra-dom sa-morng]

brake spring
  spring to be used in brake mechanism.
  ブレーキ用ばね　[bureekiyou bane]
  ブレーキ機構に用いられるばね.
  制动弹簧　[zhì dòng tán huáng]
  pegas rem　[pugasu ren]
  브레이크 스프링　[beureikeu spring]
  สปริงเบรค　[sa-pring brake]

brazing
  a process to solder metals by melting a nonferrous filler metal.
  ろう付け　[rouzuke]
  非鉄金属のろう材を溶融し，接合される部材を溶融しないで行う接合方法.
  铜焊　[tóng hàn]
  menyolder　[menyoruderu]
  납땜　[napttaem]
  การเคลือบทองเหลือง　[gaan klu-ab tong leu-ang]

breaking elongation
  an elongation at the breakage of tensile test.
  破断伸び　[hadan nobi]
  引張試験における試験片破断時の永久伸び.
  断裂延伸　[duàn liè yán shēn]
  perpanjangan pada kerusakan [purupanjangan pada kerusakan]
  파단 연신율　[padan yeonsinnyul]
  การยืดตัวของส่วนที่แตก　[gaan yuud to-a khong su-an tii taa-ek]

breaking load
  a maximum load to cause the breakage of test piece.
  破断荷重　[hadan kazyuu]
  試験片が破断するときの最大荷重.
  断裂负荷　[duàn liè fù hè]
  beban perusak　[beban purusaku]
  파단 하중　[padan hajung]
  โหลดสูงสุดที่รับได้　[load soong sud tii rab dai]

breaking stress
  stress under a breaking load.
  破断応力　[hadan ouryoku]
  破断荷重時の応力.
  断裂应力　[duàn liè yīng lì]
  tegangan perusak　[tegangan purusaku]
  파단 응력　[padan eungnyeok]
  ความเค้นจุดแตกหัก　[kwaam ken tii chud taa-ek hak]

breaking test
  a test whereby a test piece is broken.
  破断試験　[hadan siken]
  破断荷重を測定する試験. 試験片は破断される.
  破坏实验　[pò huài shí yàn]
  uji rusak　[uji rusaku]
  파단 시험　[padan siheom]
  การทดสอบการแตกหัก　[gaan tod-sorb gaan taaek hak]

bridle (of leaf spring)
  a hardware of leaf springs to hold leaves tightly.
  胴締め金具　[douzime kanagu]
  重ね板ばねのばね板の横ずれ防止金具.
  簧箍　[huáng gū]
  pengekangan　[pungekangan]
  밴딩 지그　[baending jigeu]
  คลิปรัดแหนบ　[clip rad naaep]

bright annealing
  an annealing process whereby heating and cooling is executed in an inert atmosphere to inhibit oxidation.
  光輝なまし　[kouki yakinamasi]
  保護雰囲気中で熱処理することによって高温酸化及び脱炭を防止し表面光輝状態を保持しつつ行う焼なまし熱処理.
  光亮退火　[guāng liàng tuì huǒ]

anneal sampai cemerlang [annearu sanpai chemerurangu]
광휘 풀림 [gwanghwi pullim]
การอบที่อุณหภูมิต่ำเพื่อป้องกันการเกิดอ็อกซิเดชั่น [gaan ob tii un-na-ha-poom tam pe-ua pong-gan gaan gerd oxidaiton]

Brinnel hardness
    a test to determine the hardness of material by pressing a steel ball on surface. The area of dent is measured and divided by the load. The results are expressed as Brinnell number.
    ブリネル硬さ [burineru katasa]
    試験材料に焼入鋼球を押し付けたときに生じるくぼみの表面積を，押し付けた荷重で割った値をいう．
    布氏硬度 [bù shì yìng dù]
    kekerasan brinnel [kekerasan burinneru]
    브리넬 경도 [beurinel gyeongdo]
    ความแข็งบรินเนล [kwaam khaaeng brinnel]

brittle fracture
    phenomenon where the breakdown takes place accompanied with almost no plastic distortion. In many cases, this term signifies the breakdown by hydrogen brittleness.
    ぜい性破壊 [zeisei hakai]
    塑性変形をほとんど伴わずに破壊する現象．多くは水素ぜい性（水素ぜい化）による破壊をいう．
    脆性失效 [cuì xìng shī xiào]
    retakan rapuh [retakan rapuhu]
    취성 파괴 [chwiseong pagoe]
    แตกจากความเปราะ [taaek chaak kwaam pra-or]

brittleness
    the property of a material manifested by fracture without appreciable prior plastic deformation.
    ぜい性 [zeisei]

ほとんど塑性変形しないで破壊する性質．
脆性 [cuì xìng]
kerapuhan [kerapuhan]
취성 [chwiseong]
ความเปราะ [kwaam pra-or]

brush spring
    a spring to hold the brush in the electric motor, power generator, and dynamo.
    ブラシスプリング [burasi supuringu]
    電動機，発電機，ダイナモなどにブラシを保持しているばねで，通常は圧縮コイルばねが用いられている．
    电刷压簧 [diàn shuā yā huáng]
    garis terputus-putus [garisu teruputusu putusu]
    브러시 스프링 [beureosi spring]
    สปริงแปรง [sa-pring praaeng]

buckle
    framed metal component to clamp the leaves of a leaf spring. (see Fig B-8)
    胴締め [douzime]
    重ね板ばねのばね板を締め付ける枠形金具．（Fig B-8 参照）
    卡箍 [kǎ gū]
    lengkung [rengukungu]
    버클 [beokeul]
    คลิปรัดแหนบ [clip rad naaep]

buckling
    phenomenon where the axis of a coil suddenly bends in a waved- or spiral-shape when the compressive load or the torsional moment applied to the coil spring exceeds the limit.
    座屈 [zakutu]
    コイルばねに加わる圧縮荷重又はねじりモーメントがある大きさに達すると，コイルの軸線が急に波形，らせん形などに曲がる現象．

buckling — build up edge

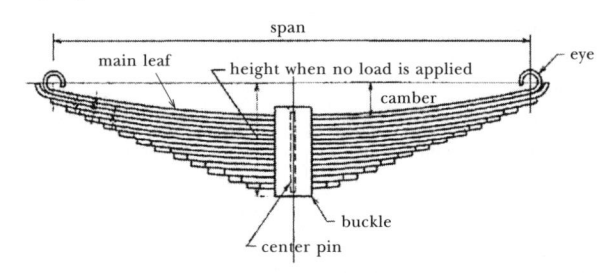

Fig B-8 buckle

(labels in figure: span, main leaf, height when no load is applied, eye, camber, buckle, center pin)

失稳　[shī wěn]
melengkung　[merengukungu]
좌굴　[jwagul]
การยุบตัวลงด้านข้างในสปริงขด　[gaan yub toa long daan khaang nai sa-pring khod]

buckling length
   length of a long beam (long spring) when it suddenly buckles under an increasing axial load.
座屈長さ　[zakutu nagasa]
細長い柱（細長い圧縮コイルばね）に軸荷重を加えていくと突然柱が横に変位するときの長さをいう．
失稳高度　[shī wěn gāo dù]
panjang kelengkungan　[panjangu kerengukungan]
좌굴 길이　[jwagul giri]
ระยะยุบตัวด้านข้าง　[ra-ya yub toa long daan khaang]

buckling load
   a minimum compressive load where a buckling occurs.
座屈荷重　[zakutu kazyuu]
座屈を起こさせる最小の圧縮荷重をいう．臨界荷重ともいう．
失稳临界负荷　[shī wěn lín jiè fù hè]
beban melengkung　[beban merengukungu]
좌굴 하중　[jwagul hajung]
โหลดที่ทำให้เกิดการยุบตัวด้านข้าง　[load tii tam hai gerd gaan yup tu-a daan khaang]

buckling strength
   strength to withstand a buckling.
座屈強さ　[zakutu tuyosa]
座屈に耐え得る強さ．
失稳强度　[shī wěn qiáng dù]
kuat melengkung　[kuato merengukungu]
좌굴 강도　[jwagul gangdo]
กำลังต้านการยุบตัวด้านข้าง　[gam-lang taan gaan yub toa daan khaang]

buckling stress
   stress under buckling load.
座屈応力　[zakutu ouryoku]
圧縮コイルばねが外力により座屈状態に達する直前に生じている応力．
失稳应力　[shī wěn yīng lì]
tegangan kelengkungan　[tegangan kerengukungan]
좌굴 응력　[jwagul eungnyeok]
ความเค้นของการยุบตัวด้านข้าง　[kwaam ken khong gaan yub toa long daan khaang]

build up edge
   2nd edge emerged on the fresh edge of cutting tool during cutting operation.
構成刃先　[kousei hasaki]
金属切削において，切削中に被削材の一部が加工硬化により母材より著しく硬くなり，刃部にたい積凝着して新たに構成される刃先．

构成刀刃　[gòu chéng dāo rèn]
ujung bentukan　[ujungu bentukan]
구성 날 끝　[guseong nal kkeut]
การเกิดขอบขณะตัด　[gaan gerd khorb kha-na tad]

bump stroke
  stroke from the position under the commonly used load to the position of the bump stopper in the suspension spring.
  バンプストローク　[banpu sutorooku]
  懸架ばねにおける，常用荷重時からバンプストッパで止まるまでの行程．
  冲程量　[chōng chéng liàng]
  langkah sampai terantuk　[rangukahu sanpai terantuku]
  범프 스트로크　[beompeu seuteurokeu]
  ระยะการกระแทก　[ra-ya gaan gra-taaek]

burn
  heat affected area on metal surface exposed by a color change. It may be caused by a heavy cutting.
  焼け　[yake]
  金属表面の色調が熱などによって著しく変化している状態．研削時に必要以上に切削負荷が加わった場合に多く発生する．この場合，研磨焼けあるいは切削焼けという．
  烧伤　[shāo shāng]
  membakar　[menbakaru]
  연삭 소손　[yeonsak soson]
  ไหม้　[mai]

burr
  an extra material sticking out from the edge of worked area caused by cutting or hole punching process.
  ばり　[bari]
  切断，穴あけなどの加工をする際，その縁にできるまくれ上がったような形状の余肉．

毛刺　[máo cì]
bram(dalam permesinan)　[buran (daran perumesinan)]
버　[beo]
ครีบ　[kriip]

bush
  component inserted into the eye of a leaf spring and works as a bearing.
  ブシュ　[busyu]
  重ね板ばねの目玉にはめ込み，軸受の働きをする部品．
  衬套　[chèn tào]
  cincin bantalan　[chinchin bantaran]
  부시 (부싱)　[busi (busing)]
  บุช　[bush]

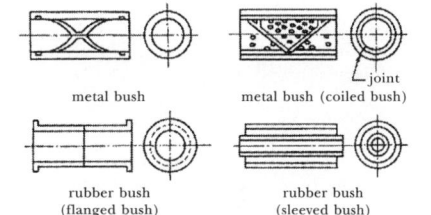

Fig B-9　bush

bushing
  process to press the bush into the eye of the leaf spring tightly.
  圧入　[atunyuu]
  重ね板ばねなどの目玉にしまりばめとなるようにブシュを押し入れる加工．
  压入　[yā rù]
  ring pemandu　[ringu pumandu]
  부시 (부싱) 압입　[busi(busing) abip]
  การอัดบุช　[gaan at-bush]

bushing test
  a test to evaluate the tightness of inserted bush.
  圧入試験　[atunyuu siken]
  ブシュの圧入荷重などを調べる試験．
  压入实验　[yā rù shí yàn]

uji ring pemandu　　[uji ringu pumandu]
부시 (부싱) 압입 하중 시험　　[busi (busing) abip hajung siheom]
การทดสอบการอัดบุช　　[gaan tod-sorb gaan at-bush]

butt welding
　　a weld to join plate type works at their edges.
突合せ溶接　　[tukiawase yousetu]
　　板を重ねずに突き合わせて行う溶接.
对焊　　[duì hàn]
las ujung　　[rasu ujungu]
맞댄 용접　　[matdaen yongjeop]
การเชื่อมปลายให้ติดกัน　　[gaan shi-um plaai hai tid gan]

C

C type retaining ring
    eccentric retaining ring shaped like letter "C", including two types: for on-shaft-use and for in-bore-use.
    C形止め輪　[siigata tomewa]
    偏心形の止め輪で，その形が文字Cに似ている止め輪．軸用及び穴用がある．
    C形挡圈　[C xíng dǎng quān]
    cincin penahan jenis C　[chinchin punahan jenisu C]
    씨 (C) 형 스냅 링　[ssihyeong seunaep ring]
    แหวนกั้นชนิด ซี　[waaen gan cha-nid C]

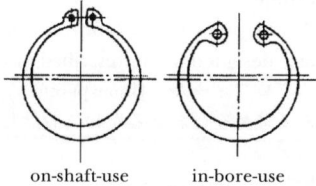

on-shaft-use   in-bore-use
Fig C-1　C type retaining ring

C type retaining ring uniform section
    concentric retaining ring shaped like letter "C", including two types: for on-shaft-use and for in-bore-use.
    C形同心止め輪　[siigata dousin tomewa]
    同心形の止め輪で，その形が文字Cに似ている止め輪．軸用及び穴用がある．
    C形同心挡圈　[C xíng tóng xīn dǎng quān]
    cincin penahan jenis C yang berpenampang sama　[chinchin punahan jenisu C yangu berupenanpangu sama]
    씨 (C) 형 동심 스냅 링　[ssihyeong dongsim seunaep ring]
    แหวนกั้นแบบแกนร่วมชนิด ซี　[waaen gan baaeb gaaen ro-um cha-nid C]

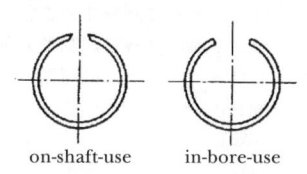

on-shaft-use   in-bore-use
Fig C-2　C type retaining ring uniform section

cable rewind spring
    a spring to retract cable after its use. A constant load clock spring are often used.
    コード巻戻し用ばね　[koodo makimodosiyou bane]
    コードを収納するとき，巻戻しするばね．
    线缆收卷弹簧　[xiàn lǎn shōu juǎn tán huáng]
    pegas penggulung kabel　[pugasu pungugurungu kaberu]
    케이블 리와인드 스프링　[keibeul riwaindeu spring]
    สปริงสำหรับดึงสายไฟกลับ　[sa-pring sam-rab dung saai-fai glab]

caliper rule
    a measuring implements like a pair of compasses to measure the dimension of machined cylindrical parts.
    キャリパー尺　[kyaripaa zyaku]
    円筒形状の径を測定するコンパス状の測定具．
    卡规　[kǎ guī]
    ukuran kaliper　[ukuran kariperu]
    캘리퍼 게이지　[kaellipeo geiji]
    กฎของคาลิเปอร์　[god khong caliper]

cam

a plate or cylinder which controls the motion of follower by means of its edge or groove while rotating around an axis.

カム [kamu]
一定軸のまわりに回転することによってエッジや溝によって従動子の動きをコントロールする板又は円筒.
凸轮 [tū lún]
nok [noku]
캠 [kaem]
แคม [cam]

cam adjustment

an adjustment to be applied to a cam mechanism.

カム調整 [kamu tyousei]
非加工物の形状を決めるために行うカム機構の調整.
凸轮调整 [tū lún tiáo zhěng]
pengaturan nok [pungaturan noku]
캠 조정 [kaem jojeong]
การปรับแคม [gaan prab cam]

cam control

a mechanism to convert a turning motion into a reciprocating motion by cam and follower.

カム制御 [kamu seigyo]
回転するカムに接触子が追従して工具の動きを制御する方法. 特殊な形状を持ったカムと, ナイフエッジ, ローラ平面などの単純な形状の接触子との直接接触によって, 従動体に所要の周期的運動を与える機構である.
凸轮控制 [tū lún kòng zhì]
pengendalian nok [pungendarian noku]
캠 제어 [kaem jeeo]
การควบคุมแคม [gaan kuob-kum cam]

cam follower

the output link of a cam mechanism.

カム従動子 [kamu zyuudousi]
カム機構の出力側リンク.
凸轮从动件 [tū lún cóng dòng jiàn]
pengikut nok [pungikuto noku]
캠 종동자 [kaem jongdongja]
การเชื่อมต่อของแคมแมคคานิซึ่ม [gaan shi-um toa khorng cam mechanism]

cam mechanism

a mechanical linkage to produce a prescribed motion by means of cam and cam follower.

カム機構 [kamu kikou]
カムとカム従動子により, ある目的の運動を行わせる機構.
凸轮机构 [tū lún jī gòu]
mekanisme nok [mekanisume noku]
캠 기구 [kaem gigu]
แคมแมคคานิซึ่ม [cam mechanism]

cam profile

the shape of the contoured cam surface designed for a prescribed motion.

カムプロフィール [kamu purofiiru]
ある運動を行わせるために設計したカムの輪郭.
凸轮外形 [tū lún wài xíng]
profile nok [purofire noku]
캠 푸로필 [kaem puropil]
รายละเอียดของแคม [raai-la-iead khong cam]

cam shape

see "cam profile".

カム形状 [kamu keizyou]
"cam profile" 参照.
凸轮形状 [tū lún xíng zhuàng]
bentuk nok [bentuku noku]
캠 형상 [kaem hyeongsang]
รูปร่างของแคม [ruup-rang khong cam]

camber

perpendicular distance measured at the center pin or bolt of leaf spring, from the tension surface of main leaf

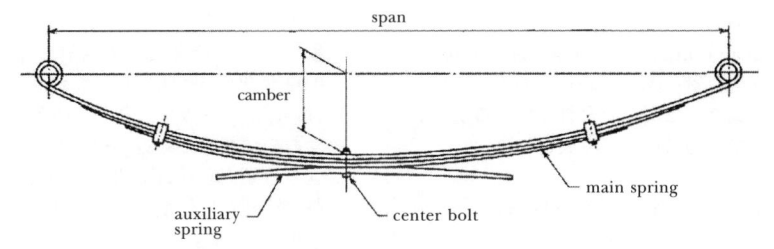

Fig C-3 camber

to the straight line connecting the centers of both eyes. (see Fig C-3)
反り　[sori]
重ね板ばねのセンタピン又はセンタボルトの位置における親板のテンション面から両目玉中心間又は荷重支持点間までを結んだ直線への垂直距離．(Fig C-3 参照)
翹曲　[qiào qǔ]
kamber　[kanberu]
캠버(휨)　[kaem beo (hwim)]
ระยะความสูงส่วนโค้งของแหนบ　[ra-ya kwaam soong su-an kong khong naaep]

camshaft
a rotating shaft to which a cam is attached.
カムシャフト　[kamu syahuto]
カムが取り付けられている回転軸．
凸轮轴　[tū lún zhóu]
poros nok　[porosu noku]
캠 축　[kaem chuk]
เพลายึดแคม　[pla-o yud cam]

cantilever
a beam with one end fixed.
片持ち梁　[katamoti hari]
一端が固定されている梁．
悬臂梁　[xuán bì liáng]
penyangga gantung　[punyangga gantungu]
외팔 보　[oepal bo]
คานตรง　[kaan trong]

cantilever length
length of a beam with one end fixed.
片持ち梁の長さ　[katamoti hari no nagasa]
一端が固定されている梁の長さ．
悬臂梁长度　[xuán bì liáng cháng dù]
panjang penyangga gantung　[panjangu punyangga gantungu]
ความยาวของคาน　[kwaam ya-o khong kaan]
외팔보 길이　[oepal bo giri]

cantilever spring
a flat spring with fixed end on one side and load carrying end on the other side.
片持ちばね　[katamoti bane]
一端を支持し，他端で荷重を支える平らなばね．
悬臂弹簧　[xuán bì tán huáng]
pegas penyangga gantung　[pugasu punyangga gantungu]
캔틸레버 스프링　[kaentillebeo spring]
สปริงคานตรง　[sa-pring kaan trong]

carbide
a compound between carbon and one or more metallic element(s).
炭化物　[tankabutu]
炭素と一つ又はそれ以上の金属元素の化合物．
碳化物　[tàn huà wù]
karbida　[karubida]
탄화물　[tanhwamul]
คาร์ไบด์　[carbide]

carbon content
    an amount of carbon percent in steel.
    炭素含有量　[tanso gan'yuuryou]
    鉄鋼などに炭素が含有している量のことで，通常百分率で表される．
    含碳量　[hán tàn liàng]
    kandungan karbon　[kandungan karubon]
    탄소 함유량　[tanso hamnyuryang]
    ปริมาณคาร์บอนในเหล็ก　[pa-ri-man carbon nai lek]

carbon fiber spring
    a spring made of CFRP (carbon fiber reinforced plastic).
    炭素繊維ばね　[tanso sen'i bane]
    CFRP を素材に用いたばね．
    碳素纤维弹簧　[tàn sù xiān wéi tán huáng]
    pegas serat karbon　[pugasu serato karubon]
    탄소 섬유 스프링　[tanso seomnyu spring]
    สปริงคาร์บอนไฟเบอร์　[sa-pring carbon fibre]

carbon potential
    a term that indicates carburizing capability of atmosphere on steel material.
    カーボンポテンシャル　[kaabon potensyaru]
    鋼を加熱する雰囲気の浸炭能力．当該雰囲気と平衡に達したときの鋼の表面の炭素濃度で表す．
    渗碳钢表面碳含量　[shèn tàn gāng biǎo miàn tàn hán liàng]
    potensi karbon　[potensi karubon]
    카본 퍼텐셜　[kabon peotensyeol]
    คาร์บอนโพเทนเชียล　[carbon potential]

carbon steel
    a steel of which carbon content is 0.02 to 2% is usually called carbon steel. Normally it contains small amount of Si, Mn, P, and S. For convenience's sake, it may be categorized by carbon content (high, middle, and low).
    炭素鋼　[tansokou]
    鉄と炭素の合金で炭素含有量が通常 0.02～2% の範囲の鋼．なお，少量のけい素，マンガン，りん，硫黄などを含むのが普通である．便宜上，炭素鋼含有分類(低,中,高)及び硬さ分類(極軟,軟,硬)に分類される場合がある．
    碳素钢　[tàn sù gāng]
    baja karbon　[baja karubon]
    탄소강　[tansogang]
    เหล็กคาร์บอน　[lek carbon]

carbon steel spring
    springs made of carbon steel.
    炭素鋼ばね　[tansokou bane]
    炭素鋼を用いたばねの総称．
    碳素钢弹簧　[tàn sù gāng tán huáng]
    pegas baja karbon　[pugasu baja karubon]
    탄소강 스프링　[tansogang spring]
    สปริงเหล็กคาร์บอน　[sa-pring lek carbon]

carbonitriding
    heat treatment to diffuse carbon and nitrogen simultaneously on steel surface in order to harden the work.
    浸炭窒化　[sintan tikka]
    鋼の表面層に炭素及び窒素を同時に拡散させる操作．
    碳氮共渗　[tàn dàn gòng shèn]
    penitridan karbon　[punitoridan karubon]
    침탄 질화　[chimtan jilhwa]
    การทำคาร์บอนไนไตรด์　[gaan tam carbonnitride]

carburizing
    operation to heat and treat the steel in the carburizing agent in order to increase the carbon content in the surface layer of steel. This operation falls into three types of solid carburizing, liquid carburizing and gas carburizing, according to the type of the carburizing agent.

浸炭　[sintan]
　鋼の表面層の炭素量を増加させるため，浸炭剤中で加熱処理する操作．浸炭剤の種類によって固体浸炭，液体浸炭及びガス浸炭に分けられる．
渗碳　[shèn tàn]
pengarbonan　[pungarubonan]
침탄　[chimtan]
การเพิ่มปริมาณคาร์บอน　[gaan perm pa-ri-man carbon]

carriage spring
　leaf spring which supports the bodies of railway cars.
担いばね　[ninai bane]
　鉄道車両などの車体を支える重ね板ばね．
车辆用悬架板弹簧　[chē liàng yòng xuán jià bǎn tán huáng]
pegas pembawa　[pugasu punbawa]
지지 스프링　[jiji spring]
สปริงค้ำลอ　[sa-pring kam lor]

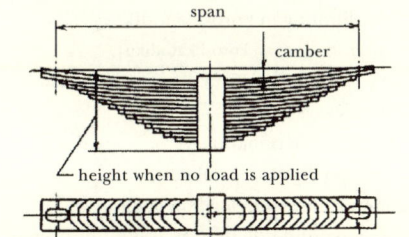

Fig C-4　carriage spring

case
　hardend surface area by case hardening.
はだ焼層　[hadayakisou]
　鋼を浸炭し，焼入焼戻しによって硬化した浸炭層．
表面渗碳硬化层　[biǎo miàn shèn tàn yìng huà céng]
lapisan yang mengalami pengerasan　[rapisan yangu mengarami pungerasan]
표면경화층　[pyomyeon gyeonghwacheung]
ผิวเหล็กที่ชุบแข็ง　[piw lek tii chub khaaeng]

case depth
　depth of hardened area by case hardening.
はだ焼深さ　[hadayaki hukasa]
　鋼を浸炭し，焼入焼戻しによって硬化した浸炭層の深さ．
表面渗碳硬化层深度　[biǎo miàn shèn tàn yìng huà céng shēn dù]
kedalaman pengerasan　[kedaraman pungerasan]
표면경화깊이　[pyomyeon gyeonghwagipi]
ความลึกของผิวเหล็กที่ชุบแข็ง　[kwaam luk khong piw lek tii chup khaaeng]

case hardening
　heat treatment to enhance the hardness of steel work by flame quenching, induction hardening or carburzing.
はだ焼き　[hadayaki]
　鋼の表面層に炭素量を増加させるため浸炭剤の中で加熱する操作．また，表面だけを焼入れ，硬化させることで，浸炭以外に高周波焼入れ，フレーム焼入れなどをいう場合もある．
表面渗碳硬化処理　[biǎo miàn shèn tàn yìng huà chǔ lǐ]
pengerasan permukaan　[pungerasan purumukaan]
표면경화　[pyomyeon gyeonghwa]
การชุบแข็ง　[gaan shup-khaaeng]

case hardening steel
　low carbon steel or low carbon steel alloy suitable for case hardening.
はだ焼鋼　[hadayakikou]
　低炭素鋼又は低炭素鋼合金で，主として浸炭焼入れによって，表面硬化させる鋼．
渗碳钢　[shèn tàn gāng]

case hardening steel — cationic electrodeposition coating

baja yang permukaannya dikeraskan
[baja yangu purumukaannya dikerasukan]
표면 경화강　[pyomyeon gyeonghwagang]
กระบวนการทำให้ผิวแข็งในเหล็กคาร์บอนต่ำ
[gra-buan gaan tam hai piw khaaeng nai lek carbon tam]

case hardness
 hardness of case hardened layer.
はだ焼硬さ　[hadayaki katasa]
　鋼を浸炭し，焼入焼戻しによって硬化
　した浸炭層の硬さ．
表面滲碳硬度　[biǎo miàn shèn tàn yìng dù]
kekerasan permukaan karburisasi
[kekerasan purumukaan karuburisasi]
침탄층 경도　[chimtancheung gyeongdo]
ความแข็งของผิวเหล็กที่ชุบแข็ง　[kwaam khaaeng khong piw lek tii chup khaaeng]

case zone
 case hardened layer.
浸炭層　[sintansou]
　鋼を浸炭し，焼入焼戻しによって硬化
　した層．
渗碳层　[shèn tàn céng]
zona karburisasi　[zona karuburisasi]
침탄층　[chimtancheung]
ชั้นผิวเหล็กที่ชุบแข็ง　[shan piw lek tii chup khaaeng]

cast steel grit
 cast steel particle used for grit blasting.
 (= cast steel shot)
鋳鋼グリット　[tyuukou guritto]
　破砕面及び稜角を持つ鋳鋼粒子．球状
　粒子を破砕して製造する．
铸钢丸　[zhù gāng wán]
butiran baja cor　[butiran baja koru]
주강 쇼트　[jugang syoteu]
กริทเหล็กหล่อ　[girt lek lor]

cast steel shot
 shot material manufactured from cast
 steel where melting steel is sprayed
 into ball-type shots.
鋳鋼ショット　[tyuukou syotto]
　溶鋼を噴霧することによって球形の
　ショットとし，さらに，熱処理のプロ
　セスによって製造された投射材．
铸钢喷丸　[zhù gāng pēn wán]
tumbukan baja cor　[tunbukan baja coru]
주강 쇼트　[jugang syoteu]
ช็อทเหล็กหล่อ　[shot lek lor]

cathodic protection
 a corrosion protection technique
 whereby coupling an active metal with
 a more noble metal to result in current
 flow and corrosion of the noble metal.
 (=electrical corrosion protection)
陰極酸化防食処理　[inkyoku sanka bousyoku syori]
　流電陽極又は外部電流を用いて金属体
　を陰極として通電し，腐食を防止する
　こと．電気防食ともいう．
阴极防腐处理　[yīn jí fáng fǔ chǔ lǐ]
perlindungan korosi katodik
[pururindungan korosi katodiku]
음극 산화 방식 처리　[eumgeuk sanhwa bangsik cheori]
การป้องกันการสึกแบบคาโธดิค　[gaan pong gan gaan suk baaep cathodic]

cationic electrodeposition coating
 a painting method to use cationic coating material.
カチオン電着塗装　[kation dentyaku tosou]
　カチオン塗料を用いて電着する塗装
　法．
阴极电镀　[yīn jí diàn dù]
penyepuhan deposisi kation
[punyepuhan deposisi kation]
카티온 전착도장　[kation jeonchakdojang]
การเคลือบโดยใช้ไฟฟ้าเป็นตัวนำ　[gaan kluab dooi chai fai-faa pen toa nam]

cavitation
   a phenomenon where in-liquid gas separates from the liquid and emerges as gas bubbles in flowing liquid when the flow locally speeds up and decreases its pressure.
   キャビテーション [kyabiteesyon]
   流体の流れの中で，速度が大となり低圧部ができるとき流体中に溶解していた気体が分離し気泡が生じること．
   气孔，气蚀 [qì kǒng, qì shí]
   rongga [ronguga]
   공동 현상 [gongdong hyeonsang]
   การเกิดฟองอากาศของก๊าซในของเหลว [gaan gerd forng aa-gaad khong gas nai shaan khong leew]

CCT-curve
   Continuous Cooling Transformation curve. It is a diagram showing steel transformation when steel is continually cooled from its austenite phase.
   連続冷却変態曲線 [renzoku reikyaku hentai kyokusen]
   鋼をオーステナイト状態から連続的に冷却した場合に生じる変態の様相を，縦軸に温度，横軸に時間（対数目盛）を取って図示した線図．
   连续冷却变化曲线 [lián xù lěng què biàn huà qū xiàn]
   kurva CCT [kuruva CCT]
   연속 냉각 변태 곡선 [yeonsok naenggak byeontae gokseon]
   เส้นโค้งซีซีที [sen kong CCT]

cellular manufacturing
   a type of manufacturing in which equipment is organized into cells.
   セル製造方式 [seru seizou housiki]
   設備が小区分された場所（セル）に配置される方式の生産方式．
   生产线 [shēng chǎn xiàn]
   manufaktur pola sel [manufakuture pora seru]
   셀 제조방식 [sel jejobangsik]
   การผลิตชนิดจัดอุปกรณ์เป็นสัดส่วนเฉพาะ [gaan pa-lit sha-nid chad up-pa-korn pen sad-suan]

cementation
   a heat treatment whereby introducing one or more elements into the surface of a metal by high temperature diffusion.
   セメンテーション [sementeesyon]
   金属材料の表面層の硬さ又は耐熱，耐食性などを向上させるため高温度の各種媒剤中で，他の元素を表面に拡散させる操作．
   渗碳 [shèn tàn]
   sementasi [sementasi]
   확산침투처리 [hwaksanchimtucheori]
   การยึดกัน [gaan yud gan]

cementite
   a compound between iron and carbon of which chemical formula is described approximately as Fe_3C.
   セメンタイト [sementaito]
   鉄と炭素の化合物で化学式が近似的に Fe_3C で示される炭化物．
   渗碳体 [shèn tàn tǐ]
   sementit [sementito]
   세멘타이트 [sementaiteu]
   เหล็กโครงสร้างซีเมนไทท์ [lek krong saang cementite]

center bolt
   fastening bolt of the leaf spring.
   センタボルト [senta boruto]
   重ね板ばねの中央部で，ばね板を締め付けるボルト．
   中心螺栓 [zhōng xīn luó shuān]
   baut pusat [bauto pusato]
   센터 볼트 [senteo bolteu]

center bolt — center spacer

เซ็นเตอร์โบลท์ [center bolt]

center bore
a hole in a leaf spring assembly for a fastening bolt.
センタ穴 [senta ana]
重ね板ばねの各リーフを締結するための穴.
中心孔 [zhōng xīn kǒng]
lubang pusat [rubangu pusato]
중심공 [jungsimgong]
รูตรงกลางแผ่นแหนบ [ruu trong glaang paaen]

center cup
protruded area around the center bolt hole to improve fatigue strength.
センタカップ [senta kappu]
疲れ強さの向上を目的として，センタボルト穴の周囲を押し出したもの.
中心密封圏 [zhōng xīn mì fēng quān]
mangkuk pusat [mangukuku pusato]
센터 배꼽 [senteo baekkop]
เบ้ากลาง [pa-o glaang]

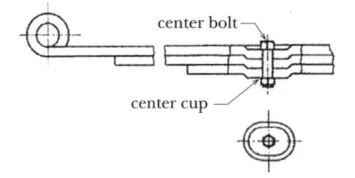

Fig C-5 center cup

center of gravity
the center position of the mass through which the gravity force is applied.
重心 [zyuusin]
物体の質量の中心位置.
重心 [zhòng xīn]
titik berat [titiku berato]
중심 [jungsim]
ศูนย์ถ่วง [soon tu-ang]

center pin
rivet to prevent the relative movement between the buckle and the leaves of a leaf spring. (see Fig C-6)
センタピン [senta pin]
重ね板ばねの胴締めとばね板相互とのずれを防ぐリベット．（Fig C-6 参照）
中心销 [zhōng xīn xiāo]
pin tengah [pin tengahu]
센터 핀 [senteo pin]
เซ็นเตอร์พิน [center pin]

center spacer
flat material inserted between the leaves in the vicinity of center bolt to prevent fretting corrosion.
センタスペーサ [senta supeesa]
フレッティングコロージョンを防ぐために，重ね板ばねの中央締付部の板間に挿入される合成樹脂製などの板状の部品.

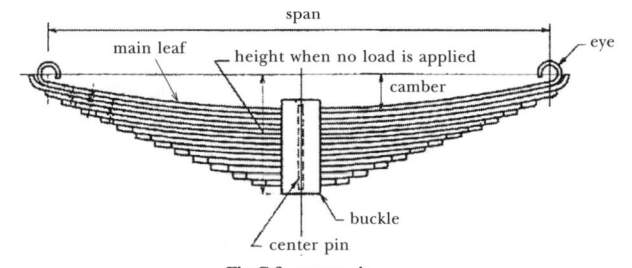

Fig C-6 center pin

内衬片　[nèi chèn piàn]
pusat spacer　[pusato supaseru]
센터 스페이서　[senteo seupeiseo]
เซ็นเตอร์สเปเซอร์　[center spacer]

centerless ground spring steel
　　a type of spring steel of which surface is ground off to diameter by a centerless grinding machine.
研削ばね鋼　[kensaku banekou]
　　熱間圧延された丸棒鋼材を所定の寸法にセンタレス研削したばね用鋼.
无心磨削弹簧钢　[wú xīn mó xuē tán huáng gāng]
baja pegas tergerinda tanpa pusat　[baja pugasu terugerinda tanpa pusato]
센터레스 연삭 스프링강　[sentareseu yeonsak spring gang]
เหล็กสปริงที่เจียร์ผิวนอกเพื่อลดขนาด　[lek sa-pring tii chia piw nork pe-ua lod kha-nard]

centrifugal force
　　force acting away from the center of a circle when an object makes circumferencial movement. It is the counter force of centripetal force.
遠心力　[ensinryoku]
　　物体が円運動するとき，円の中心から離れる方向に働く力.
离心力　[lí xīn lì]
gaya sentripetal　[gaya sentoripetaru]
원심력　[wonsimnyeok]
แรงหนีศูนย์　[rang nii soon]

centripetal force
　　force acting toward the center of a circle when an object makes circumferencial movement.
求心力　[kyuusinryoku]
　　物体が円運動するときに，円の中心に向かって働く力.
向心力　[xiàng xīn lì]
gaya sentrifugal　[gaya sentorihugaru]

구심력　[gusimnyeok]
แรงหาศูนย์　[rang haa soon]

ceramic spring
　　spring made of ceramics (sintered materials mainly comprising aluminum oxide and the like), as called by its material.
セラミックばね　[seramikku bane]
　　ばねの材料の種類による名称で，セラミック（酸化アルミニウムを主体とした焼結物及びこれに類似のもの）を用いたばね.
陶瓷弹簧　[táo cí tán huáng]
pegas keramik　[pugasu keramiku]
세라믹 스프링　[seramik spring]
สปริงเซรามิค　[sa-pring ceramic]

CFRP spring
　　a spring made of CFRP material. CFRP is an acronym of Carbon Fiber Reinforced Plastic.
カーボン繊維強化ばね　[kaabon sen'i kyouka bane]
　　CFRP 製のばね. CFRP は Carbon Fiber Reinforced Plastic の略で有機合成繊維，例えばアクリル繊維などを炭化させた繊維（carbon fiber）を強化材とするプラスチック材料.
碳纤维强化弹簧　[tàn xiān wéi qiáng huà tán huáng]
roda gerinda ikatan keramik　[roda gerinda ikatan keramiku]
탄소 섬유 강화 스프링　[tanso seomnyu ganghwa spring]
สปริงคาร์บอนไฟเบอร์　[sa-pring carbon fiber]

chamber furnace
　　a furnace used for batch type operation.
バッチ炉　[battiro]
　　被加熱物を定位置で加熱する炉.
箱式炉　[xiāng shì lú]

chamber furnace — characteristic diagram

tungku ruangan   [tunguku ruangan]
뱃치로   [baetchiro]
เตาเผาสำหรับกระบวนการแบทช์   [tao-pao sam-rab gra-buan gaan batch]

chamfered edge
an edge surface after chamfering.
面取り端面   [mentori tanmen]
面取りされた端面.
端面倒角   [duān miàn dǎo jiǎo]
tepi tirus   [tepi tirusu]
모따기 단면   [mottagi danmyeon]
ขอบที่ลบมุมแล้ว   [khorb tii lop mum laew]

chamfering
process to remove the sharp corners of materials and products.
面取り   [mentori]
素材及び製品の角を落とす工法.
倒角   [dǎo jiǎo]
tirus   [tirusu]
모따기   [mottagi]
การลบมุม   [gaan lop mum]

chamfering device
machine which chamfers the outer or inner circumference of the end of the helical compression spring. There are dedicated machines for outer or inner chamfering.
面取り装置   [mentori souti]
圧縮コイルばねの端部の外周側又は内周側の面取りを行う機械. 外面取り用及び内面取り用がある.
倒角裝置   [dǎo jiǎo zhuāng zhì]
alat penirus   [arato punirusu]
모따기 장치   [mottagi jangchi]
อุปกรณ์ที่ใช้ในการลบมุม   [up-pa-gorn tii chai nai gaan lop mum]

chamfering grinder
machine which chamfers the outer or inner circumference of the end of the helical compression spring. There are dedicated machines for outer or inner chamfering.
面取り研削盤   [mentori kensakuban]
圧縮コイルばねの端部の外周側又は内周側の面取りを行う機械. 外面取り用及び内面取り用がある.
倒角研磨机   [dǎo jiǎo yán mó jī]
mesin penirus   [mesin punirusu]
모따기 연삭반   [mottagi yeonsakban]
เครื่องเจียรลบมุม   [kru-ang chia lop mum]

characteristic curve
a curve showing load vs. deflection characteristics of a spring.
特性曲線   [tokusei kyokusen]
ばねの荷重–たわみ特性曲線.
特性曲线   [tè xìng qū xiàn]
kurva karakteristik   [kuruva karakuterisutiku]
특성 곡선   [teukseong gokseon]
กราฟแสดงโหลดและการยุบตัวของสปริง   [graph sa-daaeng load laae gaan yup to-a khong sa-pring]

characteristic curve during recoil
a characteristic curve of spring during unwinding.
巻戻し特性曲線   [makimodosi tokusei kyokusen]
ぜんまいの巻戻し時の特性曲線. 後戻り曲線ともいう.
滞后特性曲线   [zhì hòu tè xìng qū xiàn]
kurva karakteristik selama menggulung   [kuruva karakuterisutiku serama mengugurungu]
되풀림 특성 곡선   [doepulrim teukseong gokseon]
เส้นกราฟขณะม้วนซ้ำ   [sen graph kha-na muan sam]

characteristic diagram
a systematic diagram showing a relationship between a specific result and

its causes. It is also called a diagram of cause and result.
特性要因図 [tokusei youinzu]
特定の結果と原因系との関係を系統的に表した図.
特性要素图 [tè xìng yào sù tú]
diagram karakteristik [diaguramu karakuterisutiku]
특성 요인도 [teukseong yoindo]
แผนภูมิแสดงเหตุและผล [paaen poom sa-daaeng heet laae pon]

characteristic value
a numerical value showing properties or characteristics of a machine, device, or component.
特性値 [tokuseiti]
機械・器具・部品などにおいて，これらの性質・性能を表す数値.
特性值 [tè xìng zhí]
nilai karakteristik [nirai karakuterisutiku]
특성치 [teukseongchi]
ค่าตัวเลขแสดงคุณสมบัติ [kaa to-a leek sa-daaeng kun-na-som-bat]

Charpy impact test
a test conducted with a Charpy impact test machine.
シャルピー衝撃試験 [syarupii syougeki siken]
シャルピー衝撃試験機を用い，吸収エネルギーを測定する試験.
摆锤式冲击试验 [bǎi chuí shì chōng jī shì yàn]
uji tumbukan charpy [uji tunbukan charupi]
샬피 충격시험 [syalpi chunggyeoksiheom]
เครื่องทดสอบแรงกระแทกแบบชาร์พี้ [kru-ang tod-sorp rang gra-taaek baaep Charpy]

chart recorder
a recorder in which a dependent variable is plotted against an independent variables. The plot may be linear or curvilinear on a strip chart or circular chart.
チャート式記録計 [tyaatosiki kirokukei]
従属変数が独立変数に対応してプロットされる記録計．プロットは細長いチャート上の直線座標か，円形チャート上の曲線座標で行われる.
图表记录器 [tú biǎo jì lù qì]
grafik perekam [gurafiku purekan]
챠-트식 기록계 [chya-teusik girokgye]
เครื่องบันทึกชนิดแสดงผลเป็นแผนภูมิ [kru-ang ban-tuk cha-nid sa-daaeng phon pen paaen-poom]

chassis
a frame on which the body of an automobile is mounted to which driving equipment is attached.
シャシ [syasi]
自動車の車体を載せるフレームで，走行に必要な装置を取り付けたもの.
底盘 [dǐ pán]
sasis [sasisu]
새시 [syaesi]
โครงรถยนต์ประกอบกับอุปกรณ์รถยนต์ [kroong rot-yon pra-gorb gab up-pa-gorn rot-yon]

chassis spring
spring which generally supports the bodies of automobiles, railway cars or the like.
懸架ばね [kenga bane]
一般に自動車，鉄道車両などの車体を支えるばね.
悬架弹簧 [xuán jià tán huáng]
pegas sasis [pugasu sasisu]
새시 스프링 [syaesi spring]
สปริงสำหรับโครงรถยนต์ [sa-pring sam-rab krong rot-yon]

chemical composition
化学成分 [kagaku seibun]

化学成份　[huà xué chéng fèn]
komposisi kimia　[konposisi kimia]
화학 성분　[hwahag sseongbun]
องค์ประกอบทางเคมี　[ong pa-gorb taang ke-mii]

chemical nickel plating
　　a plating technique by a chemical reaction to have similar effect of electrical plating.
　化学ニッケルめっき　[kagaku nikkeru mekki]
　　化学反応によって電気ニッケルめっき同様のめっきを行う方法.成分は,ニッケル,ニッケルりん化合物で防食性,耐摩耗性などを目的とする.
　化学镀镍　[huà xué dù niè]
　pelapisan nikel secara kimia　[purapisan nikeru sechara kimia]
　화학 니켈 도금　[hwahak nikel dogeum]
　การชุบนิคเคิลโดยทางเคมี　[gaan chub nickel dooi tang ke-mii]

chemical resistance
　　resistance of paint, plating, or base metal against chemicals like acid, alkali, or a chloride.
　耐薬品性　[tai yakuhinsei]
　　塗料，めっきあるいは金属素地が酸，アルカリ，塩などに耐え得る程度.
　耐药性　[nài yào xìng]
　ketahanan zat kimia　[ketahanan zato kimia]
　내약품성　[naeyakpumseong]
　การต้านทานต่อสารเคมี　[gaan taan taan ta-o saan ke-mii]

chemical vapor deposition coating (CVD)
　　a process whereby a thin layer of substance is deposited on an object through chemically vapored substance.
　化学蒸着法（CVD）　[kagaku zyoutyaku hou (siibuidjii)]

　化学反応によって，基盤上に単結晶半導体や絶縁膜を成長させる薄膜形成法.
　化学气相蒸镀法　[huà xué qì xiāng zhēng dù fǎ]
　permintaan oksigen secara kimia [purumintaan okusigen sechara kimia]
　화학 증착법 (CVD)　[hwahak jeungchakbeop (CVD)]
　การเคลือบโดยทางเคมี (ซีวีดี)　[gaan klu-ab dooi taang ke-mii CVD]

chromating
　　a way to deposit anti-corrosion film by submerging the work in the solution of chromium acid or chromium acid salt.
　クロメート処理　[kuromeeto syori]
　　クロム酸又はクロム酸塩を主成分とする溶液中に品物を浸せきして，防せい皮膜を生成させる方法.
　镀铬　[dù gè]
　ketahanan terhadap zat kimia [ketahanan teruhadapu zato kimia]
　크로메이트처리　[keuromeiteucheori]
　การชุบโครเมียม　[gaan chub cromium]

circlip
　　generic term for circular springs, such as C type retaining ring. (see Fig C-1)
　サークリップ　[saakurippu]
　　C形止め輪のような輪状のばねの総称.（Fig C-1 参照）
　开口弹簧挡圈　[kāi kǒu tán huáng dǎng quān]
　pengkhormatan　[pungukuhorumatan]
　스냅 링　[seunaep ring]
　เซอร์คลิป　[circlip]

circular arc cam
　　a cam to have a contour line of a base circle, tip circle, and tangential lines.
　円弧カム　[enko kamu]
　　基礎円，先端円及びこれらと接する直線からなる輪郭のカム.

圆弧凸轮　[yuán hú tū lún]
klip cincin　[kuripu chinchin]
원호 캠　[wonho kaem]
แคมรูปวงกลม　[cam roop wong-glom]

circular coil spring
generic term for coil springs coiled in a circular shape.
円形コイルばね　[enkei koiru bane]
円形に巻いたコイルばねの総称.
圆柱螺旋弹簧　[yuán zhù luó xuán tán huáng]
nok busur　[noku busuru]
원형 코일 스프링　[wonhyeong koil spring]
สปริงขดรูปวงกลม　[sa-pring khod roop wong-glom]

circular section wire spring
a spring made of round cross section wire.
円形断面ばね　[enkei danmen bane]
円形断面をした素材（通称、丸線）を用いたばね.
圆截面钢丝弹簧　[yuán jié miàn gāng sī tán huáng]
pegas kawat berpenampang bulat　[pugasu kawato berupenanpangu burato]
원형 단면 선 스프링　[wonhyeong danmyeon seon spring]
สปริงชนิดลวดกลม　[luad sa-pring cha-nid luad glom]

circumferential speed
a linear speed of object in circumferential movement.
周速度　[syuusokudo]
回転部の外周の速度.
角速度　[jiǎo sù dù]
kecepatan lingkar　[kechepatan ringukaru]
원주 속도　[wonju sokdo]
ความเร็วรอบนอก　[kwaam rew rorb-nork]

cladding
a rolling process in an elevated temperature whereby different metal sheets are put together to make laminated material.
クラッド　[kuraddo]
異種金属板を密着させ高温下で圧延して合せ板とすること.
双金属材料　[shuāng jīn shǔ cái liào]
pelapisan　[purapisan]
크래드 법　[keuraedeu beop]
การรีดหุ้ม　[gaan riid hum]

clamp
fittings to hold a round object like cable, hose and pipe in place tightly.
クランプ　[kuranpu]
線、ホース、ブーツなどを固定する部品.
夹紧　[jiā jǐn]
penjepit, pemegang　[punjepito, pumegangu]
클램프　[keullaempeu]
แคลมป์　[clamp]

clamping bolt
a bolt to fasten an object tightly.
締付けボルト　[simetuke boruto]
対象物を締金とともに締め付けるボルト.
夹紧螺栓　[jiā jǐn luó shuān]
baut penjepit　[bauto punjepito]
체결 볼트　[chegyeol bolteu]
โบลท์ยึดแคลมป์　[bolt yud clamp]

clamping test
a test to hold a spring in a compressed state under certain temperature and load conditions.
締付試験　[simetuke siken]
一定の温度や荷重条件下でばねに一定のたわみを与えて保持する試験.
夹紧试验　[jiā jǐn shì yàn]
uji jepit　[uji jepito]
체결 시험　[chegyeol siheom]
การทดสอบการจับยึด　[gaan tod sorb gaan

clamping test — clip bolt

chab yud]

clean room
    a room in which elaborate facilities are employed to reduce dust particles and other contaminations in the air.
    クリーンルーム　[kuriin ruumu]
    空気中のちり粒子その他の不純物を極めて低く抑えるための設備を施した室.
    清洁室　[qīng jié shì]
    ruang bersih　[ruangu berusihu]
    청정실　[cheongjeongsil]
    ห้องคลีนรูม　[horng clean room]

clearance
    a difference between a hole diameter and a shaft diameter. It is also a gap between a punch and a die where a press die is referred.
    クリアランス　[kuriaransu]
    穴径と軸径の差．プレス金型では，パンチとダイのすきまをいう．
    间隙　[jiān xì]
    kelonggaran　[kerongugaran]
    틈새　[teumsae]
    ช่องว่าง　[chorng waang]

cleavage fracture
    a brittle fracture with no plastic deformation. The fracture surface looks glossy, since the cleavage run through a certain cristal plane.
    へき開破壊　[hekikai hakai]
    ほとんど塑性変形を伴わないぜい性破壊．特定の結晶面で破壊するので光沢のある破面が観察される．
    断裂强度　[duàn liè qiáng dù]
    kuat belah　[kuato berahu]
    벽개 파괴　[byeokgae pagoe]
    การแตกเปราะ　[gaan taa-ek pra-or]

clip
    metal component to keep the leaves of a leaf spring in position. (see Fig C-7)
    クリップ　[kurippu]
    重ね板ばねのばね板が，相互に離れること及び横ずれすることを防ぐ金具．（Fig C-7 参照）
    夹箍　[jiā gū]
    klip　[kuripu]
    클립　[keullip]
    คลิปรัดแหนบ　[clip rad naaep]

clip band
    clip body made of steel plate. (see Fig C-8)
    クリップバンド　[kurippu bando]
    クリップに用いる板状の部品．（Fig C-8 参照）
    夹紧带　[jiā jǐn dài]
    pelat klip　[perato kuripu]
    클립 밴드　[keullip baendeu]
    สายรัดแหนบ　[saai rad naaep]

clip bolt
    a bolt to fasten a clip band.
    クリップボルト　[kurippu boruto]

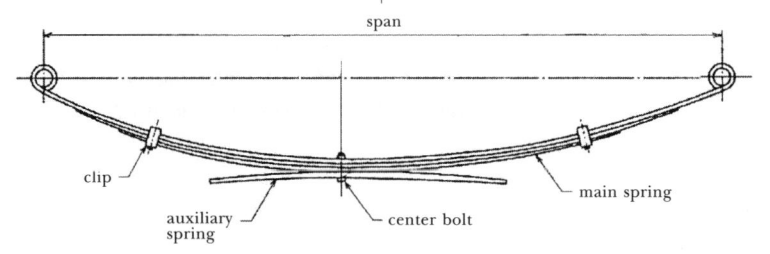

Fig C-7 clip

clip bolt — closed loop

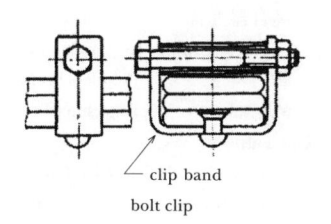

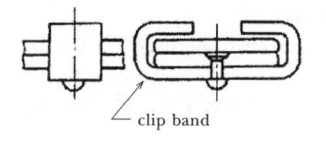

bolt clip   clinched clip

Fig C-8 clip band

重ね板ばねのクリップに用いるボルト．
夹紧螺栓 [jiā jǐn luó shuān]
baut klip [bauto kuripu]
클립 볼트 [keullip bolteu]
โบลท์ยึดคลิป [boly yud clip]

clip pipe
    a tubular type fittings for leaf spring clip.
クリップパイプ [kurippu paipu]
重ね板ばねのクリップに用いる管部品．
夹紧管 [jiā jǐn guǎn]
pipa klip [pipa kuripu]
클립 파이프 [keullip paipeu]
ไปป์ยึดคลิป [pipe yud clip]

clock spring
    a contact type power spring to provide power for clock.
時計用ぜんまいばね [tokeiyou zenmai bane]
主に時計に用いられる動力用接触形の渦巻ばね．
钟表用弹簧 [zhōng biǎo yòng tán huáng]
pegas lonceng [pugasu ronsengu]
시계용 태엽스프링 [sigyeyong taeyeop spring]
สปริงลานนาฬิกา [sa-pring laan na-ri-gaa]

closed and ground end coil
    a coil spring with its closed end ground to provide a flat bearing surface.
密着形研削座部 [mittyakugata kensaku zabu]
座部が密着しているコイルばねの研削してある端面．
两端圈并紧并磨平 [liǎng duān quān bìng jǐn bìng mó píng]
ulir ujung tertutup dan digerinda [uriru ujungu terututupu dan digerinda]
연삭 맞댐 끝 [yeonsak matdaem kkeut]
บริเวณขดปลายปิด [bo-ri-wen khod plaai pid]

closed end
    compression spring ends with coil pitch angle reduced so they are square with the spring axis. The end turn touches 2nd turn.
クローズドエンド [kuroozudo endo]
コイルばねの端末が2番コイルに接触するように巻かれている座部．
端圈并紧 [duān quān bìng jǐn]
ujung pegas ulir yang tertutup [ujungu pugasu uriru yangu terututupu]
비연삭 맞댐 끝 [biyeonsak matdaem kkeut]
ขดปลายปิด [khod plaai pid]

closed loop
    an end loop of tension springs.
すきま無しフック [sukima nasi hukku]
引張ばね端部が環状になっているもの．
无间隙簧钩 [wú jiàn xì huáng gōu]
lingkar tertutup [ringukaru terututupu]
밀폐 고리 [milpye gori]

ลูปปิด [luup pid]

clutch disk spring
    a disc spring to be used for friction clutch.
    クラッチディスクスプリング [kuratti djisuku supuringu]
    摩擦クラッチに用いられている皿ばね．
    离合器碟形弹簧 [lí hé qì dié xíng tán huáng]
    pegas cakram kopling [pugasu chakuramu kopuringu]
    클러치 디스크 스프링 [keulleochi diseukeu spring]
    สปริงแผ่นคลัช [sa-pring paaen clutch]

clutch fin spring
    a sector shaped spring used for friction clutch to hold friction plate.
    クラッチフィンスプリング [kuratti fin supuringu]
    摩擦クラッチの外周部に取り付けられている摩擦板（フェーシング）の内側に取り付けられている扇形の薄板ばね．
    离合器翼片弹簧 [lí hé qì yì piàn tán huáng]
    pegas kopling sekat [pugasu kopuringu sekato]
    클러치 휜 스프링 [keulleochi hwin spring]
    สปริงที่มีครีบสำหรับแผ่นคลัช [sa-pring tii mee kriip sam-rab paaen clutch]

clutch spring
    helical torsion spring with a comparatively long solid coil, used for transferring torque only in the winding direction using the winding force of the coiling section.
    クラッチスプリング [kuratti supuringu]
    密着巻き部分が比較的長いねじりコイルばねで，コイル部の巻締め力を利用し，巻締めの方向にだけトルクを伝達するばね．

离合器弹簧 [lí hé qì tán huáng]
pegas kopling [pegasu kopuringu]
클러치 스프링 [keulleochi spring]
สปริงคลัช [sa-pring clutch]

Cm Value
    an index defined by $Cm=T/6\sigma$ where T refers to tolerance width and σ refers to standard deviation. It represents the degree of stable run of a machine.
    機械能力指数 [kikai nouryoku sisuu]
    $Cm=T/6\sigma$ で定義される指数．Tは公差幅でσは標準偏差．本指数は機械設備の安定稼働の程度を表す．
    机械性能参数 [jī xiè xìng néng cān shù]
    nilai Cm [nirai Cm]
    기계 능력 지수 [gigye neungnyeok jisu]
    ค่าซีเอ็ม [kaa CM]

coating
    painting, chemical plating, plastic coating for the purpose of protecting springs.
    被覆処理 [hihuku syori]
    塗装，表面保護皮膜などのばねに施す処理．
    覆膜处理 [fù mó chǔ lǐ]
    pelapisan [purapisan]
    코팅 [koting]
    การเคลือบ [gaan klu-ab]

coating test
    a test to be conducted to verify the durability of coating including exposing test, weather resistance test, salt spray test, and adhesion strength test.
    コーティング耐久試験 [kootjingu taikyuu siken]
    コーティングの耐久性を調べる試験で，一般的には暴露試験，耐候性試験，塩水噴霧試験などの耐食性試験や密着性試験などがある．
    覆膜耐久试验 [fù mó nài jiǔ shì yàn]

uji kontinuitas pelapis　[uji kontinuitasu purapisu]
코팅 내구 시험　[koting naegu siheom]
การทดสอบความทนทานของการเคลือบสี　[gaan tod-sorb kwaam ton-taan khong gaan klu-ab sii]

coating thickness
thickness of coating difined by film thickness gauge.
皮膜厚さ　[himaku atusa]
膜厚計で測った塗料の厚み.
保护膜厚度　[bǎo hù mó hòu dù]
tebal pelapis　[tebaru purapisu]
피막 두께　[pimak dukke]
ความหนาของการเคลือบ　[kwaam naa khong gaan klu-ab]

cobalt alloy spring
a spring made of cobalt alloy which performs higher strength, durability, and acid resistance.
コバルト合金ばね　[kobaruto goukin bane]
高温強度, 耐酸化性, 耐食性に優れたコバルト基超合金を用いたばね.
钴合金弹簧　[gǔ hé jīn tán huáng]
pegas kobalt paduan　[pugasu kobaruto paduan]
코발트 합금 스프링　[kobalteu hapgeum spring]
สปริงโคบอลท์อัลลอย　[sa-pring cobalt alloy]

coefficient of cubic expansion
a thermal expansion coefficient applied to volume expansion. For isotropic substance, it is three times as much as liner expansion coefficient.
体膨張率　[tai boutyouritu]
体積の熱膨張に用い, 等方性固体では線膨張率の3倍である.
体积膨胀系数　[tǐ jī péng zhàng xì shù]
koefisien muai kubik　[koefisien muai kubiku]
체적 팽창 계수　[chejeok paengchang gyesu]
สัมประสิทธิ์ของการขยายตัวของปริมาตร　[sam-pra-sit khong gaan kha-yaai toa khong pa-ri-mart]

coefficient of friction
a ratio of friction force to the perpendicular force when two objects are in contact with each other.
摩擦係数　[masatu keisuu]
物体が接触しているとき, 接触面に生じる摩擦力と両面間の法線力の比をいう.
摩擦系数　[mó cā xì shù]
koefisien gesekan　[koefisien gesekan]
마찰 계수　[machal gyesu]
สัมประสิทธิ์ของแรงเสียดทาน　[sam-pra-sit khong rang-siead-taan]

coefficient of rolling friction
the ratio of the frictional force which is parallel to the surface of contact to the perpendicular force to the surface when the motion of the body is rolling around an arbor.
回転摩擦係数　[kaiten masatu keisuu]
物体が軸上で回転するとき, 接触面に平行な摩擦力と接触面に対する法線力との比をいう.
滚动摩擦系数　[gǔn dòng mó cā xì shù]
koefisien gesekan putar　[koefisien gesekan putaru]
회전 마찰 계수　[hoejeon machal gyesu]
สัมประสิทธิ์ของแรงเสียดทานหมุน　[sam-pra-sit khong rang-siead-taan mun]

coefficient of sliding friction
the ratio of the frictional force which is parallel to the surface of contact to the perpendicular force to the surface when the motion of the body is sliding

coefficient of sliding friction — coiled spring pin

on the other body.
すべり摩擦係数　[suberi masatu keisuu]
物体が他の物体上ですべり運動するとき，表面に平行な摩擦力と接触面に対する法線力の比をいう．
滑动摩擦系数　[huá dòng mó cā xì shù]
koefisien gesekan luncur　[koefisien gesekan runkuru]
미끄럼 마찰 계수　[mikkeureom machal gyesu]
สัมประสิทธิ์ของแรงเสียดทานเลื่อน　[sam-pra-sit khong rang-siead-taan le-un]

coefficient of static friction

the ratio of the frictional force which is necessary to initiate the motion and parallel to the surface of the contact to the force which is vertical to the surface.
静止摩擦係数　[seisi masatu keisuu]
静止状態の物体の運動を開始させるきに必要な表面に平行な力と，法線力の比．
静摩擦系数　[jìng mó cā xì shù]
koefisien gesekan statis　[koefisien gesekan statisu]
정지 마찰 계수　[jeongji machal gyesu]
สัมประสิทธิ์ของแรงเสียดทานสถิต　[sam-pra-sit khong rang-siead-taan sa-tit]

coefficient of thermal expansion

length or volume expansion rate per unit rise in temperature.
熱膨張率　[netu boutyouritu]
物体の温度が1℃上昇したときの長さ又は体積の膨張の割合．
热胀系数　[rè zhàng xì shù]
koefisien pemuaian　[koefisien pumuaian]
열 팽창율　[yeol paengchangyul]
สัมประสิทธิ์ของการขยายตัวของความร้อน
[sam-pra-sit khong gaan kha-yaai to-a khong kwaam rorn]

coil

コイル　[koiru]
螺旋　[luó xuán]
ulir　[uriru]
코일　[koil]
ขด　[khod]

coil end

end tip of a coil spring.
コイル端末　[koiru tanmatu]
コイルばねのコイル状に巻かれた材料の先端部．
螺旋端面　[luó xuán duān miàn]
ujung ulir　[ujungu uriru]
코일 끝　[koil kkeut]
ปลายขด　[plaai khod]

coil spring

coil-shaped spring. This includes helical compression spring, helical extension spring and helical torsion spring.
コイルばね　[koiru bane]
コイル状のばね．圧縮コイルばね，引張コイルばね，ねじりコイルばねなど．
螺旋弹簧　[luó xuán tán huáng]
pegas ulir　[pugasu uriru]
코일 스프링　[koil spring]
สปริงขด　[sa-pring khod]

coiled spring pin

spring pin coiled between 1 to 1.5 times.
二重巻きスプリングピン　[nizyuumaki supuringu pin]
スプリングピンの一種で，1巻き以上1.5巻き以下に丸めたピン．
螺旋弹簧销　[luó xuán tán huáng xiāo]
pin pegas terulir　[pin pugasu teruriru]
이중감김 롤 핀　[ijunggamgim rol pin]

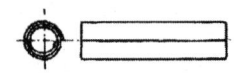

Fig C-9　coiled spring pin

พินที่มีลักษณะเป็นขด [pin tii mii lak-sa-na pen khod]

coiled spring with tapered material
    coil spring made of tapered round bar.
    テーパコイルばね [teepa koiru bane]
    材料の直径が変化しているコイルばね.
    锥形螺旋弹簧 [zhuī xíng luó xuán tán huáng]
    pegas ulir dengan bahan tirus [pugasu uriru dengan bahan tirusu]
    테이퍼 코일 선스프링 (미니부록 스프링 등) [teipeo koil seon spring (miniburok spring deung)]
    สปริงขดชนิดปลายเรียว [sa-pring khod cha-nid plaai-riew]

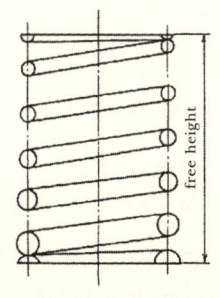

Fig C-10 coiled spring with tapered material

coiled volute friction spring
    a volute shaped spring made of plate material with a telescoping function. It has a style of cone or barrel. The friction between plates works for shock absorbing function.
    つる巻き摩擦ばね [turumaki masatu bane]
    板材を用いて渦巻形に巻いたばね．板間接触による摩擦特性を利用する．円すい形のものとたる形形状のものとがある．
    接触型截锥涡卷弹簧 [jiē chù xíng jié zhuī wō juǎn tán huáng]

pegas gesek vokute terulir [pugasu geseku vokute teruriru]
마찰형 벌류트 스프링 [machalhyeong beollyuteu spring]
สปริงขดก้นหอยเสียดทาน [sa-pring khod gon-hoi siead-taan]

coiled wave spring
    a flat wire coiled and waved in a manner where flat surface is facing each other.
    波形コイルスプリング [namigata koiru supuringu]
    主にばね用平線を幅広面が向かい合うようにコイリングし，そのコイルした円周に波形をつけてばね機能を持たせたもの．
    螺旋波形弹簧 [luó xuán bō xíng tán huáng]
    pegas gelombang terulir [pugasu geronbangu teruriru]
    파형 스프링 [pahyeong spring]
    สปริงขดรูปคลื่น [sa-pring khod ruup kluun]

coiling
    process of coiling to provide the wire or the bar with curvature and torsion to form into the coiled shape.
    コイリング [koiringu]
    コイルの加工工程をいい，線又は棒に指定の曲率曲げ及び指定のねじり率のねじりを与え，コイル状に成形する加工．
    卷簧 [juǎn huáng]
    mengulir [menguriru]
    코일링 [koilling]
    การม้วน [gaan muan]

coiling arbor
    core shaft to form the wire into a coil and to cut the wire after coiling.
    心金 [singane]
    線又は条をコイル状に成形するための心軸．

coiling arbor — coiling tool

芯軸 ［xīn zhóu］
poros pengulir　［porosu punguriru］
코일링 아버　［koilling abeo］
แกนม้วน　［gaaen muan］

coiling machine
　machine which forms coil springs.
　コイリングマシン　［koiringu masin］
　コイルばねを成形する機械.
　卷簧机　［juǎn huáng jī］
　mesin ulir　［mesin uriru］
　코일링기　［koillinggi］
　เครื่องม้วน　［kru-ang muan］

coiling mandrel
　core shaft to form the wire into a coil and to cut the wire after coiling.
　心金　［singane］
　線又は条をコイル状に成形するための心軸.
　芯軸　［xīn zhóu］
　alat bentuk pengulir　［arato bentuku punguriru］
　맨드릴　［maendeuril］
　แกนม้วน　［gaaen muan］

coiling pin
　tool which provides the material with the curvature when forming a coil spring. The coiling machine has either an one pin or a two pin coiling mechanism.
　コイリングピン　［koiringu pin］
　コイルばねをコイリングマシンで成形するとき，材料に曲率を与えるために使用する工具．コイリングマシンには，これを1本使用するタイプ及び2本使用するタイプがある．
　卷簧销　［juǎn huáng xiāo］
　pin pengulir　［pin punguriru］
　코일링 핀　［koilling pin］
　สลักในการม้วน　［sa-lak nai gaan muan］

coiling point
　the point where a coiling pin touches to the wire.
　コイリングポイント　［koiringu pointo］
　コイリングピンがワイヤと接触するポイント.
　卷簧销　［juǎn huáng xiāo］
　titik pengulir　［titiku punguriru］
　코일링 핀　［koilling pin］
　จุดม้วน　［chud muan］

coiling roller
　a roller mounted on the tip of coiling pin to reduce friction during coiling operation.
　コイリングローラ　［koiringu roora］
　コイリングするときにコイルの曲率を与えるために用いるコイリングピン先端のローラ.
　卷簧滚轮　［juǎn huáng gǔn lún］
　rol pengulir　［roru punguriru］
　코일링 롤러　［koilling rolleo］
　ลูกกลิ้งสำหรับม้วน　［luuk-gling sam-rab muan］

coiling system
　a coiling condition in hot forming or cold forming.
　コイリング方式　［koiringu housiki］
　熱間又は冷間成形でコイリングする条件.
　卷簧系统　［juǎn huáng xì tǒng］
　sistim pengulir　［sisutimu punguriru］
　코일링 방식　［koilling bangsik］
　ระบบการม้วน　［ra-bob gaan muan］

coiling tool
　tools of coiling machine like feed roller, wire guide, coiling pin, or mandrel, etc.
　コイリング工具　［koiringu kougu］
　コイリングするための工具のことで，フィードロール，ワイヤガイド，コイ

リングピン，心金などの作業工具の総称．
巻簧工具　[juǎn huáng gōng jù]
alat pengulir　[arato punguriru]
코일링 공구　[koilling gonggu]
อุปกรณ์การม้วน　[up-pa-gorn gaan muan]

coining
   a process whereby a piece of metal is given a shape in one pair of die.
コイニング　[koiningu]
彫り型のついた一対のダイの間で材料に圧力を加え表面に凸凹を作る工法．
模圧　[mó yā]
pencetakan　[punsetakan]
코이닝　[koining]
การรีดขึ้นด้านเดียว　[gaan riid khun daan dieaw]

cold drawn
   a process whereby wire is drawn through dice in room temperature.
冷間引抜き　[reikan hikinuki]
常温でダイスを通じて引抜加工をすることで，線材，鋼管，棒鋼などの成形加工に用いられる．
冷拉拔　[lěng lā bá]
ditarik dalam keadaan dingin　[ditariku daran keadaan dingin]
냉간 인발　[naenggan inbal]
การดึงขึ้นรูปเย็น　[gaan dung khun ruup yen]

cold drawn steel wire
   steel wire made by cold drawn process.
冷間引抜鋼線　[reikan hikinuki kousen]
常温でダイスを通じて引抜加工して伸線した線．硬鋼線，ピアノ線などがある．
冷拉钢丝　[lěng lā gāng sī]
kawat baja ditarik dingin　[kawato bajya ditariku dingin]
냉간 인발 강선　[naenggan inbal gangseon]
ลวดขึ้นรูปเย็น　[luad khun ruup yen]

cold formed helical compression spring
   a compression coil spring formed in room temperature.
冷間成形圧縮コイルばね　[reikan seikei assyuku koiru bane]
冷間で成形する圧縮コイルばね．
冷成形压缩螺旋弹簧　[lěng chéng xíng yā suō luó xuán tán huáng]
pegas tekan ulir bentukan dingin　[pugasu tekan uriru bentukan dingin]
냉간성형 압축 코일 스프링　[naenggan seonghyeong apchuk koil spring]
สปริงกดชนิดขึ้นรูปเย็น　[sa-pring god cha-nid khun ruup yen]

cold formed helical extension spring
   an extension coil spring formed in room temperature.
冷間成形引張コイルばね　[reikan seikei hippari koiru bane]
冷間で成形する引張コイルばね．
冷成形拉伸螺旋弹簧　[lěng chéng xíng lā shēn luó xuán tán huáng]
pegas tarik ulir bentukan dingin　[pugasu tariku uriru bentukan dingin]
냉간성형 인장 코일 스프링　[naenggan seonghyeong injang koil spring]
สปริงดึงก้นหอยชนิดขึ้นรูปเย็น　[sa-pring dung gon hoi cha-nid khun ruup yen]

cold formed spring
   spring formed in low temperature, such as the "cold formed coil compression spring" and the "cold formed coil extension spring", as called by its forming method.
冷間成形ばね　[reikan seikei bane]
ばねの成形方法による名称で，"冷間成形圧縮コイルばね"，"冷間成形引張コイルばね"など，冷間で成形するばね．
冷成形弹簧　[lěng chéng xíng tán huáng]

cold formed spring — collet

    pegas dibentuk dingin　[pugasu dibentuku dingin]
    냉간 성형 스프링　[naenggan seonghyeong spring]
    สปริงขึ้นรูปที่อุณหภูมิห้อง　[sa-pring khun ruup tii un-na-ha-poom hong]

cold forming
    a forming process in room temperature.
    冷間成形　[reikan seikei]
    常温で加工，成形すること．
    冷成形　[lěng chéng xíng]
    membentuk dalam keadaan dingin　[menbentuku daran keadaan dingin]
    냉간 성형　[naenggan seonghyeong]
    การขึ้นรูปที่อุณหภูมิห้อง　[gaan khun ruup tii un-na-ha-poom hong]

cold hardening
    hardening effect caused by cold forming.
    常温硬化　[zyouon kouka]
    常温で伸線又は圧延加工を施すことによって硬化させること．
    常温硬化　[cháng wēn yìng huà]
    pengerasan dalam keadaan dingin　[pungerasan daran keadaan dingin]
    상온 경화　[sangon gyeonghwa]
    ชุบแข็งจากการขึ้นรูปเย็น　[chup khaaeng chaak gaan khun ruup yen]

cold roll
    a rolling process in room temperature. Most of the materials for flat springs are made by this process.
    冷間圧延　[reikan atuen]
    常温で圧延すること．薄板ばね用材料のほとんどがこの工法である．
    冷压延　[lěng yā yán]
    rol dingin　[roru dingin]
    냉간 압연　[naenggan abyeon]
    การรีดที่อุณหภูมิห้อง　[gaan riid tii un-na-ha-poom hong]

cold rolled steel strip
    a steel strip made by cold roll.
    冷間圧延鋼帯　[reikan atuen koutai]
    常温で圧延した鋼帯．
    冷压带钢　[lěng yā dài gāng]
    keping baja kanal dingin　[kepingu baja kanaru dingin]
    냉간 압연 강대　[naenggan abyeon gangdae]
    แผ่นเหล็กรีดเย็น　[paaen lek riid yen]

cold setting
    intentional set given to a spring in room temperature. This operation is to prevent undesirable set during operation.
    常温セッチング　[zyouon settingu]
    常温でばねにあらかじめ使用荷重以上の荷重を加えてある程度の永久変形を生じさせばねの弾性限度を高める操作．
    常温立定　[cháng wēn lì dìng]
    set dingin　[seto dingin]
    냉간 세팅　[naenggan seting]
    เซ็ตดิ้งที่อุณหภูมิห้อง　[setting tii un-na-ha-poom hong]

cold work
    a process to induce work hardening below recrystallization temperature.
    冷間加工　[reikan kakou]
    再結晶温度以下で加工硬化を伴う加工方法．
    冷加工　[lěng jiā gōng]
    pekerjaan dalam keadaan dingin　[pukerujaan daran keadaan dingin]
    냉간 가공　[naenggan gagong]
    กระบวนการชุบแข็งแบบเย็น　[gra-buan gaan shup-khaaeng baaep yen]

collet
    a sprit, coned sleeve to hold small, cir-

cular tools or works in the nose of a lathe type machine.
コレット　[koretto]
旋盤のような機械の先端に取り付けられ，工具や加工物を保持するために用いられる円すい状ですり割りを設けたスリーブ状金具．
夹头　[jiā tóu]
kolet　[koreto]
콜릿　[kollit]
อุปกรณ์สำหรับยึดชิ้นงานในเครื่องกลึง　[up-pa-gorn sam-rab yud shin-ngaan nai kru-ang glung]

color check
a defect detection method on metal surface by infiltrating color liquid (red in usual).
カラーチェック　[karaa tyekku]
欠陥中にカラー（通常は赤色）の浸透液を浸透させ，毛管現象によって，拡大した模様を目視で観察する方法．
着色检验　[zhuó sè jiǎn yàn]
pemeriksaan warna　[pumerikusaan waruna]
칼라 체크　[kalla chekeu]
การตรวจเช็ครอยแตกด้วยสี　[gaan truat check roi taaek do-ui sii]

color marking
a small color mark put on metal parts for identification.
識別　[sikibetu]
物品の形状，形式などを目で見て分かりやすくするため，赤や青などの色を物品や梱包に塗布あるいは着色すること．
着色识别　[zhuó sè shí bié]
penandaan warna　[punandaan waruna]
마킹　[making]
การมาร์คสี　[gaan mark sii]

combination spring
combination of multiple springs to obtain the required characteristics. Springs may be combined in a serial or parallel way. A combination spring comprising two springs is called serial double spring or parallel double spring.
組合せばね　[kumiawase bane]
所要の特性を得るために，幾つかのばねを組み合わせたばね．組合せ方法には，直列法又は並列法がある．2個のばねを組み合わせたものは，直列2連ばね又は並列2連ばねという．
组合弹簧　[zǔ hé tán huáng]
pegas kombinasi　[pugasu konbinasi]
조합 스프링　[johap spring]
สปริงชนิดผสม　[sa-pring cha-nid pa-som]

combined double wire torsion spring
a helical torsion spring formed by a grooved wire and a round wire. It utilizes the friction forces between coils when torsional moment is applied.
組合せ二本線ねじりコイルばね　[kumiawase nihonsen neziri koiru bane]
溝付の素線と丸線をコイリング加工したねじりコイルばね．ねじりモーメントが加わると，線間摩擦が大きくなることを利用したばね．
双层组合扭转弹簧　[shuāng céng zǔ hé niǔ zhuǎn tán huáng]
pegas puntir kombinasi dua kawat　[pugasu puntiru konbinasi dua kawato]
이중선 조합 비틀림 코일 스프링　[ijungseon johap biteullim koil spring]
สปริงขดดึงชนิดลวดคู่　[sa-pring khod dung sha-nid kuu]

combustion engine
燃焼機関　[nensyou kikan]
燃烧发动机　[rán shāo fā dòng jī]

mesin pembakaran [mesin punbakaran]
연소 기관 [yeonso gigwan]
เครื่องยนต์ระบบเผาไหม้ [kru-ang yon ra-bob pa-o]

combustion furnace
燃焼加熱炉 [nensyou kaneturo]
燃烧加热炉 [rán shāo jiā rè lú]
tungku pembakaran [tunguku punbakaran]
연소 가열로 [yeonso gayeollo]
เตาเผาระบบเผาไหม้ [ta-o pa-o ra-bob pa-o mai]

composite fiber plastic spring
　a spring made of FRP (Fiber Reinforced Plastics).
繊維強化プラスチックばね [sen'i kyouka purasutikku bane]
　FRP (Fiber Reinforced Plastics) と呼ばれる複合材料を用いたばね。FRPは広義には繊維で補強されたプラスチックの総称。
纤维强化塑料弹簧 [xiān wéi qiáng huà sù liào tán huáng]
pegas serat komposisi [pugasu serato konposisi]
섬유 강화 프라스틱 스프링 [seomnyu ganghwa peuraseutik spring]
สปริงไฟเบอร์ [sa-pring fiber]

composite material
　material consisting of more than two material that have different natures.
複合材料 [hukugou zairyou]
　性質の異なる二つ以上の素材を組み合わせて単独の素材より優れた性質をもたせた材料。繊維強化プラスチックなどがある。
复合材料 [fù hé cái liào]
bahan campuran [bahan chanpuran]
복합 재료 [bokhap jaeryo]
วัตถุดิบชนิดวัสดุประกอบ [wat-tu-dip cha-nid was-sa-du pra-gorb]

composite spring
　spring made of composite resin materials, as called by its material.
複合材料ばね [hukugou zairyou bane]
　ばねの材料の種類による名称で，合成樹脂材料を用いたばね。
树脂弹簧 [shù zhī tán huáng]
pegas komposisi [pugasu konposisi]
복합 지료 스프링 [bokhap jiryo seupeuring]
สปริงชนิดผสม [sa-pring cha-nid pa-som]

compound corrosion test
　an accelarated corrosion test by exposing test pieces in humid and dry condition alternatively.
複合腐食試験 [hukugou husyoku siken]
　湿潤と乾燥を複合し，一定サイクルで行う腐食促進試験。
复合腐蚀试验 [fù hé fǔ shí shì yàn]
uji korosi campuran [uji korosi kanpuran]
복합 부식시험 [bokhap busiksiheom]
การทดสอบการสึกกร่อนในภาวะแห้งสลับเปียก [gaan tod-sorb gaan suk-gron nai paa-wa haaeng sa-lab pi-eak]

compound layer
　a surface layer on the metal created by a thermochemical process consisting of a chemical compound made up of a diffused element and a matrix element.
化合物層 [kagoubutu sou]
　熱化学処理により金属表面に形成された導入元素と母金属からなる化合物の層。
化合物层 [huà hé wù céng]
lapisan campuran [rapisan chanpuran]
화합물 층 [hwahammul cheung]
ชั้นของการสารประกอบ [shaan khong saan pra-gorb]

compression spring

a spring subjected mainly to a compressive force. In the narrow sense, it refers to a helical compression spring.

圧縮ばね　[assyuku bane]
　主として圧縮荷重を受けるばね．狭義には，圧縮コイルばね．
压缩弹簧　[yā suō tán huáng]
pegas tekan　[pugasu tekan]
압축 스프링　[apchuk spring]
สปริงกด　[sa-pring god]

compression spring testing machine

a testing machine to measure load and deflection of compression spring.

圧縮ばね試験機　[assyuku bane sikenki]
　圧縮ばねを試験・検査する荷重測定器，疲労寿命試験機などの試験機．
压缩弹簧试验机　[yā suō tán huáng shì yàn jī]
mesin uji pegas tekan　[mesin uji pugasu tekan]
압축 스프링 시험기　[apchuk spring siheomgi]
เครื่องทดสอบสปริงกด　[kru-ang tod-sorb sa-pring god]

compression-tension alternating stress

stress caused by reversed compression and tension load.

圧縮引張繰返し応力　[assyuku hippari kurikaesi ouryoku]
　引張り及び圧縮垂直応力の極大値と極小値の間を単純，かつ周期的に変動する応力．
拉压循环变应力　[lā yā xún huán biàn yīng lì]
tegangan bolak balik tekan tarik [tegangan boraku bariku tekan tariku]
압축인장 반복응력　[apchukinjang banbokeungnyeok]
ความเค้นสลับกด_ดึง　[kwaam ken sa-lab god / duung]

compression-tension fatigue limit

the level of reversed compression and tension load by which the spring does not experience fatigue fracture.

圧縮引張耐久限度　[assyuku hippari taikyuu gendo]
　静的破壊荷重以下の荷重であっても圧縮と引張りの繰返し荷重によって破壊する．何回繰り返しても破壊しなくなる限度荷重を圧縮引張耐久限度という．
拉压疲劳极限　[lā yā pí láo jí xiàn]
tegangan lelah tekan tarik　[tegangan rerahu tekan tariku]
압축 인장 내구한도　[apchuk injang naeguhando]
ขีดจำกัดของความล้ากดดึง　[khiid cham-gad khong kwaam laa god dung]

compression-torsion spring

a spring which has both compression and torsion function.

圧縮・ねじりコイルばね　[assyuku neziri koiru bane]
　圧縮及びねじり荷重が負荷されるばね．
压缩扭转螺旋弹簧　[yā suō niǔ zhuǎn luó xuán tán huáng]
pegas puntir/ tekan　[pugasu puntiru / tekan]
압축토션 코일 스프링　[apchuktosyeon koil spring]
สปริงสำหรับรับแรงกด / บิด　[sa-pring sam-rab rab rang god / bid]

compressive residual stress

compressive residual stress generated in a spring by shot peening. The shot peening exerts a slight increase of hardness and a tension layer on the surface. The compressive residual stress results from these facts.

圧縮残留応力　[assyuku zanryuu ouryoku]
ばねにショットピーニングなどを施すことによって，若干の加工硬化とともに表面層だけが展延される．これに起因して生じる圧縮の残留応力．
压缩残余应力　[yā suō cán yú yīng lì]
tegangan sisa tekanan　[tegangan sisa tekanan]
압축 잔류응력　[apchuk jallyueungnyeok]
ความเค้นตกค้างจากการกด　[kwaam ken tok kaang chaak gaan god]

compressive strength
maximum compressive stress without material breakage.
圧縮強さ　[assyuku tuyosa]
材料が破壊することなしに懸けられる最大荷重を断面積で除した値．
压缩强度　[yā suō qiáng dù]
kekuatan tekan　[kekuatan tekan]
압축 강도　[apchuk gangdo]
ความเค้นกดสูงสุด　[kwaam ken god soong-sud]

compressive stress
a stress that results by a compression load.
圧縮応力　[assyuku ouryoku]
圧縮荷重によって生じる応力．
压缩应力　[yā suō yīng lì]
tegangan tekan　[tegangan tekan]
압축 응력　[apchuk eungnyeok]
ความเค้นกด　[kwaan ken god]

computer aided design (CAD)
a design work with a computer operation assistance.
CAD (Computer Aided Design)　[kyado]
図形情報を介して設計者が対話して行うコンピュータ支援による設計．
计算机辅助设计　[jì suàn jī fǔ zhù shè jì]
disainbantu komputer　[disainbantu konputeru]
캐드 (CAD)　[kaedeu (CAD)]
แคด　[CAD]

computer aided manufacturing (CAM)
a manufacturing process with a computer operation assistance.
CAM (Computer Aided Manufacturing)　[kyamu]
コンピュータの支援を受けて，製作を行うこと．製作分野における総合的な効率化・自動化をねらって実施され，数値制御，生産管理なども含まれる．
计算机辅助制造　[jì suàn jī fǔ zhù zhì zào]
manufaktur bantu komputer　[manufakuture bantu konputeru]
캠 (CAM)　[kaem (CAM)]
แคม　[CAM]

computer-controlled system
a feed back control system in which a computer operates on both the input and feed back signal to effect control.
コンピュータ制御システム　[konpyuuta seigyo sisutemu]
コンピュータが入力側と出力側の双方に信号を与えて制御を行うフィードバック制御システム．
计算机控制系统　[jì suàn jī kòng zhì xì tǒng]
sistem kendali komputer　[sisutemu kendari konputeru]
컴퓨터 제어 시스템　[computer jeeo siseutem]
ระบบที่ควบคุมด้วยคอมพิวเตอร์　[ra bop tii ku-ab kum du-oi computer]

computer-integrated manufacturing
a system in which individual engineering, production, marketing, and support functions of a manufacturing enterprise are organized by a computer.
コンピュータ統合生産　[konpyuuta tougou seisan]
生産企業が持つエンジニアリング，生

産，市場開発及びこれらの支援システム機能がコンピュータによって統合的に運用される生産システム．
計算机集成制造　[jī suàn jī jí chéng zhì zào]
manufaktur terintegrasi dengan komputer　[manufakuturu terintegurasi dengan konputeru]
컴퓨터 통합생산　[computer tonghapsaengsan]
ระบบการผลิตที่รวมโดยคอมพิวเตอร์　[ra-bob gaan pa-lit tii ru-am dooi computer]

computer numerical control
a control system in which numerical values corresponding to desired tool position are generated by a computer.
コンピュータ数値制御　[konpyuuta suuti seigyo]
工具の位置を表す数値情報がコンピュータから与えられる制御システム．
数控　[shù kòng]
kendali komputer numerik　[kendari konputeru numeriku]
컴퓨터 수치제어　[computer suchijeeo]
ระบบที่ใช้คอมพิวเตอร์ควบคุมค่าตัวเลข　[ra-bob tii chai computer ku-ab kum kaa tu-a lek]

concave surface of disk spring
the under side of disk spring when placed edge down.
皿ばねの凹面　[sarabane no oumen]
皿ばね，円盤ばねなどの下面（凹面側）をいう．
碟形弹簧凹面　[dié xíng tán huáng āo miàn]
permukaan cekung dari pegas cakram [purumukaan chekungu dari pugasu chakuran]
접시스프링의 오목면　[jeopsispringui omongmyeon]
ผิวเว้าเข้าของสปริงแผ่น　[piew wa-o khao khong sa-pring paaen]

concentricity
the distance between the center line of a cylindrical body and the reference axis.
同心度　[dousindo]
基準軸線の真の位置から回転体の中心線までの距離．
同軸度　[tóng zhóu dù]
konsentisitas　[konsentisitasu]
동심도　[dongsimdo]
ระนาบเดียวกัน　[ra-naap dieaw gan]

concurrent engineering
the simultaneous design of products and related processes, including all product life-cycle aspects.
コンカレント工学　[konkarento kougaku]
製品のライフサイクル全体を考えて，これに関連する工程の統合化と製品の設計を同時に進める技術．
并行开发工程　[bìng xíng kāi fā gōng chéng]
teknik concurrent　[tekuniku chonchururento]
컨커런트 공학　[keonkeoreonteu gonghak]
วิศวกรรมคอนเคอร์เรนท์　[wis-sa-wa-gam concurrent]

cone surface
the convex surface of cone.
円すい面　[ensuimen]
直線がそれと交わる軸線の周りに回転することによって得られる回転面．
圆锥面　[yuán zhuī miàn]
permukaan kerucut　[purumukaan keruchuto]
원추면　[wonchumyeon]
ผิวนูนของกรวย　[piw noon khong gru-oi]

coned disc spring
spring shaped like a dish without bottom.
皿ばね　[sara bane]
底のない皿形のばね．

coned disc spring — connecting rod type solid stabilizer bar with flat ends

碟形弹簧　[dié xíng tán huáng]
pegas piring kerucut　[pugasu piringu keruchuto]
접시 스프링　[jeopsi spring]
สปริงแผ่นกรวย　[sa-pring paaen gru-oi]

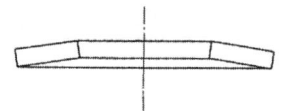

Fig C-11　coned disc spring

coned end with swivel hook

　　a hook inserted into the end of a helical extension spring.
スイベルフック　[suiberu hukku]
引張コイルばねの端部の一種で，端末がテーパ状に絞り込まれたコイル部にフックが差し込まれたもの．
带转动螺栓的圆锥端　[dài zhuǎn dòng luó shuān de yuán zhuī duān]
kait dengan mur　[kaito dengan muru]
나사박음형 고리　[nasabakeumhyeong gori]
ปลายกรวยชนิดมีห่วงร้อยโบลท์　[plaai gru-oi cha-nid mii hu-ang rooi bolt]

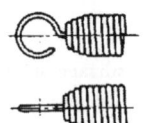

Fig C-12　coned end with swivel hook

conical helical compression spring

　　generic term for cone-shaped compression coil springs.
円すい形圧縮ばね　[ensuikei assyuku bane]
円すい形をした圧縮コイルばねの総称．
圆锥形压缩螺旋弹簧　[yuán zhuī xíng yā suō luó xuán tán huáng]
tekan heliks kerucut　[tekan herikusu keruchuto]
원추형 압축 코일 스프링　[wonchuhyeong apchuk koil spring]
สปริงกดรูปกรวย　[sa-pring god ruup gru-oi]

conical spring

　　generic term for cone-shaped springs, including conical coil springs, volute springs, and cone disk springs.
円すい形ばね　[ensuikei bane]
円すい形のばねの総称で，円すい形コイルばね，竹の子ばね，皿ばねなどがある．
圆锥形弹簧　[yuán zhuī xíng tán huáng]
pegas kerucut　[pugasu keruchuto]
원추형 스프링　[wonchuhyeong spring]
สปริงรูปกรวย　[sa-pring ruup gru-oi]

conical spring washer

　　spring washer shaped like a dish without bottom.
皿ばね座金　[sara bane zagane]
底がない皿状のばね座金．
碟形弹簧垫圈　[dié xíng tán huáng diàn quān]
ring pipih pegas kerucut　[ringu pipihu pugasu keruchuto]
접시 스프링 와셔　[jeopsi spring wasyeo]
แหวนสปริงรูปกรวย　[waaen sa-pring ruup gru-oi]

connecting rod type solid stabilizer bar with flat ends

　　a solid stabilizer bar with both ends flattened having bolt holes for connecting rods.
平つぶし中実スタビライザ　[hiratubusi tyuuzitu sutabiraiza]
腕の先端がコネクティングロッドにねじ締めしやすいように平たくつぶされ，ボルト穴を設けた中実のスタビライザ．
接杆形平端实心稳定杆　[jiē gǎn xíng píng duān shí xīn wěn dìng gǎn]

connecting rod type solid stabilizer bar with flat ends — contact spring

stabiliser solid dengan ujung datar [sutabiriseru sorido dengan ujungu dataru]
컨네팅로트형 플랫 끝 중실 스태비라이저 [keonnetingnoteuhyeong peullaet kkeut jungsil seutaebiraijeo]
เหล็กกันโคลงแบบตันปลายแบนชนิดแท่งเชื่อมต่อ [lek gan klong baaep tan plaai baaen cha-nid taaeng shi-um tou]

connecting rod type tubular stabilizer bar with flat ends

 a tubular stabilizer bar with both ends flattened having bolt holes for connecting rods.
平つぶし中空スタビライザ [hiratubusi tyuukuu sutabiraiza]
腕の先端がコネクティングロッドにねじ締めしやすいように平たくつぶされ、ボルト穴を設けたパイプ状のスタビライザ.
接杆形平端空心稳定杆 [jiē gǎn xíng píng duān kōng xīn wěn dìng gǎn]
stabiliser pipa dengan ujung datar [sutabiriseru pipa dengan ujungu dataru]
컨네팅로트형 플랫 끝 중공 스태비라이저 [keonnetingnoteuhyeong peullaet kkeut junggong seutaebiraijeo]
เหล็กกันโคลงแบบกลวงปลายแบนชนิดแท่งเชื่อมต่อ [lek gan klong baaep glu-ang plaai baaen cha-nid taaeng shi-um tou]

constant force spiral spring

 a roll of pre-stressed strip which exerts approximately constant force to resist uncoiling.
定荷重ぜんまい [teikazyuu zenmai]
たわみが変化しても力がほとんど変化しない密着巻きのぜんまい.
恒力渦卷弾簧 [héng lì wō juǎn tán huáng]
pegas ulir gaya tetapu [pugasu uriru gaya tetapu]
일정하중 스파이럴 스프링
[iljeonghajung seupaireol spring]
แถบป้องกันการคลายขดของสปริงเกลียว [taaeb pong gan gaan klaai khod khong sa-pring gli-eaw]

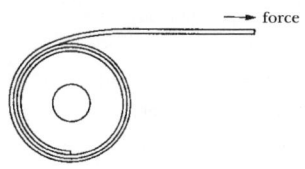

Fig C-13 constant force spiral spring

constant force spring

 spring that indicates constant load or torque despite of the deflection change.
定荷重ばね [teikazyuu bane]
たわみが変化しても、荷重及びトルクがほとんど変化しないばね.
恒力弾簧 [héng lì tán huáng]
pegas gaya tetap [pugasu gaya tetapu]
일정하중 스프링 [iljeonghajung spring]
สปริงที่มีโหลดคงที่ [sa-pring tii mee load kong-tii]

contact pressure

 a contact load of leaf tip against adjacent leaf when a leaf spring is loaded.
接触荷重 [sessyoku kazyuu]
板ばねが負荷されるとき板端が次のリーフに接触する荷重.
接触负荷 [jiē chù fù hè]
tekanan sentuh [tekanan sentuhu]
접촉 하중 [jeopchok hajung]
แรงกดจุดสัมผัส [rang god chud sam-pad]

contact spring

 a spring used for contacts in relays and switches.
接点ばね [setten bane]
リレー、スイッチなどの接点用のばね.
触点弾簧 [chù diǎn tán huáng]
pegas penghubung [pugasu

punguhubungu]
접점 스프링　[jeopjeom spring]
สปริงหน้าสัมผัส　[sa-pring naa sam-pad]

content
含有量　[gan'yuuryou]
含量　[hán liàng]
kandungan　[kandungan]
함유량　[hamnyuryang]
ปริมาณจุ　[pa-ri-maan chu]

continuous furnace
　a furnace constructed to allow work to be moved continuously or in stages from entrance to exit during furnace operation. Types include walking beam, pusher, conveyor, roller hearth, shaker hearth, rotary hearth furnace, rotary retort, strand, and tunnel types.
連続炉　[renzokuro]
　加熱中に被加工物が挿入口から取出口へ連続的に又は段階的に移動するような構造の炉．ウォーキングビーム，プッシャ，コンベヤ，ローラハース，シェーカーハース，回転炉床，回転レトルト，ストランド，トンネル式などがある．
连续炉　[lián xù lú]
tungku berkelanjutan　[tunguku berukeranjutan]
연속로　[yeonsongno]
เตาเผาชนิดต่อเนื่อง　[ta-o pa-o cha-nid to-u nu-ang]

continuous creep limit
　stage at which the strain rate becomes virtually constant in a creep test. Stress at which the steady-state creep rate becomes zero is defined as the creep limit.
定常クリープ限度　[teizyou kuriipu gendo]
　クリープ試験においてひずみ速度がほぼ一定になる段階．高温下での使用上の目安として定常クリープ速度がゼロになるような応力をクリープ限度として定義している．
恒蠕变极限　[héng rú biàn jí xiàn]
batas rambat berlanjut　[batasu ranbato beruranjuto]
정상 크-립 한도　[jeongsang keu-rip hando]
อัตราความเครียดคงที่ขณะทำครีบเทสต์　[at-traa kwaam kri-ead kong tii kha-na tam creep test]

continuous patenting
　a patenting process for an unwound steel wire executed in a continuous manner.
連続パテンチング　[renzoku patentingu]
　結束していない鋼線に施されるパテンチングで連続的に処理される．
连续铅浴淬火　[lián xù qiān yù cuì huǒ]
proses patenting kontinyu　[purosesu patentingu kontinyu]
연속 패턴팅　[yeonsok paeteonting]
กระบวนการพาเทนติ้งแบบต่อเนื่อง　[gra-buan gaan patenting baaep to-a nu-ang]

continuous resistance annealer
　continuous annealing furnace using heat generated by electrical resistance. Includes direct and indirect resistance furnaces.
抵抗式連続焼鈍炉　[teikousiki renzoku syoudonro]
　電気抵抗による発熱を利用して焼なましをする連続炉．直接抵抗炉と間接抵抗炉とがある．
电阻式连续退火炉　[diàn zǔ shì lián xù tuì huǒ lú]
penganeal tahan berkelanjutan [punganearu tahan berukeranjutan]
저항식 연속 풀림로　[jeohangsik yeonsok pullimno]

เตาอบต้านทานชนิดต่อเนื่อง [ta-o op taan taan cha-nid to-u nu-ang]

continuous shaker hearth furnace
    a furnace in which the furnace floor moves back and forth intermittently to convey work into the furnace during operation.
    炉床振動式炉 [rosyou sindousiki ro]
    ワークを加熱炉内に搬送する炉床板を間欠的に往復運動させる構造の炉．
    炉床振动式炉 [lú chuán zhèn dòng shì lú]
    tungku berguncang berkelanjutan [tunguku berugunkangu berukeranjutan]
    로상 진동식로 [rosang jindongsingno]
    เตาเผาชนิดสั่นต่อเนื่อง [ta-o pa-o cha-nid to-u nu-ang]

continuous shot peening
    shot peening performed continuously while work is being moved.
    連続式ショットピーニング [renzokusiki syotto piiningu]
    ワークを移動させながらショットピーニングを連続加工すること．
    通过式喷丸 [tōng guò shì pēn wán]
    shot peening berkelanjutan [shoto peeningu berukeranjutan]
    연속식 쇼트 피닝 [yeonsoksik syoteu pining]
    ช็อทพีนนิ่งชนิดต่อเนื่อง [shot-peening cha-nid to-u nu-ang]

control characteristic
    characteristics subjected to quality control to ensure product quality.
    管理特性 [kanri tokusei]
    製品の品質を保持するために，品質管理の対象として取り上げる特性値．
    控制特性 [kòng zhì tè xìng]
    karakteristik kontrol [karakuterisutiku kontororu]
    관리 특성 [gwalli teukseong]

คุณสมบัติที่ต้องควบคุม [kun-na-som-bat tii tong ku-ob kum]

control chart
    a chart used to check whether processes are consistent. A pair of lines indicating the control limits are drawn on the diagram, and data points representing product quality or process conditions are plotted in order to control processes.
    管理図 [kanrizu]
    工程が安定状態にあるかどうかを調べるために用いる図．管理限界を示す一対の線を引いておき，これに品質又は工程の条件などを表す点を打点し管理する．
    控制图 [kòng zhì tú]
    grafik kontrol [gurafiku kontorolu]
    관리도 [gonghak]
    แผนผังควบคุม [paaen pang ku-ob kum]

upper control limit (UCL)
center line
lower control limit (LCL)
→ time

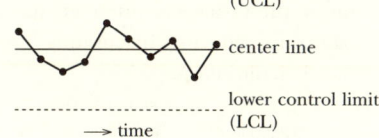

Fig C-14 control chart

control range
    limits indicating the range where the measured value exists with a high probability on the control chart when processes are statistically controlled.
    管理限界 [kanri genkai]
    工程が統計的管理状態にあるとき，管理図上で測定値の値が高い確率で存在する範囲を示す限界．
    控制范围 [kòng zhì fàn wéi]
    rentang pengendalian [rentangu pungendarian]
    관리 한계 [gwalli hangye]

control spring
ช่วงการควบคุม [cho-ang gaan ku-ob kum]
a spring to set the fluid pressure for proportioning valve (valve regulating outlet fluid pressure at constant ratio to inlet fluid pressure) in automotive brakes.
コントロールスプリング [kontorooru supuringu]
自動車ブレーキ部品のひとつであるプロポーショニングバルブ（インレット液圧に対し，アウトレット液圧を一定の比率で減圧するバルブ）の作動液圧を設定するばね．
控制弾簧 [kòng zhì tán huáng]
pegas pengendali [pugasu pungendari]
콘트롤 스프링 [konteurol spring]
สปริงควบคุม [sa-pring ku-ob kum]

controlled atmospheric furnace
a heat treatment furnace that automatically adjusts gaseous atmosphere inside the furnace as intended (e.g., oxidizing, reducing, inactivating, carburizing, nitriding).
雰囲気熱処理炉 [hun'iki netsyori ro]
炉内の雰囲気ガスを酸化性，還元性，不活性，浸炭性，窒化性などの目的にあわせて調整を行う熱処理炉．
气体热处理炉 [qì tǐ rè chǔ lǐ lú]
tungku atmosphere terkendali [tunguku atomosuhere terukendari]
분위기 열처리로 [bunwigi yeolcheoriro]
เตาเผาชนิดปรับระดับก๊าซภายใน [ta-o pa-o cha-nid prab ra-dab gas paai nai]

convection oven
an oven with a convection room.
対流炉 [tairyuu ro]
対流加熱室を設けた炉．
气体炉 [qì tǐ lú]
tungku konveksi [tunguku konbekusi]

대류로 [daeryu ro]
เตาอบชนิดถ่ายเทความร้อน [ta-o pa-o cha-nid taai tee kwaan rorn]

convex
コンベックス（凸） [konbekkusu (totu)]
凸起 [tū qǐ]
cembung [chenbungu]
볼록면 [bollong myeon]
ส่วนโค้งนูน [su-an kong nuun]

convex surface of coned disc spring
upper surface (convex face) of coned disk springs.
皿ばねの凸面 [sarabane no totumen]
皿ばねの上面（凸面側）をいう．
碟形弾簧的凸面 [dié xíng tán huáng de tū miàn]
permukaan cembung dari pegas [purumukaan chenbungu dari pugasu]
접시스프링 볼록면 [jeopsi spring bollongmyeon]
ผิวโค้งนูนของสปริงแผ่นรูปกรวย [piew kong nuun khong sa-pring paaen ruup gru-oi]

conveyer belt furnace
a furnace constructed to allow work to be moved using belt conveyor.
コンベヤ式加熱炉 [konbeyasiki kaneturo]
加熱する品物をベルトコンベヤによって移動搬送する構造の加熱炉．
网带式加热炉 [wǎng dài shì jiā rè lú]
tungku dengan ban berjalan [tunguku dengan ban berujaran]
콘베이어식 가열로 [konbeieosik gayeollo]
เตาเผาที่ใช้สายพานลำเลียง [ta-o pa-o tii chai saai phaan lam-liang]

coolant
冷却材 [reikyakuzai]
冷却材料 [lěng què cái liào]
pendingin [pundingin]
냉각재 [naenggakjae]
ตัว / สารให้ความเย็น [to-a / san hai kwaam

yen]

cooling in furnace
    cooling of steel materials in the furnace after heating and holding them at desired temperature. Primarily used for annealing to remove internal stress, reduce hardness, regulate structure, and improve machinability.
    炉冷 [rorei]
    鋼を目的の温度に加熱，保持した後，炉内で徐冷すること．主に焼なましに用いられ，目的は内部応力の除去，硬さの低下，組織の調整，被削性向上などがある．
    炉内冷却 [lú nèi lěng què]
pendinginan dalam tungku [pundinginan daran tunguku]
    로냉 [ronaeng]
การเย็นตัวของชิ้นงานในเตา [gaan yen to-a khong shin-ngaan nai ta-o]

cooling liquid
    冷却液 [reikyakueki]
    冷却液 [lěng què yè]
    cairan pendingin [chairan pundingin]
    냉각액 [naenggakaek]
ของเหลวที่ทำให้วัตถุเย็นตัว [khong lew tii tam hai wat-tu yen to-a]

cooling power
    cooling capacity of coolant used for quenching.
    冷却能 [reikyakunou]
    焼入れに用いる冷却剤の冷却能力．
    冷却功率 [lěng què gōng lǜ]
kekuatan pendinginan [kekuatan pundinginan]
    냉각능 [naenggangneung]
ความจุของความเย็นในสารหล่อเย็น [kwaam chu khong kwaam yen nai saan lo-a yen]

cooling rate
    temperature drop rate per unit of time due to radiation, conduction, or convection.
    冷却速度 [reikyaku sokudo]
    ある温度が熱放射，熱伝導及び対流によって低下する時間的割合．
    冷却速度 [lěng què sù dù]
laju pendinginan [raju pundinginan]
    냉각 속도 [naenggak sokdo]
อัตราการเย็นตัว [at-traa gaan yen to-a]

cooling temperature
    cooling temperature is one of the most important factor in heat treatment. It is sometimes as important as cooling velocity and heating temperature.
    冷却温度 [reikyaku ondo]
    冷却温度は熱処理では重要な因子の一つである．熱処理目的によって異なるが冷却速度，加熱温度などと並んで重要である．
    冷却温度 [lěng què wēn dù]
suhu pendinginan [suhu pundinginan]
    냉각 온도 [naenggak ondo]
อุณหภูมิการเย็นตัว [un-na-ha puum gaan yen to-a]

cooling time
    time interval between two temperatures during the heat treatment of metals.
    冷却時間 [reikyaku zikan]
    金属の熱処理において，二つの温度間の時間間隔．
    冷却时间 [lěng què shí jiān]
waktu pendinginan [wakutu pundinginan]
    냉각 시간 [naenggak sigan]
ระยะเวลาการเย็นตัว [ra-ya we-laa gaan yen to-a]

cooling unit
    冷却装置 [reikyaku souti]
    冷却装置 [lěng què zhuāng zhì]
satuan pendingin [satuan pundingin]
    냉각 장치 [naenggak jangchi]

อุปกรณ์การทำความเย็น [up-pa-gorn gaan tam kwaam yen]

copper alloy spring
   a spring made of copper alloy. Used in electrical equipments and measuring instruments due to its excellent electrical conduction, non-magnetism, machinability, and corrosion resistance.

銅合金ばね [dougoukin bane]
銅合金ばね材料を用いたばねのこと．電気伝導性，非磁性，加工性，耐食性などに優れていることから，電気，計測機器などに用いられている．

銅合金弹簧 [tóng hé jīn tán huáng]

pegas paduan tembaga [pugasu paduan tenbaga]

동합금 스프링 [donghapgeum spring]

สปริงโลหะผสมทองแดง [sa-pring loo-ha pa-som tong-daaeng]

copper beryllium alloys
   precipitation hardened copper alloy containing 0.4% to 2.2% beryllium and trace amounts of cobalt, nickel, and iron. Used for precision instruments in both wires and plates.

ベリリウム銅合金 [beririumu dougoukin]
銅にベリリウム 0.4～2.2% 及び少量のコバルト，ニッケル，鉄を添加した析出硬化形の合金．線，板ともに精密機器分野に用いられている．

铍青铜合金 [pí qīng tóng hé jīn]

paduan brilium tembaga [paduan buririumu tenbaga]

베릴륨 동 [berillyum dong]

สปริงโลหะผสมแร่นิเกิลและทองแดง [sa-pring loo-ha pa-som raae nin laae tong-daaeng]

core hardness
   internal hardness of a metal. It is not necessarily the same as surface hardness.

内部硬さ [naibu katasa]
金属の内部の硬さのことで，必ずしも表面硬さと同じであるとは限らない．

心部硬度 [xīn bù yìng dù]

kekerasan inti [kekerasan inti]

내부 경도 [naebu gyeongdo]

ความแข็งใจกลางแกน [kwaam-khaaeng jai klaang gaaen]

core structure
   internal structure of a metal. It is compared with surface structure for heat treatment, processing methods, and defect detection.

内部組織 [naibu sosiki]
金属表面から内側の内部組織のことで，熱処理，加工方法，異常検出などの目的で表面組織と比較される．

内部组织 [nèi bù zǔ zhī]

struktur inti [sutorukuture inti]

내부 조직 [naebu jojik]

โครงสร้างใจกลางแกน [krong-saang jai-klaaeng gaaen]

corner press
   pressing work to eliminate sharp edges of or to provide residual compressive stress to the area around the center bolt hole of a leaf spring, to improve the fatigue strength.

コーナープレス [koonaa puresu]
重ね板ばねのばね板のセンタボルト穴周囲に，疲れ強さの向上を目的として行う鋭角除去又は圧縮残留応力を付与するためのプレス加工．

圧稜角 [yā léng jiǎo]

tempa inti [tenpa inti]

코너 프레스 [koneo peureseu]

การเพรสลบขอบคม [gaan press lob khorb kom]

corner radius
コーナー半径 [koonaa hankei]

圓角半径 [yuán jiǎo bàn jìng]
jari-jari pojok [jari-jari pojoku]
코너 반경 [koneo bangyeong]
รัศมีมุม [ras-sa-mii mum]

corrected stress
    stress of coil spring calculated by multiplying the correction factor to the value obtained through a simple equation.
修正応力 [syuusei ouryoku]
コイルばねに生じる応力で，単純な計算式で算出した値に修正係数を乗じたもの．
修正应力 [xiū zhèng yīng lì]
tegangan terkoreksi [tegangan terukorekusi]
수정 응력 [sujeong eungnyeok]
ความเค้นจากการคำนวณ [kwaam-ken chaak gaan kam nu-an]

correction factor
    factor obtained from actual measurements or experience and used to correct theoretical values. Used for correction or calibration.
補正係数 [hosei keisuu]
理論値に対して実情に添うように実測値又は経験値などから得た係数で修正又は補正するのに用いる係数．
修正系数 [xiū zhèng xì shù]
faktor koreksi [fakutoru korekusi]
보정 계수 [bojeong gyesu]
องค์ประกอบในการแก้ไข [ong-pra-gorb nai gaan gaae khai]

corrective action
    steps taken to prevent recurrence of defects or failure. Corrective action is conducted for design, manufacturing, inspection, usage, and maintenance methods.
是正処置 [zesei syoti]
欠陥や故障が再発しないように手段を講じること．設計・製造・検査・使用・方法・保全方法などを対象に行う．
修正処理 [xiū zhèng chǔ lǐ]
aksi koreksi [akusi korekusi]
시정 조치 [sijeong jochi]
มาตรการแก้ไข [maat-tra-gaan gaae khai]

correlation analysis
    quantitative analysis of relationship between one continuously varying measurement and another.
相関分析 [soukan bunseki]
ある測定値の連続的な変化に対して，他の測定値が連続的に変化する場合，これらの関係を定量的に分析すること．
相关分析 [xiāng guān fēn xī]
analisa korelasi [anarisa korerasi]
상관 분석 [sanggwan bunseok]
การวิเคราะห์หาความสัมพันธ์เชิงปริมาณ [gaan wi kro-a haa kwaam sam-pan cha-ung pa-ri-man]

correlation coefficient
    an index to express the relationship between two events. Symbol ρ is usually assigned to the index. When ρ approaches to 1, the two events have a tendency of positive correlation.
相関係数 [soukan keisuu]
二つの事象の相関を表す指標で通常ρを用いて示す．ρが1に近づくとき二つの事象が正の相関を持つ割合が大きくなる．
相关系数 [xiāng guān xì shù]
fungsi korelasi [fungusi korerasi]
상관 계수 [sanggwan gyesu]
สัมประสิทธิ์ของความสัมพันธ์ [sam-pra-sit khong kwaam sam-pan]

corrosion
    phenomenon whereby surface of a

metal erodes due to chemical or electrochemical action.
腐食 [husyoku]
金属がその表面から化学的，電気的に作用されて変質していく現象．
腐蚀 [fǔ shí]
korosi [korosi]
부식 [busik]
การกัดกร่อน [gaan gad gron]

corrosion fatigue
decrease of fatigue strength which occurs in the metal materials caused by the multiple effect of corrosion and repetitive stress.
腐食疲れ [husyoku tukare]
腐食と繰返し応力との相乗作用によって，金属材料に生じる強度低下．
腐蚀疲劳 [fǔ shí pí láo]
kelelahan karena korosi [kererahan karena korosi]
부식 피로 [busik piro]
ความล้าจากการกัดกร่อน [kwaam-laa chak gaan gad gron]

corrosion pit
scores of pits on the surface of steel plate due to corrosion.
腐食荒れ [husyoku are]
鋼板などの表面に腐食によって生じるあばた状のくぼみ．
凹坑 [āo kēng]
lubang korosi [rubangu korosi]
부식 핏트 [busik pitteu]
แนวของการกัดกร่อน [naaew khong gaan gad gron]

corrosion preventing oil
oil applied to protect metals from corrosion.
さびどめ油 [sabidome abura]
金属がさびによって腐食されるのを防ぐために塗布する油．

防锈油 [fáng xiù yóu]
minyak pencegah korosi [minyaku punsegahu korosi]
방청유 [bangcheongyu]
น้ำมันป้องกันการกัดกร่อน [nam-man ponggan gaan gad gron]

corrosion protection
protection of metals from corrosion.
防食 [bousyoku]
金属が腐食するのを防止すること．
防腐蚀 [fáng fǔ shí]
perlindungan korosi [pururindungan korosi]
방식 [bangsik]
การป้องกันการกัดกร่อน [gaan pong-gan gaan gad gron]

corrosion test
a test to investigate anti-corrosion effects on steel in liquids or gases.
腐食試験 [husyoku siken]
液体や気体中での鉄鋼の腐食の起こりやすさ及び防食処理の効果を調べる試験．
耐腐蚀试验 [nài fǔ shí shì yàn]
uji korosi [uji korosi]
부식 시험 [busik siheom]
การทดสอบการกัดกร่อน [gaan tod-sorb gaan gad gron]

corrugated type mesh spring
a spring formed by compressing corrugated fine-wire mesh.
コルゲート式メッシュスプリング [korugeetosiki messyu supuringu]
小径の鋼線を編んだ金網に波形をつけて圧縮成形したばね．
波纹形网格弹簧 [bō wén xíng wǎng gé tán huáng]
pegas jaring jenis corrugate [pugasu jaringu jenisu korurugate]
콜게이트식 메시 스프링 [kolgeiteusik

mesi spring]
สปริงชนิดขดต่อกัน   [sa-pring cha-nid khod to-a gan]

cotter
    a tapered piece that can be driven in a tapered hole to hold together an assemble by machine.
    コッタ   [kotta]
    機械を固定するためにテーパ穴に打ち込まれるくさび状の金具.
    销   [xiāo]
    cotter   [kotteru]
    코터   [koteo]
    ตัวเชื่อมยึดเหล็ก   [to-a shi-um yud lek]

cottered joint
    a joint in which a clevis and a tongue are connected by a cotter.
    コッタ継手   [kotta tugite]
    クレビスとトングをコッタで連結する継手.
    开口铰链   [kāi kǒu jiǎo liàn]
    penyambung cotter   [punyanbungu kotteru]
    코터 커플링   [koteo keopeulling]
    จุดเชื่อมยึด   [chud shi-um yud]

cotter pin
    a pin that retains cotter.
    コッタピン   [kotta pin]
    コッタの抜け止め用のピン.
    开口销   [kāi kǒu xiāo]
    pin cotter   [pin kotteru]
    코터 핀   [koteo pin]
    สลักชนิดลิ่ม   [sa-lak cha-nid lim]

countersink
    a machining process to allow flush fitting of bolt head or nut seats.
    皿座ぐり   [sara zaguri]
    ボルトの首部又はナットの座のすわりを良くする加工.
    锥形沉孔   [zhuī xíng chén kǒng]

perseng   [purusengu]
카운터 싱크   [kaunteosingkeu]
การเจาะรู   [gaan cho-a ruu]

counter type full loop
    reversed full loop over center at the end of a helical extension spring.
    逆丸フック   [gyaku maruhukku]
    引張コイルばねの端部の一種で，丸フックを逆にねじった形のフック.
    反向圆钩环   [fǎn xiàng yuán gōu huán]
    kait bulat terbalik   [kaito burato terubariku]
    역원 고리   [yeokwon gori]
    ลูปเต็มวง   [loop tem wong]

Fig C-15 counter type full loop

coupling
    a component that connects drive shaft to driven shaft.
    軸継手   [zikutugite]
    駆動軸と被駆動軸とを結合する部品.
    联轴器   [lián zhóu qì]
    penghubung   [punguhubungu]
    커플링   [keopeulling]
    จานต่อเพลา   [chaan to-a pla-o]

coverage
    ratio of the area of indentation marked by shot peening to total area of the processed work indicated by percent.
    カバレージ   [kabareezi]
    ショットピーニングにおける圧こん面積と被加工物の総面積との比を百分率で表したもの.
    喷丸覆盖面积   [pēn wán fù gài miàn jī]
    cakupan   [chakupan]
    카바레이지   [kabareiji]
    การกระจายของเม็ดโลหะที่คลอบคลุมผิวโลหะ / การโคเวอร์เรจ   [gaan gra-chaai khong med

loo-ha tii kla-ob kum piw loo-ha]

coverage percentage
 a synonym for "coverage".
 カバレージ百分率　[kabareezi hyakubunritu]
  カバレージと同義語.
 喷丸覆盖率　[pēn wán fù gài lǜ]
 percentase cakupan　[purusentase chakupan]
 카바레이지 백분율　[kabareiji baekbunnyul]
 เปอร์เซ็นต์การโคเวอร์เรจ　[per-cen gaan coverage]

covered wire
 wire which is plated, coated, or covered with resin or other materials for purposes of rust prevention, insulation, etc.
 被覆線　[hihukusen]
  防せい，絶縁などの目的でめっき，塗装，樹脂などで覆った線材.
 绝缘线　[jué yuán xiàn]
 kawat berselaput　[kawato beruserapuln]
 피복선　[piboksen]
 ลวดที่ชุบแล้ว　[lu-ad tii shub laaew]

crack
 a fissure caused by localized breakage due to thermal or mechanical stress.
 割れ　[ware]
  熱の又は機械的応力のために引き起こされる，局部的な破断によって生じるき裂.
 裂纹　[liè wén]
 retak　[retaku]
 균열　[gyunnyeol]
 รอยแตก　[roi-taaek]

crack depth
 the propagated depth of crack.
 割れの深さ　[ware no hukasa]
  き裂の進行した長さ（深さ）をいう.

 裂纹深度　[liè wén shēn dù]
 kedalaman retak　[kedaraman retaku]
 균열 깊이　[gyunnyeol gipi]
 ความลึกรอยแตก　[kwaam luk roi taaek]

crack detector
 a device used to detect cracks. Examples of crack detectors include X-ray inspection, ultrasonic inspection, magnetic particle inspection, penetration testing and eddy current testing.
 割れ検出器　[ware kensyutuki]
  割れを検知する装置のこと．X線透過試験，超音波探傷，磁粉探傷，浸透探傷，渦流探傷試験などがある.
 裂纹探测器　[liè wén tàn cè qì]
 penemu retakan　[punemu retakan]
 균열 검출기　[gyunnyeol geomchulgi]
 เครื่องตรวจหารอยแตก　[kru-ang tru-at kaa roi taaek]

crack growth
 the growth of a crack under the deformation due to repeated loading.
 き裂成長　[kiretu seityou]
  繰返し荷重による変形のもとでき裂が成長すること.
 裂纹形成　[liè wén xíng chéng]
 jpertumbuhan retakan　[juperutumbuhan retakan]
 균열 성장　[gyunnyeol seongjang]
 การขยายตัวของรอยแตก　[gaan kha-yaai to-a khong roi taaek]

crack propagation
 the propagation of a crack under repeated stress.
 き裂伝ぱ　[kiretu denpa]
  繰返し応力によりき裂が伝わり広がること.
 裂纹扩展　[liè wén kuò zhǎn]
 penjalaran retakan　[punjararan retakan]
 균열 전파　[gyunnyeol jeonpa]

การแพร่กระจายของรอยแตก [gaan praae gra-chaai khong roi taaek]

creep
a phenomenon whereby strain increases with time under stress.
クリープ [kuriipu]
応力を作用させた状態においてひずみが時間とともに増大していく現象.
蠕変 [rú biàn]
rambatan [ranbatan]
크리프 [keuripeu]
การครีป (ความเครียดเพิ่มขึ้นเมื่อเวลาผ่านไป) [gaan creep]

creep forming
low temperature annealing carried out in the state after the spring is compressed to the specified height in room temperature.
クリープテンパ [kuriipu tenpa]
常温でばねを所定の高さまで締め付け，そのままの状態で行う低温焼なまし.
蠕変定形 [rú biàn dìng xíng]
pemberntukan rambatan [punberuntukan ranbatan]
크리프 뜨임 [keuripeu tteuim]
การเกิดการครีป [gaan gerd gaan creep]

creep limit
the maximum stress a given material can withstand in a given time without exceeding a specified quantity of creep.
クリープ限度 [kuriipu gendo]
指定されたクリープ量を超えることなく，指定の時間内で材料が耐え得る最大応力.
蠕変極限 [rú biàn jí xiàn]
bats rambatan [batosu ranbatan]
크리프 한도 [keuripeu hando]
ขอบเขตของการครีปที่ความเค้นสูงสุด [khob khet khong gaan creep tii kwaam ken soong-sud]

creep rate
the rate at which strain develops on a creep curve. Normally refers to a steady-state creep rate.
クリープ速度 [kuriipu sokudo]
クリープ曲線におけるひずみの速度．一般的には定常クリープ速度をいう．
蠕変速度 [rú biàn sù dù]
laju rambatan [raju ranbatan]
크리프 속도 [keuripeu sokdo]
อัตราการครีป [at-traa gaan creep]

creep resistance
the property which withstands high-temperature creep (i.e., phenomenon whereby strain increases with time at high temperature under given stress).
耐クリープ性 [tai kuriipusei]
高温において一定の応力の下でひずみが時間とともに増加する現象を高温クリープといい，これに耐え得る性質をいう．
抗蠕変性 [kàng rú biàn xìng]
tahanan rambatan [tahanan ranbatan]
내크리프성 [naekeuripeuseong]
การทนทานต่อการครีป [gaan ton-taan to-a gaan creep]

creep rupture strength
stress that causes creep rupture after a given length of time at a given temperature.
クリープ破断強さ [kuriipu hadan tuyosa]
一定温度のもとで一定時間でクリープ破断するときの応力.
蠕変失効強度（蠕変失効応力） [rú biàn shī xiào qiáng dù (rú biàn shī xiào yīng lì)]
kuat pecah karena rambatan [kuato pekahu karena ranbatan]

creep rupture strength — critical cooling rate

크리프 파단 강도　[keuripeu padan gangdo]
กำลังที่ทำให้แตกเนื่องจากการครีป　[gam-lang tii tam hai taaek nu-ang chaak gaan creep]

creep strain
　strain that has developped during creep test. It refers to the total of 1st stage (transition creep rate), 2nd stage (constant creep rate), and 3rd stage (acceralated creep rate).
クリープひずみ　[kuriipu hizumi]
　クリープ試験中に生じたひずみ．第1次クリープひずみ（遷移クリープひずみ），第2次クリープひずみ（定常クリープひずみ）及び第3次クリープひずみ（加速クリープひずみ）の合計．
蠕变应变　[rú biàn yīng biàn]
regangan rambatan　[regangan ranbatan]
크리프 변형률　[keuripeu byeonhyeongnyul]
ความเครียดครีป　[kwaam kri-ead creep]

creep strength
　stress that causes specified distortion at a specified loading time and a given temperature.
クリープ強さ　[kuriipu tuyosa]
　一定の温度の下で規定した負荷時間に規定したひずみを生じる応力．
蠕变强度　[rú biàn qiáng dù]
kuat rambatan　[kuato renbatan]
크리프 강도　[keuripeu gangdo]
กำลังของการครีป　[gam-lang khong gaan creep]

creep tempering
　a low temperature tempering given to a spring which is tightened to a designated hight.
クリープテンパ　[kuriipu tenpa]
　ばねを所定の高さまで締め付け，そのままの状態で行う低温焼なまし．

蠕变回火　[rú biàn huí huǒ]
temper rambatan　[tenperu ranbatan]
크리프 뜨임　[keuripeu tteuim]
การอบครีป　[gaan op creep]

creep test
　test to measures the time-increasing strain of a spring or a test piece by holding it in constant temperature and applying a constant load.
クリープ試験　[kuriipu siken]
　ばねや試験片を一定の温度に保持し，これに一定の荷重を加えて時間とともに増大するひずみを測定する試験．
蠕变试验　[rú biàn shì yàn]
uji rambatan　[uji ranbatan]
크리프 시험　[keuripeu siheom]
การทดสอบการครีป　[gaan tod-sorb gaan creep]

crevice corrosion
　metal corrosion caused by a concentration cell formed in narrow metal to metal or non-metal to metal gaps.
すきま腐食　[sukima husyoku]
　金属間又は金属と他の材料との間にすきまが存在する場合，すきまの内外において腐食電池が構成されて生じる金属腐食．
间隙腐蚀　[jiān xì fǔ shí]
korosi celah　[korosi chelahu]
틈새 부식　[teumsae busik]
การสึกกร่อนจากร่องแตก　[gaan suk-gron chaak rong-taaek]

critical cooling rate
　the minimum cooling rate needed to develop martensitic transformation during hardening of steel.
臨界冷却速度　[rinkai reikyaku sokudo]
　鉄鋼の焼入れの際，マルテンサイト変態を生じるのに必要な最小の冷却速度．

临界冷却速度　[lín jiè lěng què sù dù]
laju pendinginan kritis　[raju pundinginan kuritisu]
임계 냉각 속도　[imgye naenggak sokdo]
อัตราการเย็นตัววิกฤติ　[at-traa gaan yen to-a wi-grit]

cross section
断面　[danmen]
断面積　[duàn miàn jī]
penampang　[punanpangu]
단면　[danmyeon]
หน้าตัดของวัสดุ　[naa-tad khong was-sa-du]

cross section with ditch for spring plate
the cross section of a spring plate with a groove on a part of the single side.
溝付き断面　[mizotuki danmen]
片面の一部に溝を付けたばね板の断面．
沟槽断面　[gōu cáo duàn miàn]
penampang beralur　[punanpangu berarure]
홈 단면　[hom danmyeon]
หน้าตัดของวัสดุแบบมีร่อง　[naa-tad khong was-sa-du baaep mii rong]

cross section with rib for spring plate
rectangular cross-section of a leaf with a groove on one side and a rib on the opposite side at the center of the width.
リブ付き断面　[ributuki danmen]
長方形断面の幅方向の中央の，片面に溝を，片面にリブを付けたばね板の断面．
肋条断面　[lèi tiáo duàn miàn]
penampang dengan rib　[punanpangu dengan ribu]
마루골 단면　[marugol danmyeon]
หน้าตัดของวัสดุแบบแกน　[naa-tad khong was-sa-du baaep gaaen]

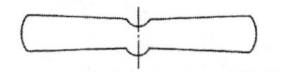

Fig C-16　cross section with rib for spring plate

crosshead
a block sliding between guides as found in testing machines.
クロスヘッド　[kurosuheddo]
試験機等に用いられるガイドの間をスライドするブロック金物．
十字头　[shí zì tóu]
crosshead　[kurosuheado]
크로스헤드　[keuroseuhedeu]
ครอสเฮด　[crosshead]

crossover loop
an end loop of an extension spring formed by crossing over the center of the coil.
クロスオーバーループ　[kurosuoobaa ruupu]
コイル内側に折り返してから成形した引張ばねの端末ループ．
压中心圆钩环　[yā zhōng xīn yuán gōu huán]
lingkaran silang　[ringukaran sirangu]
크로스 오버형 고리　[keuroseu obeohyeong gori]
ลูปชนิดครอสโอเว่อร์　[loop cha-nid crossover]

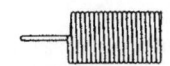

Fig C-17　crossover loop

crushing strength
the maximum load per unit area that a refractory material can withstand without crushing.
破砕強さ　[hasai tuyosa]
耐火物が押しつぶされるまでに耐える

crushing strength — curved washer

単位面積当たりの最大荷重.
粉砕強度　[fěn suì qiáng dù]
kuat remuk　[kuato remuku]
파쇄 강도　[paswae gangdo]
กำลังการชนสูงสุด　[gam-lung gaan shon soong sud]

cryogenic engineering
the study of properties of material under extremely low temperatures. Temperetures of primary interest range from about −120°C to −273°C (absolute zero).
極低温工学　[kyokuteion kougaku]
極低温下における材料特性を研究する工学．対象とする温度領域は −120°C から −273°C（絶対温度）である．
超低温工程　[chāo dī wēn gōng chéng]
teknik kriogenik　[tekuniku kuriogeniku]
극저온 공학　[geukjeoon gonghak]
วิศวกรรมครีโอเจนิค　[wis-sa-wa-gam cryogenic]

curling
process to wind the end of the leaf into a round shape. Also called eye curling.
カーリング　[kaaringu]
ばね板の端部を丸く巻込みする加工方法．目玉巻きともいう．
翻边　[fān biān]
menggulung　[mengugurungu]
귀감기 (둘째판 감기 포함)　[gwigamgi (duljjaepan gamkki poham)]
การม้วนปลายแหนบ　[gaan muan plaai naaep]

Fig C-18　curling

curvature
the reciprocal of the radius of an arc.
曲率　[kyokuritu]

円弧の半径の逆数.
曲率　[qū lǜ]
koefisien lengkung　[koefisien rengukungu]
곡률　[gongnyul]
อัตราโค้ง　[at-tra gaan kong]

curved compression spring
a helical compression spring with the curved axis line.
圧縮湾曲コイルばね　[assyuku wankyoku koiru bane]
軸線が湾曲した圧縮コイルばね.
成形压缩弹簧　[chéng xíng yā suō tán huáng]
pegas tekan lengkung　[pugasu tekan rengukungu]
압축 시 (C) 형 코일 스프링　[apchuk si (C) hyeong koil spring]
สปริงกดรูปโค้ง　[sa-pring god ruup kong]

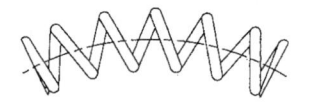

Fig C-19　curved compression spring

curved washer
a slightly bent washer to provide spring action.
波形座金　[namigata zagane]
ばね作用を持たせるためにわずかに曲げられたワッシャ.
成形垫圈　[chéng xíng diàn quān]
ring pipih lenkung　[ringu pipihu renkungu]
커브드 와셔　[keobeudeu wasyeo]

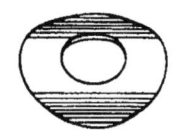

Fig C-20　curved washer

แหวนรองชนิดโค้ง [waaen rong cha-nid koong]

curving press
    a pressing machine used to provide a spring plate with curvature.
    カービングプレス [kaabingu puresu]
    ばね板に曲率を与えるプレスをいう．
    曲面成形机（滚或压） [qǔ miàn chéng xíng jī (gǔn huò yā)]
    tempa pelengkung [tenpa purengukungu]
    커빙 프레스 [keobing peureseu]
    การปั๊มโค้ง [gaan pump kong]

curving roll
    machine which provides the curvature to the leaf by rolling.
    カービングロール [kaabingu rooru]
    ばね板にロールで曲率を与える機械．
    曲面滚轧 [qǔ miàn gǔn zhá]
    rol pelengkung [roru purengukungu]
    커빙 롤 [keobing rol]
    การรีดโค้ง [gaan riid kong]

cushion
    緩衝材 [kansyouzai]
    缓冲器 [huǎn chōng qì]
    bantalan [bantaran]
    쿠션재 [kusyeonjae]
    การบดอัด [gaan bod ad]

cushion type mesh spring
    a compressed wire net spring. Generally it is used for machine's anti-vibration cushion parts.
    クッション式メッシュスプリング [kussyonsiki messyu supuringu]
    金網ばね（小径の鋼線を編んで作った金網）を圧縮成形したばね．一般には，機械設備などの振動防止に用いられる．
    缓冲式网格弹簧 [huǎn chōng shì wǎng gé tán huáng]
    pegas jaring jenis bantalan [pugasu jaringu jenisu bantaran]
    쿠션식 메시 스프링 [kusyeonsik mesi spring]
    สปริงกดชนิดเป็นตาข่าย [sa-pring god cha-nid pen taa-khaai]

cut wire shot
    shot material cut into cylindrical shape from steel wire with specified hardness. Ratio of wire diameter to the length is approximately 1:1 and this may be used for shot peening after the corners are rounded.
    カットワイヤ [katto waiya]
    所定の硬さの鋼線を切断して，円柱状にした投射材．線径と長さとの比がほぼ１：１で，ショットピーニングには，角を丸めてから使用する場合がある．
    钢丝切丸 [gāng sī qiē wán]
    kawat potongan [kawato potongan]
    커트 와이어 [keoteu waieo]
    เม็ดช็อท [med short]

cut wire size
    the length shall be approximately the same as diameter. The diameter is approximately 0.4 mm to 1.2 mm. Hardness should be close to the work.
    カットワイヤの大きさ [katto waiya no ookisa]
    直径に等しい長さの円筒状で直径が 0.4 mm から 1.2 mm 程度まである．硬さは，被加工物と同等のものが用いられる．
    钢丝切丸尺寸 [gāng sī qiē wán chǐ cùn]
    ukuran kawat potongan [ukuran kawato potongan]
    커트 와이어 크기 [keoteu waieo keugi]
    ขนาดเม็ดช็อท [kha-naad med shot]

cut-off grinding
    a process to cut a workpiece by grinding.

研削切断　[kensaku setudan]
　　工作物を研削によって切断する作業.
磨削切断　[mó xuē qiē duàn]
potong dengan gerinda　[potongu dengan gerinda]
연삭 절단　[yeonsak jeoldan]
การตัดออกด้วยวิธีเจียร์　[gaan tad-ork duoi witii chiaa]

cutting force
　　the force imposed on a tool. Cutting force is the sum of reaction force to remove chips during cutting and frictional force between tool and workpiece.
切削抵抗　[sessaku teikou]
　　切削時に切粉を除去する反力と，工具と被削材との摩擦力により工具に加わる力をいう.
切削阻力　[qiē xuē zǔ lì]
gaya pemotong　[gaya pumotongu]
절삭 저항　[jeolsak jeohang]
แรงตัด　[rang tad]

cutting speed
　　a rate of cutting expressed in mm/s at which a cutting tool cuts a workpiece in a lathe and other machine tools.
切削速度　[sessaku sokudo]
　　旋盤，その他工作機械などで刃物が工作物を削る速度を mm/s で表したもの.
切削速度　[qiē xuē sù dù]
laju pemotong　[raju pumotongu]
절삭 속도　[jeolsak sokdo]
ความเร็วการตัด　[kwaam rew gaan tad]

cutting tool
　　a cutting edge to cut the wire after coiling.
カッティングツール　[kattjingu tuuru]
　　コイリング後に線を切断するための刃状工具.
切断工具　[qiē duàn gōng jù]

alat potong　[arato potongu]
절단 날　[jeoldan nal]
อุปกรณ์ตัด　[up-pa-gorn tad]

cyanide zinc plating
　　zinc plating performed in electrolyte which contains cyanide.
シアン化亜鉛めっき　[sianka aen mekki]
　　亜鉛イオンとシアン化ナトリウム，水酸化ナトリウム，及び炭酸ナトリウムからなるめっき浴で行う亜鉛めっき.
氰化鍍锌　[qíng huà dù xīn]
pelapisan seng sianida　[purapisan sengu sianida]
시안화 아연 도금　[sianhwa ayeon dogeum]
การชุบผิวด้วยซิงค์ไซยาไนด์　[gaan chub piw duoi zinc cyanide]

cycle ratio
　　the ratio of the number of cycles of stress to the number of cycles that leads to rupture under the same stress.
繰返数比　[kurikaesisuu hi]
　　同一応力における応力の繰返数の破壊までの繰返数に対する比.
再現比　[zài xiàn bǐ]
perbandingan daur　[purubandingan dauru]
반복회수 비　[banbokhoesu bi]
อัตราการหมุน　[at-tra gaan mun]

cyclic frequency
　　number of cycles per second (unit: Hz).
振動数　[sindousuu]
　　1秒当たりのサイクル数（単位はHz）.
频率　[pín lǜ]
frekuensi daur　[furekuensi dauru]
진동수　[jindongsu]
ความถี่ในรอบ　[kwaan tii nai roab]

cyclic testing
　　a series of test on equipment at regular

intervals.
定期検査　[teiki kensa]
　一定の期間をおいて機器に実施される一連の検査．
定期检查　[dìng qī jiǎn chá]
uji berkala　[uji berukara]
정기검사　[jeonggigeomsa]
การทดสอบตามช่วงเวลาที่กำหนด　[gaan todsorb taam chu-ang wee-laa tii gan-nod]

cyclone separator
　a dust collector in which a wind spirals. Dust particles are trapped by fine meshes under centrifugal force.
遠心分離式集じん装置　[ensinbunrisiki syuuzin souti]
　風が渦を巻く構造の集じん装置．ちりの粒子は遠心力を受けて，目の細かいメッシュに捕捉される．
离心式除尘器　[lí xīn shì chú chén qì]
pemisah centrifugal　[pemisahu chentorifugaru]
원심분리식 집진장치　[wonsimbullisik jipjinjangchi]
เครื่องเก็บฝุ่นชนิดใช้แรงหนีศูนย์　[kru-ang geb fun cha-nid chaai raaeng nii suun]

cylindrical helical compression spring
　a cylindrical coil spring used to bear compressive load.
円筒形圧縮コイルばね　[entoukei assyuku koiru bane]
　円筒形の圧縮荷重を受けるコイルばね．
圆柱螺旋压缩弹簧　[yuán zhù luó xuán yā suō tán huáng]
pegas kompresi heliks silinder　[pugasu konpuresi herikusu sirinderu]
원통형 압축 코일 스프링　[wontonghyeong apchuk koil spring]
สปริงกดชนิดขดทรงกระบอก　[sa-pring god cha-nid khod song gra-borg]

cylindrical helical extension spring
　a cylindrical coil spring used to bear tensile load. The hook shape varies with the purpose of use.
円筒形引張コイルばね　[entoukei hippari koiru bane]
　円筒形の引張荷重を受けるコイルばね．フックの形状については使用目的によって形状がいろいろある．
圆柱螺旋拉伸弹簧　[yuán zhù luó xuán lā shēn tán huáng]
pegas tank heliks silinder　[pugasu tanku herikusu sirinderu]
원통형 인장 코일 스프링　[wontonghyeong injang koil spring]
สปริงดึงชนิดขดทรงกระบอก　[sa-pring dung cha-nid khod song gra-borg]

cylindrical spring
　the generic term for springs of cylindrical shape.
円筒形ばね　[entoukei bane]
　円筒形をしたばねの総称．
圆柱形弹簧　[yuán zhù xíng tán huáng]
pegas silinder　[pugasu sirinderu]
원통형 스프링　[wontonghyeong spring]
สปริงรูปทรงกระบอก　[sa-pring ruup song gra-borg]

cylindrically coiled spring
　the generic term for coil springs of cylindrical shape.
円筒コイルばね　[entou koiru bane]
　円筒形をしたコイルばねの総称．
圆柱螺旋弹簧　[yuán zhù luó xuán tán huáng]
pegas ulir silinder　[pugasu uriru sirinderu]
원통 코일 스프링　[wontong koil spring]
สปริงขดรูปทรงกระบอก　[sa-pring kod ruup gru-oi song gra-borg]

D

D-loop
    an end loop of an extension spring formed by raising half coil.
    D ループ　[djii ruupu]
    半コイルを起こしたD字形の引張コイルばね端末ループ.
    D字形半圓钩环　[D zì xíng bàn yuán gōu huán]
    lingkaran D　[ringukaran D]
    반원고리　[banwon-gori]
    ดีลูป　[D-loop]

Fig D-1　D-loop

damping
    energy consumption by the resistance to the motion of the element or its part of the vibrating system.
    減衰　[gensui]
    振動系の要素又はその一部の運動に対する抵抗力によって, 失われるエネルギー消費.
    阻尼　[zǔ ní]
    peredamam　[puredaman]
    감쇠　[gamsoe]
    การหน่วง　[gaan nu-ang]

dashpot
    a vibration attenuator.
    ダッシュポット　[dassyu potto]
    振動を減衰させる装置.
    減振器　[jiǎn zhèn qì]
    bercak　[beruchaku]
    대시 포트　[daesi poteu]
    อุปกรณ์ลดการสั่น　[up-pagorn lod gaan san]

dead turns
    difference between maximum number of turns and the upper limit number of turns actually used, when winding up the power spring.
    余裕巻き　[yoyuu maki]
    ぜんまいを巻心に巻き付ける方向に回転させるとき, その回転可能な巻数(最大巻数)と使用される範囲の巻数の上限値との差.
    无效线匝　[wú xiào xiàn zā]
    ulir mati　[uriru mati]
    여유권수　[yeoyugwonsu]
    ขดสปริงส่วนที่ไม่ได้ใช้งาน　[khod sa-pring suan tii mai dai chai ngaan]

deburr
    process to remove burrs and curling edge generated by machining and cutting.
    ばり取り　[baritori]
    材料を切断及び切削したときに生じた, ばり及びまくれを取り除く作業.
    去毛刺　[qù máo cì]
    menghilangkan bram　[menguhirangukan buran]
    버 제거　[beo jegeo]
    การขจัดครีบ　[gaan kha-chad creep]

deburred spring ends
    the ends of a coil spring from which burrs are removed by grinding.
    ばり取りされたばね端末　[baritori sareta bane tanmatu]
    コイルばねの座(巻き端末)の研削ばりが取り除かれたものをいう.
    弹簧端面去毛刺　[tán huáng duān miàn qù máo cì]

deburred spring ends — deflection

ujung pegas terhaluskan　[ujungu pugasu teruharusukan]
버제거된 스프링끝　[beo jegeodoen spring kkeut]
ปลายสปริงที่เจียร์ครีบออกแล้ว　[plaai sa-pring tii chia criip oorg laaew]

deburring equipment
　an equipment for deburring.
　ばり取り装置　[baritori souti]
　　ばりを除去するための装置及び付属装置をいう．
　去毛刺装置　[qù máo cì zhuāng zhì]
　alat penghilang bram　[arato punguhirangu buran]
　버 제거 장치　[beo jegeo jangchi]
　อุปกรณ์ขจัดครีบ　[up-pa-gorn kha-chad criip]

decarburization
　phenomenon where the carbon density decreases in the surface layer of the spring during the process of manufacturing, hot working and heat treatment of the spring. When decarburized, fatigue strength of the spring decreases.
　脱炭　[dattan]
　　材料の製造工程，ばねの熱間加工，熱処理などの工程で表層部の炭素濃度が減少する現象．脱炭すると，ばねの疲れ強さが低下する．
　脱碳　[tuō tàn]
　dekarburisasi　[dekaruburisasi]
　탈탄　[taltan]
　การสูญเสียคาร์บอนในโลหะ　[gaan soon-sia carbon nai loo-ha]

decarburization depth
　the distance from a steel surface to a location with a given decarburization rate.
　脱炭深さ　[dattan hukasa]
　　表面からある一定の脱炭率をもつ位置までの距離．

脱碳深度　[tuō tàn shēn dù]
kedalaman dekarburisasi　[kedaraman dekaruburisasi]
탈탄 깊이　[taltan gipi]
ระยะจากผิวถึงชั้นที่เสียคาร์บอนในโลหะ　[ra-ya chaak piw te-ung shan tii sia carbon nai loo-ha]

decontamination
汚染除去　[osen zyokyo]
去污　[qù wū]
penghilangan kontaminasi [punguhirangan kontaminasi]
오염 제거　[oyeom jegeo]
การเอาสิ่งปนเปื้อนออกไป　[gaan aow sing pon pe-un oorg pai]

deep drawing
　the process of working metal sheet into a deep container using hot or cold forming.
　深絞り　[hukasibori]
　　薄板から底深の成形品を熱間成形又は冷間成形により加工する操作をいう．
　深铰　[shēn jiǎo]
　tarikan dalam　[tarikan daran]
　딥 드로잉　[dip deuroing]
　การทำดีพดรออิ้ง　[gaan tam deep drawing]

deflection
　displacement or a rotation angle of a spring generated when a load or moment is applied. Actual examples are as follows: Relative displacement of both ends of the spring for a helical compression spring. Change of the vertical position of the center bolt against the line connecting both eyes for a leaf spring. Rotation angle between both ends of the spring for a torsion bar. Number of windings for a power spring.
　たわみ　[tawami]
　　ばねに荷重，モーメントなどを加えた

deflection — degradation loss

ときに発生する変位又は回転角．圧縮コイルばねの場合は，ばね両端の相対変位をいう．重ね板ばねの場合は，両目玉を結んだ線に対するセンタボルト位置からの垂線の距離の変化をいう．トーションバーの場合は，ばねの両端の回転角をいう．ぜんまいの場合は，巻き回数をいう．
挠度　[náo dù]
lendutan　[rendutan]
휨 량　[hwim lyang]
การยุบตัวของสปริง　[gaan yup to-a khong sa-pring]

deflection angle

the angle formed between the tangent to centerline at a given point of a deflected beam and the centerline of the beam before deflection occurs.
たわみ角　[tawamikaku]
梁がたわんだとき，任意の点でたわんだ梁の中心線への接線と元の梁の中心線がなす角度．
偏角　[piān jiǎo]
sudut lendutan　[suduto rendutan]
편각　[pyeongak]
มุมการยุบตัว　[mum gaan yup to-a]

deflection measurement

to measure the deflection of a spring which is loaded. It means to measure load and deflection relationship.
たわみ測定　[tawami sokutei]
荷重を受けているばねのたわみを測定すること．すなわち，荷重とたわみの関係を測定すること．
挠度测量　[náo dù cè liáng]
pengukuran lendutan　[pungukuran rendutan]
변위 측정　[byeonwi cheukjeong]
การวัดการยุบตัว　[gaan wad gaan yup to-a]

deformation

a variation of dimensions in a body caused by stresses. There are elastic and plastic deformation.
変形　[henkei]
応力によって生じる部材の寸法変化．弾性変形と塑性変形がある．
变形　[biàn xíng]
pembentukan　[punbentukan]
변형　[byeonhyeong]
การเปลี่ยนรูป　[gaan plian ruup]

deformation curve

a curve showing the relationship between the stresses or load on a structural member and the strain or deformation that resalts.
変形曲線　[henkei kyokusen]
応力又は荷重と，その結果生じるひずみ又は変形の関係を示す曲線．
变形曲线　[biàn xíng qū xiàn]
kurva pembentukan　[kuruva punbentukan]
변형 곡선　[byeonhyeong gokseon]
เส้นโค้งแสดงการผิดรูป　[sen kong sa-daaeng gaan pid ruup]

degassing

removal of gaseous components which are included in molten steel.
脱ガス　[datu gasu]
製鋼時に含有されているガス成分を除くこと．
脱气　[tuō qì]
pembuangan gas　[punbuangan gasu]
탈가스　[talgaseu]
การขจัดก๊าซ　[gaan kha-chad gas]

degradation loss

loss due to deteriorated equipment performance. For example, degradation loss is the sum of losses due to decreased production output and increased number of defects, increased operating costs, and maintenance costs incurred to maintain or recover a proper performance.

劣化損失　[rekka sonsitu]
設備の性能の低下による損失．例えば，出来高の減少や不良の増加などによる損失，運転費の増加，性能を維持・回復させるための保全費などの合計である．

劣化損耗　[liè huà sǔn hào]
kehilangan akibat degradasi　[kehirangan akibato deguradasi]

분화 손실　[bunhwa sonsil]
การเสียหายจากการเสื่อมสภาพ　[gaan sia-haai chak gaan se-um sa-paarb]

degrease

cleaning of a metal surface by removing oil or grease adhered thereto.

脱脂　[dassi]
金属表面に付着している油脂性の汚れを除去して清浄にすること．

脱脂　[tuō zhī]
penghilangan lemak　[punguhirangan remaku]

탈지　[talji]
การขจัดจาระบี　[gaan kha-chad cha-ra-bii]

degree of luster

an indicator of the degree of glossiness on the surface of an object.

光沢度　[koutakudo]
物体表面の光沢の程度を表す指標．

光洁度　[guāng jié dù]
derajat kilau　[derajato kirau]

광택도　[gwangtaekdo]
ระดับความมันเงา　[ra-dap kwaam man-ngao]

degree of rustiness

the extent to which a corrosive product develops on metal surface.

発せい度　[hasseido]
金属表面にできる腐食生成物の生成程度をいう．

锈蚀程度　[xiù shí chéng dù]
derajat kekaratan　[derajato kekaratan]

발청도　[balcheongdo]
ระดับการเกิดสนิม　[ra-dap gaan gerd sa-nim]

dehydrogenation

removal of hydrogen from steel surface. The purpose is to prevent hydrogen embrittlement. Dehydrogenation is performed by heating the work at about 200°C for several hours.

脱水素化　[dassuisoka]
鋼表面の水素を除去すること．その目的は，水素ぜい性を防止するためで通常，200°C前後の温度で数時間加熱する．

去氢处理　[qù qīng chǔ lǐ]
mengeringkan hidrogen　[mungeringukan hidorogen]

탈수소화　[talsusohwa]
การขจัดไฮโดรเจนออก　[gaan kha-chad hydrogen ork]

delayed brittle-fracture

a phenomenon whereby a high-strength steel part used under constant load (static load) in a normal environment suddenly breaks down without exhibiting any apparent change. Hydrogen invading into steel is considered the probable main cause of this type of fracture.

遅れ破壊　[okurehakai]
通常の環境下において一定負荷（静荷重）のもとで使用されている高強度鋼部品が，見かけのうえでは何の変化も示さないのにあるとき突然，破壊する現象をいう．遅れぜい性破壊ともいう．遅れ破壊は鋼中に侵入した水素が主原因と考えられている．

delayed brittle-fracture — design of experiments

延遅脆性断裂　[yán chí cuì xìng duàn liè]
retak rapuh tertunda　[retaku rapuhu terutunda]
지연 파괴　[jiyeon pagoe]
การเปราะแตกอย่างช้าๆ　[gaan pra-o taaek yaang shaa shaa]

density
密度　[mitudo]
密度　[mì dù]
kerapatan　[kerapatan]
밀도　[mildo]
ความหนาแน่น　[kwaam naaen]

deposition
the formation of a substance due to separation from a liquid or solid phase into a different solid phase in response to changes in temperature and other conditions.
析出　[sekisyutu]
温度その他の条件を変えることによってある液相又は固相から一つの物質が別の固相を生成して分離してくること.
析出，沉淀　[xī chū, chén diàn]
endapan　[endapan]
석출　[seokchul]
การเปลี่ยนสถานะ　[gaan plian sa-ta-na]

depth micrometer
a micrometer used to measure the depth of holes, slots.
デプスマイクロメータ　[depusu maikuromeeta]
穴あるいは溝の深さを測定するのに用いられるマイクロメータ.
深度測微計　[shēn dù cè wēi jì]
mikrometer pengukur kedalaman [mikurometeru pungukuru kedaraman]
깊이 마이크로 미터　[gipi maikeuro miteo]
ไมโครมิเตอร์สำหรับวัดความลึก　[micrometer sam rap wad kwaam luk]

depth of nitriding
the depth of the nitrogen diffusion layer of the steel surface. Measured by hardness or under a microscope observation.
窒化深さ　[tikka hukasa]
鉄鋼の表面層に窒素を拡散させた層の深さ．硬さや顕微鏡などで測定する．
氮化層深度　[dàn huà céng shēn dù]
kedalaman nitrasi　[kedaraman nitorasi]
질화 깊이　[jilhwa gipi]
ความลึกของการเคลือบผิวด้วยไนไตรด์
[kwaam luk khong gaan klu-ab piw du-oi nitride]

descaling
removal of oxide film on a steel product resulting from rolling or heat treatment processes.
スケール落とし　[sukeeru otosi]
鉄鋼製品の圧延加工や熱処理などに発生する酸化皮膜を除去すること．
去氧化皮　[qù yǎng huà pí]
membuang kerak　[menbuangu keraku]
스케일 제거　[seukeil jegeo]
การขัดผิวสะเก็ด　[gaan kha-chad piw sa-ged]

design load
load used at the design stage.
設計荷重　[sekkei kazyuu]
設計段階で用いる荷重．
设计负荷　[shè jì fù hè]
beban yang dirancang　[beban yangu diranchangu]
설계 하중　[seolgye hajung]
ค่าโหลดจากการดีไซน์　[kaa-load chak gaan design]

design of experiments
methods of designing an experiment by laying out each component of the experiment rationally so that the results can be analyzed in an economic

and precise manner.
実験計画法　[zikken keikakuhou]
合理的に実験を割り付けて，経済的に精度よく結果が解析できるように実験の設計をすること．
实验计划法　[shí yàn jì huà fǎ]
rancangan percobaaan　[rankangan perukobaaan]
실험 계획법　[silheom gyehoekbeop]
การออกแบบการทดลอง　[gaan ork-baaep gaan tod-long]

design standards
generally specified uniform procedures, dimensions, materials, or parts that directly affect the design of a product.
設計標準　[sekkei hyouzyun]
製品の設計に直接影響する寸法，材料，部品等を統一の様式で記述した手順書．
设计标准　[shè jì biāo zhǔn]
standar rancangan　[sutandaru rankangan]
설계표준　[seolgyepyojun]
มาตรฐานในการดีไซน์　[maat-ta-taan nai gaan design]

design stress
a permissible maximun stress to which a machine parts or structural member may be subjected.
設計応力　[sekkei ouryoku]
機械部品あるいは構造部材が受ける応力に許される最大許容値．
设计应力　[shè jì yīng lì]
tegangan rancangan　[tegangan rankangan]
설계응력　[seolgyeeungnyeok]
ค่าความเค้นจากการดีไซน์　[kaa kwaam ken chaak gaan design]

determination of yield point
a test to determine yield point of metals. Yield point is defined as the maximum stress of metal material where the load does not increase and the elongation suddenly becomes larger in tensile tests.
降伏点測定試験　[kouhukuten sokutei siken]
金属材料の引張試験において荷重が増加せず急激に伸び始めるときの応力によって降伏点とする．
屈服点测定试验　[qū fú diǎn cè dìng shì yàn]
penentuan titik mulur　[punentuan titiku murure]
항복점 측정 시험　[hangbokjeom cheukjeong siheom]
การกำหนดยิลด์พอยท์　[gaan gam-nod yield point]

developed leaf method
design method which assumes that the characteristics of a leaf spring is identical to that of a single sheet generated by developing the original leaf spring into a plane, dividing each leaf in equal halves for a symmetric spring or dividing it in proportion to the span ratio for an asymmetrical spring.
展開法　[tenkaihou]
重ね板ばねを，対称ばねの場合は長手方向に 2 等分し，非対称ばねの場合はスパン比に案分して，同一平面上に並べ直してできる 1 枚板のばね特性が，元の重ね板ばねのそれに等しいとする設計方法．
展开法　[zhǎn kāi fǎ]
metoda pengembangan　[metoda pungenbangan]
전개법　[jeongaebeop]
วิธีการพัฒนาแหนบ　[wi-tii gaan pat-ta-na naaep]

developed length

total length of a spring material when developed into straight line. This term is applied to the length of the material after rolling or bending, which differs from the original cut length.

展開長さ　[tenkai nagasa]
ばね製品の部材を平面状に展開したときの全長．素材に圧延，曲げなどの加工を施した場合，素材の切断長さと異なるのでそれと区別して呼ぶ．
展开长度　[zhǎn kāi cháng dù]
panjang yang dikembangkan　[panjangu yangu dikenbangukan]
펼친 길이　[pyeolchin giri]
ระยะที่ได้　[ra-ya tii dai]

developed length of coils spring

length of the centerline of the material of a coil spring when developed on a plane.

コイルばねの展開長さ　[koiru bane no tenkai nagasa]
　コイルばねの材料の中心線を，平面に展開したときの長さ．
螺旋弹簧的展开长度　[luó xuán tán huáng de zhǎn kāi cháng dù]
penjang material pegas koil　[punjangu materiaru pugasu koiru]
코일스프링의 펼친길이　[koilspringui pyeolchin giri]
ความยาวของสปริงขดที่ได้　[kwaam ya-o khong sa-pring khod tii dai]

deviation from circular form

the maximum radial distance between a point on a perfect circle that internally (for inner surface) or externally (for outer surface) contacts a presumed circular line and corresponding point on the latter circular line.

真円度　[sin'endo]
円形であるべき線に内接する円（内面の場合）又は外接する円（外面の場合）と，その円形であるべき線上の各点との間の最大のラジアル距離．
不圆度　[bù yuán dù]
penyimpangan dari bentuk bulat　[puninpangan dari bentuku burato]
진원도　[jinwondo]
การเบี่ยงเบนจากรูปทรงกลม　[gaan bi-ang bain chak ruup song glom]

deviation from cylindrical form

the maximum radial distance in the radial plane between a point on a perfect cylinder that internally (for inner surface) or externally (for outer surface) contacts the surface of a presumed cylinder and corresponding point on the latter cylindrical surface.

円筒度　[entoudo]
円筒状の面に内接する円筒（内面の場合）又は外接する円筒（外面の場合）と，その面上の各点との間のラジアル平面内での最大のラジアル距離．
圆柱度　[yuán zhù dù]
penympangan dari bentuk silinder　[puninpangan dari bentuku sirinderu]
원통도　[wontongdo]
การเบี่ยงเบนจากรูปทรงกระบอก　[gaan bi-ang bain chak ruup song gra-borg]

deviation from parallelism

in a pair of straight lines, a straight line and a plane, or planes parallel to each other, the magnitude of deviation of the line(s) or plane(s) from parallelism using the corresponding line or plane as reference.

平行度　[heikoudo]
平行であるべき直線同士，直線と平面又は平面同士の組合せにおいて，それらのうち一方を基準として，この基準

に対して他方の直線部分又は平面部分の狂いの大きさ．
平行度　[píng xíng dù]
penyimpangan kesejajaran
[puninpangan kesejajaran]
평행도　[pyeonghaengdo]
การเบี่ยงเบนจากการขนาน　[gaan bi-ang bain chak gaan kha-naan]

dew point hygrometer
an instrument used to measure the humidity of sample gas. It indicates the surface temperature of a cooled object under the condition that dew drops appears on the surface.
露点計　[rotenkei]
試料気体の湿度を測定する装置．気体中で物体を冷却し表面に結露が観察された状態での表面温度を表示する．
露点仪　[lù diǎn yí]
pengukur tekanan udara　[pungukuru tekanan udara]
노점계　[nojeomgye]
เครื่องวัดจุดกลั่นตัว　[kru-ang wad chud glan-toa]

dial gauge
a measuring instrument used to obtain precise length and displacement by magnifying minute displacement with gears.
ダイヤルゲージ　[daiyaru geezi]
歯車で微小変位を拡大して長さや変位を精密測定する計器．
千分表　[qiān fēn biǎo]
pengukur bentuk cakra　[pungukuru bentuku chakura]
다이얼 게이지　[daieol geiji]
ไดอัลเกจ　[dial gauge]

diamond wheel
a grinding wheel in which synthetis diamond dust is bonded as the abrasive.

ダイヤモンドホイール　[daiyamondo hoiiru]
研磨材として人工ダイヤモンド粉を結合させた円盤といし．
砂轮　[shā lún]
roda berlian　[roda berurian]
다이어몬드 숫돌　[daieomondeu sutdol]
หินเจียร์ชนิดมีผงเพชร　[hin-chia cha-nid mii pong pet]

diaphragm spring
disc spring which has multiple levers shaped in tongue flaps directing to the center on its inner circumference, generating spring action when supported at the outer circumference and the bases of levers.
ダイアフラムスプリング　[daiahuramu supuringu]
皿ばねの内周側に，中心に向かう複数の舌片状のレバーを形成し，作用時には外周及びレバー基部を支点としてばね作用をするばね．
膜片弹簧　[mó piàn tán huáng]
pegas diafragma　[pugasu diafuraguma]
다이어프램 스프링　[daieopeuraem spring]
สปริงไดอะแฟรม　[sa-pring diaphragm]

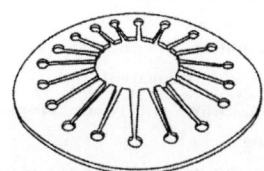

Fig D-2　diaphragm spring

die
a tool or mold used to give shapes to, or to form pattern on, materials.
金型　[kanagata]
材料にパターンを刻んだり，形状を与えるために用いられる工具あるいはひ

な型.
模具　[mó jù]
cetakan　[chetakan]
금형　[geumhyeong]
แม่พิมพ์　[maae-pim]

die block
　a tool-steel block which is bolted to the bed of a punch press working as the part of die.
　ダイブロック　[dai burokku]
　パンチプレスのベッドにボルトで組み付けられる工具鋼のブロックで，金型の一部を構成する.
　模具块　[mó jù kuài]
　tanda matres　[tanda matoresu]
　다이 블록　[dai beullok]
　ดายบล็อค　[die block]

die mark
　a tool mark that appears in stamping operation or a linear thin mark that develops along the extruding or drawing direction on extruded or drawn material.
　ダイマーク　[dai maaku]
　プレス加工における型のきず又は押出し又は引抜材表面の押出し又は引抜方向に現れる線状の細かい凹凸をいう.
　膜具标记　[mó jù biāo jì]
　pegas matres　[pugasu matoresu]
　다이 마크　[dai makeu]
　รอยจากแม่พิมพ์　[roi chaak maae-pim]

die spring
　a spring used in a die for plate metal working or die-casting. Used mainly for holding or projecting. Includes helical compression springs and disc springs.
　金型用ばね　[kanagatayou bane]
　板金加工用やダイカスト用の金型にいるばね．主に押さえや突出し用にいられ，圧縮コイルばね，皿ばねなどがある.
　膜具用弹簧　[mó jù yòng tán huáng]
　pegas matres (pegas untuk cetakan)　[pugasu matoresu (pugasu untuku chetakan)]
　금형용 스프링　[geumhyeongyong spring]
　สปริงสำหรับทำแม่แบบ　[sa-pring sam-rab tam maae baaep]

diffusion coating
　a method of heat treatment to diffuse other metal or nonmetal chemical elements on the surface.
　拡散浸透処理　[kakusan sintou syori]
　表面に他の金属元素又は非金属元素を拡散浸透させる熱処理.
　扩散渗透处理　[kuò sàn shèn tòu chǔ lǐ]
　pelapisan dengan difusi　[purapisan dengan difusi]
　확산 침투처리　[hwaksan chimtucheori]
　การเคลือบแบบกระจาย　[gaan klu-ab baaep gra-chaai]

diffusion layer
　a layer of material where the constituent chemical elements are diffused by heat treatment after thermal spraying.
　拡散浸透層　[kakusan sintousou]
　溶射した後の熱処理によって，溶射皮膜の成分元素を素材に拡散させた層.
　扩散层　[kuò sàn céng]
　lapisan difusi　[rapisan difusi]
　확산 침투층　[hwaksan chimtucheung]
　ชั้นของการแพร่กระจาย　[shan khong gaan praae gra-chaai]

digital control
　a method of control that uses a digital quantity as the main signal within a control system.
　デジタル制御　[dezitaru seigyo]
　制御系内の主要信号としてデジタル量を用いる制御.
　数字控制　[shù zì kòng zhì]

kontrol digital [kontororu digitaru]
디지털 제어 [dijiteol jeeo]
การควบคุมแบบตัวเลข [gaan ku-ob kum baaep to-a leek]

diluent
    a substance added to lower the concentration of another substances.
希釈剤 [kisyakuzai]
ある物質の濃度を低くするために加える物質.
稀释剂 [xī shì jì]
pengencer [pungenser]
희석제 [huiseokje]
สารทำละลาย [saan tam la-laai]

dimension
    a synonym of size.
次元 [zigen]
    寸法.
尺寸 [chǐ cùn]
dimensi [dimensi]
치수 [chisu]
ขนาด [kha-naad]

dip coating
    a method of coating by immersing a workpiece in paint dissolved in water or a solvent.
浸せき塗装 [sinseki tosou]
水又は溶剤に溶解した塗料中に品物を浸せきして塗装を行う方法.
浸涂 [jìn tú]
melapis dengan celupan [merapisu dengan cherupan]
디핑 도장 [diping dojang]
การเคลือบแบบจุ่ม [gaan klu-ab baaep chum]

direct cost
    the cost to manufacture a product which would not be spent if the product were not made.
直接費 [tyokusetuhi]
製品を作る費用で，もしその製品が作られなかったら発生しないものをいう.
直接成本 [zhí jiē chéng běn]
biaya langsung [biaya rangusungu]
직접비 [jikjeopbi]
ค่าใช้จ่ายในการผลิต [kaa shaai chaai nai gaan pa-lit]

direct quenching
    a method of hardening by rapidly removing the work from the furnace and immersing it in agitated oil or water.
直接焼入れ [tyokusetu yakiire]
ワークを炉から手早く取り出して直ちに撹拌されている油又は水中に浸せきして行う焼入法.
直接淬火 [zhí jiē cuì huǒ]
pengerasan langsung [pungerasan rangusungu]
직접 담금질 [jikjeop damgeumjil]
การชุบแข็งทันที [gaan shub khaaeng tan-tii]

direct stress
    the value obtained by dividing tensile load on a test piece by the original cross-sectional area of parallel parts of the test piece. Also called nominal stress.
直接応力 [tyokusetu ouryoku]
試験片に加わる引張荷重を試験片平行部の原断面積で除した値．"nominal stress"ともいう.
正应力 [zhèng yīng lì]
tegangan langsung [tegangan ransungu]
공칭 응력 [gongching eungnyeok]
ความเค้นตรง [kwaam ken trong]

direction of coiling
    a coiling direction of a coil spring in either a left or right-hand manner. If the index finger of the right hand can be bent to simulate direction of coil and end tip position, the spring is right-

direction of coiling — disc clutch

    hand wound, and vice versa.
    コイルの巻方向　[koiru no makihoukou]
    右巻あるいは左巻で表されるコイルばねの巻方向．右手の人差し指がコイルばね巻方向と一致し，コイル端末位置を示す場合は右巻であり，逆もまた同様である．
    旋向　[xuǎn xiàng]
    arah lilitan　[arahu riritan]
    코일의 감김방향　[koirui gamgimbanghyang]
    ทิศทางการม้วน　[tid-taang gaan mu-an]

direction of drawing
    the direction of drawing in drawn material. It affects moderately the mechanical property of the material.
    引抜方向　[hikinuki houkou]
    引抜材における引抜方向をいう．引抜材の機械的性質にある程度影響する．
    拉伸方向　[lā shēn fāng xiàng]
    arah menarik　[arafu menariku]
    인발 방향　[inbal banghyang]
    ทิศทางการดึง　[tid taang gaan duung]

direction of helix
    see "direction of coiling".
    巻方向　[makihoukou]
    "direction of coiling" 参照．
    卷绕方向　[juǎn rào fāng xiàng]
    arah mengulir　[arafu menguriru]
    감김 방향　[gamgim banghyang]
    ทิศทางการขด　[tid taang gaan khod]

direction of rolling
    the direction in which metal is processed by rolling into a wire rod, plate, or other products.
    圧延方向　[atuen houkou]
    金属を圧延機によって，線材，板材などの形状に成形加工するときの方向．
    压延方向　[yā yán fāng xiàng]
    arah rol　[arafu roru]
    압연 방향　[abyeon banghyang]
    ทิศทางการรีด　[tid taang gaan riid]

direction of rotation
    clockwise or counterclockwise derection of a rotation looking toward the axis of the rotation.
    回転方向　[kaiten houkou]
    回転軸に向かって，時計回り又は反時計回りの回転の向きを示す．
    回转方向　[huí zhuǎn fāng xiàng]
    arah putaran　[arafu putaran]
    회전 방향　[hoejeon banghyang]
    ทิศทางการหมุน　[tid taang gaan mun]

disc brake
    a type of brake in which discs attached to a fixed frame are pressed against discs attached to a rotating axle.
    ディスクブレーキ　[djisuku bureeki]
    回転軸に取り付けられたディスクに固定枠側のディスクが押し付けられる形式のブレーキ．
    圆盘式制动器　[yuán pán shì zhì dòng qì]
    rem cakram　[ren chakuran]
    디스크 브레이크　[diseukeu beureikeu]
    ดิสก์เบรค　[disc brake]

disc cam
    a cam shaped like an irregular disk.
    円盤カム　[enban kamu]
    変則的な円盤形状のカム．
    圆盘凸轮　[yuán pán tū lún]
    nok cakram　[noku chakuran]
    원판 캠　[wonpan kaem]
    แคมชนิดจาน　[cam cha-nid chaan]

disc clutch
    a clutch in which torque is transmitted by friction between discs with friction material attached to both sides and contact plates keyed to the inner surface of an external hub.
    円板クラッチ　[enban kuratti]

両面に摩擦材料を取り付けた円板とハブの内面に嵌合された接触板との摩擦によってトルクを伝達するクラッチ．
盘式离合器　[pán shì lí hé qì]
kopling pipih　[kopuringu pipifu]
원판 클러치　[wonpan keulleochi]
คลัทช์ชนิดแผ่น　[clutch cha-nid paaen]

disc grinder
a grinding machine that employs abrasive discs.
ディスクグラインダ　[djisuku gurainda]
円板といしを用いるグラインダ．
圆盘式磨削机　[yuán pán shì mó xuē jī]
grinda cakram　[gurinda chakuran]
디스크 그라인더　[diseukeu geuraindeo]
เครื่องเจียร์ชนิดเป็นแผ่นขัด　[kru-ang chia cha-nid pen paaen khad]

dissolution
a phenomenon where a substance is melt in liquid uniformly.
溶解　[youkai]
液体の中に他の物質が溶け込んで均一な相になること．
溶解　[róng jiě]
penguraian, pelarutan　[punguraian, purarutan]
용해　[yonghae]
การละลาย　[gaan la-laai]

distance between coils
the distance between adjacent centers of wires parallel to the coil axis in the cross-section area containing the coil axis.
コイル間ピッチ　[koirukan pitti]
コイルばねの中心線を含む断面で互いに隣り合うコイル中心線に平行な材料断面の中心距離．
螺距　[luó jù]
jarak antar ulir　[jaraku antaru uriru]
코일간 피치　[koilgan pichi]

ระยะระหว่างขด　[ra-ya ra-waang khod]
distance between inside loop edge and spring body
the dimension (distance) from the spring body of a tensile spring to the inside hook.
フック内側とコイル間長さ　[hukku utigawa to koirukan nagasa]
引張ばねのコイリング部本体からフック内側までの寸法（距離）．
钩环内侧与主体之间的尺寸　[gōu huán nèi cè yǔ zhǔ tǐ zhī jiān de chǐ cùn]
jarak antara tepi lingkar dalam dan badan pegas　[jaraku antara tepi ringkaru daramu dan badan pugasu]
고리내측과 코일본체간 거리　[gorinaecheukgwa koilbonchegan geori]
ระยะระหว่างขอบวงด้านในและตัวสปริง　[ra-ya ra-waang khorb wong daan nai laae to-a sa-pring]

distortion
deviation from specified dimensions or shape of forging products due to thermal handling after forging.
ひずみ　[hizumi]
鍛造後の熱的取扱いに起因する鍛造品の所定寸法形状からの片寄り．
变形　[biàn xíng]
distorsi　[disutorosi]
뒤틀림　[dwiteullim]
การบิดจากรูปปกติ　[gaan bid chaak ruup pak-ka-ti]

distortion due to hardening
change in dimension or shape of steel or parts that occurs after quenching.
焼入れひずみ　[yakiire hizumi]
鋼材や部品を焼入れしたときに発生する寸法変化，形状変化をいう．
淬火变形　[cuì huǒ biàn xíng]
distorsi karena pengerasan　[disutorosi

distortion due to hardening — double coil spring

karena pungerasan]
담금질 변형 [damgeumjil byeonhyeongn]
การบิดจากรูปปกติเนื่องจากการชุบแข็ง [gaan bid chaak ruup pak-ka-ti nu-ang chaak gaan shub khaaeng]

distortion due to heat

a distortion caused by thermal stress, transformation stress, or residual stress due to prior process.
熱ひずみ [netu hizumi]
熱応力，変態応力あるいは前加工の残留応力に起因するひずみ．
热变形 [rè biàn xíng]
distorsi karena panas [disutorosi karena panasu]
열변형 [yeolbyeonhyeong]
การบิดจากรูปปกติเนื่องจากความร้อน [gaan bid chaak ruup pak-ka-ti nu-ang chaak kwaam rorn]

distribution of stresses

stress distribution pattern in a spring.
応力分布 [ouryoku bunpu]
ばねに生じる応力が分布して作用する状態，又はその分布状況．
应力分布 [yīng lì fēn bù]
distribusi tegangan [disutoribusi tegangan]
응력 분포 [eungnyeok bunpo]
การกระจายของความเครียด [gaan gra-chaai khong kwamm kriead]

double action coil spring

a helical spring which has the shape of helical compression spring and screw seats at both ends to receive both tensile and compressive load.
引張・圧縮両用コイルばね [hippari assyuku ryouyou koiru bane]
圧縮コイルばねの形をしており，両端末にねじ込みの座を設け引張荷重と圧縮荷重を共用させるコイルばね．

拉压双作用螺旋弹簧 [lā yā shuāng zuò yòng luó xuán tán huáng]
pegas ulir aksi ganda (tarik-tekan) [pugasu uriru akusi (tariku-tekan)]
인장 압축 양용 코일 스프링 [injang apchuk yangyong koil spring]
สปริงขดคู่ [sa-pring khod kuu]

double-bodied torsion spring

a helical torsion spring with two coiled parts which are subjected to tortional moment.
ダブルねじりばね [daburu neziri bane]
ねじりモーメントが働くコイル部が2か所あるねじりコイルばね．
双体扭转弹簧 [shuāng tǐ niǔ zhuǎn tán huáng]
pegas puntir ganda [pugasu puntiru ganda]
이중 비틀림 스프링 [ijung biteullim spring]
สปริงแบบแรงบิด 2 คำ [sa-pring baaep rang bid song kaa]

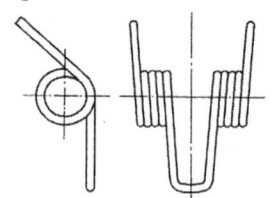

Fig D-3 double-bodied torsion spring

double coil spring

a coil spring consisting of two element wires.
二重コイルばね [nizyuu koiru bane]
2本の素線を用いたコイルばね．
双重螺旋弹簧 [shuāng chóng luó xuán tán huáng]
pegas heliks ganda [pugasu herikusu ganda]
이중 코일 스프링 [ijung koil spring]
สปริงขดประกอบด้วยลวด 2 ชนิด [sa-pring

khod pra-gorb du-oi song cha-nid]

double conical helical extension spring

an extension spring whose coil diameter is gradually reduced toward the end of the coil.

端末絞り引張ばね [tanmatu sibori hippari bane]

コイル端部のコイル径を順次小さくした引張ばね．

端部錐形拉伸弾簧 [duān bù zhuī xíng lā shēn tán huáng]

pegas tarikan heliks kerucut ganda [pugasu tarikan herikusu kerukuto ganda]

원추형 인장 스프링 [wonchuhyeong injang spring]

สปริงรับแรงดึงชนิดขดก้นหอยคู่ [sa-pring rab rang duung cha-nid khod gon hoi kuu]

double leg torsion spring

a torsion spring with straight parts to transmit momentum to both ends.

両腕付きねじりばね [ryouudetuki neziri bane]

両端にモーメントを伝えるための直線部があるねじりばね．

双臂扭簧 [shuāng bì niǔ huáng]

pegas puntir berkaki ganda [pugasu puntiru berukaki ganda]

이중 암 토션스프링 [ijung am tosyeonseupeuring]

สปริงรับแรงบิดขาคู่ [sa-pring rab rang bid kha-kuu]

double loop

a double-ring shaped part at an end of an extension spring.

二重フック [nizyuu hukku]

二重環形状をした引張ばねの取付用端部．

双钩环 [shuāng gōu huán]

lingkaran ganda [ringukaran ganda]

이중 고리 [ijung gori]

ลูปคู่ที่ปลายสปริงดึง [loop kuu tii plaai sa-pring dung]

Fig D-4 double loop

double peening

shot peening carried out in two stages, where the first comparatively strong shot peening is followed by the second shot peening with different grain size and projection speed. It is to obtain an effective distribution of residual stress in the surface layer. Also called two-staged peening.

ダブルピーニング [daburu piiningu]

比較的強度の高いショットピーニングの後，ショットの粒度及び投射速度を変えて，2段階で行うショットピーニング．表面層の効果的な残留応力分布を得るために行う加工．2段ピーニングともいう．

二次喷丸 [èr cì pēn wán]

peen ganda [peen ganda]

이단 피닝 [idan pining]

การยิงเม็ดโลหะ 2 ครั้ง [gaan ying med looha song krang]

double spindle spring end grinding machine

a double spindle machine used to grind the end faces of a helical compression spring.

2軸端面研削盤 [niziku tanmen kensakuban]

圧縮コイルばねの端面を研削するのに用いる2軸の機械．

双端面磨簧机 [shuāng duān miàn mó huáng jī]

mesin gerinda ujung pegas poros ganda [mesin gerinda ujungu pugasu porosu ganda]

double spindle spring end grinding machine — downturned eye

이축단면 연삭기 [ichuk danmyeon yeonsakgi]
เครื่องเจียร์ปลายสปริงแบบ 2 เพลา [kru-ang chiaa plaai sa-pring baaep song pla-o]

double stroke feed mechanism
  a mechanism to feed workpieces in double stroke action.
2段行程送り機構 [nidan koutei okuri kikou]
被加工物の搬送を2行程で行う機構．
双行程供料机构 [shuāng xíng chéng gòng liào jī gòu]
mekanisme umpan langkah ganda [mekanisume unpan rangkahu ganda]
이단 행정 이송 기구 [idan haengjeong isong gigu]
กลไกป้อนคู่ [gon gai porn kuu]

double support transverse leaf spring
  a set of two leaf springs placed parallel to the wheel axle.
横置き2連重ね板ばね [yokooki niren kasaneita bane]
車軸と平行に設置した二つの重ね板ばね．
双支承重叠板弹簧 [shuāng zhī chéng chóng dié bǎn tán huáng]
pegas daun melintang berpenyangga ganda [pugasu daun merintangu berupenyanguga ganda]
횡치식 2 련 겹판 스프링 [hoengchisik iryeon gyeop pan spring]
สปริงแบบแผ่นชนิดยึดทแยง [sa-pring baaep paaen cha-nid yuud ta-yaaeng]

double wire extension spring
  helical extension spring made of two parallel wires.
二本線引張コイルばね [nihonsen hippari koiru bane]
2本の素線からなる引張コイルばね．
双线拉簧 [shuāng xiàn lā huáng]

pegas tarik berkawat ganda [pugasu tariku berukawato ganda]
이중선 인장 스프링 [ijung seon injang spring]
สปริงดึงแบบลวดคู่ [sa-pring duung baaep lu-ad kuu]

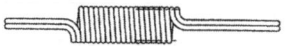

Fig D-5 double wire extension spring

double wire torsion spring
  helical torsion spring made of two parallel wires.
二本線ねじりばね [nihonsen neziri bane]
2本の素線からなるねじりコイルばね．
双线扭转弹簧 [shuān xiàn niǔ zhuǎn tán huáng]
pegas puntir berkawat ganda [pugasu puntiru berukawato ganda]
이중선 비틀림 스프링 [ijung seon biteullim spring]
สปริงรับแรงบิดลวดคู่ [sa-pring rab rang bid lu-ad kuu]

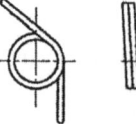

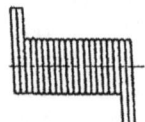

Fig D-6 double wire torsion spring

downturned eye
  an eye of a leaf spring in which the end part of the main leaf is rounded downwards into a loop.
下巻き目玉 [sitamaki medama]
板ばねの親板端部を下向きに丸めた目玉をいう．
下卷耳 [xià juǎn ěr]
gelang terputar kebawah [gerangu teruputaru kebawahu]

아랫방향 귀 [araetbanghyang gwi]
หูแหนบชนิดม้วนลง [hoo-naaep cha-nid muan long]

drawability
the easiness in drawing a workpiece. Good drawability refers to a workpiece which has a good surface condition free of stress cracking after drawing.
引抜加工性 [hikinuki kakousei]
引抜加工を行うときの加工しやすさ．加工後に応力割れがなく表面状態のよいものを加工性がよいという．
拉伸加工性 [lā shēn jiā gōng xìng]
mampu tarik [manpu tariku]
인발 가공성 [inbal gagongseong]
ความสามารถในการดึง [kwaam sa-maad nai gaan duung]

drawbar spring
a compression spring with hooks on both sides to function as an extension spring. It has a feature of having a solid stop which an ordinary extension spring does not provide.
ドローバーばね [doroobaa bane]
両端のフックによって引張ばねの挙動をする圧縮コイルばね．普通の引張ばねでは得られない密着ストッパ機能を有するのが特徴．
拉伸用圧縮弾簧 [lā shēn yòng yā suō tán huáng]
pegas batang tarik [pegasu batangu tariku]
인장 막대 스프링 [injang makdae spring]
สปริงกดชนิดติดตะขอที่ปลาย 2 ด้าน [sa-pring god cha-nid tid ta-khor tii plaai song daan]

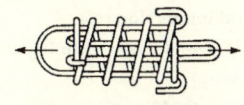

Fig D-7 drawbar spring

drawing speed
the rate of drawing wire in a drawing process. An appropriate rate is determined depending on the material, shape, and reduction ratio, etc.
引抜速度 [hikinuki sokudo]
引抜加工のときの素線を引っ張る速度．材質，形状，減面率などによって適切な速度が定まる．
拉拔速度 [lā bá sù dù]
kecepatan tarikan [kechepatan tarikan]
인발 속도 [inbal sokdo]
อัตราการดึง [at-tra gaan dung]

drawn wire
a drawn wire, including hard steel wire and piano wire.
引抜線 [hikinuki sen]
引抜加工をした線．硬鋼線，ピアノ線などがある．
拉拔线 [lā bá xiàn]
kawat hasil tarikan [kawato hasiri tarikan]
신선 재료 [sinseon jaeryo]
ลวดที่ผ่านการดึง [lu-ad tii paan gaan duung]

dressing
the process of smoothing or cleaning performed to a grinding stone recover the grinding capability.
ドレッシング [doressingu]
といしの研削力を回復するためにといしの表面を清掃し平面度を元に戻すこと．
砂轮修正 [shā lún xiū zhèng]
pencucian [punchuchian]
드레싱 [deuresing]
กระบวนการทำความสะอาดหินเจียร์ใหม่ [gra-bunn gaan tam kwaam sa-aad hin chia mai]

dressing attachment
a device used to grind the blunt surface of a grinding wheel, so as to allow new abrasive grains to emerge.

dressing attachment — dryness and immersion test

目立て装置　[metate souti]
といし車の切れ味が鈍ったとき，といしの表面を削って新しいと粒面を得るための工具装置をいう．
修正装置　[xiū zhèng zhuāng zhì]
alat koreksi gerinda　[arato korekusi gerinda]
드레싱 장치　[deuresing jangchi]
อุปกรณ์แต่งหินเจียร์　[up-pa-gorn taaeng hin chiaa]

drive shaft

a shaft used to transmit power from a motor or engine to the rest of the machine.
駆動軸　[kudou ziku]
原動機の動力を伝達する軸．
驱动轴　[qū dòng zhóu]
unsur penggerak　[unsuru pungugeraku]
구동 축　[gudong chuk]
เพลาขับ　[pla-o khab]

drop forging press

a type of forging press whereby kinetic enegy of dropping ram is used to perform forging operation.
ドロップ鍛造機　[doroppu tanzouki]
ラムが下降するときの運動エネルギーによって鍛造成形を行うプレス．
落锤式锻造机　[luò chuí shì duàn zào jī]
penempa jatuhan　[punenpa jatuhan]
낙하 단조기　[nakha danjogi]
การกดขึ้นรูป　[gaan god khun ruup]

dry drawing

a method of drawing with the aid of a dry lubricant.
乾式引抜き　[kansiki hikinuki]
乾式潤滑材を使用する引抜加工．
干式拉伸　[gàn shì lā shēn]
penarikan kering　[punarikan keringu]
건식 인발　[geonsik inbal]
การดึงแบบแห้ง　[gaan duung baaep haaeng]

dry grinding

a method of grinding without using a liquid lubricant.
乾式研削　[kansiki kensaku]
液状の研削剤を使用しない研削加工．
干式磨削　[gàn shì mó xuē]
penggerindaan kering　[pungugerindaan keringu]
건식 연삭　[geonsik yeonsak]
การเจียร์แบบแห้ง　[gaan chiaa baaep haaeng]

dry honing

surface processing method to spray the abrasive grains directly by compressed air. By this process, surface smoothing and peening effect is obtained.
ドライホーニング　[dorai hooningu]
と粒を直接圧縮空気で吹き付ける表面加工方法．表面の平滑化，ピーニング効果などが得られる．
干式研磨　[gàn shì yán mó]
honing kering　[honingu keringu]
건식 호닝　[geonsik honing]
การลับแบบแห้ง　[gaan lab haaeng]

dry oven

an oven used to remove moisture, solvent, etc. from an object at about 200°C.
乾燥炉　[kansouro]
200℃程度で物体から水分又は溶剤などを取り除くための炉．
干燥烘箱　[gàn zào hōng xiāng]
tungku pengering　[tunguku pungeringu]
건조로　[geonjoro]
เตาอบแห้ง　[ta-o ob haaeng]

dryness and immersion test

乾湿交互浸せき試験　[kansitu kougo sinseki siken]
干湿交互浸渍试验　[gàn shī jiāo hù jìn zì shì yàn]
uji serapan dan pengeringan　[uji serapan

dan pungeringan]
건식 침적 교차 시험 [geonsik chimjeok gyocha siheom]
การทดสอบด้วยการจุ่มและทำแห้ง [gaan todsorb do-ui gaan chum laae haaeng]

dual pitch coil spring
a coil spring wound to have two pitches. It has two-stage spring rate.
二段ピッチコイルばね [nidan pitti koiru bane]
二つのピッチで巻かれたコイルばね. ばね定数が2段階に変化する.
二段节距螺旋弹簧 [èr duàn jié jù luó xuán tán huáng]
pegas ulir pitch ganda [pugasu uriru pitti ganda]
이단 피치 코일 스프링 [idan pichi koil spring]
สปริงขดที่มีพิทช์คู่ [sa-pring khod tii mii pitch kuu]

ductile fracture
phenomenon where the breakdown takes place after large plastic distortion.
延性破壊 [ensei hakai]
大きな塑性変形の後に破壊する現象.
塑性破坏 [sù xìng pò huài]
retakan lunak [retakan runaku]
연성 파괴 [yeonseong pagoe]
การแตกหักจากการยืด / ดึง [gaan taaek hak chaak gaan yuud / duung]

ductility
a property of a material to expand before breaking when external force is applied.
延性 [ensei]
物体に外力が加わるとき, 破壊の前に伸びることのできる性質.
可塑性 [kě sù xìng]
kelunakan [kerunakan]

연성 [yeonseong]
คุณสมบัติในการยืด [kun-na-som-bat nai gaan yuud]

duplex torsion spring
a helical torsion spring with duplex coils. A spring made by turning back the end of a helical torsion spring and then winding the outer coil with the outer diameter of the first coil as the inner diameter of the second coil.
重ね巻きねじりばね [kasanemaki neziri bane]
コイル部を二重にした, ねじりコイルばね. ねじりコイルばねの端末を折り返して, 最初のコイル外径を内径側として, 二重にコイリングしたばね.
双重扭转弹簧 [shuāng chóng niǔ zhuǎn tán huáng]
pegas puntiran rankap [pugasu puntiran rankapu]
이중감김 비틀림 스프링 [ijunggamgim biteullim spring]
สปริงรับแรงบิด 2 ทาง [sa-pring rab rang bid song taang]

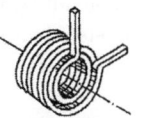

Fig D-8 duplex torsion spring

duplex rectangular wire torsion spring
a duplex torsion spring made of a rectangular wire.
重ね巻き角形断面ねじりばね
[kasanemaki kakugata danmen neziri bane]
角線を用いた重ね巻きねじりコイルばね.
双重方形钢丝截面扭转弹簧 [shuāng chóng fāng xíng gāng sī jié miàn niǔ zhuǎn tán huáng]

pegas puntiran rankap [pugasu puntiran rankapu]
이중감김 각형 단면 비틀림 스프링
[ijunggamgim gakhyeong danmyeon biteullim spring]
สปริงรับแรงบิด 2 ทางชนิดลวดเหลี่ยม [sapring rab raaeng bid song taang cha-nid lu-ad liam]

durability
a property that allows a product to maintain its required functions if it is subjected to stress during normal use.
耐久性 [taikyuusei]
製品が通常使用中にストレスの経過を受けても，所定の機能を保持できる性能．
耐久性（寿命） [nài jiǔ xìng (shòu mìng)]
ketahanan [ketahanan]
내구성 [naeguseong]
ความคงทน [kwaam kong ton]

dust collector
a device used to collect and separate soot, dust, etc. from smoke or polluted air by using gravity, inertia, centrifugal force, heat, diffusion adhesive force, sound wave force, electric force, etc. There are dry type and wet type collectors.
集じん装置 [syuuzin souti]
煤煙又は汚染空気から，煤じん，粉じんなどを分離補集するために，重力，慣性力，遠心力，熱，拡散付着力，音波力，電気力などを利用した装置．乾式と湿式とがある．
集尘器 [jí chén qì]
penyaring debu [punyaringu debu]
집진장치 [jipjinjangchi]
เครื่องเก็บฝุ่น [kru-ang geb fun]

dynamic characteristic test
a test to measure the load and displacement of a spring when step, sinusoidal, random or simulated oscillation is applied.
動的特性試験 [douteki tokusei siken]
ばねにステップ振動，正弦波振動，ランダム振動又は実際に加わる振動を入力し，そのときのばねの動きや荷重を計測する試験．
动特性试验过程 [dòng tè xìng shì yàn guò chéng]
uji karakter dinamis [uji karakuteru dinamisu]
동적 특성 시험 [dongjeok teukseong siheom]
การทดสอบลักษณะการทำงานแบบไดนามิค [gaan tod-sorb lak-sa-na gaan tam ngaan baeep dynamic]

dynamic damper
a device consisting of weight and rubber spring attached to the large amplitude section of a leaf spring to suppress resonance. (see Fig D-9)
ダイナミックダンパ [dainamikku danpa]
重ね板ばねの共振抑制のために，振幅の大きい部分に付けるダンパで，おもり，ゴムばねなどで構成される部品．(Fig D-9 参照)
动力阻尼器 [dòng lì zǔ ní qì]
peredaman dinamis [puredaman dinamisu]
매스 댐퍼 [maeseu daempeo]
ไดนามิคแดมเปอร์ [dynamic damper]

dynamic load
time dependent load, such as repeatedly varying load or oscillating load.
動的荷重 [douteki kazyuu]
繰り返し変動する繰返し荷重，大きさと向きが変わる交番荷重など，時間的に変動する荷重．
动态负荷 [dòng tài fù hé]

dynamic load — dynamic strength

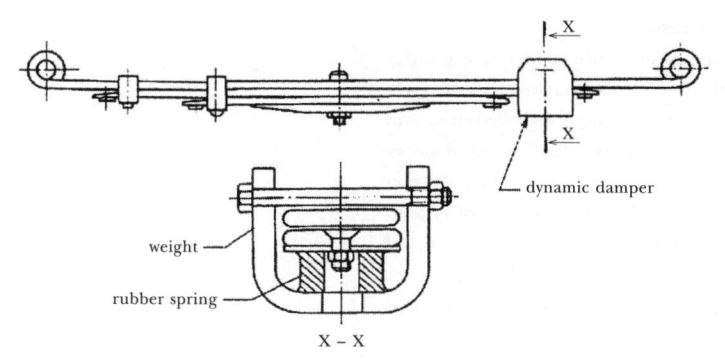

Fig D-9 dynamic damper

beban dinamis   [beban dinamisu]
동적 하중   [dongjeok hajung]
การโหลดแบบไดนามิค   [gaan load baaep dynamic]

dynamic measurement
a measurement of instantaneous value of a fluctuating quantity.
動的測定   [douteki sokutei]
変動する量の瞬間値の測定．
动态测量   [dòng tài cè liáng]
pengukuran dinamis   [pungukuran dinamisu]
동적 측정   [dongjeok cheukjeong]
การวัดแบบไดนามิค   [gaan wad baaep dynamic]

dynamic spring constant
spring rate against the dynamic force, indicating the spring characteristics in actual vibrating condition relating to constantly vibrating springs such as suspension springs for automobiles.
動ばね定数   [dou bane zyousuu]
自動車用懸架ばねなどのように定常的に振動しているばねに関し，実際の振動状態におけるばね特性を表すもので，動的荷重に対するばね定数．
动态弹簧刚度   [dòng tài tán huáng gāng dù]

tetapan pegas dinamis   [tetapan pugasu dinamisu]
동 스프링 정수   [dong spring jeongsu]
ค่าสปริงเรทของแรงไดนามิค   [kaa sa-pring rate khong raaeng dynamic]

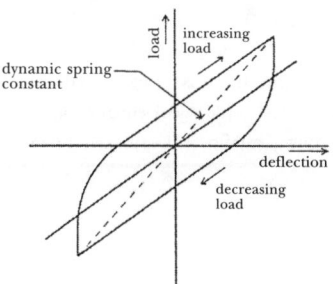

Fig D-10 dynamic spring constant

dynamic strength
strength to withstand breakage resulting from dynamic load.
動的強さ   [douteki tuyosa]
動的負荷による破壊に耐え得る強さ．
动态强度   [dòng tài qiáng dù]
kuat dinamis   [kuato dinamisu]
동적 강도   [dongjeok gangdo]
กำลังต้านการแตกเนื่องจากไดนามิคโหลด   [gam-lang taan gaan taaek ne-uang chaak dynamic load]

dynamic stress

dynamic stress
a type of stress whose magnitude varies with time. Stress that simply and cyclically varies between the maximum and the minimum is called repeated stress. Repeated stress that fluctuates between positive and negative stress (of equivalent absolute values) is called reversed stress. Repeated stress that fluctuates between a given stress and 0 is called pulsating stress.

動的応力　[douteki ouryoku]
大きさが時間的に変動する応力．極大値と極小値の間を単純，かつ周期的に変動する応力を繰返し応力，絶対値が等しい正負の応力の間を変動する繰返し応力を両振応力，一定の応力と0との間を変動する繰返し応力を片振り応力という．

动态应力　[dòng tài yīng lì]

tegangan dinamis　[tegangan dinamisu]

동적 응력　[dongjeok eungnyeok]

ความเค้นไดนามิค　[kwaam-ken dynamic]

E

E modulus
 a ratio of perpendicular stress generated in a cross-section of a bar to the perpendicular strain. Commonly designated by E. Unit is MPa or N/mm^2.
縦弾性係数　[tate dansei keisuu]
棒の断面に働く垂直応力と単位長さ当たりの伸び又は縮み（垂直ひずみ）との比．量記号：E，単位記号：MPa 又は N/mm^2.
弾性模量　[tán xìng mó liàng]
modulus E　[modulusu E]
종탄성계수　[jongtanseonggyesu]
อี โมดูลัส　[E modulus]

E type retaining ring
 a retaining ring shaped like letter "E" for on-shaft-use.
E 形止め輪　[iigata tomewa]
軸用の止め輪で，その形が文字 E に似ている止め輪．
E 形挡圈　[E xíng dǎng quān]
cincin penahan bentuk E　[chinchin punahan bentuku E]
이 (E) 형 스냅 링　[ihyeong seunaep ring]
แหวนกันชนิด อี　[waaen gan cha-nid E]

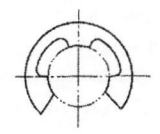

Fig E-1　E type retaining ring

earplug
 a piece of pliable material to fit the ear used to protect the ear from harmful noise.
耳栓　[mimisen]
耳に合うように柔らかい材料で作られた栓で，有害な騒音から耳を保護するために用いられる．
挂耳　[guà ěr]
penyumbat telinga　[punyunbato teringa]
귀마개　[gwimagae]
ปลั๊กอุดหู　[plug ud-huu]

eccentric cam
 a cylindrical cam with the shaft displased from the geometric center.
偏心カム　[hensin kamu]
回転軸が幾何学的中心からずれている円筒カム．
偏心凸轮　[piān xīn tū lún]
nok eksentris　[noku ekusentorisu]
편심 캠　[pyeonsim kaem]
แคมเยื้องศูนย์　[cam yu-ang soon]

eccentric load
 a load imposed on a spring off the center.
偏心荷重　[hensin kazyuu]
ばねの中心から外れた点に加わる荷重．
偏心负荷　[piān xīn fù hè]
beban eksentris　[beban ekusentorisu]
편심하중　[pyeonsimhajung]
โหลดเยื้องศูนย์　[load yu-ang soon]

eccentricity
 the deviation of the geometric center of a revolving body from the axis of rotation.
偏心　[hensin]
中心の片寄りのこと．
偏心　[piān xīn]
eksentrisitas　[ekusentorisitasu]
편심　[pyeonsim]

eccentricity — edge friction

เยื้องศูนย์ [yu-ang soon]

economic life
the number of years after which a capital good should be replaced in order to minimize running cost.

耐用寿命 [taiyou zyumyou]
稼働コストを最小にするために資本財を交換するべき年数.

耐久寿命 [nài jiǔ shòu mìng]
umur ketahanan [umuru ketahanan]

내용수명 [naeyongsumyeong]

อายุการใช้งาน [aa-yu gaan shai ngaan]

economic tool life
the total time during which a tool performs its required function under the most efficient cutting conditions.

工具耐用寿命 [kougu taiyou zyumyou]
工具が最適研削条件のもとで必要な機能を果たす合計時間.

工具耐久寿命 [gōng jù nài jiǔ shòu mìng]
umur ketahanan alat [umuru ketahanan arato]

공구 내용수명 [gonggu naeyongsumyeong]

อายุการใช้งานของอุปกรณ์ [aa-yu gaan shai ngaan khong up-pa-gorn]

eddy current sensor
a proximity sensor which uses an alternating magnetic field to create eddy currents in nearby objects. The currents are used to detect the presence of the objects.

渦電流センサ [kadenryuu sensa]
付近の物体に渦電流を生じさせる交流磁界を利用した近接センサ. このときの渦電流が物体の存在を検知するのに用いられる.

电涡流传感器 [diàn wō liú chuán gǎn qì]
sensor arus berlebihan [sensoru arusu berurebihan]

에디커렌트 센서 [edikeorenteu senseo]
เอ็ดดี้โคเรนท์เซนเซอร์ [eddy current sensor]

eddy current test
nondestructive testing which applies time varying magnetic field (by alternating current) to an electric conductive spring or its material. The defects are detected by the variation of generated eddy current caused by the defects.

渦流探傷試験 [karyuu tansyou siken]
電気的導体であるばね材料又はばねへ時間的に変化する磁場（交流など）を与え，それに生じた渦電流が，欠陥によって変化するのを検出する非破壊試験方法.

涡流探伤试验 [wō liú tàn shāng shì yàn]
pengujian arus eddy [pungujian arusu edi]

와류 탐상 시험 [waryu tamsang siheom]

การทดสอบการแตกด้วยแม่เหล็ก [gaan todsorb gaan taaek do-ui maae-lek]

edge crack
a crack or fissure of the edge of a leaf spring. Edge crack is caused by bending process or heat treatment.

端面割れ [tanmen ware]
板ばね端面の割れ，き裂. 曲げ加工あるいは熱処理で発生する.

端面裂纹 [duān miàn liè wén]
retakan tepi [retakan tepi]

단면 균열 [danmyeon gyunnyeol]

การแตกที่ขอบ [gaan taaek tii khorb]

edge friction
friction that occurs when a load is imposed on the contact edge (top and bottom) of a disk spring.

端面摩擦 [tanmen masatu]
皿ばねの上下の接触端面において荷重がかかるとき発生する摩擦.

端面摩擦　[duān miàn mó cā]
gesekan tepi　[gesekan tepi]
단면 마찰　[danmyeon machal]
ความฝืดที่ขอบ　[kwaam phuud tii khorb]

effective length
    a span serving as the basis for calculating spring characteristics of a leaf spring. It includes the effects of bands and U-bolts.
    有効スパン　[yuukou supan]
    重ね板ばねのばね特性を計算する基礎となるスパン．胴締め，Uボルトなどの影響を考慮したもの．
    有效长度　[yǒu xiào cháng dù]
    panjang efektif　[panjangu efekutifu]
    유효 스팬　[yuhyo seupaen]
    ระยะที่มีผล　[ra-ya tii mii pon]

egg-shaped wire helical spring
    coil spring made of the material the cross-section of which is egg-shaped, which is one of the noncircular wire coil springs.
    卵形断面ばね　[tamagogata danmen bane]
    異形断面ばねの一種で，材料の断面形状が卵形のコイルばね．
    卵形钢丝截面弹簧　[luǎn xíng gāng sī jié miàn tán huáng]
    pegas ulir berpenampang telur　[pugasu uriru berupenanpangu teruru]

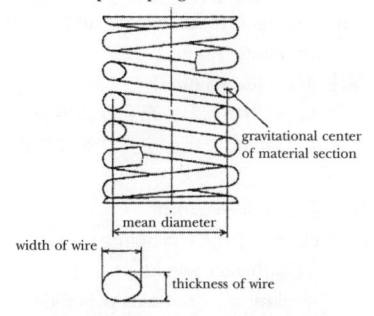

Fig E-2　egg-shaped wire helical spring

난형단면 스프링　[nanhyeongdanmyeon spring]
สปริงลวดเกลียวรูปไข่　[sa-pring lu-ad gliaw ruup khai]

ejector
    a device used to eject forged products, press-formed products, etc. from dies.
    エジェクタ　[ezyekuta]
    鍛造品，プレス成形品などを金型から排出する装置．
    脱模器（顶出器）　[tuō mó qì (dǐng chū qì)]
    penyembur　[punyenburu]
    이젝터　[ijekteo]
    ตัวฉีด　[to-a shiid]

elastic buckling
    an abrupt increase in the lateral deflection of a slender compression spring at a critical load while the stress are within elastic range.
    弾性座屈　[dansei zakutu]
    細長い圧縮コイルばねに荷重を加えていくとき，応力が弾性域の限界荷重で突然横たわみが大きくなること．
    弹性失稳　[tán xìng shī wěn]
    lenkungan elastis　[renkungan erasutisu]
    탄성좌굴　[tanseongjwagul]
    การยุบตัวในช่วงยืดหยุ่น　[gaan yup to-a nai chu-ang yuud-yun]

elastic deformation
    among all deformations of an object that may occur under a given stress, a deformation whereby the object immediately returns to its original shape when stress is removed.
    弾性変形　[dansei henkei]
    物体に応力を加えたとき起きる全変形のうち，応力を取り除くと直ちに復元する変形をいう．
    弹性变形　[tán xìng biàn xíng]
    pembentukan elastis　[punbentukan

erasutisu]
탄성 변형 [tanseong byeonhyeong]
การผิดรูปจากการคืนตัว [gaan pid ruup chaak gaan kuun to-a]

elastic limit
a stress limit under which permanent strain remains in the material after the external load is removed.
弾性限度 [dansei gendo]
外力を除いても永久ひずみが残留するときの応力限界.
弹性极限 [tán xìng jí xiàn]
batas kekenyalan [batasu kekenyaran]
탄성 한도 [tanseong hando]
ลิมิตของความเค้นเมื่อเอาโหลดออก [limit khong kwaam ken me-ua aow load oorg]

elastic ratio
the ratio of the elastic limit to the ultimate strength of materials.
弾性比率 [dansei hiritu]
材料における弾性限界の破壊強度に対する比率.
弹性比率 [tán xìng bǐ lǜ]
daerah elastis [daerahu erasutisu]
탄성비율 [tanseongbiyul]
อัตราการยืดหยุ่น [at-traa gaan yuud-yun]

elastic region
a range within which stress and strain with respect to the load imposed on an object are in direct proportion. Within this range, the strain disappears when load is removed.
弾性域 [danseiiki]
物体に作用する荷重に対する応力とひずみとが正比例する範囲. この範囲では荷重を取り除くとひずみは消える.
弹性区域 [tán xìng qū yù]
perbandingan elastis [purubandingan erasutisu]
탄성 영역 [tanseongyeongyeok]

บริเวณที่ยืดหยุ่น [bo-ri-wen tii yuud-yun]

elastic spike
a nail used to pin down a tie plate to a sleeper.
ばね釘 [bane kugi]
レール押え板を枕木に打ち込んで止める釘.
弹簧钉 [tán huáng dìng]
paku elastis [paku erasutisu]
탄성 스파이크 [tanseong seupaikeu]
ตะปู (ราง) ยืด [ta-poo(rang) yud]

elastic strain energy
the work being done by external force in deforming a solid body within elastic range.
弾性ひずみエネルギー [dansei hizumi enerugii]
外力によって物体が弾性変形を受けるときになされる仕事.
弹性变形能 [tán xìng biàn xíng néng]
energi regangan elastis [enerugi regangan erasutisu]
탄성변형 에너지 [tanseongbyeonhyeong eneoji]
งานที่ใช้แรงภายนอกทำการเปลี่ยนรูป [ngaan tii chai raaeng paai nork tam gaan plian ruup]

elastic theory
theory of the relations between the forces acting on a body and the resulting changes in dimensions and stresses within elastic range.
弾性理論 [dansei riron]
物体に外力が加わったときの外力と寸法変化の関係や応力を弾性領域で扱う理論.
弹性理论 [tán xìng lǐ lùn]
teori elastis [teori erasutisu]
탄성이론 [tanseongiron]
ทฤษฎีการยืดหยุ่น [trid-sa-dii gaan yuud-yun]

elasticity

a property whereby a solid object that is deformed by the application of a force returns to its original shape immediately after the stress is removed.

弾性　[dansei]
固体の物体に力を加え変形させた後、応力を取り除くと直ちに原形に戻る性質.

弾性　[tǎn xìng]
elastisitas　[erasutisitasu]
탄성　[tanseong]
การยืดหยุ่น　[gaan yuud-yun]

elastomer spring

a spring made of polymeric material (e.g., natural rubber, various types of synthetic rubber, acrylic resin). Also referred to as resin spring.

高分子ばね　[koubunsi bane]
高分子材料(天然ゴム、各種合成ゴム、アクリル系樹脂など)を用いたばね.単に樹脂ばねという場合もある.

高分子橡胶弹簧　[gāo fēn zǐ xiàng jiāo tán huáng]
pegas elastomer　[pugasu erasutomeru]
고분자 스프링　[gobunja spring]
สปริงอีลาสโตเมอร์　[sa-pring elastomer]

elastoplasticity

the state of a body subjected to a stress greater than its elastic limit but not exceeding a rapture stress.

弾塑性　[dansosei]
物体が弾性域を越えてはいるが、破壊応力を越えない範囲の応力を受けている状態.

弾塑比　[tǎn sù bǐ]
elastoplastik　[erasutopurasutiku]
탄소성　[tansoseong]
อีเลคโตพลาสติกซิตี้　[elec to plas tic city]

electrical conductivity

a substance-specific value that represents the ease of electric current to flow. The reciprocal of resistivity.

導電率　[doudenritu]
電流の流れやすさを示す、物質固有の値. 比抵抗の逆数.

导电率　[dǎo diàn lǜ]
konduktifitas listrik　[kondukutifitasu risutoriku]
전도율　[jeondoyul]
การนำไฟฟ้า　[gaan nam fai-faa]

electro-galvanizing

a process of zinc plating by electrolysis. The object to be galvanized is to be a cathode.

電気亜鉛めっき　[denki aen mekki]
めっきされる対象物を陰極として、その表面に電解によって亜鉛を被せる操作をいう.

电镀　[diàn dù]
galbani listrik　[garubani risutoriku]
전기 아연 도금　[jeongi ayeon dogeum]
การชุบซิงค์ด้วยไฟฟ้า　[gaan shub zinc du-oi fai-faa]

electrolytic polishing

a method of electrochemically polishing metal surface in an acid solution.

電解研磨　[denkai kenma]
酸溶液中で電気化学的に表面を研磨する方法.

电解抛光　[diàn jiě pāo guāng]
pemolesan elektrolitik　[pumoresan erekutororitiku]
전해 연마　[jeonhae yeonma]
การขัดเงาด้วยไฟฟ้าและกรด　[gaan chad ngao du-oi fai-faa laae grod]

electrophoretic coating

a painting method whereby the product is placed to be an anode in water-

electrophoretic coating — elliptically wound arc spring

soluble paint, and voltage is applied to deposit paint on surface using electrophoretic migration.
電着塗装　[dentyaku tosou]
解離可能な水溶性塗料中で被加工物を陽極として陰極との間に直流電圧を印加し, 電気泳動によって塗装する方法.
电镀　[diàn dù]
pelapisan elektropoleik　[purapisan erekutoroporeiku]
전착 도장　[jeonchak dojang]
การชุบฟอสเฟตด้วยไฟฟ้า　[gaan shub fos-fate du-oi fai-faa]

electroplating

a method of plating in which the thin film of metal is deposited electrochemically on the surface of a metallic or non-metallic product.
電気めっき　[denki mekki]
金属又は非金属製品の表面に他の金属薄層を電気化学的に析出させるめっき方法.
电镀(技术)　[diàn dù (jì shù)]
pelapisan listrik　[purapisan risutoriku]
전기 도금　[jeongi dogeum]
การชุบด้วยไฟฟ้า　[gaan shub du-oi fai-faa]

electrostatic painting

a coating method in which the object to be coated is set as anode and the spraying device as cathode, and the atomized coating material which is electrostatically-charged by high voltage is electrically attracted to the object.
静電塗装　[seiden tosou]
被塗装物を陽極, 噴霧装置を陰極とし, 高電圧をかけて静電気を帯電させた噴霧状の塗料を, 対象物に電気的に引き付けて塗装する方法.
静电镀膜　[jìng diàn dù mó]
pengecatan elekrostasis　[pungesatan erekurosutasisu]
정전 도장　[jeongjeon dojang]
การเคลือบสีด้วยไฟฟ้า　[gaan klu-ab sii du-oi fai-faa]

Eligo spring

a metal coil spring coated with rubber. Used as axle springs for railway cars or supporting springs for vibrating conveyors and similar vibrating machines.
エリゴばね　[erigo bane]
ゴムで金属のコイルばねを被覆したばね. 鉄道車両の軸ばね, 振動コンベヤなど各種振動機械の支持ばねとして用いられている.
埃里哥弹簧　[āi lǐ gē tán huáng]
pegas eligo　[pugasu erigo]
에리고 스프링　[erigo spring]
สปริงอีลิโก　[sa-pring Eligo]

elliptically coiled spring

a coil spring which is wound in ellipse shape. It is one of the irregular shaped coil springs.
だ円コイルばね　[daen koiru bane]
異形コイルばねの一種で, コイル横断面がだ円又はこれと類似の形状のコイルばね.
椭圆螺旋弹簧　[tuǒ yuán luó xuán tán huáng]
pegas ulir elips　[pugasu uriru eripusu]
타원 코일 스프링　[tawon koil spring]
สปริงขดรูปไข่　[sa-pring khod ruup khai]

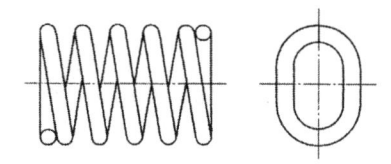

Fig E-3　elliptically coiled spring

elliptically wound arc spring

an elliptically coiled spring whose axis line is formed into an arch shape.

だ円アークばね [daen aaku bane]
  だ円コイルばねの形状で，軸線が弓形に成形されたばね．
  轴线呈弧形的椭圆形螺旋弹簧 [zhóu xiàn chéng hú xíng de tuǒ yuán xíng luó xuán tán huáng]
  pegas busur elips [pugasu busuru eripusu]
  타원형 아크 스프링 [tawonhyeong akeu spring]
  สปริงกดรูปไข่ [sa-pring god ruup khai]

elongation
  the increase in gage length of a test specimen before and after tensile testing.
  伸び [nobi]
  引張試験片の引張り後の標点間の長さと元の標点距離との差．
  伸长 [shēn cháng]
  perpanjangan [purupanjangan]
  연신 [yeonsin]
  การยืด [gaan yuud]

elongation parcentage
  a percentage of elongation length of tensile test specimen obtained during tensile testing to its original length.
  伸び率 [nobiritu]
  引張試験時の伸び量の元の長さに対する百分率．
  伸长率 [shēn cháng lǜ]
  persentase perpanjangan [purusentase purupanjangan]
  연신율 [yeonsinnyul]
  เปอร์เซ็นต์การยืด [percent gaan yuud]

elongation rate
  the degree of elongation of ferrite grains due to steel processing. The elongation rate is derived by dividing the number of grains per unit length vertical to the direction of elongation by the number of grains per unit length parallel to the direction of elongation.
  展伸度 [tensindo]
  鋼の加工によってフェライト結晶粒が伸展された度合い．伸展度は垂直方向の結晶粒数を平行方向の結晶粒数で除したものである．
  延伸率 [yán shēn lǜ]
  laju perpanjangan [raju purupanjangan]
  전신도 [jeonsindo]
  อัตราการยืด [at-traa gaan yuud]

embossing
  producing a raised pattern on the surface by a mechanical process.
  エンボス [enbosu]
  表面に機械的方法によって，凸凹模様を付けること．
  使凸起（压纹） [shǐ tū qǐ (yā wén)]
  embos [enbosu]
  엠보싱 [embosing]
  การทำให้นูนเป็นลวดลาย [gaan tam hai noon pen lu-ad laai]

embrittlement
  a loss of ductility or toughness in a metal with little chage in other mechanical properties.
  ぜい化 [zeika]
  金属において，他の機械的性質がほとんど変化せず，じん性あるいは延性が失われること．
  脆化 [cuì huà]
  kerapuhan [kerapuhan]
  취화 [chwihwa]
  เปราะง่าย [pra-or ngai]

encoder
  a combinational circuit for converting the decimal number data to the binary number data.
  エンコーダ [enkooda]
  10進数のデータを2進数に変換するための組合せ回路．

encoder — end of helical compression spring

编码器　[biān mǎ qì]
encoder　[enkoderu]
인코더　[inkodeo]
เอ็นโค้ดเดอร์　[en-coder]

end block
　a piece of hardware used to support a load, provided at both ends of the main leaf of a full- or semi-elliptic spring.
端受け　[hasiuke]
だ円ばね又は半だ円ばねの親板の両端で荷重を支える金具.
端面支承座　[duān miàn zhī chéng zuò]
blok ujung　[broku ujungu]
받침쇠　[badchimsoe]
อุปกรณ์ที่ใช้ในการโหลด　[up-pa-gorn tii chai nai gaan load]

end coil
　an end turn of compression coil spring. Number of the turn is not counted in spring rate calculation.
座巻　[zamaki]
圧縮コイルばねの端部コイル．ばね定数計算時に巻数として算入しない．
端圈　[duān quān]
uir ujung　[uiru ujungu]
자리감김　[jarigamgim]
ขดสุดท้ายของสปริงขด　[khod sud-taai khong sa-pring khod]

end coil not ground
　a helical compression spring whose ends are not ground. Mostly used for a helical spring made of fine wire or a spring which does not require a high degree of stability.
無研削コイル端末　[mukensaku koiru tanmatu]
コイルの端末を研削しない圧縮コイルばね．一般的には，細線のコイルばねや安定度を気にしないばねに多い．

端圈不磨　[duān quān bù mó]
ulir ujung tidak digerinda　[uriru ujungu tidaku digerinda]
무연삭 코일끝단　[muyeonsak koil kkeutdan]
ขดสุดท้ายของสปริงขดที่ไม่ได้เจียร์　[khod sud-taai khong sa-pring khod tii mai dai chiaa]

end grinding
　process to grind the end face of the helical compression spring. When both ends of the spring are ground, it is called both-end grinding.
端面研削　[tanmen kensaku]
主として，圧縮コイルばねの端面を研削する加工．ばねの両端を端面研削する場合を，両端研削という．
端面磨削　[duān miàn mó xuē]
penggerinda ujung　[pungugerinda ujungu]
단면 연삭　[danmyeon yeonsak]
การเจียร์ปลาย　[gaan chiaa plaai]

end of helical compression spring
　an end part of a helical compression spring. There are various types: for example, closed end, open end, pigtail end and tangent tail end. (see Fig E-4)
圧縮コイルばね端部　[assyuku koiru bane tanbu]
圧縮コイルばねの端末のことで，クローズドエンド，オープンエンド，ピッグテールエンド，タンジェントテールエンドなどがある．(Fig. E-4 参照)
压缩螺旋弹簧端部　[yā suō luó xuán tán huáng duān bù]
ujung pegas ulir tekan　[ujungu pugasu uriru tekan]
압축 코일 스프링의 끝부분　[apchuk koil springui kkeutbubun]
ปลายของสปริงกดรูปก้นหอย　[plaai khong sa-pring god ruup gon hoi]

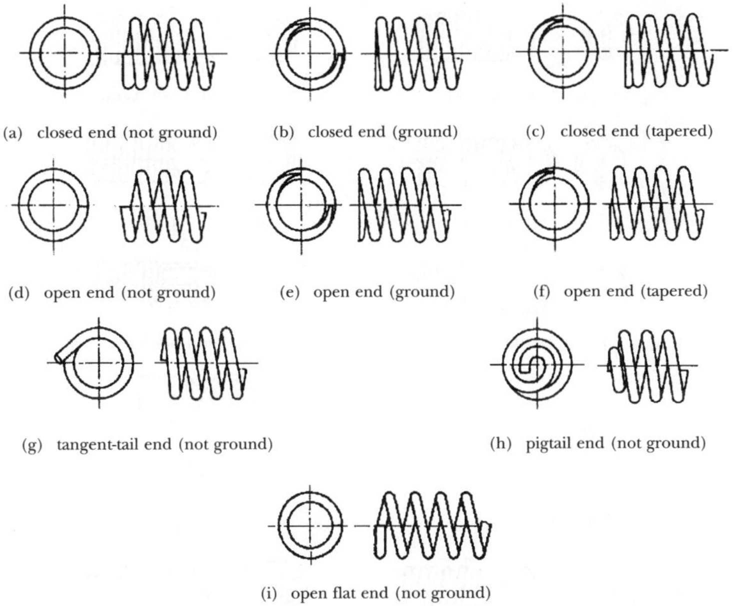

end of helical extension spring — end of power spring

(a) closed end (not ground)   (b) closed end (ground)   (c) closed end (tapered)

(d) open end (not ground)   (e) open end (ground)   (f) open end (tapered)

(g) tangent-tail end (not ground)   (h) pigtail end (not ground)

(i) open flat end (not ground)

Fig E-4 end of helical compression spring

end of helical extension spring
    winding end of a helical extension spring. (see Fig E-5)
    引張コイルばね端部 [hippari koiru bane tanbu]
    引張コイルばねの巻端の部分．(Fig E-5 参照)
    拉伸螺旋簧端部 [lā shēn luó xuán huáng duān bù]
    ujung pegas ulir tarik [ujungu pugasu uriru tariku]
    인장 코일 스프링의 끝부분. [injang koil springui kkeutbubun]
    ปลายของสปริงดึงรูปก้นหอย [plaai khong sa-pring duung ruup gon hoi]

end of helical torsion spring
    winding end of a helical torsion spring. (see Fig E-6)
    ねじりコイルばね端部 [neziri koiru bane tanbu]
    ねじりコイルばねの巻端の部分．(Fig E-6 参照)
    扭转弹簧的端部 [niǔ zhuǎn tán huáng de duān bù]
    ujung pegas puntir [ujungu pugasu puntiru]
    비틀림 스프링 끝부분. [biteullim spring kkeutbubun]
    ปลายสปริงก้นหอยทอร์ชั่น [plaai sa-pring gon hoi torsion]

end of power spring
    inner or outer end of a power spring.
    ぜんまい端部 [zenmai tanbu]
    ぜんまいの内外の巻端の部分．
    涡卷弹簧端部 [wō juǎn tán huáng duān bù]
    ujung pegas spiral [ujungu pugasu supiraru]
    태엽 스프링 끝 [tae-yeobtbseupeuling

end of helical extension spring — end of helical torsion spring

(a) half loop over

(b) full loop over center

(c) counter type full loop

(d) full loop at side

(e) rectangular hook

(f) long round end hook over center

(g) V-hook over center

(h) tapered end with circular hook

plug-type   plate-type
(i) threaded plug to fit plain end spring

(j) inclined side hook

Fig E-5 end of helical extension spring

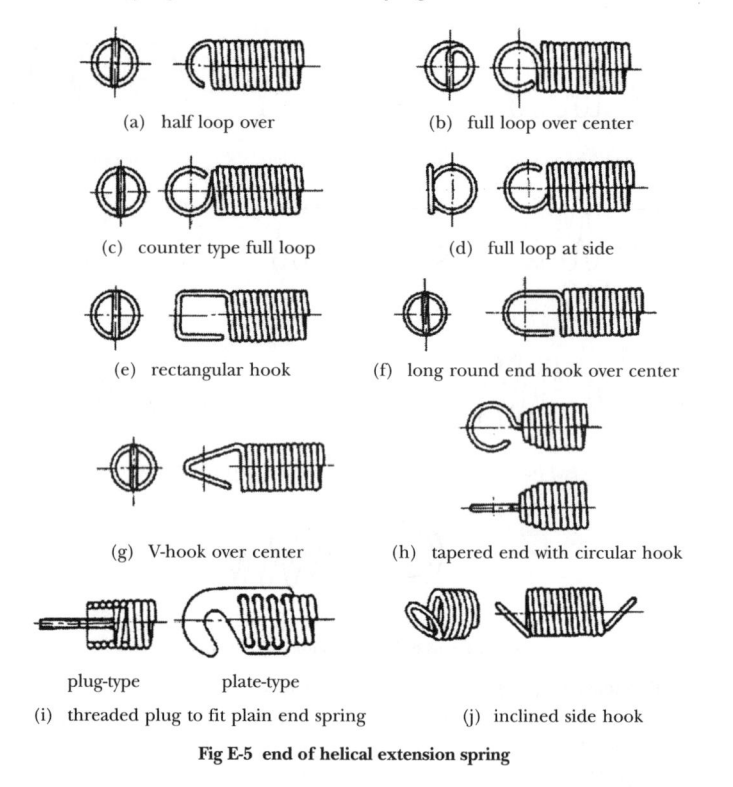

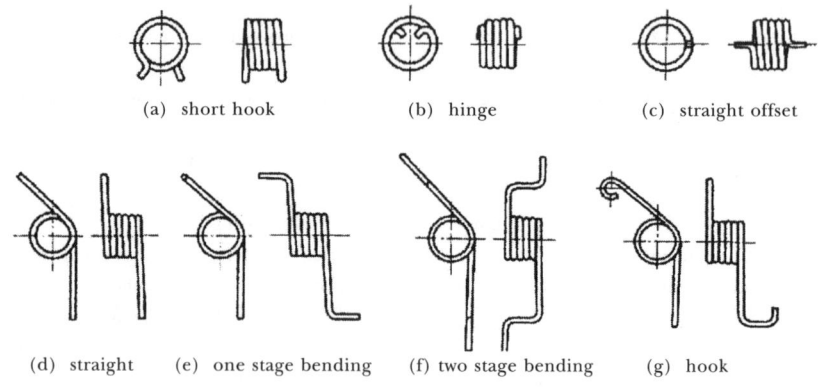

(a) short hook

(b) hinge

(c) straight offset

(d) straight   (e) one stage bending   (f) two stage bending   (g) hook

Fig E-6 end of helical torsion spring

kkeud]
ปลายของสปริงพาวเวอร์ [plaai sa-pring power]

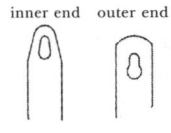

inner end   outer end

Fig E-7 end of power spring

end piece of spring
a piece of hardware used to stabilize the seat of a helical spring.
ばねの端末具 [bane no tanmatugu]
コイルばねの座部の安定のために用いる金具.
弹簧端部的紧固件 [tán huáng duān bù de jǐn gù jiàn]
bagian ujung pegas [bagian ujungu pugasu]
스프링 끝부분 부품 [spring kkeutbubun bupum]
อุปกรณ์ที่ใช้ประคองซีทของสปริงขด [uppa-gorn tii chai pra-kong seat khong sa-pring khod]

end plane
a plane including one turn of coiling at the end of a helical spring.
端面 [tanmen]
コイルばね端部のひと巻きを含む平面.
端面 [duān miàn]
permukaan ujung [purumukaan ujungu]
단면 [danmyeon]
ปลายสปริงขด [plaai sa-pring khod]

end shape
端末形状 [tanmatu keizyou]
端头形状 [duān tóu xíng zhuàng]
bentuk ujung [bentuku ujungu]
단말 형상 [danmal hyeongsang]

รูปร่างของปลายสปริงขด [ruup rang plaai sa-pring khod]

end trimmed with diamond point
triangularly trimmed end of a back-up leaf.
三角開先 [sankaku kaisaki]
板ばねの子板端部の形状種類で，三角形状にトリミングしたもの.
三角形坡口端面 [sān jiǎo xíng pō kǒu duān miàn]
ujung terpotong dengan pucuk wajik [ujungu terupotongu dengan pukuku wajiku]
삼각 끝 [samgak kkeut]
ตัดขอบด้วยเพชร [tad khorb du-oi pet]

end turn
an end turn of compression coil spring. Number of the turn is not counted in spring rate calculation.
座巻 [zamaki]
圧縮コイルばねの端部コイル．ばね定数計算時に巻数として算入しない．
座(支承)圈 [zuò (zhī chéng) quān]
ujung lenkungan [ujung lenkungan]
자리감김 [jarigamgim]
ขดสุดท้าย [khod sud-taai]

endurance limit
a life time whereby an item can not be repaired or over-hauled to meet the required standard any longer.
耐久限度 [taikyuu gendo]
アイテムを合格基準に適合するまで修理又はオーバーホールすることが，もはや物理的経済的にできないアイテムの耐久限度（寿命）をいう．
疲劳极限 [pí láo jí xiàn]
batas ketahanan [batasu ketahanan]
내구한도 [naeguhando]
พิกัดความทนทาน [pi-gad kwaam ton-taan]

endurance test
a test to examine the effects of stress

and time on product performance.
耐久性試験　[taikyuusei siken]
製品の性能がストレスと時間の経過によってどのような影響を受けるかを調べる試験．
疲劳试验　[pí láo shì yàn]
uji ketahanan　[uji ketahanan]
내구성 시험　[naeguseong siheom]
การทดสอบความทนทาน　[gaan tod-sorb kwaam ton-taan]

energy saving
an activity to promote the efficient use of energy and reduce energy consumption.
省エネルギー　[syouenerugii]
エネルギーの効率的利用と消費の低減を図る活動．
节能　[jié néng]
hemat energi　[hemato enerugi]
에너지 효율화　[eneoji hyoyulhwa]
การประหยัดพลังงาน　[gaan pra-yad pa-lang-ngaan]

Engler degree
a unit of viscosity for industrial use, measured with the Engler viscometer. It is derived by dividing time during which a sample flows out by time during which the same volume of water flows out.
エングラー度　[enguraa do]
エングラー粘度計による工業用粘度単位．試料が流出する時間を同体積の水が流出する時間で除した値．
恩氏度（粘度指数）　[ēn shì dù (nián dù zhǐ shù)]
derajat engler　[derajato engureru]
엥글라도　[enggeulla do]
ค่าของอิงเลอร์　[kaa khong engler]

English hook
a hook that rises from the center of a helical extension spring coil.
英式フック　[eisiki hukku]
引張コイルばねのコイルの中心部より立ち上げたフック．
英式钩环　[yīng shì gōu huán]
lingkaran inggris　[ringukaran ingurisu]
잉그리쉬 고리　[inggeuriswi gori]
อิงลิชฮุก　[english hook]

enlarged hook
a hook with an outer diameter greater than the coil diameter of a helical extension spring.
拡大フック　[kakudai hukku]
引張ばねのコイル径より大きな外径のフック．
加大钩环　[jiā dà gōu huán]
kait diperbesar　[kaito diperubesaru]
확대 고리　[hwakdae gori]
ตะขอที่มีเส้นผ่าศูนย์กลางนอกใหญ่กว่าขด　[ta-khor tii mii sen-pha-soon-glaang norg yaai gwaa khod]

enriched gas
a gas obtained by adding hydrocarbon gas to a carrier gas so as to increase carbon potential.
エンリッチガス　[enritti gasu]
カーボンポテンシャルを上げるためにキャリヤガスに炭化水素ガスを添加したガス．
浓缩气体　[nóng suō qì tǐ]
memperkaya gas　[menperukaya gasu]
인리치 가스　[illichi gaseu]
ก๊าซที่มีความเข้มข้นสูง　[gas tii mii kwaam khem-kon suung]

environmental assessment
preliminary survey, forecast and evaluation of the influence of environment.
環境影響評価　[kankyou eikyou hyouka]
環境影響について，事前に調査，予測，評価すること．

环境影响评价 [huán jìng yǐng xiǎng píng jià]
penilaian pengaruh lingkungan
[puniraian pungaruhu ringukungan]
환경 영향 평가 [hwangyeong yeonghyang pyeongga]
การประเมินทางสิ่งแวดล้อม [gaan pra-mern taang sing-waaed-lorm]

environmental deterioration
 deterioration of initial performance as a result of chemical action due to environmental conditions.
环境劣化 [kankyou rekka]
環境条件による化学的作用の結果，初期性能が劣化すること．
环境污染 [huán jìng wū rǎn]
penuaan karena lingkungan [punuaan karena ringukungan]
환경 악화 [hwangyeong akhwa]
การเสื่อมลงในด้านสิ่งแวดล้อม [gaan se-um long nai daan sing-waaed-lorm]

environmental engineering
 the technology conserned with the reduction of pollution, contamination, and deterioration of human surroundings.
環境工学 [kankyou kougaku]
人間の環境の変化，汚染，公害を減少させるための工学．
环境工程学 [huán jìng gōng chéng xué]
teknik lingkungan [tekuniku ringukungan]
환경공학 [hwangyeonggonghak]
วิศวกรรมสิ่งแวดล้อม [wis-sa-wa-gam sing-waaed-lorm]

environmental impact analysis
 predetermination of the extent of pollution or environmental degradation which will be involved in a new developping or industrial project.
環境影響調査 [kankyou eikyou tyousa]

新規の鉱工業を開発するにあたり，公害や環境悪化の程度を事前に予測すること．
环境影响调查 [huán jìng yǐng xiǎng diào chá]
analisa pengaruh lingkungan [anarisa pungaruhu ringukungan]
환경영향 조사 [hwangyeongyeonghyang josa]
การวิเคราะห์ผลกระทบด้านสิ่งแวดล้อม [gaan wi ka-or pon gra-top daan sing-waaed-lorm]

environmental stress cracking
 stress cracking accelerated by environmental conditions.
環境応力き裂 [kankyou ouryoku kiretu]
環境条件によって促進される応力き裂のこと．
环境应力裂纹 [huán jìng yīng lì liè wén]
retak tegangan lingkungan [retaku tegangan ringukungan]
환경 응력 균열 [hwangyeong eungnyeok gyunnyeol]
การแตกเนื่องจากความเค้นที่สัมพันธ์กับสิ่งแวดล้อม [gaan taaek ne-uang chaak kwaam ken tii sam-pan gab sing-waaed-lorm]

environmental test
 a laboratory test to determine the functional performance of a product under conditions that simulate the real environment.
環境試験 [kankyou siken]
実際の環境を模した実験室の環境下で行う製品の機能試験．
环境试验 [huán jìng shì yàn]
uji lingkungan [uji ringukungan]
환경시험 [hwangyeongsiheom]
การทดสอบเกี่ยวกับสิ่งแวดล้อม [gaan tod-sorb giaew gab sing-waaed-lorm]

equalizing spring
 a spring used to reduce the difference in load between a pair of axles of a ve-

hicle.
釣合いばね [turiai bane]
車両の車軸間にかかる荷重差を減らすために用いるばね.
平衡弾簧 [píng héng tán huáng]
pegas pengimbang [pugasu punginbangu]
평형 스프링 [pyeonghyeong spring]
สปริงใช้ในการปรับสมดุลของโหลดที่ต่างกัน [sa-pring chai nai gaan prab som-dun khong load tii taang gan]

equivalent length
    the equivalent length of torsion bar used for calculation assuming that the torsion bar including both ends is a round bar which has the same diameter of the central part of the torsion bar.
等価長さ [touka nagasa]
両端を含めたトーションバーを中央部の直径を有する単純丸棒として計算する場合に想定する長さ.
等效长度 [děng xiào cháng dù]
panjang ekuivalen [panjangu ekuivaren]
등가 길이 [deungga giri]
ความยาวเทียบเท่า [kwaam ya-o ti-eab ta-o]

ergonomics
    the study of human capability and psychology in relation to the working environment and the operability of equipments.
人間工学 [ningen kougaku]
機器の操作性や作業環境に関連して人間の心理や能力を研究する学問.
人类工程学 [rén lèi gōng chéng xué]
ergonomik [erugonomiku]
인간공학 [ingangonghak]
สมรรถยะศาสตร์ [sa mad ta ya sard]

error limit
    a limit value of every pressumed error.
誤差限界 [gosa genkai]
推定した総合誤差の限界の値.
误差极限 [wù chà jí xiàn]
batas kesalahan [batasu kesarahan]
오차 한계 [ocha hangye]
ขอบเขตของความผิดพลาด [khorb-khet khong kwaam pid-plaad]

etching
    the process of corroding a metallic or non-metallic surface by using a chemical or electrochemical method.
エッチング [ettingu]
金属又は非金属表面を化学的又は電気化学的に腐食する方法.
蚀刻 [shí kè]
etching [ettingu]
에칭 [eching]
การกัดกร่อนที่ผิว [gaan gad-gron tii piw]

Euler force
    the greatest force that a slender column carry without buckling. The force is calculated by Euler formura.
オイラーの力 [oiraa no tikara]
細長い柱が座屈せずに耐え得る最大の力. この力はオイラーの公式により計算される.
欧拉力 [ōu lā lì]
tenaga euler [tenaga eureru]
오일러의 힘 [oilleoui him]
แรงของยูเลอร์ [rang khong Euler]

Euler's theorem
    a theory derived by Euler who analyzed buckling phenomena whereby a long, slender column destroys due to cross bending under compressive load.
オイラー定理 [oiraa teiri]
断面に比して長さの大きい柱が圧縮荷重を受けたとき, 曲げによって破壊される座屈現象を解析したオイラーの理論.
欧拉变形理论 [ōu lā biàn xíng lǐ lùn]

teori euler [teori eureru]
오일러 정리 [oilleo jeongni]
ทฤษฎีของยูเลอร์ [tis-sa-dii khong Euler]

eutectoid
    a phenomenon whereby two or more solid phases are formed from one solid solution during the cooling process.
共析 [kyouseki]
冷却の過程で，一つの固溶体から二つ以上の固相が密に混合した組織へ変態する現象．
共析 [gòng xī]
eutectoid [eutekutoido]
공석 [gongseok]
เทคทอยด์ [eutectoid]

eutectoid steel
    a type of steel that consists of a texture resulting from eutectoid transformation, contains 0.9% carbon. It has entirely perlite phase.
共析鋼 [kyousekikou]
共析変態で生じた組織の鋼で，0.9%の炭素を含み組織がすべてパーライトからなるものをいう．
共析钢 [gòng xī gāng]
baja eutectoid [baja eutekutoido]
공석강 [gongseokgang]
เหล็กยูเทคทอยด์ [lek eutectoid]

evaporation rate of heating surface
    a heat quantity transferred per unit hour and unit area.
伝熱面熱負荷 [dennetumen netuhuka]
単位伝熱面積について，1時間当たりに伝えられる熱量．
传热面热负荷 [chuán rè miàn rè fù hè]
laju evaporasi permukaan perpindahan panas [raju ebaporasi purumukaan purupindahan panasu]
전열면 열 부하 [jeonnyeolmyeon yeol buha]

อัตราการระเหยของผิวร้อน [at-traa gaan ra-heri khong piw rorn]

even stress leaf spring
    a leaf spring tapered toward either end. It ensures the uniform distribution of stress in the longitudinal direction of the leaf.
テーパリーフスプリング [teepa riihu supuringu]
長手方向にテーパが施してある板ばね．板の長手方向の応力分布が均一になる．
等应力板簧 [děng yīng lì bǎn huáng]
pegas tegangan rata [pugasu tegangan rata]
등응력 스프링 (테이퍼 판 스프링) [deungeungnyeok spring (teipeo pan spring)]
แหนบที่มีความเค้นเท่ากัน [naaep ti mii kwaam khen ta-o gan]

event
    a specified accomplishment in a sequential program at a particular time.
事象 [zisyou]
一連のプログラムの過程で，特定の時点で達成された出来事．
事件 [shì jiàn]
kejadian [kejadian]
사상 (事象) [sasang]
เหตุการณ์ [heed-gaan]

event recorder
    a recorder that plots event information against time indicating occurrence time and duration.
事象記録計 [zisyou kirokukei]
事象の発生した時刻と存続時間等の事象情報を時間軸に対してプロットする記録計．
事件记录仪 [shì jiàn jì lù yí]
perekam kejadian [purekan kejadian]
사상 기록계 [sasang girokgye]

event recorder — extension spring

อุปกรณ์บันทึกข้อมูล [up-pa-gorn ban-tuk kho-moon]

event tree
a graphical representation of the possible sequence of events that might triggers an accident.
事故発生系統図 [ziko hassei keitouzu]
事故を引き起こす可能性のある事象のつながりを図で表現したもの.
事件发生系统图 [shì jiàn fā shēng xì tǒng tú]
sistematika kejadian [sisutematika kejadian]
사고발생 계통도 [sagobalsaeng gyetongdo]
แผนภูมิรูปแบบการเกิดเหตุการณ์ [paaen-poom ruup baaep gaan gerd heed-gaan]

expander spring
a spring used to give uniform pressure to a packing ring in a brake chamber.
エキスパンダスプリング [ekisupanda supuringu]
ブレーキチャンバのパッキンに均一な張りを与えるために用いられるばね.
扩张弹簧 [kuò zhāng tán huáng]
pegas pemuai [pugasu pumuai]
익스팬더 스프링 [ikseupaendeo spring]
สปริงที่ใช้ในการดึงขยาย [sa-pring tii chai nai gaan duung kha-yaai]

expansion coefficient
a rate of increase in the length or volume of a heated object per unit temperature. Increase in length is called linear expansion; increase in volume called volumetric expansion.
膨張係数 [boutyou keisuu]
物体が熱せられたとき,長さ又は体積が増大する単位温度当たりの割合. 前者を線膨張, 後者を体積膨張という.
膨胀系数 [péng zhàng xì shù]
koefisien muai [koefisien muai]

팽창 계수 [paengchang gyesu]
สัมประสิทธิ์การขยายตัว [sam-pra-sit gaan kha-yaai to-a]

extended length
see "developed length".
展開長さ [tenkai nagasa]
"developed length" 参照.
展开长度 [zhǎn kāi cháng dù]
panjang tambahan [panjangu tanbahan]
펼친 길이 [pyeolchin giri]
ระยะดึง [ra-ya duung]

extended round loop
"U" shaped hook at the end of a helical extension spring.
U フック [yuu hukku]
引張コイルばねの端部の一種で, U字形のフック.
U 形钩 [U xíng gōu]
kait bentuk huruf U [kaito bentuko hurufu U]
유 (U) 형 고리 [Uhyeong gori]
ยูคลิป [U clip]

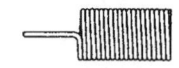

Fig E-8 extended round loop

extension spring
spring subjected mainly to an extensive force. In the narrow sense, helical extension spring.
引張ばね [hippari bane]
主として引張荷重を受けるばね. 狭義には, 引張コイルばね.
拉伸弹簧 [lā shēn tán huáng]
pegas tarik [pugasu tariku]
인장 스프링 [injang spring]
สปริงดึง [sa-pring duung]

extension spring machine
　forming machine of coils and hooks for the helical extension spring.
　テンションマシン　[tensyon masin]
　主として，引張コイルばねのコイル部及びフック部を成形する機械．
　拉伸弹簧成形码机　[lā shēn tán huáng chéng xíng mǎ jī]
　mesin pegas tarik　[mesin pugasu tariku]
　인장 스프링 코일링기　[injang spring koillinggi]
　เครื่องม้วนสปริงดึง　[kru-ang muan sa-pring duung]

external radius
　外半径　[sotohankei]
　外圈半径　[wài quān bàn jìng]
　jari-jari luar　[jari-jari ruaru]
　바깥 반지름　[bakkat banjireum]
　รัศมีนอก　[ras-sa-mii nork]

external rataining ring
　a retaining ring which is assembled over a circumferential groove on a shaft.
　軸用C形止め輪　[zikuyou siigata tomewa]
　軸円周上の溝に組み付ける止め輪．
　轴向挡圈　[zhóu xiàng dǎng quān]
　cincin pengunci luar　[chinchin pungunchi ruaru]
　외측 멈춤 링　[oecheuk meomchum ring]
　แหวนกั้นด้านนอก　[waaen gan daan nork]

eye
　turned section at the end of a leaf.
　目玉　[medama]
　ばね板の端部を丸く巻いた部分．
　卷耳　[juǎn ěr]
　gelang　[gerangu]
　귀　[gwi]
　หูแหนบ　[huu-naaep]

eye bush
　eye of a bar stabilizer with bush.
　アイブシュ　[ai busyu]
　バースタビライザ目玉部形状の一種で，孔にブシュを通して固定する形式．
　卷耳套　[juǎn ěr tào]
　bush gelang　[busuhu gerangu]
　아이 부시　[aibusi]
　หูของสตาบิไลเซอร์ที่ไม่ใส่บุช　[huu khong stabilizer tii mai sai bush]

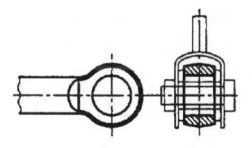

Fig E-9　eye bush

eye leaf
　leaf with an eye other than main leaf.
　アイリーフ　[ai riihu]
　重ね板ばねに組み込んだ目玉付きのばね板．
　吊耳片　[diào ěr piàn]
　gelang pegas daun　[gerangu pugasu daun]
　아이 리-푸　[ai ri-pu]
　หูแหนบ　[huu-naaep]

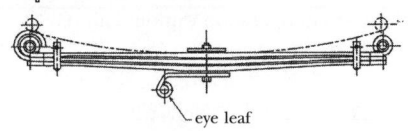

eye leaf
Fig E-10　eye leaf

eye rolling machine
　machine which forms the eyes of a leaf.
　目玉巻き機　[medama makiki]
　ばね板の目玉を成形する機械．
　卷耳成形机　[juǎn ěr chéng xíng jī]
　mesin pembentuk gelang　[mesin punbentuku gerangu]
　귀 감기 기계　[gwigamgi gigye]
　เครื่องม้วนหูแหนบ　[kru-ang muan huu naaep]

failure analysis

a systematic research or study to determine corrective action for failure by examining the possible causes of product failure, rate of occurrence, and effects of failure.

故障解析　[kosyou kaiseki]
製品の故障原因，発生率及び故障の影響を検討し，是正処置を決定するための系統的な調査・研究．

故障分析　[gù zhàng fēn xī]
analisis kegagalan　[anarisisu kegagaran]
고장 해석　[gojang haeseok]
การวิเคราะห์ความล้มเหลวเพื่อหาวิธีการแก้ไข
[gaan wi kro-a kwaam lom-leew pe-ua haa wi-tii gaae-khai]

failure rate

the probability of failure per next unit of time for an apparatus which has been in operation without failure.

故障率　[kosyouritu]
ある時点まで故障なしで運転してきた機器が次の単位時間に故障を起こす確率．

故障率　[gù zhàng lǜ]
laju kerusakan　[raju kerusakan]
고장율　[gojangyul]
อัตราความล้มเหลว　[at-traa kwaam lom-leew]

fan type lock washer

a type of toothed washer. The teeth are shaped like a fan.

ファンタイプ座金　[fan taipu zagane]
歯付座金の一種で，歯部が扇状に切り出されている座金．

扇形齿垫圈　[shàn xíng chǐ diàn quān]
ring pipih kunci jenis kipas　[ringu pipihu kunsi jenisu kipasu]
팬형 잠금 와셔　[paenhyeong jamgeum wasyeo]
แหวนล็อคชนิดฟันรูปพัด　[waaen lock cha-nid fan ruup pad]

fastener

a device for joining two separate machine parts.

ファスナ　[fasuna]
二つの別々の機械部品を締結する金具．

拧紧工具　[nǐng jǐn gōng jù]
pengencang　[pungenchangu]
파스너　[paseuneo]
ตัวรัดเข้าด้วยกัน　[to-a rad khao-do-ui gan]

fastener spring

various shapes of spring for fastening.

ファスナばね　[fasuna bane]
締結を目的とした，各種形状のばね．

紧固件弹簧　[jǐn gù jiàn tán huáng]
pegas pengikat　[pugasu pungikato]
파스너 스프링　[paseuneo spring]
สปริงฟาสเทนเนอร์　[sa-pring fastener]

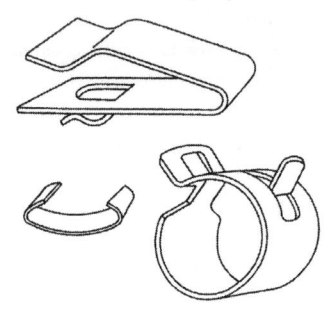

Fig F-1　fastener spring

fatigue

failure of a material by cracking result-

ing from repeated or cyclic stress.
疲労　[hirou]
　　繰返し応力あるいは周期的応力に起因
　　し，割れが進展して破壊すること．
疲労　[pí láo]
kelelahan　[kererahan]
피로　[piro]
ความล้าในวัตถุ　[kwaam laa nai wat-tu]

fatigue crack
　　a crack that occurs in a material subject to repeated stress.
疲労き裂　[hirou kiretu]
　　材料が繰返し荷重を受けたとき発生するき裂．
疲劳裂纹　[pí láo liè wén]
retak kelelahan　[retaku kererahan]
피로 균열　[piro gyunnyeol]
การแตกเนื่องจากความล้า　[gaan taaek nu-ang chaak kwaam laa]

fatigue durability
　　the number of repeated stress cycles a material can endure until fatigue fracture results.
疲労寿命　[hirou zyumyou]
　　疲労破壊を生じるまでの応力繰返し回数．
疲劳寿命　[pí láo shòu mìng]
ketahanan terhadap kelelahan　[ketahanan teruhadapu kererahan]
피로 수명　[piro sumyeong]
ความทนทานต่อการล้า　[kwaam ton-taan to-u gaan laa]

fatigue fracture
　　a phenomenon whereby cracks generated in a material under repeated load grows to cause material breakage.
疲労破壊　[hirou hakai]
　　材料が繰返し荷重を受けて発生した割れが進展して破壊に至る現象．
疲劳破坏　[pí láo pò huài]

retak karena kelelahan　[retaku karena kererahan]
피로 파괴　[piro pagoe]
การร้าวเนื่องจากการใส่โหลดซ้ำ ๆ　[gaan ra-o nu-ang chaak sai load sam sam]

fatigue fracture surface
　　the surface of fractured section as a result of breakage or damage due to fatigue.
疲労破面　[hirou hamen]
　　疲労によって破損又は損傷した破面をいう．
疲劳破坏面　[pí láo pò huài miàn]
permukaan retak kelelahan
[purumukaan retaku kererahan]
피로 파면　[piro pamyeon]
ผิวแตกจากการล้า　[piw taaek chaak gaan laa]

fatigue life
　　the number of repeated stress cycles a material can endure until fatigue fracture occurs.
疲れ寿命　[tukare zyumyou]
　　疲労破壊を生じるまでの応力繰返し回数．
疲劳寿命　[pí láo shòu mìng]
usia kelelahan　[usia kererahan]
피로 수명　[piro sumyeong]
จำนวนของความเค้นสูงสุดที่ทนได้ (ก่อนแตก)
[cham-nuan khong kwaam ken suung sud tii ton dai]

fatigue limit
　　the maximum stress that a material can endure for an unlimited number of stress cycles without breaking.
疲れ限度　[tukare gendo]
　　無限回数の繰返しに耐える応力の上限値．
疲劳极限　[pí láo jí xiàn]
batas kelelahan　[batasu kererahan]
피로 한도　[piro hando]

fatigue limit — fatigue strength under bending stresses

พิกัดของการล้า,ความเค้นสูงสุด [pi-gad khong gaan laa]

fatigue limit diagram

diagram which shows the way the fatigue limit varies depending on the combination of the stress amplitude and mean stress. There are a number of diagrams presented by Goodman, Haigh and Smith.

疲れ限度線図 [tukare gendo senzu]
疲れ限度が応力振幅と平均応力との組合せ方によって，変化する様子を示す線図．グッドマン，ヘイ及びスミス，それぞれの疲れ限度線図がある．

疲劳强度极限图 [pí láo qiáng dù jí xiàn tú]
bagan batas lelah [bagan batasu rerahu]
내구한도 선도 [naeguhando seondo]
แผนภูมิความสัมพันธ์ขอบเขตการล้าและความเค้น [paaen poom kwaam sam-pan khob khet gaan laa laae kwaam ken]

fatigue limit under pulsating stress

the limit of fatigue subject to repeated stress in which the maximum or minimum value of stress is zero.

片振り疲れ限度 [kataburi tukare gendo]
片方の応力が 0 となる繰返し応力が作用するときの疲れ限度．

脉动循环疲劳极限 [mài dòng xún huán pí láo jí xiàn]
batas kelelahan pada tegangan getar [batasu kererahan pada tegangan getaru]
편진 피로 한도 [pyeonjin piro hando]
พิกัดของการล้าสลับ (ความเค้นสูงสุด-ต่ำสุด) [pi-gad khong gaan laa sa-lab]

fatigue notch factor

a value of dividing the fatigue strength of an unnotched test piece by the fatigue strength of a notched test piece.

切欠き係数 [kirikaki keisuu]
平滑試験片の疲れ強さを切欠き試験片の疲れ強さで除した値．

应力集中系数 [yīng lì jí zhōng xì shù]
faktor tarik kelelahan [fakutoru tariku kererahan]
피로 놋치계수 [piro notchigyesu]
แฟกเตอร์ของรอยบากความล้า [factor khong roi-bark kwaam-laa]

fatigue strength

maximum stress level to withstand unlimitted number of repetitive stress.

疲れ強さ [tukare tuyosa]
疲れ限度及び時間強さの総称，又は反復する応力によって生じる，破壊に耐え得る性質．

疲劳强度 [pí láo qiáng dù]
kuat lelah [kuato rerahu]
피로 강도 [piro gangdo]
กำลังของการล้า [gam-lang khong gaan laa]

fatigue strength diagram

diagram drawn by taking the stress amplitude on the ordinate and the mean stress on the abscissa to indicate the life (number of repetition times to the breakdown) under arbitrary repetition condition.

時間強度線図 [zikan kyoudo senzu]
任意の繰返し応力条件下での寿命（折損までの繰返数）を見るために，縦軸に応力振幅，横軸に平均応力をとり，適当な寿命間隔で，等寿命線を引き表した線図．

疲劳强度线图 [pí láo qiáng dù xiàn tú]
bagan kuat lelah [bagan kuato rerahu]
피로 강도 선도 [piro gangdo seondo]
แผนภูมิกำลังของการล้า [paaen poom gam lang khong gaan laa]

fatigue strength under bending stresses

a fatigue limit of a material subject to bending stress.

曲げ応力疲労限度 [mage ouryoku hirou

gendo]
　曲げ応力が作用するときの疲労限度．
弯曲应力疲劳强度　[wān qǔ yīng lì pí láo qiáng dù]
kuat lelah pada tegangan lentur　[kuato rerahu pada tegangan renture]
굽힘 응력 피로 한도　[gupim eungnyeok piro hando]
พิกัดของความล้าจากความเค้นโค้งงอ　[pi-gad khong kwaam laa chaak kwaam ken koong ngor]

fatigue strength under completely reversed bending stresses
　a fatigue limit of a material subject to bending stress of which the absolute value of max. and min. stress is equal.
両振曲げ応力疲労限度　[ryouburi mage ouryoku hirou gendo]
　絶対値の等しい正負の曲げ応力が作用するときの疲労限度．
循环弯曲应力下的疲劳强度　[xún huán wān qǔ yīng lì xià de pí láo qiáng dù]
kuat lelah pada tegangan lentur terbalik [kuato rerahu pada tegangan renture terubariku]
양진 굽힘 응력 피로 한도　[yangjin gupim eungnyeok piro hando]
พิกัดของความล้าจากความเค้นโค้งงอที่ค่าความเค้นสูงสุด=ต่ำสุด　[pi-gad khong kwaam laa chaak kwaam ken koong ngor tii kaa kwaam ken suung sud ta-o gab tam sud]

fatigue strength under fluctuating stresses
　a fatigue limit of a material subject to repeated stress cycles having different absolute values for the maximum and minimum stress.
変動応力疲労限度　[hendou ouryoku hirou gendo]
　絶対値の異なる最大値と最小値との間を繰り返す応力が作用するときの疲労限度．
循环应力疲劳强度　[xún huán yīng lì pí láo qiáng dù]
kuat lelah pada teganngan berubah-ubah [kuato rerahu pada teganngan berubahu-ubahu]
변동 응력 피로 한도　[byeondong eungnyeok piro hando]
พิกัดของการล้าความเค้น (สูงสุด=ต่ำสุด)　[pi-gad khong kwaam laa suung sud ta-o gab tam sud]

fatigue test
　test to determine the fatigue life or limit of a spring or a test piece by applying repetitive stress or varying stress. This test is divided into torsional-, axial load-, rotating bending- and plane bending-fatigue test according to the type of the test stress.
疲労試験　[hirou siken]
　ばね又は試験片に，繰返し応力又は変動応力を加えて，疲労寿命，限度などを求める試験．試験応力の種類に応じて，ねじれ，軸荷重，回転曲げ，平面曲げ疲労試験などに分類される．
疲劳试验　[pí láo shì yàn]
uji kelelahan　[uji kererahan]
피로 시험　[piro siheom]
การทดสอบการล้า　[gaan tod-sorb gaan laa]

fault analysis
　the detection and diagnosis malfunctions in machine system.
故障解析　[kosyou kaiseki]
　機械システムの不具合要因を発見して分析すること．
故障分析　[gù zhàng fēn xī]
analisa kerusakan/kesalahan　[anarisa kerusakan / kesarahan]
고장해석　[gojanghaeseok]
การวิเคราะห์ความผิดปกติ　[gaan wi ka-or

fault tree
    a graphical representation of faults caused by equipment failure, human error, or environmental events.
過失系統図　[kasitu keitouzu]
環境要因，人の過失，機器の故障などから引き起こされる故障を系統的に図示したもの．
故障树　[gù zhàng shù]
sistematika kelelahan　[sisutematika kererahan]
고장의 트리　[gojangui teuri]
แผนภูมิรูปแบบการผิดปกติ　[paaen-poom ruup baaep gaan pid pa-ka-ti]

feasibility study
    a study to determine whether a plan is capable of being accomplished successfully.
実行可能性調査　[zikkou kanousei tyousa]
ある計画が成功裡に達成できるか否かを決める調査．
可行性研究　[kě xíng xìng yán jiū]
studi fisibilitas　[sutudi fisibiritasu]
실행 가능성 조사　[silhaeng ganeungseong josa]
การศึกษาความเป็นไปได้　[gaan suk-saa kwaam pen-pai-dai]

feasibility test
    a test conducted to obtain data in support of a feasibility study.
実行可能性試験　[zikkou kanousei siken]
ある計画の実行可能性調査を裏付けるデータを得る目的で実施する試験．
可行性试验　[kě xíng xìng shì yàn]
uji fisibilitas　[uji fisibiritasu]
실행 가능성 시험　[silhaeng ganeungseong siheom]
การทดสอบความเป็นไปได้　[gaan tod-srob kwaam pen-pai-dai]

feed accuracy
    accuracy of wire feed length in coiling operation.
送り精度　[okuri seido]
コイリングにおける線送り長さの精度．
进给精度　[jìn gěi jīng dù]
ketelitian unpan　[keteritian unpan]
이송 정밀도　[isong jeongmildo]
ความแม่นยำในการป้อน　[kwaam maaen-yam nai gaan porn]

feed clutch
    mechanical component which provides on and off material feed, in the spring forming machine.
フィードクラッチ　[fiido kuratti]
線，帯鋼などを材料としてばねを成形する成形機の一部で，材料の送りを断続する機構部品．
自动进退刀离合器　[zì dòng jìn tuì dāo lí hé qì]
kopling unpan　[kopuringu unpan]
피드 클러치　[pideu keulleochi]
ฟีดคลัชท์　[feed clutch]

feed roller
    rollers driven in pairs to feed the material to the coiling section of the coiling machine.
フィードローラ　[fiido roora]
コイルばねをコイリングマシンで成形するとき，材料を送るために対をなして駆動するローラ．
进给滚轮　[jìn gěi gǔn lún]
roda pengumpan　[roda pungunpan]
피드 롤러　[pideu rolleo]
ฟีดโรลเลอร์　[feed roller]

feedback control
    a system in which the value of some output quantity is controlled by feeding back the value of the controlled

quantity and using it to manipulate an input quantity so as to bring the value of the controlled quantity closer to a desired value.
フィードバック制御　[fiidobakku seigyo]
出力側の制御量の一部を入力側にフィードバックして入力量を制御し、その結果当該制御量が目的値に近似するようにする管理システム．
反馈控制　[fǎn kuì kòng zhì]
kontrol feed-back　[kontororu feedo-bakku]
피드백 제어　[pideubaek jeeo]
การควบคุมการฟีดแบ็ค　[gaan kuob-kum feedback]

feeding device
　a device for supplying objects (transferring, transporting). There are various machine types such as electronic, vibrational, magnetic, pressurized.
供給装置　[kyoukyuu souti]
物体を供給（移送，運搬など）するための装置．機械式，電気式，振動式，磁気式，空圧式などがある．
送料装置　[sòng liào zhuāng zhì]
sarana pengumpan　[sarana pungunpan]
공급 장치　[gonggeup jangchi]
อุปกรณ์สำหรับป้อนชิ้นงาน　[up-pa-gorn sam-rab porn shin-ngaan]

feeding length
　the length of a workpiece which is fed per stroke.
送り長さ　[okuri nagasa]
1ストローク当たりの長さ．
送料长度　[sòng liào cháng dù]
panjang umpan　[panjangu unpan]
이송 길이　[isong giri]
ระยะการป้อน　[ra-ya gaan porn]

feeler gauge
　a tool with many blades of different thickness used to establish clearance between parts.
すきまゲージ　[sukima geezi]
既知の厚みを持つ薄鋼板ゲージで，すきまや平面度の検査に用いる．シックネスゲージともいう．
塞規　[sāi guī]
pengukur perasa　[pungukuru purasa]
틈새 게이지　[teumsae geiji]
เกจสำหรับทำช่องว่างระหว่างชิ้นงาน　[gauge sam-rab tam chong waang ra-waang shin-ngaan]

ferrite
　metallographical name for α iron solid solution.
フェライト　[feraito]
α鉄固溶体につけた金属組織上の名称．
铁素体　[tiě sù tǐ]
ferit　[ferito]
페라이트　[peraiteu]
เหล็กโครงสร้างเฟอร์ไรท์　[lek krong-sang ferrite]

ferromagnetic materials
　materials that are strongly attracted to a magnet.
強磁性体　[kyouziseitai]
磁石に強く引き付けられる物質．
铁磁体　[tiě cí tǐ]
bahan feromagnet　[bahan feromaguneto]
강자성체　[gangjaseongche]
วัตถุดิบเหล็กที่มีความสามารถเป็นแม่เหล็กได้สูง　[wat-tu-dip lek tii mii kwaam sa-maad pen maae-lek soong]

fiber flow of fracture
　a state of flow of microstructures at a fractured surface.
破面の繊維流れ　[hamen no sen'i nagare]
破面における組織の流れ状態をいう．
断面纤维组织状态　[duàn miàn xiān wéi zǔ zhī zhuàng tài]

fiber flow of fracture — finish grinding

aliran serat dari keretakan　[ariran serato dari keretakan]
파면 섬유 흐름　[pamyeon seomnyu heureum]
ทิศทางแนวเส้นร้าว　[tid-tang naaew sen ra-o]

fiber reinforced plastic leaf spring
　leaf spring made of fibre reinforced plastics. Compared to steel, this spring is lighter and has larger specific elasticity energy and higher corrosion resistance.
　FRP 板ばね　[ehuaarupii itabane]
　繊維強化プラスチックを用いた板ばね．鋼に比べ軽量で，比弾性エネルギーが大きく耐食性が優れる．
　FRP 板弹簧　[FRP bǎn tán huáng]
　pegas daun plastik diperkuat serat [pugasu daun purasutiku diperukuato serato]
　에푸알피 (FRP) 판 스프링　[epualpi (FRP) pan seupeuring]
　แหนบชนิด FRP　[naaep cha-nid FRP]

fine blanking
　high precision blanking process.
　ファインブランキング　[fain burankingu]
　精度の高い打抜加工．
　精密冲裁　[jīng mì chōng cái]
　bentuk awal yang halus　[bentuku awaru yangu harusu]
　파인 블랭킹　[pain beullaengking]
　ปั๊มละเอียด　[pump la-iead]

fine grain
　an austenite grain size number of 5 or above. Coarse grain refers to grain size numbers less than 5.
　細粒　[sairyuu]
　オーステナイト結晶粒度番号5以上を細粒という．また，5未満を粗粒という．
　细晶粒　[xì jīng lì]
　butiran halus　[butiran harusu]

미립자　[miripja]
เม็ดเกรนละเอียด　[med grain la-iead]

fine grained steel
　steel having the austenite grain size number of 5 or above.
　細粒鋼　[sairyuukou]
　オーステナイト結晶粒度番号5以上の鋼．
　细晶粒钢　[xì jīng lì gāng]
　baja berbutiran halus　[baja berubutiran harusu]
　미립강　[miripgang]
　เหล็กเม็ดเกรนละเอียด　[lek med grain la-iead]

finger spring
　a washer to which spring action is given by cutting out small flap-like pieces around the circumference and raising them by bending.
　フィンガスプリング　[finga supuringu]
　外径側に円周に沿って切り出した部分を曲げ起こし，ばね作用させる座金．
　指针弹簧　[zhǐ zhēn tán huáng]
　pegas jari　[pugasu jari]
　핑거 스프링　[pinggeo spring]
　สปริงฟิงเกอร์　[sa-pring finger]

finger washer
　a washer which combined the flexibility of fingers located around the circle which provides distributed loading points.
　フィンガ座金　[finga zagane]
　円周上のフィンガのばね作用を組み合わせた働きを持つワッシャで，円周上に分布された荷重点がある．
　指形垫圈　[zhǐ xíng diàn quān]
　ring pipih jari　[ringu pipihu jari]
　핑거 와셔　[pinggeo wasyeo]
　แหวนรองฟิงเกอร์　[waaen rong finger]

finish grinding
　grinding procedure done at the final

finishing stage.
仕上げ研磨　[siage kenma]
最終仕上げの段階で行う研削作業.
精磨　[jīng mó]
penggerindaan akhir　[pungugerindaan akuhiru]
마무리 연마　[mamuri yeonma]
การเจียร์สุดท้าย　[gaan chiaa sud-taai]

finished size
仕上げ寸法　[siage sunpou]
成品尺寸　[chéng pǐn chǐ cùn]
ukuran akhir　[ukuran akuhiru]
마무리 치수　[mamuri chisu]
ขนาดสำเร็จ　[kha-naad sam-ret]

finishing
processing such as deburring and polishing to meet required appearance specifications.
仕上げ　[siage]
製品のばり取り，つや出しなど外観の要求仕様を満足させる加工.
精加工　[jīng jiā gōng]
pengerjaan akhir　[pungerujaan akuhiru]
마무리 가공　[mamuri gagong]
กระบวนการเก็บความเรียบร้อยสุดท้าย　[grabuan gaan geb kwaam riab roi sud taai]

finite element method
a method of analysis based on numerical calculation. A continuum with infinite degrees of freedom is divided into discrete elements with finite degrees of freedom, and then each element is solved numerically by simultaneous equations to approximate an entire system.
有限要素法　[yuugen youso hou]
無限の自由度を有する連続体を離散化して有限の自由度を持つ要素に分割し，個々の要素を連立方程式に組み立てて全体の系を数値計算する方法.

有限元方法　[yǒu xiàn yuán fāng fǎ]
metoda unsur terbatas　[metoda unsuru terubatasu]
유한 요소법　[yuhan yosobeop]
วิธีการวิเคราะห์โดยการคำนวณ　[wi-tii gaan wika-or dooi gaan kam-nuan]

fissure
a defect such as fracture and crack caused by bending, drawing, etc.
き裂　[kiretu]
曲げ，絞り加工などの工程で生じる割れ，き裂等の不具合.
亀裂　[guī liè]
celah　[cherahu]
균열　[gyunnyeol]
การแตกฉีกขาด　[gaan taaek shiik khaad]

fittings
accessories attached to a leaf spring such as a bolt, clip, and buckle.
付属金具　[huzoku kanagu]
重ね板ばねに取り付けるボルト，クリップ，胴締め金具などの付属品.
装配件　[zhuāng pèi jiàn]
kelengkapan　[kerengukapan]
판스프링 부품　[panspring bupum]
ตัวยึด　[to-a yud]

fixed arm length
the length of an arm that carries moment to torsion bar body.
固定腕長さ　[kotei ude nagasa]
トーションバーにモーメントを与える腕の長さ.
固定杆长　[gù dìng gǎn cháng]
panjang lengan yang tetap　[panjangu rengan yangu tetapu]
고정팔 길이　[gojeongpal giri]
ระยะแขนยึด　[ra-ya khaaen yud]

fixture tempering
tempering process for parts restrained by a fixture to improve dimensional

control.
拘束テンパ [kousoku tenpa]
寸法管理を改善するために治具で拘束して行うテンパ処理.
定形回火 [dìng xíng huí huǒ]
tempering pengekang bentuk
[tenperingu pungekangu bentuku]
구속 뜨임 [gusok tteuim]
การอบฟิกซ์เจอร์ [gaan op fixture]

flame hardening
a method of hardening using direct heating with a flame. Mainly used to harden steel surface.
火炎焼入れ [kaen yakiire]
炎で直接加熱して行う焼入れ. 主に鉄鋼の表面の焼入れに用いる.
火焰淬火 [huǒ yàn cuì huǒ]
pengerasan percikan api [pungerasan purusikan api]
화염 담금질 [hwayeom damgeumjil]
ชุบแข็งด้วยเปลวไฟ [shub khaaeng du-oi pleew fai]

flange
a removable fitting used to attach a grinding wheel to an arbor.
フランジ [huranzi]
研削といしをアーバに取り付けるための取外し可能なといし取付具.
法兰盘 [fǎ lán pán]
flange [frange]
플랜지 [peullaenji]
อุปกรณ์สำหรับติดหินเจียร์เข้ากับเพลา [up-pagorn san-rab tid hin chia khaow gab pla-o]

flat bar
a slender steel bar rolled on all sides into a flat rectangular section. Typically the width is at least double the thickness and the maximum width is about 200 mm.
平鋼 [hirakou]
長方形の断面に四面とも圧延された細長い鋼材. 通常, 幅が厚さの2倍以上で, 幅の最大は約200 mm.
扁钢 [biǎn gāng]
batang rata [batangu rata]
평강 [pyeonggang]
เหล็กแผ่นยาว [lek paaen ya-o]

flat position
the state of a loaded leaf spring with its loaded point and supporting point horizontally balanced.
水平位置 [suihei iti]
重ね板ばねに荷重を加えたとき, 荷重点と支持点が水平になった状態.
水平位置 [shuǐ píng wèi zhì]
kedudukan rata [kedudukan rata]
수평 위치 [supyeong wichi]
ตำแหน่งแนวระนาบ [tam-naaeng naaew ra-naap]

flat ring spring
a flattened doughnut-shaped thin metal. The spring action is achieved by supporting the inner and outer rim against perpendicular force to the plane.
円盤ばね [enban bane]
外周及び内周に隔て輪を入れ, 一方を支え他方に軸線方向の荷重を加えて使用する円盤状のばね.
环行片弹簧 [huán xíng piàn tán huáng]
pegas cincin datar [pugasu chinchin dataru]
플랫 링 스프링 [peullaet ring seupeuring]
สปริงแผ่นบางรูปแหวน [sa-pring paaen baang ruup waaen]

flat spring
various shapes of spring made of thin flat materials, as called by its material.
薄板ばね [usuita bane]
ばねの材料の種類による名称で, 薄い

板状の材料を用いた各種形状のばね．
片弾簧　[piǎn tán huáng]
pegas rata　[pugasu rata]
박판 스프링　[bakpan spring]
แฟลตสปริง　[flat sa-pring]

flat wire
a wire with a flat rectangular cross-section obtained by rolling or drawing a wire rod.
平線　[hirasen]
線材に圧延又はダイス引きを施して加工した平板状の線．
扁钢丝　[biǎn gāng sī]
kawat rata　[kawato rata]
각선　[gakseon]
ลวดแผ่นเรียบ　[lu-ad paaen riab]

flatness
the degree of deviation from a geometrical straightness of a plain surface.
平面度　[heimendo]
平面部分の幾何学的直線からの狂いの大きさをいう．
平面度　[píng miàn dù]
kerataan　[kerataan]
평면도　[pyeongmyeondo]
การแบนเรียบ　[gaan baaen riab]

flaw detector
a device used to detect flaw and crack of a material or product with eddy current test, magnetic test, X-ray transmission test, ultrasonic test, etc.
きず検出器　[kizu kensyutuki]
渦流探傷，磁気探傷，X線透過試験機，超音波探傷器などで素材又は製品のきずや割れを検出する機器．
探伤仪　[tàn shāng yí]
penemu cacat　[penemu chachato]
결함 검출기　[gyeolham geomchulgi]
อุปกรณ์ตรวจจับรอยบนวัตถุดิบ　[up-pa-gorn truat chab roi bon wat-tu-dip]

flex test
a test to determine the number of repeated bending operations performed until a material breaks, by bending the material with one end fixed in place and the other end moved a given radius along an arc.
繰返し曲げ試験　[kurikaesi mage siken]
材料の一端を固定して他端を特定半径の円弧を描き繰り返し曲げて，破断までの繰返し回数を調べる試験．
循环弯曲试验　[xún huán wān qǔ shì yàn]
uji bengkok bolak balik　[uji bengukoku boraku bariku]
반복 굽힘시험　[banbok gupimsiheom]
การทดสอบการโค้งซ้ำ　[gaan tod-sorb gaan kong sam]

flexibility
capability of being flexed or bent repeatedly.
可撓性　[katousei]
繰返しの曲げに耐えられる状態又は性質をいう．
挠性　[náo xìng]
fleksibel　[frekusiberu]
유연성　[yuyeonseong]
ความยืดหยุ่น　[kwaam yuud yun]

flexible manufacturing system
a form of computer-integrated manufacturing system to achieve the flexibility of manufacturing.
フレキシブル生産システム　[hurekisiburu seisan sisutemu]
生産のフレキシビリティを高レベルで可能とするために設計されたコンピュータ統合化生産システム．
柔性制造系统　[róu xìng zhì zào xì tǒng]
sistem produksi fleksibel　[sisutemu purodukusi furekusiberu]
플랙시블 생산 시스템　[peullaeksibeul

flexible manufacturing system — fluid spring

saengsan siseutem]
ระบบการผลิตที่มีความยืดหยุ่นสูง   [ra-bop gaan pa-lit tii mii gaan yuud yun soong]

flexible shaft

a shaft that transmit rotary motion to accommodate a small amount of angled alignment by using flexible materials.

可撓軸   [katouziku]
小さい角度を持った軸線をつないで回転を伝達するように，可撓性を持つ材料で作られた軸.

挠性轴   [nǎo xìng zhóu]
poros fleksibel   [porosu furekusiberu]
플랙시블 샤프트   [peullaeksibeul syapeuteu]
เพลายืดหยุ่น   [pla-o yuud-yun]

flexural strength

the value of the maximum bending moment that a sample can withstand divided by the modulus of a section of the sample.

曲げ強度   [mage kyoudo]
供試体の耐えられる最大曲げモーメントを供試体の断面係数で除した値.

弯曲强度   [wān qū qiáng dù]
kuat bengkokan   [kuato bengukokan]
굽힘 강도   [gupim gangdo]
กำลังการโค้งงอ   [gam-lang gaan kong-ngor]

flow chart

a graphical representation of the progress of a system in which symbols are used to show operations. Lines and arrows are used to indicate the relationship of operations.

フローチャート   [hurootyaato]
システムの進行状態をオペレーションを表す記号を用いて図示したチャート．線と矢線がオペレーション相互の関係を示す．

流程图   [liú chéng tú]
diagram alur   [diaguramu aruru]
순서도   [sunseodo]
โฟล์ชาร์ท   [flow chart]

flow chart symbol

symbols used to represent operations in flow charts.

フローチャート記号   [hurootyaato kigou]
フローチャート中でオペレーションを表すのに用いられる記号．

流程图标记   [liú chéng tú biāo jì]
simbol diagram alur   [sinboru diaguramu aruru]
순서도 기호   [sunseodo giho]
สัญญลักษณ์บนโฟล์ชาร์ท   [san-ya-luk bon flow-chart]

fluctuating load

a load whose magnitude varies irregularly with time.

変動荷重   [hendou kazyuu]
時間的に大きさが不規則に変化する荷重．

动负荷   [dòng fù hè]
beban berubah-ubah   [beban berubahu-ubahu]
변동 하중   [byeondong hajung]
โหลดเป็นจังหวะ   [load pen chang-wa]

fluid clutch

a fluid coupling that transmits or cuts off power depending on whether a fluid is filled or discharged.

流体クラッチ   [ryuutai kuratti]
流体の充てん又は排出によって動力を伝達・遮断する流体継手．

流体离合器   [liú tǐ lí hé qì]
kopling cairan   [kopuringu chairan]
유체 클러치   [yuche keulleochi]
คลัชน้ำมัน   [clutch naam-man]

fluid spring

spring using the elasticity of gas or liquid, as called by its material.

流体ばね　[ryuutai bane]
　ばねの材料の種類による名称で，気体及び液体の弾性を利用するばね．

流体弾簧　[liú tǐ tán huáng]

pegas fluida　[pugasu furuida]

유체 스프링　[yuche spring]

สปริงน้ำมัน　[sa-pring naam-man]

fluorescence penetrant testing
　nondestructive testing where drips of penetrant is applied to open scratches on the surface of the test piece of a spring or its material, and the observation of the flaw indication can be carried out with ease in the magnified image. This is divided into visible dye penetrant testing and fluorescence penetrant testing.

蛍光浸透探傷試験　[keikou sintou tansyou siken]
　ばね材料，ばねなどの試験体表面に，開口しているきずに浸透液を浸透させた後，拡大した像の指示模様としてきずを観察する非破壊試験方法．染色浸透探傷試験及び蛍光浸透探傷試験がある．

荧光渗透探伤试验　[yíng guāng shèn tòu tàn shāng shì yàn]

uji penetrasi fluoresensi　[uji punetorasi furuoresensi]

형광 침투 탐상 시험　[hyeonggwang chimtu tamsang siheom]

การทดสอบคุณสมบัติในการเรืองแสง　[gaan tod-sorb kun-na-som-bat nai gaan re-ung saaeng]

flute
　the groove of a drill or a reamer.

溝　[mizo]
　ドリルやリーマなどの刃の溝部のこと．

沟, 槽　[gōu, cáo]

galur　[garuru]

홈　[hom]

ร่อง　[rong]

force
　action exerted on or by a spring in order to produce or modify motion, or to maintain a system of forces in equilibrium.

力　[tikara]
　運動を起こすもしくは緩和するために，又は系の釣合いを保つために，ばねに加わる作用又はばねから生じる作用．

力　[lì]

gaya　[gaya]

힘　[him]

แรง　[rang]

forced breakage
　destruction of an object by exerting external force on it.

強制破壊　[kyousei hakai]
　外力を加えて物体を破壊させること．

强制破坏　[qiáng zhì pò huài]

kerusakan yang dipaksakan　[kerusakan yangu dipakusakan]

강제 파괴　[gangje pagoe]

การแตกจากแรง　[gaan taaek chaak raaeng]

forced vibration
　vibration maintained by cyclic external force.

強制振動　[kyousei sindou]
　周期的な外力によって持続される振動．

强迫振动　[qiǎng pò zhèn dòng]

getaran gaya　[getaran gaya]

강제 진동　[gangje jindong]

การสั่นสะเทือนจากแรงที่มากระทำ　[gaan san sa-teun chaak raaeng tii ma gra-tam]

formability
the degree of ease in forming material into a specified shape.
成形性　[seikeisei]
所要の形状に成形するときの成形の容易さを表す概念．
成形性　[chéng xíng xìng]
kemmpuan bentuk　[kenmupuan bentuku]
성형성　[seonghyeongseong]
คุณสมบัติในการขึ้นรูป　[kun-na-som-bat nai gaan khun ruup]

formed wire spring
various shapes of springs made of wire, as called by its material.
線細工ばね　[senzaiku bane]
ばねの材料の種類による名称で，線状の材料を用いた各種形状のばね．
线成形弹簧　[xiàn chéng xíng tán huáng]
pegas kawat yang dibentuk　[pugasu kawato yangu dibentuku]
선세공 스프링　[seonsegong spring]
ลวดสปริงขึ้นรูป　[lu-ad sa-pring khun ruup]

forming
the process of permanently transforming a material to the specified shape and dimensions.
成形　[seikei]
素材を永久変形させて所定の形状及び寸法を与える加工．
成形　[chéng xíng]
pembentukan　[punbentukan]
성형　[seonghyeong]
การขึ้นรูป　[gaan khun ruup]

forming machine
forming machine of wire with tools sliding in from various directions.
成形機　[seikeiki]
多方向からスライドする工具によって線又は条を成形する機械．
成形机　[chéng xíng jī]
mesin pembentuk　[mesin punbentukn]
성형기　[seonghyeonggi]
เครื่องขึ้นรูป　[kru-ang khun ruup]

forming press
a punch press for forming metal parts.
成形プレス　[seikei puresu]
金属部品を成形するパンチプレス．
冲压成形　[chōng yā chéng xíng]
pres pembentuk　[puresu punbentuku]
성형 프레스　[seonghyeong peureseu]
กดขึ้นรูป　[god khun ruup]

forming rolls
rolls contoured to give a desired shape to materials passing between them.
成形ロール　[seikei rooru]
ロール加工される材料に所与の形状を与えるための断面形状を持つロール．
成形滚轮　[chéng xíng gǔn lún]
roll pembentuk　[roru punbentuku]
성형 롤러　[seonghyeong rolleo]
รีดขึ้นรูป　[riid khun ruup]

Fourier analysis
determination of the characteristics of periodic vibration with regard to individual harmonic wave (or a sinusoidal amount that has the frequency given by an integral multiple of the fundamental frequency). Also called harmonic analysis.
フーリエ解析　[huurie kaiseki]
周期振動の特性を各調波（基本振動数の整数倍の振動数を持つ正弦量）について求めること．調和分析ともいう．
傅利叶分析(变换)　[fù lì yè fēn xī (biàn huàn)]
analisi fourier　[anarisi fourieru]
푸리에 해석　[purie haeseok]
การวิเคราะห์ฟอร์เรียร์　[gaan wi kro-a Fourier]

four-wheel drive
a drive mechanism in which the drive

shaft act on all four wheels of the automobile.
四輪駆動　[yonrin kudou]
自動車の四輪全部に駆動シャフトが働く駆動機構.
四轮驱动　[sì lún qū dòng]
kemudi empat roda　[kemudi enpato roda]
사(四)륜 구동　[saryun gudong]
การขับเคลื่อน 4 ล้อในรถยนต์　[gaan khab kluan sii lor nai rot yon]

fractography
the microscopic examination of fractured metal surfaces.
フラクトグラフィ　[hurakutogurafi]
破面観察により破壊の状態を調べる学問.
断口金相学　[duàn kǒu jīn xiāng xué]
fraktograpi　[furakutogurapi]
프랙토 그래피　[peuraekto geuraepi]
การตรวจสอบรอยร้าวบนผิวโลหะ　[gaan tru-ad sorb roi raao bon piw loo-ha]

fracture
破損　[hason]
破损　[pò sǔn]
keretakan　[keretakan]
파손　[pason]
การแตกร้าว　[gaan taaek raao]

fracture face
a surface of destroyed portion of an object.
破面　[hamen]
材料が破壊した後に生じる表面.
破损面　[pò sǔn miàn]
permukaan retakan　[permukaan retakan]
파면　[pamyeon]
ผิวที่แตกร้าว　[piw taaek raao]

fracture mechanics
a theoretical study to elucidate the mechanism of how an object breaks under external forces.
破壊力学　[hakai rikigaku]
外力による物体の破壊機構を解明する理論を伴った学問.
断裂力学　[duàn liè lì xué]
mekanika retakan　[mekanika retakan]
파괴 역학　[pagoe yeokhak]
การศึกษากลไกของการแตก　[gaan suk saa gon gai khong gaan taaek]

fragility
the likelihood of breakage of an object under external forces.
もろさ　[morosa]
物体が外力によって破損しやすい性質.
易碎性　[yì suì xìng]
kerapuhan　[kerapuhan]
취성　[chwiseong]
ความเปราะ　[kwaam pra-or]

free angle
relative angle between both ends of a helical torsion spring when no load is applied.
自由角度　[ziyuu kakudo]
無荷重時のねじりコイルばねの，コイル両端部の相対角度.
自由角度　[zì yóu jiǎo dù]
sudut bebas　[suduto bebasu]
자유 각도　[jayu gakdo]
มุมอิสระ　[mum is-sa-ra]

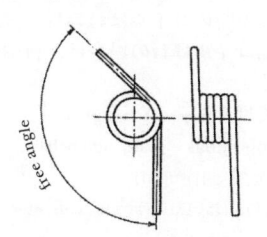

Fig F-2　free angle

free height

height of a coil spring when no load is applied. This is called free length in helical extension spring.

自由高さ　[ziyuu takasa]
無荷重時のコイルばねの高さ．ただし，引張コイルばねの場合は，自由長という．
自由高度　[zì yóu gāo dù]
tinggi bebas　[tingi bebasu]
자유고　[jayugo]
ความสูงอิสระในสปริงขดขณะไม่มีโหลด
[kwaam soong is-sa-ra]

Fig F-3　free height

free length

length of a coil spring when no load is applied. This is called free height in helical compression spring.

自由長さ　[ziyuu nagasa]
無荷重時のコイルばねの長さ．ただし，圧縮コイルばねの場合は，自由高さという．
自由长度　[zì yóu cháng dù]
panjang bebas　[panjangu bebasu]
자유 길이　[jayugiri]
ความยาวอิสระในสปริงขดขณะไม่มีโหลด
[kwaam ya-o is-sa-ra]

frequency

the number of repeated applications of stress to a metal test piece per unit time during a fatigue test.

繰返し速度　[kurikaesi sokudo]
疲れ試験中，金属片にかかる単位時間当たりの応力繰返数．
频率　[pín lǜ]
frekuensi　[furekuensi]
진동수　[jindongsu]
ความถี่　[kwaam tii]

frequency analysis

to conduct a spectral analysis on a complicated sound or vibration.

周波数分析　[syuuhasuu bunseki]
複雑な音又は振動について，その成分の大きさを周波数の関数として求めること．
频率分析　[pín lǜ fēn xī]
analisa frekuensi　[anarisa furekuensi]
주파수 분석　[jupasu bunseok]
การวิเคราะห์ความถี่　[gaan wi-kra-or kwaam tii]

frequency distribution

the graphical representation of the number of times for the respective values in a measurement session.

度数分布　[dosuu bunpu]
測定値の中に同じ値が繰り返し現れる場合，各値の出現度数を図に表したもの．
频率分布　[pín lǜ fēn bù]
distribusi frekuensi　[disutoribusi furekuensi]
도수분포　[dosubunpo]
การกระจายความถี่　[gaan gra-chaai kwaam tii]

frequency of vibration

the number of cycles per unit time.

周波数　[syuuhasuu]
単位時間当たりのサイクル数．
振动频率　[zhèn dòng pín lǜ]

frekuensi getaran [hurekuensi getaran]
진동 주파수 [jindong jupasu]
ความถี่ของการสั่น [kwaam tii khong gaan san]

fretting corrosion
    frictional corrosion caused by the high-pressure contact and minute reciprocal relative displacement between the springs or the spring and other components.
    フレッチングコロージョン [hurettingu koroozyon]
    ばね同士又は他の部材との高い面圧の接触，及び微小な相対的往復変位によって生じる摩耗腐食現象．
    微振磨損腐蚀 [wēi zhèn mó sǔn fǔ shí]
    korosi karena goresan [korosi karena goresan]
    마모 부식 [mamo busig]
    การสึกกร่อนจากการเสียดสี [gaan suk-gron chaak gaan siead-sii]

friction
    a counter force against sliding when two surfaces contact each other. There are two different frictions. One is a static friction and the other is a dynamic friction.
    摩擦 [masatu]
    互いに接触する二つの面が滑りに対して示す抵抗力．静摩擦と動摩擦に区別される．
    摩擦 [mó cā]
    gesekan [gesekan]
    마찰 [machal]
    ความเสียดทาน [kwaam siad taan]

friction between plates
    friction caused by the sliding between the leaves of a leaf spring.
    板間摩擦 [bankan masatu]
    重ね板ばねのリーフ間のしゅう動に伴って生じる摩擦．
    板间摩擦 [bǎn jiān mó cā]
    gesekan antar plat [gesekan antaru purato]
    판간 마찰 [pangan machal]
    แรงเสียดทานระหว่าง 2 แผ่น [rang siead-taan ra-wang song paaen]

friction clutch
    a clutch that connects the driving side and driven side with frictional force.
    摩擦クラッチ [masatu kuratti]
    駆動側と非駆動側とを摩擦力によって連結するクラッチ．
    摩擦离合器 [mó cā lí hé qì]
    kopling gesekan [kopuringu gesekan]
    마찰 클러치 [machal keulleochi]
    ความฝืดในคลัตช์ [kwaam fuud nai clutch]

friction coil spring
    a type of helical spring that utilizes friction between the inside and outside circumferences of two helical compression springs (one large, one small) combined on the same axis.
    摩擦コイルばね [masatu koiru bane]
    大小二つの圧縮コイルばねを同軸に組み合わせて内径と外径の摩擦を利用するコイルばね．
    摩擦螺旋弹簧 [mó cā luó xuán tán huáng]
    pegas ulir gesekan [pugasu uriru gesekan]
    마찰 코일 스프링 [machal koil spring]
    สปริงขดความฝืด [sa-pring khod kwaam-fuud]

friction force
    the force required to overcome continuous static friction force.
    摩擦力 [masaturyoku]
    摩擦で静止し続ける力に打ち勝つ力．
    摩擦力 [mó cā lì]
    gaya gesek [gaya geseku]
    마찰력 [machallyeok]

แรงเสียดทาน [rang siead-taan]

friction press
> a pressing machine that drives slides with frictional force and thread mechanism.
>
> 摩擦プレス [masatu puresu]
> 摩擦力とねじ機構によってスライドを駆動するプレス.
>
> 摩擦挤压 [mó cā jǐ yā]
> tekanan gesek [tekanan geseku]
> 마찰 프레스 [machal peureseu]
> เครื่องเพรสที่ใช้แรงเสียดทาน [hru-ang press tii chai raaeng siead taan]

friction ring spring
> a ring spring with a frictional surface where the outer and inner rings contact, with the outer ring edge sloping inward and the inner ring edge outward.
>
> 摩擦輪ばね [masatu wabane]
> 外輪は内側に, 内輪は外側に傾斜して, 互いに接触する摩擦面をもつ輪状のばね.
>
> 摩擦环形弹簧 [mó cā huán xíng tán huáng]
> pegas cincin gesek [pugasu chinchin geseku]
> 마찰 링 스프링 [machal ring spring]
> สปริงแหวนผิวฝืด [sa-pring waaen piw fuud]

friction spring
> generic term for springs for generating damping using friction. This includes volute springs, ring springs and double helical springs made of materials of irregular cross-section.
>
> 摩擦ばね [masatu bane]
> 摩擦を利用して減衰を生じさせるばねの総称. 竹の子ばね, 輪ばね, 異形断面を用いた二重巻のコイルばねなどがある.
>
> 摩擦弾簧 [mó cā tán huáng]

pegas gesek [pugasu geseku]
마찰 스프링 [machal spring]
สปริงที่ใช้ความฝืดทำให้เกิดการหน่วง [sa-pring tii chai kwaam-fuud tam hai gerd kwaam nu-ang]

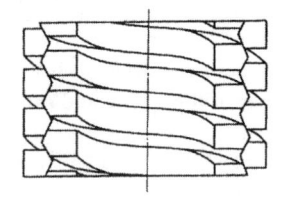

Fig F-4 friction spring

frictional corrosion
> corrosion accelerated by friction.
>
> 摩擦腐食 [masatu husyoku]
> 摩擦によって促進される腐食.
>
> 摩擦腐蚀 [mó cā fǔ shí]
> korosi karena gesekan [korosi karena gesekan]
> 마모 부식 [mamo busik]
> การสึกกร่อนจากความฝืด [gaan suk-gron chaak kwaam-fuud]

frictional force during recoil
> frictional force that occurs between the coils of a spiral spring during unwinding.
>
> 巻戻し摩擦力 [makimodosi masaturyoku]
> ぜんまいばねを巻き戻すときに発生するコイルの接触による摩擦力.
>
> 反卷摩擦力 [fǎn juǎn mó cā lì]
> gaya gesek selama menggulung [gaya geseku serama mengugurung]
> 리코일시 마찰력 [rikoilsi machallyeok]
> แรงเสียดทานขณะคลายขด [rang siead-taan kha-na klaai khod]

frictional resistance
> resistance due to friction that occurs when two objects move in contact.
>
> 摩擦抵抗 [masatu teikou]

二つの物体が接触運動を行うときに生じる摩擦による抵抗.
摩擦阻力　[mó cā zǔ lì]
tahanan gesek　[tahanan geseku]
마찰 저항　[machal jeohang]
การต้านทานเนื่องจากความฝืด　[gaan taan-taan ne-uang chaak kwaam-fuud]

frictional torque
  torque caused by a frictional force.
  摩擦トルク　[masatu toruku]
  　摩擦力によるトルク.
  摩擦力矩　[mó cā lì jù]
  momen puntir gesek　[momen puntiru geseku]
  마찰 토-크　[machal to-keu]
  แรงทอร์คเนื่องจากแรงเสียดทาน　[raaeng torque ne-uang chaak raaeng siead-taan]

fulcrum
  the point of support for a lever.
  支点　[siten]
  　てこの支点.
  支点　[zhī diǎn]
  titik topang　[titiku topangu]
  지점　[jijeom]
  จุดหมุน　[chud mun]

full annealing
  heat treatment process whereby an object is heated and maintained at a temperature equal to or higher than the A_3 transformation point (for semi-eutectoid steel) or A_1 transformation point (for hyper-eutectoid steel) and then cooled to room temperature slowly.
  完全焼まなし　[kanzen yakinamasi]
  　A_3 変態点（亜共析鋼）又は A_1 変態点（過共析鋼）以上の温度に加熱し，その温度に加熱保持後室温まで徐冷すること.
  完全退火　[wán quán tuì huǒ]

aneal penuh　[anearu punuhu]
완전 풀림　[wanjeon pullim]
การอบอ่อนสมบูรณ์　[gaan ob-orn som-boon]

full center hook
  the hook of an extension spring whose center is on the axis line of the coil.
  丸フック　[maruhukku]
  　引張ばねのフック中心がコイル軸線上にあるフック.
  全中心鉤环　[quán zhōng xīn gōu huán]
  kait tengah penuh　[kaito tengahu penuhu]
  풀센터 고리　[pulsenteo gori]
  ห่วงที่อยู่ตรงกลางขด　[hu-ong tii yuu trong glaang khod]

full hardening
  the process of hardening by rapidly cooling a steel work from the austenitizing temperature to below Ms temperature.
  完全焼入れ　[kanzen yakiire]
  　オーステナイト化温度から Ms 点以下に急冷して内部まで完全に硬化させる操作.
  完全淬火　[wán quán cuì huǒ]
  pengerasan penuh　[pungerasan punuhu]
  완전 담금질　[wanjeon damgeumjil]
  การชุบแข็งสมบูรณ์　[gaan chub khaaeng som-boon]

full length leaf
  leaf longer than the span.(see Fig F-5)
  全長板　[zentyouban]
  　スパン以上の長さをもつばね板．(Fig F-5 参照)
  全长板　[quán chǎng bǎn]
  daun panjang penuh　[daun panjangu punuhu]
  전장판 (온길이 판)　[jeonjangpan (ongiri pan)]
  แผ่นแหนบที่มีความยาวมากกว่าความยาวสแปน

full length leaf — full-elliptic spring

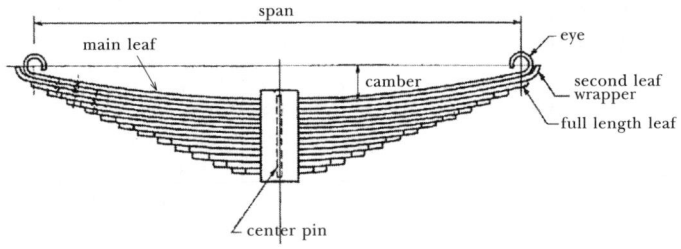

Fig F-5 full length leaf

[paaen nnaep tii ii kwaam ya-o maak kwaa span]

full load
the maximum allowable load under a given condition.
全負荷　[zenhuka]
所定の条件における最大の負荷．
全负荷　[quán fù hè]
beban penuh　[beban punuhu]
전부하　[jeonbuha]
รับโหลดเต็มที่　[rab load tem tii]

full loop at side
circular hook formed at the side of coil circle of a helical extension spring.
側面丸フック　[sokumen maruhukku]
引張コイルばねの端部の一種で，コイル外周部からコイル軸線方向に成形された丸形のフック．
侧面圆环　[cè miàn yuán huán]
lengkung penuh pada sisi　[rengukungu punuhu pada sisi]
측면 원형 고리　[cheungmyeon wonhyeong gori]
ห่วงกลมด้านข้าง　[hu-ong glom daan khaang]

Fig F-6 full loop at side

full loop over center
circular hook at the end of a helical extension spring.
丸フック　[maruhukku]
引張コイルばねの端部の一種で，丸形のフック．
圆环　[yuán huán]
kait bulat　[kaito burato]
원형 고리　[wonhyeong gori]
ห่วงกลมด้านปลาย　[hu-ong glom daan plaai]

Fig F-7 full loop over center

full-elliptic spring
leaf spring shaped like an ellipse. Combination of two or three of this spring is called dual spring or triple spring, respectively.
だ円ばね　[daen bane]
だ円のような形状をした重ね板ばね．二組組み合わせたものを二連ばね，三組組み合わせたものを三連ばねなどという．
椭圆弹簧　[tuǒ yuán tán huáng]
pegas elips　[pugasu eripusu]

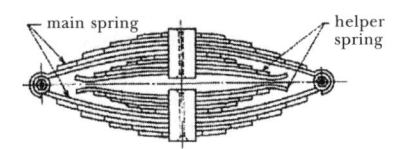

Fig F-8 full-elliptic spring

타원 스프링 [tawon spring]

สปริงรูปวงรี [sa-pring ruup wong-rii]

functional test
>a test to determine whether a function is performed as intended.

機能試験 [kinou siken]
>目的機能を果たしているかどうかを確認するための試験.

性能试验 [xìng néng shì yàn]

uji funsional [uji funsionaru]

기능 시험 [gineung siheom]

การทดสอบการทำงาน [gaan tod-sorb gaan tam-ngaan]

furnace
>an apparatus in which heat is transfered directly or indirectly to the work. Heat sources are heavy oil, kerosene, gas, electricity, etc.

炉 [ro]
>重油，灯油，ガス，電気などを熱源として熱処理や鍛造用材料を直接，又は間接的に加熱する装置.

炉 [lú]

tungku [tunguku]

로 [ro]

เตาเผา [ta-o pa-o]

G

gage
the thickness of a metal sheet, a road, or a wire.
ゲージ [geezi]
線の太さ，棒鋼の径，鋼鈑の厚さの標準寸法．
量具 [liàng jù]
gage [geji]
게이지 [geiji]
เกจ [guage]

gage block
a steel block having two flat, parallel surfaces with the distance between them being marked on the block to a guaranteed accuracy.
ブロックケージ [burokku geezi]
平行面を持つ鋼製のブロックで，平行面間の距離精度を保証して表示してあるもの．
量块 [liàng kuài]
blok gage [buroku geji]
블록 게이지 [beullok geiji]
เกจบล็อค [gauge block]

gage length
original length of the portion having constant cross sectional area in a test specimen.
ゲージ長さ [geezi nagasa]
試験片の断面積が同じ部分の長さ．
量具长 [liǎng jù cháng]
panjang pengukur, pembanding [panjangu pungukuru, punbandingu]
표점 거리 [pyojeom geori]
ความยาวเกจ [kwaam ya-o guage]

gage pressure
the amount of pressure calculated by the deduction of ambient atmospheric pressure from the total absolute pressure.
ゲージ圧 [geeziatu]
絶対圧から大気圧を差し引いた値．
气压计 [qì yā jì]
pengukur tekanan [pungukuru tekanan]
게이지 압력 [geiji amnyeok]
ความดันเกจ [kwaam dan guage]

gap between plate
the clearance between adjacent two leaves of a leaf spring before being fastened with a center bolt or other fittings.
板間すきま [bankan sukima]
重ね板ばねにおいて，センタボルトなどで締め付ける前の隣接するばね板とばね板とのすきま．
板间间隙 [bǎn jiān jiān xì]
celah anatara plat [cherahu anatara prato]
판 사이틈 [pan saiteum]
ช่องว่างระหว่างแผ่น [shong-waang ra-waang paaen]

gap of snap ring
止め輪の開口部 [tomewa no kaikoubu]
挡圈开口部 [dǎng quān kāi kǒu bù]
celah cincin penyumbat [cherahu chinchin punyunbato]
스냅링의 개구부 [seunaemningui gaegubu]
ช่องว่างระหว่างแหวนกันหลุด [shong-waang ra-waang waaen gan-lud]

garter spring
spring made by joining both ends of a close-coiled spring into a ring and used for holding.

ガータースプリング [gaataa supuringu]
　密着巻きのコイルばねの両端を連結し，環状にして締付けに用いるばね．
緊箍弾簧 [jǐn gū tán huáng]
pegas penahan [pugasu punahan]
가터 스프링 [gateo spring]
สปริงสำหรับยึด [sa-pring sam-rab yud]

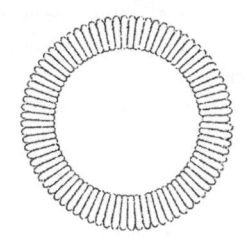

Fig G-1　garter spring

gas curtain
　the flow of gas that insulates the interior atmosphere of a heating furnace from the exterior atmosphere.
ガスカーテン [gasu kaaten]
　加熱炉の内部雰囲気と外部雰囲気を遮断するガスの流れ．
气幕 [qì mù]
tirai gas [tirai gasu]
가스 커튼 [gaseu keoteun]
แผ่นกั้นก๊าซในเตาอบ [paaen gan gas nai ta-o ob]

gas nitriding
　operation to harden the surface layer of steel by diffusing nitrogen. This operation includes gas nitriding by resolved ammonia gas and liquid nitriding by cyanide compounds.
ガス窒化 [gasu tikka]
　鉄鋼の表面層に窒素を拡散させ，表面層を硬化する操作．方法には，アンモニア分解ガスによるガス窒化及び青酸塩による液体窒化がある．
气体氮化 [qì tǐ dàn huà]

penitritan gas [punitoritan gasu]
가스 질화 [gaseu jilhwa]
การชุบแข็งที่ผิวเหล็กโดยก๊าซไนโตรเจน [gaan shup khaaeng tii piw lek dooi gas nitrogen]

gas spring
　a kind of fluid springs where gas is encapsulated in elastic body such as rubber to provide elasticity.
気体ばね [kitai bane]
　気体をゴムなどで包んで弾性のある構造にしたばね．
气弾簧 [qì tán huáng]
pegas gas [pugasu gasu]
가스 스프링 [gaseu spring]
สปริงก๊าซ [sa-pring gas]

geometrical moment of inertia
断面2次モーメント [danmen nizi moomento]
截面惯性矩 [duàn miàn guàn xìng jù]
momen geometrik [momen geometoriku]
단면 2차 모멘트 [danmyeon 2cha momenteu]
โมเมนต์หน้าตัดสเตป 2 [moment naa tad step song]

geometrical tolerance
　allowance for geometrical deviation (deviation and fluctuation from specified shape, alignment and position).
幾何公差 [kika kousa]
　幾何偏差（形状，姿勢及び位置の偏差並びに振れ）の許容差．
几何公差 [jǐ hé gōng chā]
toleransi geometrik [toreransi geometoriku]
기하학적 공차 [gihahakjeok gongcha]
พิกัดความเผื่อทางเรขาคณิต [pi-gad kwaam phuu taang re-kha-ka-nit]

German hook
　see "machine loop".
ジャーマンフック [zyaaman hukku]

German hook

"machine loop" 参照.

不压中心 (德国) 钩环　[bù yā zhōng xīn (déguó) gōu huán]

kait german　[kaito geruman]

저먼 후크 (측면 원형고리)　[jeomeon hukeu (cheungmyeon wonhyeonggori)]

เยอรมันฮุก　[German hook]

glass beads

a spherical shot made of glass. In shot peening operations, glass beads are used when steel shots bring too mach deformation.

ガラスビーズ　[garasu biizu]
ガラスを素材にした球状のショット. スチールショットによるショットピーニング作業では変形が大きすぎるものに適用されている.

玻璃珠　[bō lí zhū]

bead glas　[beado gurasu]

글라스 비 - 드 (쇼트)　[geullaseu bi-deu (syoteu)]

ลูกแก้วสำหรับช็อตพีนนิ่ง　[luuk gaaew sam-rap shot peening]

glass fiber reinforced plastic leaf spring

leaf spring made of glass fiber reinforced plastics.

GFRP 板ばね　[ziiehuaarupii itabane]
ガラス繊維を強化材とした繊維強化プラスチックを用いた板ばね.

玻璃纤维强化弹簧　[bō lí xiān wéi qiáng huà tán huáng]

pegas dengan penguat serat gelas [pugasu dengan punguato serato gerasu]

유리 섬유강화플라스틱 (GFRP) 스프링 [yuri seomnyuganghwapeullaseutik (GFRP) spring]

สปริงใยแก้วเสริมพลาสติก　[sa-pring sen yai gaaew serm plastic]

glass fiber spring

a plastic spring reinforced with glass fiber.

ガラス繊維ばね　[garasu sen'i bane]
ガラス繊維で補強されたプラスチックばね.

玻璃纤维弹簧　[bō lí xiān wéi tán huáng]

pegas penguat gelas　[pugasu punguato gerasu]

유리 섬유 스프링　[yuri seomnyu spring]

สปริงใยแก้ว　[sa-pring sen yai gaaew]

globular carbide

a spherical carbide included in steel.

球状炭化物　[kyuuzyou tankabutu]
鉄鋼に含まれる球状の炭化物.

球状碳化物　[qiú zhuàng tàn huà wù]

karbid bola　[karubido bora]

구상 탄화물　[gusang tanhwamul]

คาร์ไบด์ทรงกลมในเหล็ก　[carbide song glom nai lek]

globular inclusion

spherical inclusions in steel.

球状含有物　[kyuuzyou gan'yuubutu]
鉄鋼に含まれる球状物質.

球状夹杂物　[qiú zhuàng jiā zá wù]

masukan berbentuk bola　[masukan berubentuku bora]

구상 함유물　[gusang hamnyumul]

สารประกอบทรงกลมในเหล็ก　[saan-pra-gorb song glom nai lek]

go gage

a limit gage designed to allow the test piece fit in place.

通りゲージ　[toori geezi]
テストピースが通るように設計された限界ゲージ.

通规　[tōng guī]

pembanding lolos　[penbandingu rorosu]

통과 게 - 지　[tonggwa ge-ji]

โก เกจ　[go guage]

Gohner coefficient

a coefficient of stress correction in

Gohner's formula for helical springs.
ゲーナの応力修正係数 [geena no ouryoku syuusei keisuu]
コイルばね応力計算公式中で，ゲーナの推奨する応力修正係数．
剪切应力修正系数 [jiǎn qiē yīng lì xiū zhèng xì shù]
koefisien gohner [koefisien gohoneru]
괴너 응력 수정 계수 [goeneo eungnyeok sujeong gyesu]
สัมประสิทธิ์โกเนอร์ [sam-pra-sit Gohner]

Goodman fatigue limit diagram
see "fatigue limit diagram".
グッドマンの耐久限度線図 [guddoman no taikyuu gendo senzu]
"fatigue limit diagram" 参照．
古德曼疲劳极限图 [gǔ dé màn pí láo jí xiàn tú]
bagan goodman [bagan guddoman]
굿맨 내구 한도 선도 [gunmaen naegu hando seondo]
แผนภูมิพิกัดความล้ากู๊ดแมน [paaen poom pi-gad kwaam-laa Goodman]

governor
a device actuated by the centrifugal force rotating weights countered a spring force to provide automatic control of rotational speed.
調速機 [tyousokuki]
ばね力とバランスさせた重りの遠心力で作動させる装置で，回転速度の自動制御を行う．
调速器 [tiáo sù qì]
pengatur kecepatan [pungaturu kesepatan]
조속기 [josokgi]
ตัวควบคุมความเร็วหมุนของเครื่องจักร [tu-a ku-ab kum kwaam rew mun khong kreuang-chak]

grain boundary
the boundary of adjoining crystal grains.
粒界 [ryuukai]
隣接する結晶の境界．
晶粒界 [jīng lì jiè]
batas butiran [batasu butiran]
입계 [ipgye]
ขอบระหว่างเม็ดเกรน [khorb ra-waang med grain]

grain coarsening
growing of crystal grains due to overheating or prolonged heating. Grain coarsening causes reduction of tensile strength and elongation.
結晶粗大化 [kessyou sodaika]
加熱時にオーバーヒート又は時間超過により結晶粒が粗大化すること．引張強さや伸びが低下する．
晶粒粗大化（因过热） [jīng lì cū dà huà (yīn guò rè)]
pengasaran butiran [pungasaran butiran]
결정립 조대화 [gyeoljeongnip jodaehwa]
เม็ดเกรนหยาบขึ้นจากการโอเว่อร์ฮีท [med grain yaab khun chaak gaan overheat]

grain growth
a phenomenon whereby crystal grains grow larger due to heating polycrystalline material to a high temperature.
結晶粒成長 [kessyouryuu seityou]
多結晶材を高温に加熱することにより，結晶粒が大きくなる現象．
晶粒粗大 [jīng lì cū dà]
pertumbuhan butiran [purutunbuhan butiran]
결정립 성장 [gyeoljeongnip seongjang]
การขยายตัวของเม็ดเกรน [gaan kha-yaai to-a khong med grain]

grain refining
the process of making fine crystal

grain refining — grinding mark

grains by addition of an element(s) to molten steel or subsequent processing.
結晶粒微細化　[kessyouryuu bisaika]
　鉄鋼材料の溶解時の添加元素，あるいは二次加工によって結晶粒を微細にする操作．
晶粒细化　[jīng lì xì huà]
penghalusan butiran　[punguharusan butiran]
결정립 미세화　[gyeoljeongnip misehwa]
ทำให้เม็ดเกรนละเอียดลงมาก　[tam hai med grain la-iead long maak]

grain size
　the size of a crystal grain of polycrystalline material.
結晶粒度　[kessyouryuudo]
　多結晶材における結晶粒の大きさ．
晶粒度　[jīng lì dù]
ukuran butiran　[ukuran butiran]
결정 입도　[gyeoljeong ipdo]
ขนาดของเม็ดเกรน　[kha-naad khong med grain]

grain size distribution
　the distribution of the sizes of crystal grains.
結晶粒度分布　[kessyouryuudo bunpu]
　結晶粒の大きさの分布状態のこと．
晶粒分布　[jīng lì fēn bù]
distribusi ukuran butiran　[disutoribusi ukuran butiran]
결정 입도 분포　[gyeoljeong ipdo bunpo]
การกระจายของขนาดเม็ดเกรน　[gaan grachaai khong kha-naad med grain]

graphical symbol
　a symbol to represent a picture of an element used in a diagram.
図記号　[zukigou]
　要素の形状を表す記号で，図表に用いられる．
图号　[tú hào]

simbol gambar　[sinboru ganbaru]
기호도　[gihodo]
สัญลักษณ์ในแผนภูมิ　[san-ya-luk nai paaen-poom]

green chromate coating
　a chromate coating process for galvanized material, resulting in a green finish.
グリーンクロメート処理　[guriin kuromeeto syori]
　緑色の仕上げとなる亜鉛めっきのクロメート処理．
环保镀铬处理　[huán bǎo dù gè chǔ lǐ]
pelapisan kromat hijau　[purapisan kuromato hijau]
그린 크로맷 처리　[geurin keuromet cheori]
การชุบกรีนโครเมท　[gaan shub green chromate]

grinding
　removing a portion of workpiece surface with a grinding wheel.
研削　[kensaku]
　工作物の表面をといしで削りとること．
磨削　[mó xuē]
penggerrindaan　[pungugerurindaan]
연삭　[yeonsak]
การเจียร์　[gaan chiaa]

grinding machine
　a machine tool used to grind and polish a workpiece.
研磨（研削）機　[kenma (kensaku) ki]
　工作物に研磨・研削加工を施す工作機．
磨簧机　[mó huáng jī]
mesin gerinda　[mesin gerinda]
연마 (연삭) 기　[yeonma (yeonsak) gi]
เครื่องเจียร์　[kru-ang chiaa]

grinding mark
　surface pattern made by the roughness

of the grinding stone, grinding speed, material to be processed and the feed direction.
研磨模様 [kenma moyou]
といしの粗さ，研磨速度，被加工物の材質，送り方向などによって現れる模様．
磨纹（痕） [mó wén (hén)]
tanda penggerindaan [tanda pungugerindaan]
연마 모양 (흔적) [yeonma moyang (heunjeok)]
รูปแบบสำหรับเจียร์ผิว [ruup baaep sam0rap chiaa piw]

grinding pressure
pressure imposed on a grinding wheel during grinding.
研削圧力 [kensaku aturyoku]
研削加工時のといしにかかる圧力．
磨削压力 [mó xuē yā lì]
tekanan gerinda [tekanan gerinda]
연삭 압력 [yeonsak amnyeok]
แรงกดขณะเจียร์ [rang god kha-na chiaa]

grinding wheel change
といし交換 [toisi koukan]
砂轮更换 [shā lún gèng huàn]
pergantian roda gerinda [purugantian roda gerinda]
숫돌 교환 [sutdol gyohwan]
การเปลี่ยนหินเจียร์ [gaan plian hin chiaa]

grinding wheel compensation
see "dressing".
といし修整 [toisi syuusei]
"dressing" 参照．
砂轮修整 [shā lún xiū zhěng]
pengimbangan roda gerinda [punginbangan roda gerinda]
숫돌 수정 [sutdol sujeong]
การชดเชยหินเจียร์ที่สึก [gaan shod-shoi hin chiaa tii suk]

grinding wheel condition
requirements for a grinding wheel which are determined by the material of a workpiece, grinding roughness, working conditions, and other factors. Conditions selected typically include the type of abrasive grain, grain size, grade, the type of binder, and structure. Also considered are the equipment to be used, rotational speed, and feed rate.
といしの条件 [toisi no zyouken]
加工物の研削代，材質，研削粗さ，加工条件などによって決まるといしの要件．砥粒の種類，粒度，結合度，結合剤，組織などが選択される．また，使用機種，回転速度，送り速度も条件の一つである．
砂轮状况 [shā lún zhuàng kuàng]
kondisi roda gerinda [kondisi roda gerinda]
숫돌 요건 [sutdol yogeon]
สภาพหินเจียร์ [sa-paab hin chiaa]

grinding wheel consumption
the localized wear-out of a grinding wheel surface.
といし消耗 [toisi syoumou]
といしの使用面が局部的に摩耗すること．
砂轮损耗 [shā lún sǔn hào]
pemakaian roda gerinda [pumakaian roda gerinda]
숫돌 소모 [sutdol somo]
การสึกหรอของหินเจียร์ [gaan suk-ror khong hin chiaa]

grinding wheel quality
factors to be specified are as follows; grain material, grain size, bonding strength, bonding material shape and size.

grinding wheel quality — grooved section

といしの品質　[toisi no hinsitu]
　といしの品質は，砥粒，粒度，結合度，組織，結合剤などで，さらに形状，縁形，寸法などで規格で規定されている．
砂轮质量　[shā lún zhì liàng]
mutu roda gerinda　[mutu roda gerinda]
숫돌 품질　[sutdol pumjil]
คุณภาพหินเจียร์　[kun-na-paab hin chiaa]

grindstone
　tool to grind and polish the end face of the helical compression spring. Two types are found in the tool. One is a natural stone type, and the other is a sintered artificial stone type.
といし　[toisi]
　圧縮コイルばねの端末などの研削及び研磨を行う工具．天然石のもの及び人造研磨材を焼結したものがある．
砂轮　[shā lún]
batu gerinda　[batu gerinda]
연삭 숫돌　[yeonsak sutdol]
หินขัด　[hin khad]

grip band
　a plate part used to prevent the leaves of a leaf spring from displacement.
グリップバンド　[gurippu bando]
　重ね板ばねのばね板のずれ防止のために用いる板状の部品．
夹紧箍　[jiā jǐn gū]
sabuk pencengkram　[sabuku punsengukuran]
그립 밴드　[geurip baendeu]
สายรัดแหนบ　[saai-rad naaep]

grip feeder
　component which holds and feeds the material to the forming section in the spring forming machine.
フィードグリッパ　[fiido gurippa]
　線，帯鋼などを材料としてばねを成形する成形機の一部で，材料を保持して成形部に送り出す部品．
夹钳送料　[jiā qián sòng liào]
umpan penncengkram　[unpan punsengukuran]
그리퍼 이송　[geuripeo isong]
ตัวยึดขณะป้อนวัตถุดิบ　[to-a yud kha-na porn wat-tu-dip]

grip ring
　retaining ring used for shafts without grooves.
グリップ止め輪　[gurippu tomewa]
　軸用の止め輪で，溝を付けていない軸に使用する止め輪．
轴用挡圈　[zhóu yòng dǎng quān]
cincin pencengkram　[chinchin punsengukuran]
그립 링　[geurip ring]
แหวนรัด　[waaen-rad]

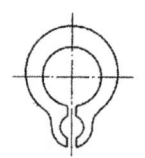

Fig G-2　grip ring

grooved section
　rectangular cross-section of a leaf with a groove on one side partially at the center of the width.
溝付き断面　[mizotuki danmen]
　長方形断面の幅方向の中央の，片面の一部に溝を付けたばね板の断面．
沟槽断面　[gōu cáo duàn miàn]
penampang dengan celah　[punanpangu dengan cerahu]
홈 단면　[hom danmyeon]

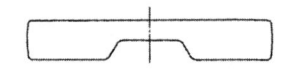

Fig G-3　grooved section

บริเวณร่องตัดบนแหนบ [bo-ri-wen rong tad bon naaep]

grooved spring pin

    spring pin formed cylindrically so that the edges of the slot do not touch each other when inserted into a hole.

溝付きスプリングピン [mizotuki supuringu pin]

    スプリングピンの一種で，穴に打ち込んだとき，円筒状に丸めた両端が接合しない程度に成形したピン．

带槽弹簧销 [dài cáo tán huáng xiāo]

pegas pin dengan celah [pugasu pin dengan cerahu]

그루브드 스프링 핀 [geurubeudeu seupeuring pin]

สปริงพินทรงกระบอก [sa-pring pin song gra-borg]

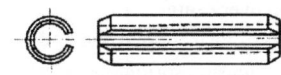

Fig G-4 grooved spring pin

ground face

    ground end surface of a helical compression spring.

研削端面 [kensaku tanmen]

    圧縮コイルばねの端面研削された面．

磨削端面 [mó xuē duān miàn]

permukaan gerinda [purumukaan gerinda]

연삭 끝면 [yeonsak kkeunmyeon]

พื้นผิวเจียร์ของปลายสปริงกด [phuun piw chiaa khong plaai sa-pring god]

guide cylinder

    a cylindrical part used to prevent a spring from buckling. There are two types of guide cylinders: one is mounted inside the spring and the other is mounted outside the spring.

案内筒 [annaitou]

    ばねの座屈防止のために用いる円筒状の部品．内径ガイドや外径ガイドのものがある．

导向筒 [dǎo xiàng tǒng]

cilindeer pemandu [chirindeeru pumandu]

안내 실린더 [annae sillindeo]

ตัวนำทรงกระบอก [to-a nam song gra-bork]

guide length

    the length of a guide cylinder or other type of guiding piece.

ガイド長さ [gaido nagasa]

    案内筒あるいは他の形式のガイド部品の長さ．

导向长度 [dǎo xiàng cháng dù]

panjang pemandu [panjangu pumandu]

안내 길이 [annae giri]

ความยาวตัวนำ [kwaam ya-o to-a nam]

guide pin

    a pin used to set the relative positions of male and female dies.

ガイドピン [gaido pin]

    金型の雄型と雌型の相対位置を決めるために用いるピン．

导向销 [dǎo xiàng xiāo]

pin pemandu [pin pumandu]

안내핀 [annaepin]

พินสำหรับกำหนดตำแหน่งของดาย [pin sam-rab gam-nod tam-naaeng khong die]

H

Haigh diagram
see "fatigue limit diagram".
ヘイの耐久限度曲線 [hei no taikyuu gendo kyokusen]
"fatigue limit diagram" 参照.
哈夫疲劳极限图 [hā fū pí láo jí xiàn tú]
bagan Haigh [bagan hai]
헤이 내구 한도 곡선 [hei naegu hando gokseon]
แผนภูมิเฮย์ [paaen-poom Haigh]

hair crack
a fine crack as thin as hair, resulting from improper shrinkage, forging, drawing, heat treatment, grinding, welding, etc.
微細割れ [bisai ware]
髪の毛のような小さなき裂で収縮, 鍛造, 引抜き, 熱処理, 研削, 溶接などが不適切に行われるとき発生する.
微裂纹 [wēi liè wén]
retak rambut [retaku ranbuto]
미세 균열 [mise gyunnyeol]
ลายแตกขนแมว [laai taaek khon maaew]

hair spring
noncontact type small and precise spiral spring, especially for instruments.
ひげぜんまい [hige zenmai]
計器用などの小型で精密な非接触形ぜんまい.
精細弾簧 [jīng xì tán huáng]
pegas rambut [pugasu ranmbuto]

Fig H-1　hair spring

헤어 스프링 (미소 태엽 스프링) [heeo spring (miso taeyeop spring)]
สปริงเส้นผม [sa-pring sen pom]

hairpin spring
a term for hairpin shaped spring clip.
ヘヤピンスプリング [heyapin supuringu]
ヘヤピンの形状を持つスプリングクリップ.
发卡式弹簧 [fā kǎ shì tán huáng]
pegas penjepit rambut [pugasu punjepito ranbuto]
헤어핀 스프링 [heeopin spring]
สปริงรูปร่างที่หนีบผม [sa-pring ruup-rang tii niip pom]

half loop over center
semi-circular loop at the end of a helical extension spring.
半丸フック [hanmaruhukku]
引張コイルばねの端部の一種で, 半円形のフック.
半圆形钩 [bàn yuán xíng gōu]
setengah kait [setenganu kaito]
반원 고리 [banwon gori]
ตะขอครึ่งวงกลม [ta0khor kre-ungg wong glom]

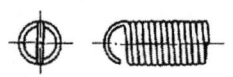

Fig H-2　half loop over center

handling device
an operation device activated by fingers, hands, or feet. The operation force is usually given through a push button, a lever, a pedal, etc.
手動装置 [syudou souti]
指, 手又は足による操作装置. 通常,

handling device — hardened and tempered spring wire

押しボタン，レバー又はペタルなどを介して操作力が与えられている．
手動装置　[shǒu dòng zhuāng zhì]
sarana penanganan　[sarana punanganan]
수동 장치　[sudong jangchi]
อุปกรณ์ควบคุมด้วยมือ　[up-pa-gorn kuob kum du-oi muu]

hard chrome plating
a type of chrome plating that provides hardness of 600 to 800 H_V. It has high resistance to wear and glossy appearance.
硬質クロムめっき　[kousitu kuromu mekki]
硬さが 600 〜 800 H_V 付近で硬く光沢があり，耐摩耗性に富んでいるクロムめっき．
硬质镀铬　[yìng zhì dù gè]
pelapisan(krom)keras　[perapisan (kuromu) kerasu]
경질 크롬 도금　[gyeongjil keurom dogeum]
ชุบแข็งโครเมียม　[shub khaaeng chromium]

hard drawn steel wire
a steel wire finished by cold-work after heat treatment.
硬鋼線　[koukousen]
熱処理後伸線など冷間加工して仕上げられた鋼線．
硬钢丝　[yìng gāng sī]
baja keras hasil tarikan　[baja kerasu hasiru tarikan]
경강선　[gyeonggangseon]
ลวดเหล็กดึงแข็ง　[luad lek duung khaaeng]

hard drawn wire
hard wire as rolled or drawn at a high reduction ratio.
硬引線　[katabikisen]
圧延，引抜加工などにおいて高い減面率で加工したままの硬い線をいう．

硬钢丝（冷拉钢丝）　[yìng gāng sī (lěng lā gāng sī)]
kawat ditarik keras　[kawato ditariku kerasu]
경인선　[gyeonginseon]
ลวดดึงแข็ง　[luad duung khaaeng]

hardenability
the degree of ease in hardening steel. In other words, hardenability is performance that governs the hardened depth and distribution of hardness.
焼入性　[yakiiresei]
鉄鋼を焼入れ硬化させる場合の焼きの入りやすさ．すなわち，焼きの入る深さと硬さの分布を支配する性能．
淬硬性　[cuì yìng xìng]
dapat diperkeras　[dapato diperukerasu]
담금질성　[damgeumjilseong]
ความสามารถในการชุบแข็ง　[kwaam saa-maad nai gaan shub khaaeng]

hardened and tempered spring wire
spring wire which is heat-treated after being processed or drawn to specified dimensions.
熱処理したばね用線　[netusyori sita baneyou sen]
所定の寸法に伸線加工あるいは引抜加工をした後，熱処理を施したばね用の線．
经过调质处理的弹簧钢丝　[jīng guò tiáo zhì chǔ lǐ de tán huáng gāng sī]
kawat pegas dikeraskan dan disepuh keras　[kawato pugasu dikerasukan dan disepuhu kerasu]
열처리한 스프링용 선　[yeolcheorihan springyong seon]
ลวดสปริงชุบแข็งและอบแล้ว　[lu-ad sa-pring shub khaaeng laae ob laaew]

hardened and tempered steel wire after drawing

steel wire which is quenched and tempered after being drawn to specified dimensions.
伸線加工後の焼入焼戻し鋼線　[sinsen kakougo no yakiire yakimodosi kousen]
所定の寸法に伸線加工あるいは引抜加工をしたのち，焼入焼戻しを施された鋼線．
拉拔后调质钢丝　[lā bá hòu tiáo zhì gāng sī]
dikeraskan dan sepuh keras setelah penarikan　[dikerasukan dan sepuhu kerasu seterahu punarikan]
신선 가공후 담금질．뜨임　[sinseon gagonghu damgeumjil.tteuim]
การชุบแข็งและอบหลังดึงขึ้นรูป　[gaan shub khaaeng laae ob lang duung khun ruup]

hardening

a process to increase hardness through a combination of aging, heating, and cooling. Includes age hardening, quenching, case hardening, and work hardening.
硬化　[kouka]
時効，加熱，冷却などの処理などで硬さを増す操作．時効析出，焼入れ，はだ焼き，加工硬化などがある．
硬化　[yìng huà]
pengerasan　[pungerasan]
경화　[gyeonghwa]
การชุบแข็ง　[gaan shub khaaeng]

hardening furnace

a furnace for hardening process. The term consists of a continuous type and a batch type with different heat sourses like electricity, gas, and salt bath.
焼入炉　[yakiirero]
焼入れするための加熱炉．形式として連続式とバッチ式があり，電気炉，ガス炉，塩浴炉などがある．
淬火炉　[cuì huǒ lú]
tungku pengerasan　[tunguku pungerasan]
담금질로　[damgeumjillo]
เตาชุบแข็ง　[ta-o shub khaaeng]

hardening medium

a coolant used for hardening. Though hardening medium is typically oil, water, air, and salt are also used depending on the material to be hardened.
焼入剤　[yakiirezai]
焼入れを行うための冷却媒体．一般的には油が用いられているが，材質によって水，空気，溶融塩などが用いられる．
淬火冷却剂　[cuì huǒ lěng què jì]
zat pengerasan　[zato pungerasan]
경화제　[gyeonghwaje]
สารหล่อเย็นที่ใช้ในการชุบแข็ง　[saan lo-r yen tii nai gaan shub khaaeng]

hardening oil

vegetable or mineral oil used for hardening high-carbon steel. Typically a mixture of mineral oil and additives is used. The temperature range between 60 and 80°C is appropriate, but a temperature from 90 to 120°C may be used depending on the quality and shape of the material to be hardened.
焼入油　[yakiire abura]
高炭素鋼の焼入れに用いられる植物性油又は鉱物性油．一般には，鉱物性油に添加物を混ぜたものが多く 60〜80°Cが適当であるが，材質・形状などによっては 90〜120°Cで行っている．
淬火油　[cuì huǒ yóu]
minyak pengerasan　[minyaku pungerasan]
담금질유　[damgeumjiryu]
น้ำมันชุบแข็ง　[naam-man shub khaaeng]

hardmetal
    material made of the minute particle of tungstein carbide or titanium carbide together with metal powder to be sintered into super hard metal block.
超硬合金　[tyoukougoukin]
炭化タングステン，炭化チタンなどの非常に硬い化合物の粉末とコバルト金属粉末を結合剤として高圧で圧縮し，金属が溶けない程度の高温に加熱し焼結・成形させたもの．
超硬质合金　[chāo yìng zhì hé jīn]
logam paduan super keras　[rogan paduan superu kerasu]
초경합금　[chogyeonghapgeum]
เหล็กแกร่ง　[lek-graaeng]

hardness
    a relative numerical value to represent a resistance to permanent deformation when pressed by a hard tip.
硬さ　[katasa]
試験片に，他の硬い物体を押し込んだ際に生じる永久変形の抵抗性を表す相対的な数値．
硬度　[yìng dù]
kekerasan　[kekerasan]
경도　[gyeongdo]
ความแข็ง　[kwaam khaaeng]

hardness test
    a test conducted on a hardness tester to measure the hardness of material, by pressing a hard indenter onto the surface of a test piece or product under a given load, or by dropping a hammer from a given height. The test methods include Vickers, Rockwell, Brinell, and Shore.
硬さ試験　[katasa siken]
硬さ試験機を用い試験片又は製品の表面に一定の荷重で一定形状の硬質圧子を押し込むか，又は一定の高さからハンマを落下させるなどの方法で硬さを測定する試験．ビッカース，ロックウエル，ブリネル，ショアーなどの試験法がある．
硬度试验　[yìng dù shì yàn]
uji kekerasan　[uji kekerasan]
경도 시험　[gyeongdo siheom]
การทดสอบความแข็ง　[gaan tod-sorb kwaam khaaeng]

harmonic analysis
    an analysis of vibration to find each harmonic constituent (the multiples of sinusoidal fundamental vibration).
調和分析　[tyouwa bunseki]
周期振動の特性を調波（基本振動数の整数倍の振動数を持つ正弦波）について求めること．
谐波分析　[xié bō fēn xī]
analasisa harmonik　[anarasisa harumoniku]
조화 분석　[johwa bunseok]
การวิเคราะห์ความสอดคล้อง　[gaan wi-kra-or kwaam sord-klong]

hazard
    any risk to which a worker is subjected to in work environment.
危険源　[kikengen]
職場環境で作業者がさせられる可能性のある危険要因．
危险源　[wēi xiǎn yuán]
sumber bahaya　[sunberu bahaya]
위험원　[wiheomwon]
การเสี่ยงต่ออันตราย　[gaan sei-ang to-a an-ta-raai]

hazardous event
    an dangerous event which triggers harm.
危険事象　[kiken zisyou]
危害を生じさせる危険な一つの出来

事.
危险现象　[wēi xiǎn xiàn xiàng]
kejadian berbahaya　[kejadian berubahaya]
위해요소　[wihaeyoso]
ภาวะเสี่ยงต่ออันตราย　[paa-wa sei-ang to-a an-ta-raai]

HDD suspension spring
 a flat spring made of thin, stainless steel to support the magnetic head of a hard disk drive. The magnetic head flies above a disk by virtue of air pressure caused by the disk rotating at high speed reaching equilibrium with the spring force.
HDD サスペンションスプリング
[ettidjiidjii sasupensyon supuringu]
 ハードディスク装置の磁気ヘッドを支えるステンレス鋼薄板のばね．磁気ヘッドはディスクの高速回転による空気圧で浮上し，ばねの支持力と釣り合う．
HDD 悬架弹簧　[HDD xuán jià tán huáng]
pegas suspensi HDD　[pugasu susupensi HDD]
에치디디 (HDD) 서스펜션 스프링
[HDD seoseupensyeon spring]
สปริงซัสเพนชั่นสำหรับหัวอ่านฮาร์ดดิสก์　[sa-pring suspension sam-rab hu-a aan harddisk]

head profile
 the shape of head part of bolts. There's a round, a hexagon, a square, etc. in the form.
頭部形状　[toubu keizyou]
 ボルトの頭部の形状をいう．丸，六角，四角などがある．
头部形状　[tóu bù xíng zhuàng]
bentuk kepala　[bentuku kepara]
머리 형상　[meori hyeongsang]
ลักษณะของหัวโบลท์　[lak-sa-na khong khong hu-a]

heat conduction
 a phenomenon whereby heat transfers from a higher-temperature section to a lower-temperature section of an object.
熱伝導　[netudendou]
 物質内を熱が高温部から低温部に移動する現象.
热传导　[rè chuán dǎo]
pembawa kalor　[punbawa karoru]
열전도　[yeoljeondo]
การนำความร้อน　[gaan nam kwaam rorn]

heat resistance
 the property of a material that resists oxidation and retains mechanical strength at high temperature.
耐熱性　[tainetusei]
 高温において対酸化性に優れ，また高温強度に優れている性質.
耐热性　[nài rè xìng]
tahan panas　[tahan panasu]
내열성　[naeyeolseong]
การทนความร้อน　[gaan ton kwaam rorn]

heat setting
 a process to prerelax a spring in an elevated temperature in order to improve stress relaxation resistance in service.
熱間セッチング　[nekkan settingu]
 使用中の応力緩和抵抗性を改善するために，熱間であらかじめばねに初期ひずみを与える処理.
热定型处理　[rè dìng xíng chǔ lǐ]
pemantapan secara panas　[pumantapan sechara panasu]
열간 세팅　[yeolgan seting]
การเซ็ทแบบร้อน　[gaan set baaeb rorn]

heat treatment
 the process of heating and cooling metal material to give it required properties. Includes quenching, tempering, normalizing, and annealing.

熱処理　[netusyori]
　金属材料へ所要の性質を付与することを目的に行う加熱・冷却の操作．焼入れ，焼戻し，焼ならし，焼なましなどがある．
热処理　[rè chǔ lǐ]
olah kalor　[orahu karoru]
열처리　[yeolcheori]
การทำฮีททรีดเมนต์　[gaan tam heat treatment]

heat up time
　the length of time required to heat an object to a specified temperature.
加熱時間　[kanetu zikan]
　所定の温度まで温度を上げるための時間．
加热时间　[jiā rè shí jiān]
waktu pemanasan　[wakutu pumanasan]
가열 시간　[gayeol sigan]
เวลาในการให้ความร้อน　[ra-ya-wee-laa nai gaan hai kwaam rorn]

heatcapacity
　the quantity of heat required to raise the temperature of an object by one unit degree.
熱容量　[netuyouryou]
　物体の温度を単位温度だけ上昇させるのに必要な熱量．
热容量　[rè róng liàng]
kapasitas panas　[kapasitasu panasu]
열용량　[yeoryongnyang]
ความจุความร้อน　[kwaam-chu kwaam rorn]

heating test
　a test to examine whether a heated sample may cause rust, scale, cracking, peel, etc.
加熱試験　[kanetu siken]
　試料を加熱して，さび，スケール，割れ，はく離などの発生状態を調べる試験．
加热试验　[jiā rè shì yàn]

uji pemanasan　[uji pumanasan]
가열시험　[gayeolsiheom]
การทดสอบความผิดปกติเนื่องจากกระบวนการฮีทติ้ง　[gaan tod-sorb kwaam pid-pa-ga-ti nu-ang chaak gra-buan gaan heating]

heat-resisting steel
　an alloy steel that retains resistance to oxidation and corrosion, and maintains strength at high temperature.
耐熱鋼　[tainetukou]
　高温環境で，耐酸化性，耐高温腐食性，高温強度を保持する合金鋼．
耐热钢　[nài rè gāng]
baja tahan panas　[baja tahan panasu]
내열강　[naeyeolgang]
เหล็กกล้าทนความร้อน　[lek glaa ton kwaam rorn]

height adjustment
　adjustment of the installation height of a spring so as to adjust load on the spring.
高さ調整　[takasa tyousei]
　ばねの荷重を調節するために取付高さを調節することをいう．
高度调整　[gāo dù tiáo zhěng]
pengaturan tinggi　[pungaturan tingugi]
높이 조정　[nopi jojeong]
การปรับความสูง　[gaan prab kwaam soong]

height of twisted tooth
　the height of the twisted tooth for a toothed washer.
歯付座金の歯の高さ　[hatuki zagane no ha no takasa]
　歯付座金の歯部ねじり高さ．
扭齿垫圈齿高　[niǔ chǐ diàn quān chǐ gāo]
tinggi gigi bantalan　[tingugi gigi bantaran]
이붙이 와셔의 이의 높이　[ibuchi wasyeoui iui nopi]
ความสูงของฟันที่บิด　[kwaam soong khong fan tii bid]

height test — helical compression spring, barrel shaped, made of tapered wire

height test
 test which applies a load to a spring and measures the length (height).
 長さ試験 [nagasa siken]
  ばねに荷重を加え，そのときの長さ（高さ）を測定する試験．
 高度试验 [gāo dù shì yàn]
 uji ketinggian [uji ketingugian]
 길이 시험 [giri siheom]
 การทดสอบความสูง [gaan tod-sorb kwaam soong]

helical compression spring
 coil spring mainly subjected to a compressive force.
 圧縮コイルばね [assyuku koiru bane]
  主として，圧縮荷重を受けるコイルばね．
 螺旋压缩弹簧 [luó xuán yā suō tán huáng]
 pegas tekan heliks [pugasu tekan herikusu]
 압축 코일 스프링 [apchuk koil spring]
 สปริงขดรับแรงกดทรงก้นหอย [sa-pring khod rab rang god song gon-hoi]

helical compression spring made of tapered wire
 a helical compression spring made of spring wire whose diameter gradually decreases. The shapes of spring include barrel, hourglass and cylinder.
 テーパ線圧縮コイルばね [teepasen assyuku koiru bane]
  材料の直径が逓減している線を用いた圧縮コイルばね．円筒形のほかたる形，つづみ形などがある．
 变截面钢丝螺旋压缩弹簧 [biàn jié miàn gāng sī luó xuán yā suō tán huáng]
 pegas tekan heliks, dengan diameter batang tidak konstan [pugasu tekan herikusu, dengan diameteru batangu tidaku konsutan]
 테이퍼 선재 압축코일 스프링 [teipeo seonjae apchukkoil spring]
 สปริงขดรับแรงกดทรงก้นหอยที่มีปลายเรียว [sa-pring khod rab rang god song gon-hoi tii mii plaai rieaw]

helical compression spring with variable pitch
 a helical compression spring coiled to unequal pitches.
 不等ピッチ圧縮コイルばね [hutou pitti assyuku koiru bane]
  ピッチが一定でない圧縮コイルばね．
 不等距螺旋压缩弹簧 [bù děng jù luó xuán yā suō tán huáng]
 pegas tekanan helik, dengan ragam jarak antara [pugasu tekanan heriku, dengan ragan jaraku antara]
 부등 피치 압축 코일 스프링 [budeung pichi apchuk koil spring]
 สปริงขดรับแรงกดทรงก้นหอยที่มีระยะพิทช์ต่างๆ [sa-pring khod rab rang god song gon-hoi tii mii ra-ya pitch taang taang]

helical compression spring, barrel shaped, made of tapered wire
 a helical compression spring that is barrell-shaped and made of tapered steel bar. This design permits a lower solid height than otherwise possible, eliminates wire-to-wire contact and provides nonlinear characteristics. Also called mini-block spring.
 テーパ線たる形圧縮コイルばね [teepasen tarugata assyuku koiru bane]
  材料の直径が逓減している線を用いた樽形の圧縮コイルばね．密着高さが低くでき，線間接触がなく，非線形特性が得られる．ミニブロックスプリングともいう．
 变截面钢丝中凸形螺旋压缩弹簧 [biàn jié miàn gāng sī zhōng tū xíng luó xuán yā suō tán

helical compression spring, barrel shaped, made of tapered wire — helical extension spring

huáng]

pegas tekan heliks, dengan diameter batang tidak konstan　[pugasu tekan herikusu , dengan diameteru batangu tidaku konsutan]

테이퍼 원단면 미니블럭 코일 스프링 [teipeo wondanmyeon minibeulleok koil spring]

สปริงขดรับแรงกดทรงก้นหอยรูปร่างบาร์เรลที่มีปลายเรียว　[sa-pring khod rab rang god song gon-hoi ruup rang barrel tii mii plaai rieaw]

helical compression spring, hourglass shaped, made of tapered wire

　　a helical compression spring that has an hourglass shape and is made of tapered steel bar. This design permits a lower solid height than otherwise possible, eliminates wire-to-wire contact, provides nonlinear characteristics and prevent rattle noises. Also called a maxi-block spring.

テーパ線つづみ形圧縮コイルばね [teepasen tuzumigata assyuku koiru bane]

材料の直径が逓減している線を用いたつづみ形の圧縮コイルばね．密着高さが低くでき，線間接触がなく，非線形特性が得られ，線間たたき音を生じる心配がない．マキシブロックスプリングともいう．

变截面钢丝中凹形螺旋压缩弹簧　[biàn jié miàn gāng sī zhōng āo xíng luó xuán yā suō tán huáng]

pegas tekan heliks, dengan diameter batang tidak konstan (pegas maxi-blok)　[pugasu tekan herikusu, dengan diameteru batangu (pugasu makisi-buroku)]

테이퍼 원단면 장고꼴 코일 스프링 [teipeo wondanmyeon janggokkol koil spring]

สปริงขดรับแรงกดทรงก้นหอยรูปร่างอาร์วกลาสที่มีปลายเรียว　[sa-pring khod rab rang god song gon-hoi ruup rang hourglass tii mii plaai rieaw]

helical conical spring

　　cone-shaped coil spring.

円すいコイルばね　[ensui koiru bane]

円すい形のコイルばね．

圆锥螺旋弹簧　[yuán zhuī luó xuán tán huáng]

pegas kerucut heliks　[pugasu kerukuto herikusu]

원추 코일 스프링　[wonchu koil spring]

สปริงขดรูปกรวย　[sa-pring khod ruup gru-oi]

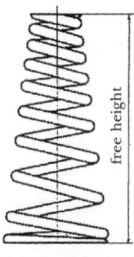

Fig H-3　helical conical spring

helical extension spring

　　coil spring mainly subjected to an extensive force.

引張コイルばね　[hippari koiru bane]

主として，引張荷重を受けるコイルばね．

拉伸螺旋弹簧　[lā shēn luó xuán tán huáng]

pegas perpanjangan heliks　[pugasu purupanjangan herikusu]

인장 코일 스프링　[injang koil spring]

สปริงขดชนิดดึงทรงก้นหอย　[sa-pring khod cha-nid duung song gon hoi]

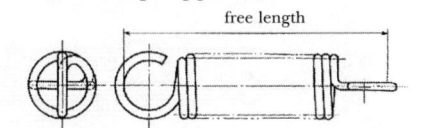

Fig H-4　helical extension spring

helical spring — hexagonal cross section torsion bar spring

helical spring
 the generic term for helically coiled springs.
 コイルスプリング　[koiru supuringu]
 ら旋状のばね.
 螺旋弹簧　[luó xuán tán huáng]
 pegas heliks　[pugasu herikusu]
 코일 스프링　[koil spring]
 สปริงขด　[sa-pring khod]

helical torsion spring
 coil spring mainly subjected to a twisting moment.
 ねじりコイルばね　[neziri koiru bane]
 主として，ねじりモーメントを受けるコイルばね.
 扭转弹簧　[niǔ zhuǎn tán huáng]
 pegas punti heliks　[pugasu punti herikusu]
 비틀림 코일 스프링　[biteullim koil spring]
 สปริงขดชนิดดึงทรงก้นหอย　[sa-pring khod cha-nid duung song gon hoi]

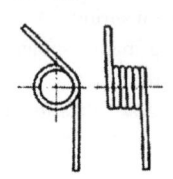

Fig H-5 helical torsion spring

helper spring
 spring which supports the load additionally after the load has increased in a 2 stage progressive leaf spring. (see Fig H-6)
 子ばね　[kobane]
 親子重ねばねにおいて，荷重が増加した後に補助的に働くばね．(Fig H-6 参照)
 補助弾簧　[fǔ zhù tán huáng]
 pegas bantu　[pugasu bantu]
 보조 스프링　[bojo spring]
 สปริงช่วย　[sa-pring shu-oi]

hexagon bar
 六角棒鋼　[rokkaku boukou]
 六角钢　[liù jiǎo gāng]
 pegas hexagon　[pugasu hekisagon]
 육각봉강　[yukgak bonggang]
 แทงหกเหลี่ยม　[taaeng hok-liam]

hexagon head
 the terminal part of a torsion bar with a hexagonal-shaped cross section.
 六角頭部　[rokkaku toubu]
 六角形のトーションバー端部.
 六角头部　[liù jiǎo tóu bù]
 kepala hexagon　[kepara hekisagon]
 육각 머리　[yukgak meori]
 หัวหกเหลี่ยม　[hu-oa hok liam]

hexagonal cross section torsion bar spring
 a torsion bar which was made of material with a hexagonal-shaped cross section.
 六角断面タイプトーションバー
 [rokkaku danmen taipu toosyon baa]
 断面が六角形の素材を用いたトーショ

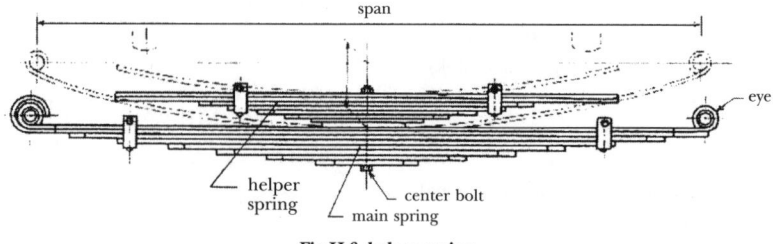

Fig H-6 helper spring

ンバー．
六角截面扭杆　[liù jiǎo jié miàn niǔ gǎn]
pegas puntir berpenampang hexagonal
[pugasu puntiru berupenanpangu hekisagonaru]
육각단면형 토션바
[yukgakdanmyeonhyeong tosyeonba]
ทอร์ชั่นบาร์หน้าตัดรูปหกเหลี่ยม　[torsion bar naa-tad ruu hok liam]

high carbon steel wire rod
　　carbon steel wire containing 0.24 to 0.86% carbon. Used to make hard drawn steel wire, oil tempered wire, etc.
硬鋼線材　[koukou senzai]
　　炭素含有量 0.24 〜 0.86% の炭素鋼線材．硬鋼線，オイルテンパ線などの製造に用いられる．
硬钢丝盘条　[yìng gāng sī pán tiáo]
kawat baja tinggi karbon (baja keras)
[kawato baja tingugi karubon (baja kerasu)]
경강선재　[gyeonggangseonjae]
แท่งลวดเหล็กชนิดคาร์บอนสูง　[taaeng lu-ad lek cha-nid carbon soong]

high polymer spring
　　a spring made of natural or synthetic polymer materials.
高分子材ばね　[koubunsizai bane]
　　天然高分子材，合成高分子材を用いたばね．
高分子弹簧　[gāo fēn zǐ tán huáng]
pegas polimer　[pugasu porimeru]
고분자재료 스프링　[gobunjajaeryo spring]
สปริงไฮโพลีเมอร์　[sa-pring high polymer]

high speed steel
　　a tool steel obtained by adding relatively large amounts of alloying elements such as Cr, Mo, W, B, and Co to high-carbon steel.
高速度鋼　[kousokudokou]

高炭素鋼に Cr, Mo, W, B, Co などの合金元素を比較的多量に添加した工具鋼．
高速钢　[gāo sù gāng]
baja kecepatan tingg　[baja kesepatan tingu]
고속도강　[gosokdogang]
เหล็กกล้ารอบสูง　[lek glaa rorb soong]

high temperature clamping test
　　test where a spring is clamped to a constant height and held for a certain period in constant high temperature. Creep test is a similar test.
高温締付試験　[kouon simetuke siken]
　　一定の高温雰囲気内で，ばねを一定の高さに締め付け保持する試験．類似のものにクリープ試験がある．
高温定型試験　[gāo wēn dìng xíng shì yàn]
uji pengencangan suhu tinggi　[uji pungenkangan suhu tingugi]
고온 체결 시험　[goon chegyeol siheom]
การทดสอบการยึดจับที่อุณหภูมิสูง　[gaan tod-sorb gaan yud chab tii un-na-ha-puum soong]

high temperature corrosion test
　　a test which inspects extent of corrosion in high temperature environment.
高温腐食試験　[kouon husyoku siken]
　　高温環境で，腐食の程度を調べる試験．
高温腐蚀試験　[gāo wēn fǔ shí shì yàn]
uji korosi suhu tinggi　[uji korosi suhu tingugi]
고온 부식 시험　[goon busik siheom]
การทดสอบการกัดกร่อนที่อุณหภูมิสูง　[gaan tod-sorb gaan gad gron tii un-na-ha-puum soong]

high temperature creep characteristic
　　a material property where the strain increases with the time lapse in high temperature under the constant load.
高温クリープ特性　[kouon kuriipu

high temperature creep characteristic — histogram

tokusei]
　高温において一定の荷重の下でひずみが時間とともに増加する現象で，その温度及び荷重の変化する過程の特性．

高温蠕変特性 [gāo wēn rú biàn tè xìng]
karakteristik rambat suhu tinggi
[karakuterisutiku ranbato suhu tingugi]
고온 크립 특성　[goon keurip teukseong]
ลักษณะการครีปที่อุณหภูมิสูง　[lak-sa-na gaan creep tii un-na-ha-puum soong]

high temperature strength
　a strength of material at high temperature, or a strength which endures high temperature.

高温強度 [kouon kyoudo]
　材料の高温時の強さあるいは高温に耐え得る強さのこと．

高温強度 [gāo wēn qiáng dù]
kekuatan suhu tinggi　[kekuatan suhu tingugi]
고온 강도　[goon gangdo]
ความแข็งแกร่งที่ทนต่อความร้อนสูงได้　[kwaam khaaeng-graaeng tii ton to-a kwaam rorn soong]

high tensile strength steel
　steel with higher tensile strength than normal steel, intended for use in architecture, bridges, ships, vehicles, other structures and pressure vessels, and manufactured with special consideration given to weldability, notch toughness, and workability. For steel plates used for automobiles, high tensile strength steel refers to steel with tensile strength of 343 N/mm^2 or higher.

高張力鋼 [koutyouryokukou]
　建築，橋，船舶，車両その他の構造物用及び圧力容器用として，普通鋼より高い引張強さを保持するとともに溶接性，切欠きじん性及び加工性も重視して製造された鋼板．自動車用鋼板では，引張強さ 343 N/mm^2 以上を高張力鋼という．

高强度钢 [gāo qiáng dù gāng]
baja berkekuatan tarik tinggi　[baja berukekuatan tariku tinggi]
고장력강　[gojangnyeokgang]
เหล็กแรงดึงสูง　[lek rang duung soong]

high tensile washer
　a flat washer made of heat treated carbon or alloy steel, or high tensile strength steel plate. Typically hardness ranges from 350 to 500 H$_V$.

高張力座金 [koutyouryoku zagane]
　炭素鋼又は合金鋼に熱処理を施したもの及び高張力鋼板を用いた平座金．硬さは 350～500 H$_V$ 程度のものが多い．

高强度垫圈 [gāo qiáng dù diàn quān]
washer kekuatan tarik tinggi　[wasuheru kekuatan tariku tingugi]
고장력 와셔　[gojangnyeok wasyeo]
แหวนแรงดึงสูง　[waaen rang duung soong]

hinge end torsion spring
　a type of torsion spring that has a hinge hook formed toward the center of the coil.

ヒンジフック形ねじりコイルばね
[hinzihukkugata nezirikoiru bane]
　コイル中心部に向けて成形されたヒンジフックを持つねじりコイルばね．

两端内铰形扭转弹簧 [liǎng duān nèi jiǎo xíng niǔ zhuǎn tán huáng]
pegas puntir hinge　[pugasu puntiru hinge]
힌지 고리형 비틀림 코일 스프링　[hinji gorihyeong biteullim koil spring]
สปริงทอร์ชั่นที่ปลายมีตะขอพับ　[sa-pring toesion tii plaai mii ta-khor pab]

histogram
　a diagram for frequencies of appear-

ance of measurement values, consisting of columns whose height is proportional to the frequency and whose width corresponds to the range of measurement values for the divided section of interest.
ヒストグラム　[hisutoguramu]
測定値の存在する範囲をいくつかの区間に分け，各区間を底辺として，その区間に属する測定値の出現度数に比例する高さを持つ柱を並べた図．
直方图　[zhí fāng tú]
histogram　[hisutoguramu]
히스토그램　[hiseutogeuraem]
แผนภูมิแสดงค่าที่วัดได้ ณ ความถี่ต่าง ๆ
[paaen puum sa-daaeng kaa tii wat dai na kwaam tii taang taang]

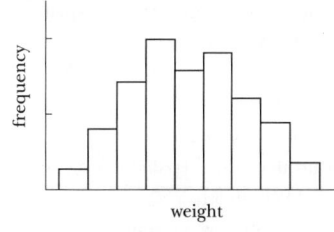

Fig H-7 histogram

hob
a rotary cutting tool with cutting edges arranged along helical thread on the outer surface of cylinder.
ホブ　[hobu]
円筒外周のら旋に沿って切刃を付けた回転刃物．
滚齿　[gǔn chǐ]
pemotong gerigi　[pumotongu gerigi]
호브　[hobeu]
อุปกรณ์ตัดชนิดหมุน　[up-pa-gorn tad cha-nid mun]

holding temperature
保持温度　[hozi ondo]
保温　[bǎo wēn]
suhu tahan　[suhu tahan]
유지 온도　[yuji ondo]
อุณหภูมิคงที่　[un-na-ha-poom kong tii]

holding time
保持時間　[hozi zikan]
保持时间　[bǎo chí shí jiān]
waktu tahan　[wakutu tahan]
유지 시간　[yuji sigan]
ระยะเวลาคงที่　[ra-ya we-laa kong tii]

hole diameter
穴径　[anakei]
孔径　[kǒng jìng]
diameter lubang　[diameteru rubangu]
구멍 지름　[gumeong jileum]
เส้นผ่าศูนย์กลางรู　[sen-pa-soon-glaang ruu]

homogenizing
a type of annealing performed by heating steel to an appropriate temperature equal to or higher than the A_3 or A_{cm} point so as to homogenize it by means of diffusion.
拡散焼なまし　[kakusan yakinamasi]
鋼を拡散によって均質化するために A_3 点又は A_{cm} 点以上の適当な温度に加熱して行う焼なまし．
扩散退火　[kuò sàn tuì huǒ]
penghomogenan　[punguhomogenan]
확산 풀림　[hwaksan pullim]
เป็นเนื้อเดียวกัน　[pen ne-uu dieaw gan]

honing
the process of removing a small amount of material from a cylindrical surface by abrasive stone to obtain close dimensional tolerance.
ホーニング　[hooningu]
精密な寸法を得るためにといしで円筒内面を研削すること．
珩磨　[héng mó]
honing　[honingu]

honing — hook shape

호닝 가공 [honing gagong]
กระบวนการเอาเศษเล็ก ๆ ออกจากผิวไซลินเดอร์โดยการขัด [gra-buan gaan aow sed lek lek oorg chaak piw cy-lin-der dooi gaan khad]

hook
    hook-shaped end of a helical extension spring to exert a load.
    フック [hukku]
    引張コイルばねの荷重をかけるかぎ状の端末部分．
    钩，环 [gōu, huán]
    kait [kaito]
    고리 [gori]
    ตะขอ [ta-khor]

hook height
    the length of the hook of a helical extension spring measured from the coil end.
    フックの高さ [hukku no takasa]
    引張ばねのフックのコイル端部から測った長さ．
    钩环高度 [gōu huán gāo dù]
    tinggi kait [tingugi kaito]
    고리 높이 [gori nopi]
    ความสูงตะขอ [kwaam soong ta-khor]

hook opening
    the length of clearance between the hook tip of a helical extension spring and the coil end.
    フック開口 [hukku kaikou]
    引張ばねのフックの先端とコイル端部のすきま長さ．
    钩环开口 [gōu huán kāi kǒu]
    mulut kait [muruto kaito]
    고리 개구 [gori gaegu]
    ระยะของช่องว่างจากปลายตะขอถึงปลายขด [ra-ya chong waang chaak plaai ta-khor tu-ung plaai khod]

hook plate with coil holes
    plate hook with pitched holes for coils of a helical extension spring to screw in.
    ねじ込みフックプレート [nezikomi hukku pureeto]
    引張コイルばねの端部の一種で，コイル部にプレートがねじ込まれたフック．
    带螺纹孔的钩头垫板 [dài luó wén kǒng de gōu tóu diàn bǎn]
    kait datar dengan lubang ulir [kaito dataru dengan rubangu uriru]
    나사형 후크 프레이트 [nasahyeong hukeu peureiteu]
    แผ่นตะขอชนิดมีเกลียว [paaen ta-khor cha-nid mii gliaew]

hook position
    an angle between the hooks at both ends of a helical extension spring.
    フック位置 [hukku iti]
    引張コイルばねの両端フックが互いになす角度．
    钩环位置 [gōu huán wèi zhì]
    kedudukan kait [kedudukan kaito]
    고리 위치 [gori wichi]
    ตำแหน่งของมุมระหว่างตะขอที่ปลายทั้งสองของสปริงดึง [tam-naaeng khong mum ra-waang ta-khor tii plaai tang song khong sa-pring dung]

hook shape
    the shape of a hook of a helical extension spring, including half loop over center, full loop over center, counter type full loop, full loop at side, rectangular hook, long round end hook over center, V-hook over center, tapered end with circular hook, inclined side hook, and threaded plug to fit plain end spring.
    フック形状 [hukku keizyou]
    引張コイルばねのフック形状のことで，半丸，丸，逆丸，側面丸，角，U，V，

絞り丸，斜め丸，ねじ込みフックなどがある．
鉤环形状　[gōu huán xíng zhuàng]
bentuk kait　[bentuko kaito]
고리 형상　[gori hyeongsang]
รูปร่างตะขอ　[roop rang ta-khor]

Hooke's law
　the law that strain of an object is directly proportional to the load applied to it until the load reaches certain limit value.
　フックの法則　[hukku no housoku]
　物体に負荷を加えると荷重がある限度に達するまでは荷重とひずみは正比例関係にあるという法則．
　虎克定律　[hǔ kè dìng lǜ]
　Hukum hooke　[Hukumu hooke]
　후크 법칙　[hukeu beopchik]
　กฎของฮุก　[god khong Hooke]

hooking machine
　a machine used to form the hooks of helical extension springs.
　フッキングマシン　[hukkingu masin]
　引張コイルばねのフック部を成形する機械．
　鉤环机　[gōu huán jī]
　mesin kait　[mesin kaito]
　후킹 머신　[huking meosin]
　เครื่องทำตะขอ　[kru-ang tam ta-khor]

hose clamp
　spring made by forming the plate or wire into a ring and used for tightening the connection of the hoses using the elasticity of the material.
　ホースクランプ　[hoosu kuranpu]
　板状又は線状の材料を用いて環状に成形し，材料の弾性を利用してホース差込み接続部を締め付けるばね．
　套管夹　[tào guǎn jiā]
　penjepit selang　[penjepito serangu]

호스 클램프　[hoseu keullaempeu]
สปริงขันเชื่อมท่อ　[sa-pring khun shi-um to-r]

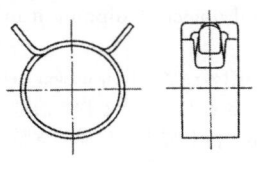

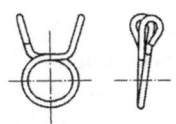

Fig H-8　hose clamp

hose protection coil
　a slender tube-shaped coil that covers a hose for feed water, oiling, etc. to protect from rupture.
　ホース保護用コイル　[hoosu hogoyou koiru]
　給水，給油などのホースの破裂防止のために被せる細長い管に成形したコイル．
　软管保护弹簧　[ruǎn guǎn bǎo hù tán huáng]
　ulir pelindung selang　[uriru purindungu serangu]
　호스 보호용 코일　[hoseu bohoyong koil]
　ลวดขดสำหรับป้องกันท่อ　[khod lu-ad sam-rab pong gan tor]

hot bending
　bending process at high temperature for a material that is difficult to cold bending.
　熱間曲げ　[nekkan mage]
　冷間曲げが困難な材料に対する高温曲げ加工．
　热弯曲　[rè wān qǔ]
　benkokan panas　[benkokan panasu]
　열간 굽힘　[yeolgan gupim]
　การดัดอดด้วยความร้อน　[gaan dad ngor du-oi kwaam rorn]

hot dip galvanize
making zinc film on the surface of a metal object by dipping it in molten zinc.
溶融亜鉛めっき　[youyuu aen mekki]
　めっきしようとする物を溶融した亜鉛の中に浸して表面に亜鉛皮膜を作ること.
热浸镀锌　[rè jìn dù xīn]
pegas tekan heliks jam pasir　[pugasu tekan herikusu jamu pasiru]
용융 아연 도금　[yongyung ayeon dogeum]
การชุบสังกะสี　[gaan shub sang-ga-sii]

hot dip tin
making tin film on the surface of a metal object by dipping it in molten tin.
溶融すずめっき　[youyuu suzu mekki]
　めっきしようとする対象物を溶融すず中に浸して表面にすず皮膜を作ること.
热浸镀锡　[rè jìn dù xī]
timah celup panas　[timahu cherupu panasu]
용융 주석 도금　[yongyung juseok dogeum]
การชุบดีบุก　[gaan shub dii-bug]

hot forging
a forging process at a certain elevated temperature to form a metal object into shape by applying pressure to it.
熱間鍛造　[nekkan tanzou]
　金属を一定の温度に加熱して，加圧力をその金属に加えて成形する鍛造作業.
热锻造　[rè duàn zào]
tempa panas　[tenpa panasu]
열간 단조　[yeolgan danjo]
การตีขึ้นขณะร้อน　[gaan tii khun ruup kha-na rorn]

hot formed helical spring
a coil spring made of thick gage wire that is formed in hot process.
熱間成形コイルばね　[nekkan seikei koiru bane]
　熱間で成形される太物のコイルばね.
热成形螺旋弹簧　[rè chéng xíng luó xuán tán huáng]
pegas bentukan panas　[pugasu bentukan panasu]
열간 성형 코일 스프링　[yeolgan seonghyeong koil spring]
สปริงขดขึ้นรูปร้อน　[sa-pring khod khun ruup kha-na rorn]

hot formed leaf spring
a leaf spring that is formed in hot process.
熱間成形重ね板ばね　[nekkan seikei kasaneita bane]
　熱間で成形される重ね板ばね.
热成形钢板弹簧　[rè chéng xíng gāng bǎn tán huáng]
pegas daun bentukan panas　[pugasu daun bentukan panasu]
열간 성형 겹판 스프링　[yeolgan seonghyeong gyeoppan spring]
สปริงแหนบขึ้นรูปร้อน　[sa-pring naaep khun ruup rorn]

hot forming
a process to form an object into shape above room temperature.
熱間成形　[nekkan seikei]
　室温より高い温度下で行う加工.
热变形　[rè biàn xíng]
pembentukan panas　[punbentukan panasu]
열간 성형　[yeolgan seonghyeong]
การขึ้นรูปร้อน　[gaan khun roop kha-na rorn]

hot peening
shot peening carried out on a spring in

hot peening — hour glass helical compression spring

a warm temperature. Fatigue strength is improved more than those by ordinary shot peening.
ホットピーニング　[hotto piiningu]
ばねに温間で行うショットピーニング．通常のショットピーニングよりも，疲れ強さが向上する．
热喷丸　[rè pēn wán]
peen panas　[peen panasu]
온간 피닝　[ongan pining]
การทำพีนนิ่งขณะร้อน　[gaan tam peening kha-na rorn]

hot roll
rolling process done at an elevated temperature.
熱間圧延　[nekkan atuen]
高温環境下での圧延．
热轧　[rè zhá]
roll panas　[roru panasu]
열간 압연　[yeolgan abyeon]
การรีดร้อน　[gaan riid rorn]

hot rolled bar
a steel bar that is rolled at high temperature typically around 800°C (equal to or higher than the A_3 transformation point).
熱間圧延棒鋼　[nekkan atuen boukou]
通常800°C付近（A_3変態点以上）の高温で圧延された棒鋼．
热轧棒料　[rè zhá bàng liào]
batang dirol panas　[batangu diroru panasu]
열간 압연 환강　[yeolgan abyeon hwangang]
แท่งเหล็กรีดร้อน　[taaeng lek riid rorn]

hot rolled strip steel
steel strip or steel plate rolled at high temperature typically around 800°C (equal to or higher than the A_3 transformation point).
熱間圧延帯鋼　[nekkan atuen obikou]
800°C付近（A_3変態点以上）の高温で圧延された帯鋼又は鋼板をいう．
热轧带钢　[rè zhá dài gāng]
baja kepingan dirol panas　[baja kepingan diroru panasu]
열간 압연 강대　[yeolgan abyeon gangdae]
เหล็กแถบรีดร้อน　[lek taaeb riid rorn]

hot setting
setting carried out in the temperature as high as low temperature annealing. Also called warm setting.
ホットセッチング　[hotto settingu]
低温焼なまし温度程度で行うセッチング．温間セッチングともいう．
热定型　[rè dìng xíng]
pengaturan saat panas　[pungaturan saato panasu]
핫세팅　[hatseting]
การทำเซ็ทดิ้งขณะร้อน　[gaan tam setting kha-na rorn]

hot working
process to work metal or an alloy into shape by rolling, forging, drawing, bending, punching, etc. at high temperature.
熱間加工　[nekkan kakou]
金属又は合金を加熱して高温のうちに圧延，鍛造，引抜き，曲げ，打抜きなどの加工を行う操作．
热加工　[rè jiā gōng]
pengerjaan panas　[pungerujaan panasu]
열간 가공　[yeolgan gagong]
กระบวนการที่ใช้ความร้อน　[gra bu-an gaan tii chai kwaam rorn]

hour glass helical compression spring
hourglass-shaped coil spring.
つづみ形圧縮コイルばね　[tuzumigata assyuku koiru bane]
中央がくびれたつづみ形のコイルば

hour glass helical compression spring — hydrogen embrittlement

ね.
中凹形螺旋圧簧 [zhōng āo xíng luó xuán yā huáng]
pegas tekan heliks jam pasir [pugasu tekan herikusu jamu pasiru]
장고꼴 압축 코일 스프링 [janggokkol apchuk koil spring]
สปริงขดรับแรงกดทรงก้นหอยอาร์วกลาส [sa-pring khod rab rang god song gon hoi hour glass]

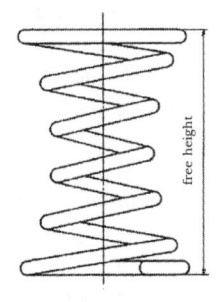

Fig H-9 hour glass helical compression spring

humidity cabinet test
a test to see corrosion, rust, deterioration of the surface of test piece in a chamber where atmospheric condition is controlled with high humidity and constant temperature.
湿潤試験 [situzyun siken]
高湿度に調整した恒温装置内に試験片を入れて湿潤状態に保ち,腐食,さび,劣化などの状態を調べる試験.
湿润试验 [shī rùn shì yàn]
uji kelembaban kabinet [uji kerenbaban kabineto]
습윤 시험 [seubyun siheom]
การทดสอบการกัดกร่อน, สนิม ในตู้ความชื้นสูง [gaan tod-sorb gaan gad-gorn, sa-nim nai tuu kwaam shi-uun suung]

hydraulic press
a press machine actuated by hydraulic pressure.
油圧プレス [yuatu puresu]
油圧によって駆動力を与えるプレス.
液压 [yè yā]
pres hidrolik [puresu hidororiku]
유압 프레스 [yuap peureseu]
เครื่องเพรสระบบไฮโดรลิค [kru-ang press ra-bob hydraulic]

hydraulic pressure spring
a spring utilizing the resilience of oil.
油圧ばね [yuatu bane]
油の弾性を利用するばね.
液压弹簧 [yè yā tán huáng]
pegas tekanan hidrolik [pugasu tekanan hidororiku]
유압 스프링 [yuap spring]
สปริงไฮโดรลิคเพรสเชอร์ [sa-pring hydraulic pressure]

hydrogen determination test
a test to determine the quantity of hydrogen contained in deposited or welded metal.
水素量試験 [suisoryou siken]
溶着金属又は溶接金属の含有水素の定量を行う試験.
氢量试验 [qīng liàng shì yàn]
uji kandungan hidrogen [uji kandungan hidorogen]
수소량 시험 [susoryang siheom]
การทดสอบปริมาณไฮโดรเจน [gaan tod-sorb pa-ri-maan hydrogen]

hydrogen embrittlement
a phenomenon whereby the ductility or toughness of steel material is reduced by hydrogen absorbed in the steel. This often occurs during pickling and electroplating.
水素ぜい化 [suiso zeika]
鋼中に吸収された水素によって鋼材に生じる延性又はじん性が低下する現

hydrogen embrittlement — hysteresis curve

象. この現象は, 酸洗い, 電気めっきなどの場合に生じることが多い.
氢脆　[qīng cuì]
kerapuhan hidrogen　[kerapuhan hidorogen]
수소 취성　[suso chwiseong]
การเปราะแตกจากไฮโดรเจน　[gaan pra-or taaek chaak hydrogen]

hygroscopic

a property whereby a material absorbs moisture in air.
吸湿性　[kyuusitusei]
空気中から水分を吸収する性質.
吸湿性　[xī shī xìng]
higroskopik　[higurosukopiku]
흡습성　[heupseupseong]
คุณสมบัติในการดูดความชื้น　[kun-na-som-bat nai gaan duud kwaam shi-un]

hyper eutectoid steel

carbon steel that contains 0.8% or more carbon and its structure consists of cementite and pearlite.
過共析鋼　[kakyousekikou]
炭素鋼のうち0.9%以上の炭素を含み組織が初析セメンタイトとパーライトからなるものをいう.
过共析钢　[guò gòng xī gāng]
baja hyper eutectic　[baja haiperu eutekutiku]
과공석강　[gwagongseokgang]
เหล็กกล้าไฮเปอร์ยูเต็กตอยด์　[lek glaa hyper eutectoid]

hypo eutectoid steel

carbon steel that contains 0.8% or less carbon and its structure consists of ferrite and pearlite.
亜共析鋼　[akyousekikou]
炭素鋼のうち0.9%以下の炭素を含み組織が初析フェライトとパーライトからなるものをいう.
亚共析钢　[yà gòng xī gāng]
baja hipo eutektoid　[baja hipo eutekutoido]
아공석강　[agongseokgang]
เหล็กกล้าไฮโปยูเต็กตอยด์　[lek-glaa hypo eutectoid]

hysteresis

phenomenon where the spring characteristic has different paths depending on increase or decrease of the applied load, caused by the frictional resistance between the springs themselves or the spring and the adjacent components.
ヒステリシス　[hisuterisisu]
ばね同士又はばねと周辺部品との摩擦抵抗によって, ばね特性が加荷時と減荷時とで異なる現象.
磁滞(現象)　[cí zhì (xiàn xiàng)]
histeresis　[hisuteresisu]
히스테리시스　[hiseuterisiseu]
คุณสมบัติที่ต่างกันของสปริงตามปริมาณโหลดที่ใส่　[kun-na-som-bat tii taang gan khong sa-pring taam load tii sai]

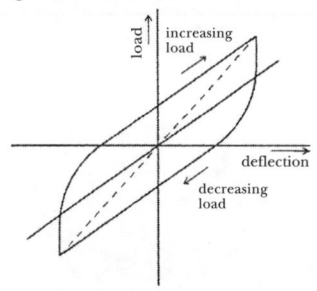

Fig H-10 hysteresis

hysteresis curve

a curve showing the difference in load deflection curve of a spring in loading and releasing conditions.
ヒステリシス曲線　[hisuterisisu kyokusen]
ばねの荷重増加時と減少時の変化を示

hysteresis curve

ヒステリシス曲線.

磁滞曲线 [cí zhì qǔ xiàn]

kurva histeresis [kuruva hisuteresisu]

히스테리시스 곡선 [hiseuterisiseu gokseon]

เส้นโค้งฮีสเทอร์เรซีส [sen kong hysteresis]

I

immersion process
    a process of immersing a work piece in a chemist bath for the purpose of painting, cleaning, removal of grease, etc.
    浸せき工程　[sinseki koutei]
    塗装，洗浄，脱脂などの目的で品物を薬剤槽に浸せきする工程．
    浸洗过程　[jìn xǐ guò chéng]
    proses pencelupan　[purosesu punserupan]
    침적 공정　[chimjeok gongjeong]
    กระบวนการล้าง　[gra bu-an gaan laang]

impact angle
    an angle made between the moving direction of a shot and the surface to be worked.
    投射角　[tousyakaku]
    ショットの運動方向と加工面とのなす角度．
    投射角度　[tóu shè jiǎo dù]
    sudut benturan　[suduto benturan]
    투사 각도　[tusa gakdo]
    มุมของการกระแทก　[mum khong gaan gra-taaek]

impact energy
    energy necessary to fracture the material under a dynamic load condition.
    衝撃エネルギー　[syougeki enerugii]
    動的荷重条件下で材料を破壊するのに要するエネルギー．
    冲击能　[chōng jī néng]
    energi benturan　[enerugi benturan]
    충격 에너지　[chunggyeok eneoji]
    พลังงานของการกระแทก　[pa-lang-ngaan khong gaan gra-taaek]

impact fatigue testing machine
    the test equipment used to apply a given magnitude of repeated impact to the test specimen.
    衝撃疲労試験機　[syougeki hirou sikenki]
    材料に一定の繰返し衝撃を加える試験装置をいう．
    冲击疲劳试验机　[chōng jī pí láo shì yàn jī]
    mesin uji kelelahan benturan　[mesin uji kererahan benturan]
    충격 피로 시험기　[chunggyeok piro siheomgi]
    เครื่องทดสอบความล้าจากการกระแทก　[kruang tod-sorb kwaam laa chaak gaan gra-taaek]

impact load
    a load which act on a body when an another object collides with the body.
    衝撃荷重　[syougeki kazyuu]
    物体に他の物体が衝突する際に生じる荷重．
    冲击负荷　[chōng jī fù hè]
    beban benturan　[beban benturan]
    충격 하중　[chunggyeok hajung]
    โหลดของการกระแทก　[load khong gaan gra-taaek]

impact speed
    the speed at which one mass collides with another.
    衝撃速度　[syougeki sokudo]
    質量と質量がぶつかり合う速さ．
    冲击速度　[chōng jī sù dù]
    laju benturan　[raju benturan]
    충격 속도　[chunggyeok sokdo]
    ความเร็วของการกระแทก　[kwaam rew khong gaan gra-taaek]

impact strength

the magnitude of resistance of a material against impact, expressed as the amount of energy needed to break a test piece at a single impact.

衝撃強さ　[syougeki tuyosa]
一回の打撃により試験片を破壊するのに必要なエネルギーで表された材料の抵抗の大きさ．

冲击强度　[chōng jī qiáng dù]
kuat benturan　[kuato benturan]
충격 강도　[chunggyeok gangdo]
ความแข็งแรงจากการกระแทก　[kwaam khaaeng rang chaak gaan gra-taaek]

impact stress

stress in a test piece when an impact test is conducted.

衝撃応力　[syougeki ouryoku]
衝撃試験時に試験片に生じる応力．

冲击应力　[chōng jī yīng lì]
tegangan benturan　[tegangan benturan]
충격 응력　[chunggyeok eungnyeok]
ความเค้นจากการกระแทก　[kwaam ken chaak gaan gra-taaek]

impact test

a test that is conducted to examine the toughness or brittleness of a material. Evaluation is made based on the amount of energy consumed, appearance of a fractured cross-section, deformation behavior, the development of cracking, and other factors when the test piece is broken with a impact load.

衝撃試験　[syougeki siken]
材料のじん性又はぜい性などを調べるため，試験片に衝撃荷重を加えて破壊し，要したエネルギーの大小，破面の様相，変形挙動，き裂の進度挙動などによって評価する試験．

冲击试验　[chōng jī shì yàn]
uji benturan　[uji benturan]
충격 시험　[chunggyeok siheom]
การทดสอบการกระแทก　[gab tod-sorb gaan gra-taaek]

impurity

an undesirable foreign material in a substance.

不純物　[huzyunbutu]
ある物質の中に含まれている主物質にとって不必要な物質．

杂质　[zá zhì]
kekotoran　[kekotoran]
불순물　[bulsunmul]
สิ่งเจือปน　[sing chi-uu pon]

incipient fatigue crack

a microscopic crack found in the early stage of fatigue fracture.

初期疲労き裂　[syoki hirou kiretu]
疲労破壊の初期段階で発生する微細なクラック．

初始疲劳裂纹　[chū shǐ pí láo liè wén]
retak kelelahan awal　[retaku kererahan awaru]
초기 피로 균열　[chogi piro gyunnyeol]
รอยแตกเริ่มต้นจากการล้า　[roi taaek rerm-ton chaak gaan laa]

inclination

the degree of squareness between coil circle plane and coil axis of a helical compression spring. It is shown by horizontal displacement from vertical line or the inclined angle of the uppermost edge of the outer side surface when the spring is set on the flat plate.

倒れ　[taore]
圧縮コイルばねの座面とコイル軸とが直角でない現象．倒れは，ばねを定盤上に立てたときの外側面最上端の横ずれ量又は傾斜角度で表す．

倾斜（偏角）　[qīng xié (piān jiǎo)]

inklinasi [inkurinasi]
경사 [gyeongsa]
การเอียง [gaan-iang]

inclined side hook
    hook formed by raising end coil halfway at the end of a helical extension spring.
斜め丸フック [naname maru hukku]
引張コイルばねの端部の一種で，コイル外周部から斜めに成形したフック．
傾斜钩环 [qīng xié gōu huán]
kait samping (lingkaran) mereng
[kaito sanpingu (ringukaran) mererengu]
경사 원형 고리 [gyeongsa wonhyeong gori]
ตะขอด้านที่เอียง [ta-khor daan tii iang]

Fig I-1 inclined side hook

inclusion
    a nonmetallic foreign material that precipitates or is accidentally trapped in molten steel during solidification.
介在物 [kaizaibutu]
金属の凝固過程で鋼中に析出又は巻き込まれた非金属の介在物．
夹杂物 [jiā zá wù]
masukan [masukan]
개재물 [gaejaemul]
สารประกอบ [saan pra-gorb]

index of cleanliness of steel
    a value to indicate how extensively nonmetallic inclusion exists in steel, expressed as the percentage of an area occupied by nonmetallic inclusion as observed under the microscope.
清浄度 [seizyoudo]
鋼中において，非金属介在物が含まれる度合い．顕微鏡視野内で，非金属介在物が占める面積百分率．

清洁度 [qīng jié dù]
tingkat pembersihan baja [tingukato punberusihan baja]
청정도 [cheongjeongdo]
ระดับความสะอาด [ra-dab kwaam sa-aad]

indexing plate
    a plate used for indexing an angle of a workpiece being sliced.
割出板 [waridasiban]
スライス加工の際，工作物の角度の割出しに使用されるプレート．
分度盘 [fēn dù pán]
plat pengindeks [purato pungindekusu]
분할판 [bunhalpan]
แผ่นแสดงดัชนี [paaen sa-daaeng dad-cha-nee]

indirect cost
    a cost that is not directly chargeable to a specific product.
間接費 [kansetuhi]
個々の製品に直接に賦課されない費用．
间接成本 [jiān jiē chéng běn]
biaya tidak langsung [biaya tidaku rangusungu]
간접비 [ganjeopbi]
ค่าใช้จ่ายทางอ้อม [kaa shai-chaai taang oam]

indirect labor
    labor not directly engaged in the actual production of the product.
間接工賃 [kansetu koutin]
製品を実際に生産することに直接関係しない労務費用．
间接工资 [jiān jiē gōng zī]
buruh tidak langsung [buruhu tidaku rangusungu]
간접 공임 [ganjeop gongim]
ค่าแรงทางอ้อม [kaa rang taang oam]

indirect measurement
    obtaining a measurement by mesuring

substituting property.
間接測定　[kansetu sokutei]
　　測定量と一定の関係にある量について測定を行って，測定値を間接的に導き出すこと．
间接测量　[jiān jiē cè liáng]
pengukuran tak langsung　[pungukuran taku rangusungu]
간접측정　[ganjeopcheukjeong]
การวัดทางอ้อม　[gaan wad taang-orm]

induction furnace
　　a furnace that utilizes heat generated by induced current.
誘導炉　[yuudouro]
　　誘導電流によって生じる熱で加熱する構造の炉．
感应电炉　[gǎn yīng diàn lú]
tungku induksi　[tunguku indukusi]
유도로　[yudoro]
เตาเผาแบบเหนี่ยวนำกระแสไฟฟ้า　[ta-o pa-o baaep niew-nam gra-saae fai-faa]

induction hardening
　　quenching utilizing induction heating capability of high frequency current. It is mainly used for quenching arbitrary surfaces or areas of steel.
高周波焼入れ　[kousyuuha yakiire]
　　高周波電流による誘導加熱作用で加熱して行う焼入れ．主に，鉄鋼の任意の表面又は部分を焼入れする場合に用いる．
高频淬火　[gāo pín cuì huǒ]
pengerasan induksi　[pungerasan indukusi]
고주파 열처리　[gojupa yeolcheori]
การชุบแข็งแบบเหนี่ยวนำกระแสไฟฟ้า　[gaan shub khaeng baaep niew-nam gra-saae fai-faa]

induction heating
　　heating steel with heat induced by high frequency current.

高周波加熱　[kousyuuha kanetu]
　　高周波電流による誘導加熱作用で鉄鋼材料を加熱すること．
高频加热　[gāo pín jiā rè]
pemanasan induksi　[pumanasan indukusi]
고주파 가열　[gojrpa gayeol]
การเผาโดยการใช้คลื่นความถี่สูง　[gaan pa-o du-oi gaan chai kluun kwaam tii soong]

industrial engineering
　　an engineering of industrial production concerned with the integrated systems of people, materials, and equipment.
生産工学　[seisan kougaku]
　　人，材料，設備を統合して工業生産を検討する工学．
工业工程　[gōng yè gōng chéng]
teknik industri　[tekuniku indusutori]
생산공학　[saengsan-gonghak]
วิศวกรรมอุตสาหการ　[wis-sa-wa-gam us-sa-ha-gaan]

industrial waste
　　waste resulting from industrial production activity. Includes metal particles, sludge, used oil, used acids and alkali, and waste plastics.
産業廃棄物　[sangyou haikibutu]
　　工業生産活動で生じる廃棄物．金属粉，汚泥，廃油，廃酸，廃アルカリ，廃プラスチックなどがある．
工业废弃物　[gōng yè fèi qì wù]
limbah industri　[rinbahu indusutori]
산업폐기물　[saneoppyegimul]
ขยะอุตสาหกรรม　[kha-ya ut-sa-ha-gam]

inert gas
　　a gas with extremely low reactivity. Includes argon, helium, and nitrogen.
不活性ガス　[hukassei gasu]
　　反応性が極めて低い気体．アルゴン，ヘリウム，窒素などがある．

惰性気体 [duò xìng qì tǐ]
gas mulia [gasu muria]
불활성 가스 [bulhwalseong gaseu]
ก๊าซเฉื่อย [gas chi-uui]

infiltration rusting
the progression of corrosion from a metal surface (such as steel) to its inside.
腐食浸透 [husyoku sintou]
腐食が鋼などの金属表面から内部へ進行すること．
腐蚀渗透 [fǔ shí shèn tòu]
pengaratan infiltrasi [pungaratan infirutorasi]
부식 침투 [busik chimtu]
สนิมซึมลึก [sa-nim sum luk]

inflection point
point determined for designing convenience to obtain the characteristics of a nonlinear characteristics spring. For example, on the load-deflection diagram of a two-stage pitch coil spring or a progressive leaf spring, this means the cross point of the extended characteristics curves of the first and the second stage.
交会点 [koukaiten]
非線形特性ばねの特性を決めるため，設計の便宜上決める点で，例えば，2段ピッチコイルばね及びプログレッシブ重ね板ばねの荷重-たわみ線図において，1段目及び2段目の特性の延長上の交点．
变形点 [biàn xíng diǎn]
titik perubahan kurva [titiku purubahan kuruva]
변곡점 [byeongokjeom]
จุดเปลี่ยนการโค้งงอ [chud-plian gaan kong-ngor]

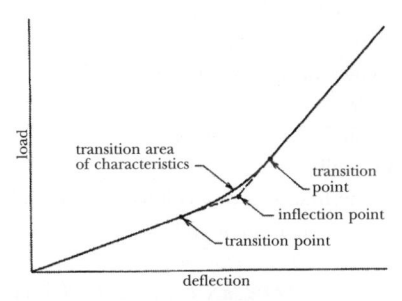

Fig I-2 inflection point

infrared drying equipment
a drying equipment that utilizes the aid of infrared radiation. Used in coating process etc.
赤外線乾燥装置 [sekigaisen kansou souti]
放射赤外線を利用して乾燥する装置．塗装工程などで用いられている．
红外干燥设备 [hóng wài gān zào shè bèi]
alat pengering inframerah [arato pungeringu infuramerahu]
적외선 건조장치 [jeokoeseon geonjojangchi]
อุปกรณ์ทำให้แห้งโดยรังสีอินฟราเรด [up-pa-gorn tam hai haaeng du-oi rang-sii in-fa-red]

ingress and egress
a driver's behavior to enter and leave the car.
乗降 [zyoukou]
乗用車の運転手が乗り降りする行動．
进出 [jìn chū]
naik turun [naiku turun]
승강 [seunggang]
เข้าๆออกๆ [khao khao oorg oorg]

inhibitor
a substance capable of stopping or retarding rapid or localized chemical or electrochemical reaction.
抑制剤 [yokuseizai]
化学反応又は電気化学反応の急激なも

しくは局部的な進行を妨げる物質.
防腐蝕剤　[fáng fǔ shí jì]
penghambat　[punguhanbato]
억제제　[eokjeje]
สารยับยั้ง　[saan yab-yang]

initial coil
the first turn of a coil where material wire begins to be coiled.
初巻きコイル　[syomaki koiru]
コイルばねを巻くとき，材料の直線部が円形に成形される最初のコイル部.
初卷　[chū juǎn]
ulir awal　[uriru awaru]
코일링 시작부　[koilling sijakbu]
ขดแรก　[khod raaek]

initial failure
a failure of equipment which takes place at a relatively early stage after the commencement of operation.
初期故障　[syoki kosyou]
使用開始後の比較的早期に機器等に生じる故障.
初期故障　[chū qī gù zhàng]
kegagalan awal　[kegagaran awaru]
초기 고장　[chogi gojang]
ความล้มเหลวในระยะแรก　[kwaam lom-leew nai ra-ya raaek]

initial load
load exerted on a spring when built into a machine.
初荷重　[syokazyuu]
ばねを機械に取り付けたとき，ばねに加わる荷重.
初始负荷　[chū shǐ fù hè]
beban awal　[beban awaru]
초하중　[chohajung]
โหลดเริ่มต้น　[load rerm-ton]

initial stress
torsion stress generated in the coil material by the initial tension of a helical extension spring.
初応力　[syouryoku]
引張コイルばねの初張力によって，コイル材料に生じるねじり応力.
初始应力　[chū shǐ yīng lì]
tegangan awal　[tegangan awaru]
초응력　[choeungnyeok]
ความเค้นเริ่มต้น　[kwaam-ken rerm-ton]

initial stress of stress peening
static stress generated by deflecting the spring in the direction to be used before performing stress peening.
ストレスピーニングの初応力　[sutoresu piiningu no syouryoku]
ストレスピーニングを行う場合に，あらかじめばねを使用する方向にたわませて生じさせる静的応力.
应力强化的初始应力　[yīng lì qiáng huà de chū shǐ yīng lì]
tegangan awal dari tegangan peen　[tegangan awaru dari tegangan peen]
스트레스 피닝 초기응력　[seuteureseu pining chogieungnyeok]
ความเค้นเริ่มต้นของการทำสเตรสพีนนิ่ง　[kwaam-ken rerm-ton khong gaan tam stree-peening]

initial tension
internal force of a helical extension spring closing tightly when no load is applied.
初張力　[syotyouryoku]
無荷重時に密着している，引張コイルばねの内力.
初拉力　[chū lā lì]
tarikan awal　[tarikan awaru]
초장력　[chojangnyeok]
แรงดึงเริ่มต้น　[rang-duung rerm-ton]

inner coil diameter
内ばね径　[utibanekei]
内弹簧径　[nèi tán huáng jìng]

diameter ulir dalam　[diameteru uriru daran]
내측 스프링 지름　[naecheuk springjireum]
เส้นผ่าศูนย์กลางของขดด้านใน　[sen pa-soon-glaang khong khod daan nai]

inner diameter
内径　[naikei]
内径　[nèi jìng]
diameter dalam　[diameteru daran]
안지름　[anjireum]
เส้นผ่าศูนย์กลางภายใน　[sen pa-soon-glaang paai-nai]

inner guide
　a guide provided to the inner side of a disc spring or a helical compression spring so as to secure positioning of the spring.
内面ガイド　[naimen gaido]
　皿ばねや圧縮コイルばねなどの位置確保のために内径側に設けるガイド.
内导向面　[nèi dǎo xiàng miàn]
pemandu dalam　[pumandu daran]
내측 가이드　[naecheuk gaideu]
ตัวนำด้านใน　[to-a nam daan-nai]

inner length
　the distance between inner surfaces of extension spring hook.
うちのり　[utinori]
　引張ばねの両端フックの内側間隔寸法.
内側面长度　[nèi cè miàn cháng dù]
panjang dalam　[pangjangu daran]
내부 길이 (자유길이)　[naebu giri (jayugiri)]
ระยะด้านใน　[ra-ya daan-nai]

inner radius
内側半径　[utigawa hankei]
内侧半径　[nèi cè bàn jìng]
jari-jari dalam　[jari-jari daran]

내측 반경　[naecheuk bangyeong]
รัศมีด้านใน　[ras-sa-mii daan-nai]

inner ring
　the ring on the inner side of the frictional surface of the two rings comprising a ring spring.
内輪　[utiwa]
　輪ばねを構成する内外の輪のうちで摩擦面の内側の輪をいう.
内环　[nèi huán]
cincin dalam　[chinchin daran]
내륜　[naeryun]
แหวนตัวใน　[waaen to-a nai]

inner spring
　the inner coil spring of a double coil spring.
内側コイルばね　[utigawa koiru bane]
　二重コイルばねにおいて，内側に配置されるコイルばね.
内弹簧　[nèi tán huáng]
pegas dalam　[pugasu daran]
내측 코일 스프링　[naecheuk koil spring]
สปริงขดด้านใน　[sa-pring khod daan-nai]

inorganic compound
無機化合物　[muki kagoubutu]
无机化合物　[wú jī huà hé wù]
senyawa anorganik　[senyawa anoruganiku]
무기화합물　[mugihwahapmul]
สารอนินทรีย์　[saan a-nin-sii]

inorganic material spring
　a spring made of inorganic material. Ceramic spring is included.
無機材ばね　[mukizai bane]
　無機物からなる材料を用いたばね. セラミックばねがある.
无机材料弹簧　[wú jī cái liào tán huáng]
pegas bahan anorganik　[pugasu bahan anoruganiku]
무기재료 스프링　[mugijaeryo spring]
สปริงที่ทำจากวัสดุอนินทรีย์　[sa-pring tii tam

inorganic material spring — intercrystalline fracture

chaak wat-sa-du a-nin-sii]

inorganic zinc coating
無機亜鉛塗装　[muki aen tosou]
涂覆锌粉　[tú fù xīn fěn]
lapisan seng anorganik　[rapisan sengu anoruganiku]
무기 아연 도장　[mugi ayeon dojang]
การเคลือบสังกะสีโดยใช้สารอนินทรีย์　[gaan klu-ab sang-ka-sii du-oi chai saan a-nin-sii]

inside coil
　a coil spring placed inside for a set of combined spring.
内ばね　[utibane]
組合せばねの内側のコイルばねのこと.
内弾簧　[nèi tán huáng]
ulir dalam　[uriru daran]
내측 스프링　[naecheuk spring]
ขดที่อยู่ด้านใน　[khod tii yuu daan-nai]

inside diameter of coil
　inner diameter of a coil spring.
コイル内径　[koiru naikei]
コイルばねの内径.
弹簧内径　[tán huáng nèi jìng]
diameter dalam ulir　[diameteru daran uriru]
코일 안지름　[koil anjireum]
เส้นผ่าศูนย์กลางด้านในของขด　[sen-pa-soon-glang daan-nai khong khod]

inter leaf
　plate of rubber or plastics inserted between the leaves of a leaf spring in full length or partially.
インターリーフ　[intaa riihu]
重ね板ばねのばね板間に, 全長又は部分的に挿入するゴム製, 合成樹脂製などの板.
内片　[nèi piàn]
antar daun　[antaru daun]
인터 리프　[inteo ripeu]

แหนบแผ่นกลาง　[naaep paaen klaang]

interchangeability
　a characteristic that parts can be replaced mutually.
互換性　[gokansei]
部品相互が互いに交換できること.
互換性　[hù huàn xìng]
mampu saling mengganti　[manpu saringu menguganti]
호환성　[hohwanseong]
ความสามารถในการสับเปลี่ยน　[kwaam samaad nai gaan sub-plian]

intercrystalline crack
　a crack that forms along or across the grain boundaries of a metal or alloy.
粒界割れ　[ryuukai ware]
結晶粒界に発生するか, 又はそこを通過した割れ.
晶界裂纹　[jīng jiè liè wén]
retak antara kristal　[retaku antara kurisutaru]
입계균열　[ipgye-gyunnyeol]
การแตกในผลึกเนื่องจากความร้อน　[gaan taaek nai pa-luk neu-ang chaak kwaam rorn]

intercrystalline fracture
　phenomenon where cracks are generated along the boundary of the crystal grains of metal and propagate to cause the breakdown.
粒界破壊　[ryuukai hakai]
金属の結晶粒の界面に沿ってき裂が発生し, 伝ぱして破壊する現象.
晶界破坏　[jīng jiè pò huài]
retakan antar kristal　[retakan antaru kurisutaru]
입계 파괴　[ipgye pagoe]
การแตกร้าวของเม็ดผลึกเนื่องจากแรงภายนอก　[gaan taaek ra-o ra-waang med pa-luk neu-ang chaak rang paai-nork]

intergranular corrosion
> corrosion growing selectively along the boundary of the crystal grains of metal.
> 粒界腐食 [ryuukai husyoku]
> 金属の結晶粒界に選択的に生じる腐食.
> 晶界腐蚀 [jīng jiè fǔ shí]
> korosi antar [korosi antaru]
> 입계 부식 [ipgye busik]
> การกัดกร่อนระหว่างเม็ด [gaan gad gorn rawaang med]

interlock
> a device that prevents activation of a piece of equipment when a protective cover is open or some other hazard exists.
> インターロック [intaarokku]
> 保護カバーが開いているときや，他の危険が存在するとき，機器の起動を防止する装置．
> 互锁 [hù suǒ]
> kunci dalam [kunsi daran]
> 인터로크 장치 [inteorokeu jangchi]
> อุปกรณ์ล็อคแบบสลับ [up-pa-gorn lock baaep sa-lab]

intermetallic compound
> a compound in which two metal bonds together in specific arrangement, thus acquiring different characteristics from those of the original metals.
> 金属間化合物 [kinzokukan kagoubutu]
> 複数の金属が特定の配置で結合し，元の金属とは異なる特性を有する化合物．
> 金属化合物 [jīn shǔ huà hé wù]
> campuran logam dan bukan logam [chanpuran rogan dan bukan rogan]
> 금속간 화합물 [geumsokgan hwahammul]
> สารประกอบโลหะอื่นที่ปนอยู่ในเหล็ก [saan-pra-gorb loo-ha u-un tii pon yuu nai lek]

intermittent testing
> a test conducted periodically at a given interval.
> 断続試験 [danzoku siken]
> 一定間隔で周期的に試験・検査すること．
> 间断试验 [jiān duàn shì yàn]
> pengujian terputus-putus [pungujian teruputusu putusu]
> 단속 시험 [dansok siheom]
> การทดสอบเป็นช่วงๆ [gaan tod-sorb pen shu-ong shu-ong]

internal combustion engine
> an engine in which the fuel is burned within the engine cylinder.
> 内燃機関 [nainen kikan]
> 燃料が気筒内で燃焼するタイプのエンジン．
> 内燃机 [nèi rán jī]
> mesin pembakaran internal [mesin punbakaran interunaru]
> 내연기관 [naeyeongigwan]
> เครื่องยนต์ระบบเผาไหม้ภายใน [kru-ang yon ra-bob pa-o mai paai nai]

internal friction
> a phenomenon where an applied mechanical vibration is attenuated loosing its energy as heat in solid material.
> 内部摩擦 [naibu masatu]
> 物体に加えられた機械的振動エネルギーが内部の欠陥や内部構造の変化で熱として失われ，振動が減衰する現象．
> 内部摩擦 [nèi bù mó cā]
> penggesekan dalam [pungugesekan daran]
> 내부 마찰 [naebu machal]
> การเสียดทานภายใน [gaan siead-taan paai-nai]

internal retaining ring
> a retaining ring assembled to circum-

internal retaining ring — irregular shaped coil spring

ferencial groove in a cylinder.
穴用Ｃ形止め輪　[anayou siigata tomewa]
円筒内面の溝に組み付けられる止め輪．
内挡圈　[nèi dǎng quān]
cincin penahan dalam　[chinchin punahan daran]
내측 멈춤링　[naecheuk meomchumning]
แหวนกั้นด้านใน　[waaen gan daan nai]

internal stress

a stress that occurs within a material due to a cause other than external forces.
内部応力　[naibu ouryoku]
材料内部に外力以外の原因によって生じる応力をいう．
内部应力　[nèi bù yīng lì]
tegangan dalam　[tegangan daran]
내부 응력　[naebu eungnyeok]
ความเค้นภายใน　[kwaam ken paai-nai]

inverter

a device that converts direct current into variable-frequency alternating current.
インバータ　[inbaata]
直流を可変周波数交流に変換する装置．
变换器　[biàn huàn qì]
inverter　[inveruteru]
인버터　[inbeoteo]
อุปกรณ์เปลี่ยนกระแสตรงเป็นกระแสสลับ　[uppa-gorn plian gra-saae trong pen gra-saae sa-lup]

iron-carbon diagram

a diagram showing how an iron-carbon alloy in a given percent composition changes in response to temperature changes, or state of the alloy in a given temperature range. The vertical axis represents temperature; the horizontal the amount of carbon.
鉄−炭素平衡状態図　[tetu-tanso heikou zyoutaizu]
ある成分比率の鉄−炭素合金が温度の変化によって状態がどのように変わるか，またある温度範囲ではどのような状態になっているかを表す線図．縦軸に温度，横軸に炭素量をとって表す．
铁−碳状态图（相图）　[tiě- tàn zhuàng tài tú (xiāng tú)]
diagram besi-karbon　[diaguramu besi-karubon]
철 - 탄소 평형 상태도　[cheol-tanso pyeonghyeong sangtaedo]
แผนภาพของเหล็กและคาร์บอน　[paaen paab khong lek laae carbon]

ironing

a method of forging or pressing to reduce wall thickness by squeezing a drawn object, thereby increasing the length and improving diameter precision.
アイアニング　[aianingu]
鍛造又はプレス加工法の一つ．絞った部品などをしごくことによって側面肉厚を薄くし，長さを増し，径精度などを向上させる．
压薄　[yā báo]
ironing　[ironingu]
아이어닝　[aieoning]
กระบวนการทำให้เหล็กบริสุทธิ์ขึ้น　[gra-buan gaan tam hai lek bo-ri-sut khun]

irregular shaped coil spring

generic term for the coil spring which is wound in non-circular manner.
異形コイルばね　[ikei koiru bane]
コイルの形状によるばねの総称で，円筒，円すい，つづみ形及びたる形以外の形のコイルばね．
异形螺旋弹簧　[yì xíng luó xuán tán huáng]

pegas ulir berbentuk tak beraturan [pugasu uriru berubentuku taku beraturan]
이형 코일 스프링 [ihyeong koil spring]
สปริงขดชนิดรูปร่างไม่ปกติ [sa-pring khod cha-nid ruup-rang mai pak-ka-ti]

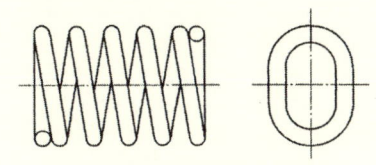

Fig I-3 irregular shaped coil spring

irreversible temper brittleness

a phenomenon whereby hardened steel held at a tempering temperature or annealed from a tempering temperature is more likely to cause brittle fracture. High-temperature temper brittleness refers to primary temper brittleness that occurs during tempering at about 500°C, and secondary temper brittleness that occurs during tempering at above 500°C. Low-temperature temper brittleness refers to temper brittleness resulting from tempering at about 300°C.

不可逆焼戻しぜい性 [hukagyaku yakimodosi zeisei]
焼入れした鉄鋼をある焼戻温度に保持した場合又は焼戻温度から徐冷した場合,ぜい性破壊が生じやすくなる現象.500℃前後の焼戻しで生じる一次焼戻ぜい性及び更に高い温度の焼戻し後の焼戻しで生じる二次焼戻ぜい性を高温焼戻ぜい性といい,300℃前後の温度に焼き戻した場合に見られる焼戻ぜい性を低温焼戻ぜい性という.
不可逆回火脆性 [bù kě nì huí huǒ cuì xìng]
kerapuhan sepuh keras [kerapuhan sepuhu kerasu]

불가역 뜨임 취성 [bulgayeok tteuim chwiseong]
ความเปราะจากการอบแบบย้อนกลับไม่ได้ [kwaam pra-o chaak gaan ob baaep yorn-glab mai-dai]

isoelastic spring alloy

an alloy material for spring in which variations of the elastic modulus in response to temperature change is minimized so that the spring constant maintains constant against changes of ambient temperature.

恒弾性ばね合金 [koudansei bane goukin]
環境温度変化に対して,常にばね定数を均一に保つため,弾性係数の温度変化を抑制したばね用の合金材料.
恒弾性弾簧合金材料 [héng tán xìng tán huáng hé jīn cái liào]
paduan pegas isoelastik [paduan pugasu isoerasutiku]
항탄성 스프링 합금 [hangtanseong spring hapgeum]
โลหะผสมชนิดสปริงไอโซอิลาสติค [lo-ha pa-som cha-nid sa-pring isoelastic]

isothermal annealing

a method of annealing whereby austenite is heated to a temperature higher than A_3 or A_1 transformation, then cooled down to a temperature below A_1 at which pearlite transformation progresses relatively rapidly in a relatively short period of time, and held at the temperature to transform into ferrite and carbide. Isothermal annealing is characterized by its relatively short process time.

等温焼なまし [touon yakinamasi]
A_3変態又はA_1変態以上の温度に加熱した後,A_1点以下の比較的急速にパーライト変態の進む温度まで急冷し,

その温度に保持してオーステナイトをフェライトと炭化物とに変態させ比較的短時間に軟化する焼なまし．
等温退火　[děng wēn tuì huǒ]
aneal isotermal　[anearu isoterumaru]
등온 풀림　[deungon pullim]
การอบแห้งที่อุณหภูมิเดียว　[gaan pob-haaeng tii un-na-ha-poom di-aw]

isothermal transformation

a transformation that occurs when steel is rapidly cooled from the austenite state to a given temperature equal to or below the A_r point, and then held at that temperature. Heat treatment intended for isothermal transformation is called austempering.

等温変態　[touon hentai]
鉄鋼をオーステナイト状態から A_r 点以下の任意の温度まで急冷し，その温度に保持した場合に生じる変態．等温変態を目的とする熱処理をオーステンパ処理と呼んでいる．
等温変化　[děng wēn biàn huà]
transformasi isotermal　[toransuforumasi isoterumaru]
등온 변태　[deungon byeontae]
การเปลี่ยนรูปที่อุณหภูมิเดียว　[gaan plian ruup tii un-na-ha-poom diaew]

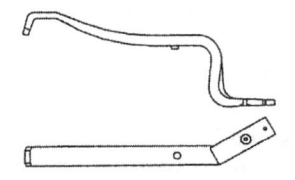

J

J leaf spring
    leaf spring shaped in letter "J".
    Jリーフ　[zyei riihu]
    J字形の板ばね.
    J形片弹簧　[J xíng piàn tán huáng]
    pegas daun J　[pugasu daun J]
    제이 (J) 형 판 스프링 (에어서스펜션)
    [J hyeong pan spring (eeoseoseupensyeon)]
    แหนบชนิด เจ　[naaep cha-nid J]

Fig J-1　**J leaf spring**

jig
    a device to position and hold parts for inspection or machining, and to guide the cutting tool.
    治具　[zigu]
    機械加工や検査のために部品を位置決めしたり、あるいは保持する金具．工具の案内をする金具の場合もある．
    夹具　[jiā jù]
    jig　[jigu]
    지그　[jigeu]
    อุปกรณ์　[up-pa-gorn]

Jominy curve
    a graph showing the relationship between changes in hardness and distance from a cooled end obtained in the Jominy test.
    ジョミニー曲線　[zyominii kyokusen]
    ジョミニー式一端焼入方法において求めた冷却端面からの距離と硬さの変化曲線.
    顶端淬透性曲线　[dǐng duān cuì tòu xìng qū xiàn]
    kurva Jominy　[kuruva Jomini]
    조미니 커브　[jomini keobeu]
    เส้นโค้งโจมีนี่　[sen kong Jominy]

journal bearing
    a cylindrical bearing which supports a rotating shat in radial direction.
    ジャーナル軸受　[zyaanaru zikuuke]
    回転軸を半径方向に支える円筒形の軸受.
    径向轴承　[jìng xiàng zhóu chéng]
    bearing jurnal　[bearingu jurunaru]
    저널 베어링　[jeoneol beeoring]
    แบริ่งของปลอกสวมเพลา　[bearing khong plork su-am pla-o]

just-in-time
    an approach to operate a whole manufacturing system so that the least amount of resources is expended in producing the final product.
    ジャストインタイム　[zyasuto in taimu]
    最終製品を生産するのに最小限の経営資源が消費されるように生産の仕組み全体を運営する考え方.
    正时　[zhèng shí]
    tepat waktu　[tepat waktu]
    저스트 인 타임　[jeoseuteu in taim]
    ระบบจัสท์อินไทม์　[ra-bop just in time]

K

kanban
an inventory control system for tracking the flow of in-process materials through the various operations of a just-in-time production.
カンバン [kanban]
ジャストインタイムの生産過程の種々の工程を通して仕掛品の流れを追跡し，在庫を管理する仕組み．
看板 [kàn bǎn]
kanban [kanban]
간판 [ganpan]
คัมบัง [kam-ban]

Kanigen nickel plating
electroless nickel plating by means of chemical reduction.
無電解ニッケルめっき [mudenkai nikkeru mekki]
電気を使用せずに化学的な還元によって行われるニッケルめっき．
非电解镀镍 [fēi diàn jiě dù niè]
pelapisan nikel Kanigen [purapisan nikeru kanigen]
무전해 니켈 도금 [mujeonhae nikel dogeum]
การชุบนิคเคิลแบบคานิเกน-นิกเคิล [gaan shub nickel baaep Kanigen]

keystone section
trapezoidal cross section.
台形断面 [daikei danmen]
断面が台形をした断面．
梯形截面 [tī xíng jié miàn]
penampang dasar [punanpangu dasaru]
사다리꼴 단면 [sadarikkol danmyeon]
ส่วนตัดรูปสี่เหลี่ยมคางหมู [suan tad ruup sii-liam-kaang muu]

killed steel
thoroughly deoxidized steel by ferro-silicon or aluminum during ingot making process.
キルド鋼 [kirudokou]
インゴットを作るときにフェロシリコンやアルミニウムで十分に脱酸を行った鋼．
镇静钢 [zhèn jìng gāng]
baja killed [baja kirudo]
킬드강 [kildeugang]
เหล็กบริสุทธิ์จากการถลุง [lek bo-ri-sut chaak gaan ta-lung]

kinetic energy
energy possessed by a moving object.
運動エネルギー [undou enerugii]
物体が運動しているときに持っているエネルギー．
动能 [dòng néng]
energi kinetik [enerugi kinetiku]
운동 에너지 [undong eneoji]
พลังงานจลน์ [pa-lang-ngaan jon]

kink
a local twist of wire by mishandling.
キンク [kinku]
不適切な取扱いに起因する局部的な線のねじれ．
扭结 [niǔ jié]
kekusutan [kekusutan]
꼬임 (킹크) [kkoim (kingkeu)]
ปม [pom]

kink test of steel wire
a test on steel wire conducted by pulling both ends of hooped steel in opposite directions to determine whether the wire breaks, how it breaks, and how

much force is required for breakage.

鋼線のキンク試験 [kousen no kinkusiken]
鋼線を輪状にして，両端を引っ張り，破断の有無，破断状況，所要の力の大小を調べる試験．

钢丝的扭结试验 [gāng sī de niǔ jié shì yàn]

uji kekusutan kawat baja [uji kekusutan kawato baja]

강선의 킹크 시험 [gangseonui kingkeu siheom]

การทดสอบปมในเหล็กเส้น [gaan tod-sorb pom nai lek-sen]

knife-edge

a sharp narrow edge to be operative as the fulcrum for a lever arm in a measuring instrument.

ナイフエッジ [naihu ezzi]
測定器具の天秤アームの支点として働く鋭いエッジ．

飞边 [fēi biān]

ujung pisau [ujungu pisau]

나이프 에지 [naipeu eji]

ขอบมีด [khorb-miid]

Knoop hardness test

a test to measure hardness of a sample by pressing a pyramid-shaped diamond indenter (with dihedral angles of 170°30' and 130°) onto the sample's test surface with a given test load, and determining sample hardness based on dimensions of the resulting rhombic permanent dent. This is one of the microhardness test methods.

ヌープ硬さ試験 [nuupu katasa siken]
二つの対りょう角が172°30'と130°の四角すいダイヤモンド圧子を一定の試験荷重で材料の試験面に押し込み，生じたひし形の永久くぼみの大きさから試料の硬さを測定する試験．微小硬さ試験法の一つ．

努普硬度测试 [nǔ pǔ yìng dù cè shì]

uji kekerasan Knoop [uji kekerasan noopu]

누-프 경도 시험 [nu-peu gyeongdo siheom]

การทดสอบความแข็งแบบนูป [gaan tod-sorb kwaam khaeng baaep Knoop]

knuckle-joint press

a press that drives a slide using a single knuckle joint.

ナックルプレス [nakkuru puresu]
単一のナックルジョイントによって，一個のスライドを駆動するプレス．

曲柄连杆式压力机 [qū bǐng lián gǎn shì yā lì jī]

kempa hubungan buku-jari [kenpa hubungan buku-jari]

너클 조인트 푸레스 [neokeul jointeu pureseu]

เครื่องเพรสแบบคอม้า [kru-ang press baaep kor-maa]

L

labor cost
the part of the cost of goods and services attributed to wages.
労務費　[roumuhi]
製品やサービスの費用の一部で賃金に帰する部分．
劳务费　[láo wù fèi]
biaya buruh　[biaya buruhu]
노무비　[nomubi]
ค่าจ้างแรงงาน　[kaa chaang rang ngaan]

lamellar carbide
flake or layered pearlite that is inferior in ductility and impact resistance. Also called lame since its fructured surface glistens.
層状炭化物　[souzyou tankabutu]
片状パーライト又は層状パーライトを指す．延性や衝撃値に劣る．破面を見るとキラキラと光るところからラメの異名もある．
片状碳化物　[piàn zhuàng tàn huà wù]
karbid bersisik　[karubido berusisiku]
층상 탄화물　[cheungsang tanhwamul]
คาร์ไบด์ละเอียดเป็นชั้น　[car-bide la-iad pen shan]

laminated cross section torsion bar spring
a torsion bar spring made from laminated plates and wire rods.
積層断面タイプトーションバー　[sekisou danmen taipu toosyon baa]
板材や線材を重ね合わせた材料を用いたトーションバー．
积层截面扭杆　[jī céng jié miàn niǔ gǎn]
pegas puntir penampang berlapis　[pugasu puntiru punanpangu berurapisu]
적층단면형 토션바　[jeokcheungdanmyeonghyeong tosyeonba]
ทอร์ชั่นบาร์ชนิดซ้อนทับขวางกัน　[torsion bar cha-nid sorn tab kwaang gan]

laminated damping steel sheet
a composite panel that absorbs vibration energy by combining a steel sheet and viscoelastic polymer.
制振鋼板　[seisin kouhan]
鋼板と粘弾性高分子材料との組合せによって，振動エネルギーを吸収する複合型板形状の材料．
积层缓冲钢板　[jī céng huǎn chōng gāng bǎn]
baja lembaran berlapis　[baja renbaran berurapisu]
제진 강판　[jejin gangpan]
เหล็กแผ่นบางที่ใช้รองรับการสั่นสะเทือน　[lek paaen baang tii chai rong-rab gaan san sa-tuan]

laminated leaf spring
leaf spring comprising laminated leaves. (see Fig L-1)
重ね板ばね　[kasaneita bane]
ばね板を重ね合わせて構成された板ばね．（Fig L-1 参照）
重叠板弹簧　[chóng dié bǎn tán huáng]
pegas daun berlapis　[pugasu daun berurapisu]
겹판 스프링　[gyeoppan spring]
แหนบแผ่นบางชนิดซ้อนกันหลายชั้น　[naaep paaen bang cha-nid sorn gan laai shan]

lateral deflection
a deflection which takes place laterally when a coil spring is compressed with no guide equipment.
横方向たわみ　[yokohoukou tawami]
コイルばねがガイド機構なしの状態

lateral deflection — leaf center

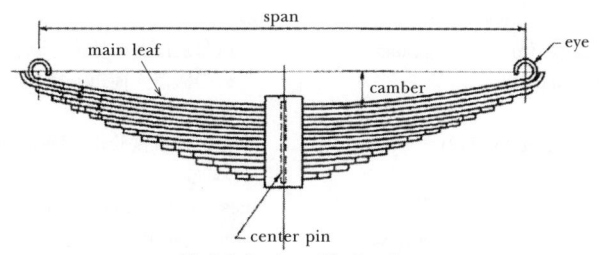

Fig L-1 laminated leaf spring

で圧縮されるときに生じる横方向たわみ．
横向挠曲　[héng xiàng náo qǔ]
defleksi menyamping　[defurekusi munyanpingu]
횡방향 변위　[hoengbanghyang byeonwi]
การยุบตัวด้านข้าง　[gaan yub to-a daan-khaang]

lateral load
　a load applied in perpendicular direction to the axis of spring.
横方向荷重　[yokohoukou kazyuu]
コイルばねの中心線に対して直角の方向に入る荷重．
横向负载　[héng xiàng fù zǎi]
beban samping　[beban sanpingu]
횡방향 하중　[hoengbanghyang hajung]
โหลดด้านข้าง　[load daan-khaang]

layer of plastic deformation
　a deformed layer in a metal piece which went through the deformation in elastoplastic process.
塑性変形層　[sosei henkeisou]
弾塑性過程を経て変形した金属片の塑性変形層．
塑性变形层　[sù xìng biàn xíng céng]
lapisan perubahan plastik　[rapisan perubahan purasutiku]
소성 변형층　[soseong byeonhyeongcheung]
ชั้นของพลาสติคที่ผิดรูป　[shaan khong plastic tii pid-roup]

layer thickness
　the thickness of coating film, plated layer, carburized layer, decarburized layer and layer of metallurgical structure.
膜厚　[makuatu]
塗装やめっきなどの厚み又は浸炭，脱炭，金属組織などの層の厚み．
膜厚　[mó hòu]
tebal lapisan　[tebaru rapisan]
피막 두께　[pimak dukke]
ความหนาของชั้นสี　[kwaam-naa khong shan sii]

leaf
　leaf constituting the leaf spring.
ばね板　[baneita]
重ね板ばねを構成する板．
弹簧板片　[tán huáng bǎn piàn]
daun　[daun]
스프링 판　[spring pan]
แผ่นแหนบ　[paaen naaep]

leaf center
　the center of a leaf spring. Normally a center bolt or band is provided here.
ばね板中心　[baneita tyuusin]
重ね板ばねの中心のことで，一般的にはこの部にセンタボルトあるいは胴締めなどが施されている．
弹簧板片中心　[tán huáng bǎn piàn zhōng xīn]
pusat daun　[pusato daun]

leaf center — leaf spring with eyes

스프링 판 중심　[spring pan jungsim]
ศูนย์กลางของแผ่นแหนบ　[soon-glaang khorng paaen-naaep]

leaf end
　　end of a leaf. (see Fig L-2)
　　ばね板端部　[baneita tanbu]
　　ばね板の端部．（Fig L-2 参照）
　　弹簧板片端部　[tán huáng bǎn piàn duān bù]
　　ujung daun　[ujungu daun]
　　스프링 판 끝　[spring pan kkeut]
　　ปลายของแผ่นแหนบ　[plaai khong paaen-naaep]

leaf length
　　the length of each single left of multi-leaf spring.
　　リーフ長さ　[riihu nagasa]
　　重ね板ばねの個々のリーフの長さ．
　　弹簧板片长度　[tán huáng bǎn piàn cháng dù]
　　panjang daun　[panjangu daun]
　　판 길이　[pan giri]
　　ความยาวของแผ่นแหนบ　[kwaam ya-o khong paaen-naaep]

leaf spring testing machine
　　a device used to test and inspect performance of leaf springs. Includes load tester, fatigue tester, corrosion tester and hardness tester.
　　重ね板ばね用試験機　[kasaneita baneyou sikenki]
　　重ね板ばねの性能を試験・検査するための機器．荷重，疲れ，耐食，硬さ試験機などがある．
　　重叠板弹簧试验机　[chóng dié bǎn tán huáng shì yàn jī]
　　mesin penguji pegas daun　[mesin punguji pugasu daun]
　　판 스프링용 시험기　[pan springyong siheomgi]
　　เครื่องทดสอบแหนบ　[kru-ang tod-sorb naaep]

leaf spring with eyes
　　a leaf spring having an eye at both ends. The eye comes in various shapes (e.g., upturned eye, downturned eye and berlin eye).
　　目玉付き重ね板ばね　[medamatuki kasaneita bane]
　　両端末に目玉を持つ重ね板ばね．目玉の形状は上巻目玉，下巻目玉，ベルリンアイなどがある．
　　带吊耳孔重叠板弹簧　[dài diào ěr kǒng chóng dié bǎn tán huáng]
　　pegas daun dengan gelang　[pugasu daun dengan gerangu]
　　아이부가 있는 겹판 스프링　[aibuga inneun gyeoppan spring]
　　แหนบชนิดมีหู　[naaep cha-nid mii huu]

flat end

diamond end

tapered end

silencer

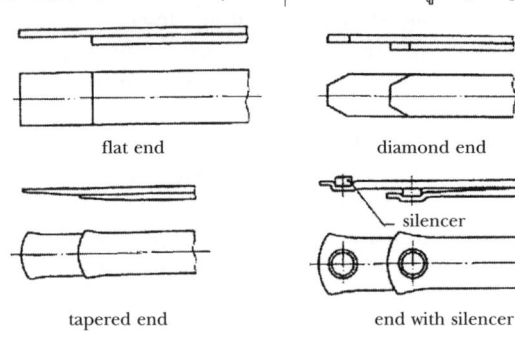

end with silencer

Fig L-2　leaf end

leaf spring with rubber cushion

a leaf spring where a columnar rubber spring bears the wind-up deformation that occurs during braking. The rubber spring is placed near the center pin of a normal leaf spring.

ゴムばね併用重ね板ばね [gomubane heiyou kasaneita bane]

一般的な重ね板ばねのセンタピン付近に柱状のゴムばねを用いて，制動時に生じるワインドアップ変形をゴムばねに負担させる形式の重ね板ばね．

带橡胶软垫重叠板弹簧 [dài xiàng jiāo ruǎn diàn chóng dié bǎn tán huáng]

pegas daun dengan bantalan karet [pugasu daun dengan bantaran kareto]

고무 스프링 혼용 겹판 스프링 [gomu spring honnyong gyeoppan spring]

แหนบชนิดมียางกันสะเทือน [naaep cha-nid mii yaang gan sa-te-un]

leaf spring with sliding ends

a leaf spring designed so that the load bearing point can slide by processing the leaf ends into a flat or curved shape instead of an eye. Also used on railway vehicles.

重ね板ばねの滑面端末 [kasaneita bane no katumen tanmatu]

重ね板ばねの端末部を目玉にせずに平面あるいは曲面形状にして荷重支持点を滑らせるように設計された重ね板ばね．鉄道車両などで用いられている．

端面滑动的重叠板弹簧 [duān miàn huá dòng de chóng dié bǎn tán huáng]

pegas daun dengan ujung geser [pugasu daun dengan ujungu geseru]

판스프링 미끄럼 끝면 [panspring mikkeureom kkeunmyeon]

แหนบชนิดที่ปลายเลื่อนได้ [naaep cha-nid tii plaai le-un dai]

leaf thickness

thickness of one leaf.

ばね板の厚さ [baneita no atusa]

ばね板の一枚の板厚をいう．

弹簧板厚度 [tán huáng bǎn hòu dù]

tebal daun [tebaru daun]

스프링 판 두께 [spring pan dukke]

ความหนาแผ่นแหนบ [kwwam-naa paaen naaep]

leaf width

the strip width of a spring leaf.

ばね板の幅 [baneita no haba]

ばね板の板幅．

弹簧板宽度 [tán huáng bǎn kuān dù]

lebar daun [rebaru daun]

판 스프링 판폭 [pan spring panpok]

ความกว้างแผ่นแหนบ [kwwamn-gwaang paaen naaep]

left-hand coil

turning direction of coiling like the left-hand thread.

左巻 [hidarimaki]

左ねじと同じ方向のコイルの巻方向．

左旋 [zuǒ xuán]

uliran arah kiri [uriru arahu kiri]

왼쪽 감기 [oenjjok gamgi]

สปริงขดชนิดหมุนซ้าย [sa-pring khod cha-nid mun saai]

left-hand rotation

counterclockwise rotation looking toward the axis of the rotation.

左旋回 [hidarisenkai]

出力側からみて時計方向の逆方向に回転すること．

左旋向 [zuǒ xuán xiàng]

putaran arah kiri [putaran arahu kiri]

좌회전 [jwahoejeon]

การหมุนซ้าย [gaan mun saai]

leg angle

an angle between one leg and another

leg angle
  leg for a stabilizer, torsion spring, etc.
  脚の角度　[asi no kakudo]
  スタビライザやねじりばねなどの脚部と他方の脚部とのなす角をいう．
  扭臂杆的角度　[niǔ bì gǎn de jiǎo dù]
  sudut kaki　[sudutu kaki]
  팔 각도　[pal gakdo]
  มุมของขา　[mum khong khaa]

leg length
  the length of the arm of a torsion bar or stabilizer.
  脚長　[kyakutyou]
  トーションバーやスタビライザなどの腕部の長さ．
  扭臂杆长度　[niǔ bì gǎn cháng dù]
  panjang kaki　[panjangu kaki]
  팔 길이　[pal giri]
  ความยาวขา　[kwaam ya-o khaa]

length of buckle
  the length of frame-like hardware that fastens the leaves of a leaf spring.
  胴締め長さ　[douzime nagasa]
  重ね板ばねのばね板を締め付ける枠形金具部の長さ．
  卡箍长度　[kǎ gū cháng dù]
  panjang pengikat　[panjangu pungikato]
  밴드의 길이　[baendeuui giri]
  ระยะโก่งงอ　[ra-ya kong ngor]

length of disc spring stack
  the length of a set of stacked disk springs.
  皿ばねの積み重ね長さ　[sarabane no tumikasane nagasa]
  皿ばねを重ね合わせた一組の長さ．
  碟形弹簧叠加长度　[dié xíng tán huáng dié jiā cháng dù]
  panjang tumpukan pegas cakram [panjangu tunpukan pugasu kakuran]
  접시 스프링 적층 길이　[jeopsi spring jeokcheung giri]

length of inside hook
  length between the inside of the hooks of a helical extension spring.
  引張コイルばねの長さ　[hippari koiru bane no nagasa]
  引張コイルばねの，フックの内側間の長さ．
  拉伸弹簧钩环内侧长度　[lā shēn tán huáng gōu huán nèi cè cháng dù]
  panjang kait dalam　[panjangu kaito daran]
  고리 내측길이　[gori naecheukgiri]
  ความยาวของขอด้านใน　[kwaam ya-o ta-khor daan-nai]

leverage
  the multiplication of force or motion achieved by a lever.
  てこ装置　[teko souti]
  レバーによって力又は動きを増幅させる装置．
  旋转中心　[xuán zhuǎn zhōng xīn]
  pengaruh　[pungaruhu]
  지렛대　[jiretdae]
  จุดหมุน　[chud mun]

life test
  a test to determine the service life of a part or finished product under specified conditions. For example, service life test, time to failure test, and shelf life test.
  寿命試験　[zyumyou siken]
  ある規定条件下における部品又は完成品の寿命に関する試験．耐用寿命試験，故障寿命試験，保管寿命試験などがある．
  寿命试验　[shòu mìng shì yàn]
  uji usia　[uji usia]
  수명 시험　[sumyeong siheom]

การทดสอบการใช้งาน　[gaan tod-sorb gaan chai-ngaan]

limit load

the maximum load applied to a structural part under normal service conditions.

限界荷重　[genkai kazyuu]

運用状態で構造部分に負荷させ得る最大の荷重．

极限负荷　[jí xiàn fù hè]

beban batas　[beban batasu]

한계 하중　[hangye hajung]

โหลดจำกัด　[load cham-gad]

limit of proportionality

the maximum stress that direct proportionality between load and strain holds when load is applied to an object.

比例限度　[hirei gendo]

物体に荷重を加えると変形して応力とひずみを生じる．この両者は応力がある値に達するまで，正比例する．この正比例関係が保たれている応力の最大限をいう．

比例极限　[bǐ lì jí xiàn]

batas perimbangan　[batasu purinbangan]

비례한도　[biryehando]

ขีดจำกัดของสัดส่วน　[khiid cham-gad khong sad-suan]

limit of stress

the limit of stress generated within an object when external force is applied to the object. There are compressive stress, shear stress and tensile stress.

应力限界　[ouryoku genkai]

物体に外力が加わるとき，その物体内に生じる応力の限界をいう．その種類は圧縮，せん断，引張応力がある．

应力极限　[yīng lì jí xiàn]

batas tegangan　[batasu tegangan]

응력 한계　[eungnyeok hangye]

ขีดจำกัดของความเค้น　[khiid cham-gad khong kwaan ken]

limit switch

a switch designed to cut off power automatically at or near the travel of a moving object.

リミットスイッチ　[rimitto suitti]

移動物が動く範囲の終端又は近傍で自動的に電源を切るスイッチ．

限位开关　[xiàn wèi kāi guān]

switch batas　[switti batasu]

리밋 스위치　[rimit seuwichi]

ลิมิตสวิทซ์　[limit switch]

linear characteristic

spring characteristics where the relationship between the load and the deflection is linear.

線形特性　[senkei tokusei]

荷重とたわみとの関係が直線となるばね特性．

线性特性　[xiàn xìng tè xìng]

karakteristik linier　[karakuterisutiku rinieru]

선형 특성　[seonhyeong teukseong]

ลักษณะเชิงเส้น　[lak-sa-na che-arng sen]

linearly acting spring

spring the deflection of which is linear to the load applied.

線形特性ばね　[senkei tokusei bane]

荷重とたわみとの関係が直線となるばね．

线性弹簧　[xiàn xìng tán huáng]

pegas yang bertindak linier　[pugasu yangu berutindaku rinieru]

선형 특성 스프링　[seonhyeong teukseong spring]

สปริงที่มีคุณสมบัติรับโหลดเชิงเส้น　[sa-pring tii mii kun-na-som-bat rab load che-arng sen]

liner

rubber or plastics tip attached at the

liner — load

end of the leaves to suppress squeak noise and rattle noise.

ライナ　[raina]
きしみ音，たたき音などの抑制のために，重ね板ばねの先端部に付ける，又はコイルばねとばね座との間に入れるゴム製，合成樹脂製などの部品．

衬垫　[chèn diàn]

pelapis　[purapisu]

라이너　[raineo]

แผ่นรอง　[paaen-rong]

link of stabilizer bar

link which connects the bar stabilizer and the body.

スタビリンク　[sutabirinku]
バースタビライザの目玉部と車体側とを連結するリンク．

稳定杆连结器　[wěn dìng gǎn lián jié qì]

penyambung batang stabiliser [punyanbungu batangu sutabiriseru]

스테비 링크　[seutebi ringkeu]

สตาบิลิงค์　[stabi link]

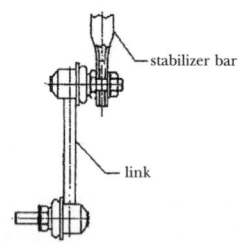

Fig L-3　link of stabilizer bar

liquid horning

process to clean the surface of a spring by spraying water containing minute polishing agents and/or appropriate corrosion retardant, while improving the fatigue strength by the similar principles to shot peening. This process can also be applied to the case the shot peening may cause too much distortion.

液体ホーニング　[ekitai hooningu]
微細な研磨材を加えた水又はそれに適当な腐食抑制剤を加えたものを吹き付けて，ばね表面を清浄にするとともに，ショットピーニングと同様な原理によって疲れ強さを向上させる加工．ショットピーニングでは，変形が大きすぎるものに適用する．

液体珩磨　[yè tǐ héng mó]

horning cair　[horuningu chairu]

액체 호닝　[aekche honing]

การลับผิวแบบเปียก　[gaan lab piw baaep piaek]

liquid penetrant testing

nondestructive testing where drips of penetrant is applied to open scratches on the surface of the test piece of a spring or its material, and the observation of the flaw indication can be carried out with ease in the magnified image. This is divided into visible dye penetrant testing and fluorescence penetrant testing.

浸透探傷試験　[sintou tansyou siken]
ばね材料，ばねなどの試験体表面の，開口しているきずに浸透液を浸透させた後，拡大した像の指示模様としてきずを観察する非破壊試験方法．染色浸透探傷試験及び蛍光浸透探傷試験がある．

浸透探伤试验　[jìn tòu tàn shāng shì yàn]

uji penetrasi cairan　[uji punetorasi chairan]

침투 탐상 시험　[chimtu tamsang siheom]

การทดสอบด้วยการสอดของเหลว　[gaan todsorb do-ui gaan sord khong-leew]

load

荷重　[kazyuu]

负荷　[fù hè]

beban [beban]
하중 [hajung]
โหลด [load]

load at rupture
    the load that breaks a component when applied.
破壊荷重 [hakai kazyuu]
部材に荷重が加えられるとき，その部材を破壊する荷重．
破坏负荷 [pò huài fù hè]
beban saat pecah [beban saato pechahu]
파괴 하중 [pagoe hajung]
โหลดสุดท้ายขณะหัก [load sud-taai kha-na hak]

load axis
    the centerline of a spring in the direction of an applied load.
荷重軸線 [kazyuu zikusen]
荷重が負荷される方向に引かれたばねの中心線．
负荷轴线 [fù hè zhóu xiàn]
sumbu beban [sunbu beban]
하중 축선 [hajung chukseon]
แกนที่รับโหลด [gaaen tii rab load]

load capacity
    the load endured by a component without permanent deformation.
許容荷重 [kyoyou kazyuu]
部材が永久変形することなく耐えられる荷重．
许用负荷 [xǔ yòng fù hè]
kapasitas beban [kapasitasu beban]
허용 하중 [heoyong hajung]
ปริมาณที่รับโหลดได้ [pa-ri-man tii rab load dai]

load extension curve
    a curve showing the relationship between load applied to a test sample and the resulting elongation for the entire process of a tensile test.
荷重-伸び曲線 [kazyuu-nobi kyokusen]
引張試験の全過程における試験片に加えた荷重とそれに伴う伸びとの関係を表す曲線．
负荷-拉伸曲线 [fù hè - lā shēn qū xiàn]
kurva perpanjanagn beban (pegas tarik) [kuruva purupanjanagun beban (pugasu tariku)]
하중-신률 곡선 [hajung-sillyul gokseon]
เส้นโค้งแสดงโหลดและการยืดตัวของสปริง [sen-kong sa-daaeng load laae gaan yuud to-a khong sa-pring]

load increase
荷重増加 [kazyuu zouka]
负荷增加 [fù hè zēng jiā]
pertambahan beban [purutanbahan beban]
하중 증가 [hajung jeungga]
การเพิ่มโหลด [gaan perm load]

load limit
荷重限度 [kazyuu gendo]
负荷极限 [fù hè jí xiàn]
batas beban [batasu beban]
하중 한도 [hajung hando]
ขนาดโหลดที่รับได้ [kha-naad load tii rab dai]

load loss
    phenomenon where the spring load decreases owing to the time lapse, temperature and cyclic load when kept at a constant height (length).
荷重低下 [kazyuu teika]
一定の高さ（長さ）であるにもかかわらず，時間，温度，繰返しなどによって，ばねがへたって荷重が減少する現象．
负荷损耗 [fù hè sǔn hào]
penurunan beban [punurunan beban]
하중 저하 [hajung jeoha]
การสูญเสียโหลด [gaan soon-siaa load]

load measurement
    measuring of spring load, torque,

load measurement — loading speed

spring rate, etc.
荷重測定 [kazyuu sokutei]
ばねの荷重，トルク，ばね定数を各種測定機器で測定すること．
負荷測定 [fù hè cè dìng]
pengukuran beban [pungukuran beban]
하중 측정 [hajung cheukjeong]
การวัดโหลด [gaan wad load]

load measuring device
a machine or device that measures spring load, torque, and spring constant. A digital device or a computer controlled measuring instrument are available nowadays.
荷重測定装置 [kazyuu sokutei souti]
ばねの荷重，トルク，ばね定数を測定する機器・装置で，最近ではデジタル式やコンピュータ付きの各種測定機器がある．
负荷測定装置 [fù hè cè dìng zhuāng zhì]
sarana pengukur beban [sarana pungukuru beban]
하중 측정 장치 [hajung cheukjeong jangchi]
เครื่องวัดโหลด [kru-ang wad load]

load test
a test of spring load. It may be a static or dynamic test.
荷重試験 [kazyuu siken]
ばねの荷重を調べる試験で，静的試験と動的試験とがある．
负荷试验 [fù hè shì yàn]
uji beban [uji beban]
하중 시험 [hajung siheom]
การทดสอบโหลด [gaan tod-sorb load]

load/deflection curve
curve showing the relationship between the load and the deflection, indicating the spring characteristics. There are linear and nonlinear categories in spring characteristics. The curve implies nonlinear type.
荷重-たわみ曲線 [kazyuu-tawami kyokusen]
ばね特性を表す，荷重とたわみの関係を表した曲線．ばね特性には，線形特性と非線形特性とがあるが，曲線の場合は後者をいう．
负荷－变形曲线 [fù hè - biàn xíng qǔ xiàn]
kurva beban/lenturan [kuruva beban/renturan]
하중-변위 곡선 [hajung-byeonwi gokseon]
เส้นโค้งแสดงโหลดและการยุบตัวของสปริง [sen-kong sa-daaeng load laae gaan yub to-a khong sa-pring]

load-eccentricity
off-center load applied to a coil spring in the direction of the coil axis.
偏心荷重 [hensin kazyuu]
コイルばねに作用する荷重の作用線がコイル軸からずれている荷重．
偏心负荷 [piān xīn fù hè]
penyimpangan beban [punimupangan beban]
편심 하중 [pyeonsim hajung]
โหลดเยื้องศูนย์ [load ye-ung soon]

loading device
负荷装置 [huka souti]
加载装置 [jiā zǎi zhuāng zhì]
sarana pembebanan [sarana punbebanan]
부하 장치 [buha jangchi]
อุปกรณ์ในการโหลด [up-pa-gorn nai gaan load]

loading speed
the speed of the load transfer unit (chuck or crosshead) when applying load on a test sample.
加圧速度 [kaatu sokudo]
試験片に荷重を加える際の荷重伝達部（チャック又はクロスヘッド）の速度．

加載速度　[jiā zǎi sù dù]
kecepatan pembebanan　[kesepatan punbebanan]
가압 속도　[gaap sokdo]
ความเร็วในการโหลด　[kwaam rew nai gaan load]

lock spring
　　a spring used in locks. Includes a torsion spring, helical compression spring, and flat spring.
錠前ばね　[zyoumae bane]
錠前に用いられるばね．ねじりばね，圧縮コイルばね，薄板ばねが多い．
锁用弹簧　[suǒ yòng tán huáng]
pegas pengunci　[pugasu pungunsi]
잠금 스프링　[jamgeum spring]
สปริงที่ทำหน้าที่ล็อค　[sa-pring tii tam naa-tii lock]

lock washer
　　a bolt tightenning washer to utilize spring action.
ばね座金　[bane zagane]
ばね作用を利用して，ゆるみどめをさせる座金の総称．
弹簧垫圈　[tán huáng diàn quān]
ring pipih pengunci pegas　[ringu pipihu pungunsi pugasu]
스프링 와셔　[spring wasyeo]
โบลท์ยึดแหวนรอง　[bolt yud waaen rong]

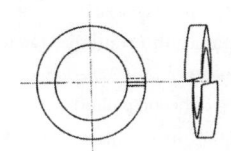

Fig L-4　lock washer

long round end hook over center
　　"U" shaped hook at the end of a helical extension spring.
U フック　[yuu hukku]

引張コイルばねの端部の一種で，U字形のフック．
U 形钩　[U xíng gōu]
kait bentuk huruf U　[kaito bentuko hurufu U]
유 (U) 형 고리　[Uhyeong gori]
ยูคลิป　[U clip]

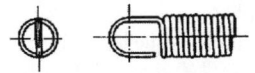

Fig L-5　long round end hook over center

longitudinal crack
　　a crack that propagates parallel to the drawing direction of wire rods.
縦割れ　[tateware]
線材の伸線方向に平行に生じる割れ．
纵向裂纹　[zòng xiàng liè wén]
retakan memanjang　[retakan memanjangu]
길이 방향 균열　[giri banghyang gyunnyeol]
การแตกตามแนวยาว　[gaan taaek taam naaew ya-o]

longitudinal leaf spring
　　a leaf spring mounted perpendicularly to an axle connecting the left and right wheels.
縦置き板ばね　[tateoki itabane]
左右の車輪を結ぶ車軸に対して直角に取り付けられる板ばね．
纵置板弹簧　[zòng zhì bǎn tán huáng]
pegas daun memanjang　[pugasu daun memanjangu]
종치식 판스프링　[jongchisik panspring]
แหนบแผนแนวยาว　[naaep-paaen naaew ya-o]

lot
　　a quantity of items which are manufactured under identical conditions.
ロット　[rotto]
同一の条件で生産されたアイテムの数量．

批　　[pī]
lot　　[roto]
로트　　[roteu]
ล็อต　　[lot]

lot number
identification number assigned to a particular lot of items from a single supplier.
ロット番号　　[rotto bangou]
単一の供給者から供給される特定のロットに与えられる識別番号.
批号　　[pī hào]
nomor lot　　[nomoru roto]
로트 번호　　[roteu beonho]
ล็อตนัมเบอร์　　[lot number]

low alloy steel
steel containing about 0.3% silicon, manganese, chromium, and nickel. It has high strength and is used for components requiring heat resistance.
低合金鋼　　[teigoukinkou]
Si, Mn, Cr, Ni などを 0.3% 程度加えた鋼の総称. 強じんで耐熱部品などに用いる.
低合金钢　　[dī hé jīn gāng]
baja paduan rendah　　[baja paduan rendahu]
저합금 강　　[jeohapgeum gang]
เหล็กโลหะผสมต่ำ　　[lek lo-ha pa-som tam]

low carbon steel
steel containing about 0.12% to 0.22% carbon.
低炭素鋼　　[teitansokou]
炭素量 0.12 〜 0.22% 程度の鋼.
低碳钢　　[dī tàn gāng]
baja karbon rendah　　[baja karubon rendahu]
저탄소강　　[jeotansogang]
เหล็กกล้าคาร์บอนต่ำ　　[lek glaa carbon tam]

low load hardness test
a hardness test conducted using a small load. Micro Vickers hardness testers with a small load are used to avoids any effects on the reverse side of thin test samples.
低荷重硬さ試験　　[teikazyuu katasa siken]
小さな荷重で硬さを試験すること. 薄い材料の硬さ試験では裏面まで影響するので小さな荷重であるマイクロビッカース硬さ試験機が用いられる.
低负荷硬度试验　　[dī fù hè yìng dù shì yàn]
penguji kekerasan beban rendah　　[punguji kekerasan beban rendahu]
저하중 경도시험　　[jeohajung gyeongdosiheom]
เครื่องวัดความแข็งด้วยโหลดต่ำ　　[kru-ang wad kwaam khaaeng du-oi load tam]

low temperature annealing
low temperature heating process to eliminate inner stress, and to improve various properties of the materials such as elastic limit, yield strength or fatigue strength, or to stabilize the form.
低温焼なまし　　[teion yakinamasi]
内部応力の除去又は材料の弾性限度, 耐力, 疲れ強さなどの諸特性の改善及び形状の安定化を目的として行う低温加熱処理.
低温退火　　[dī wēn tuì huǒ]
anneal pada suhu rendah　　[annearu pada suhu rendahu]
저온 풀림　　[jeoon pullim]
การอบที่อุณหภูมิต่ำ　　[gaan ob tii un-na-ha-puum tam]

low temperature brittleness
the phenomenon whereby the impact strength of steel decreases dramatically at room temperature or lower temperatures, and become brittle.

低温ぜい性　[teion zeisei]
　室温付近又はそれ以下の温度で鋼の衝撃値が急激に低下しもろくなる現象．

低温脆性　[dī wēn cuì xìng]
kerapuhan pada suhu rendah
[kerapuhan pada suhu rendahu]

저온 취성　[jeoon chwiseong]
การเปราะแตกที่อุณหภูมิต่ำ　[gaan pra-or taaek tii un-na-ha-puum tam]

low temperature furnace
　an annealing furnace operating at temperatures up to about 400°C.

低温炉　[teionro]
　加熱温度が 400°C 以下程度の熱処理用焼なまし炉．

低温炉　[dī wēn lú]
tungku suhu rendah　[tunguku suhu rendahu]

저온로　[jeoonno]
เตาอุณหภูมิต่ำ　[ta-o un-na-ha-puum tam]

low temperature treatment
　heat treatment performed at lower elevated temperature to eliminate internal stress, improve such properties as elastic limit, yield strength and fatigue strength, and stabilize the shape.

低温処理　[teion syori]
　内部応力の除去又は材料の弾性限度，耐力，疲れ強さ等の特性の改善と形状の安定化を目的として行う低温加熱処理．

低温处理　[dī wēn chǔ lǐ]
olahan suhu rendah　[orahan suhu rendahu]

저온 처리　[jeoon cheori]
การให้ความร้อนที่อุณหภูมิต่ำ　[gaan hai kwam rorn tii un-na-ha-puum tam]

lower grinding wheel
　the grinding and polishing wheel mounted on the underside of a double-ended spring grinding machine.

下面といし　[kamen toisi]
　ばね両面研削盤の下側に取り付ける研削といし．

下侧砂轮　[xià cè shā lún]
roda gerinda bawah　[roda gerinda bawahu]

하면 그라인더　[hamyeon geuraindeo]
ล้อหินเจียร์ด้านล่าง　[lor hin-chiaa daan-lang]

M

machine loop
    a hook that rises up from the outer circumference of a helical extension spring.
    マシンループ　[masin ruupu]
    引張ばねのコイル外径から立ち上がるフック．
    外置偏心圆钩环　[wài zhì piān xīn yuán gōu huán]
    mesin rup　[mesin rupu]
    머신 루프　[meosin rupeu]
    แมชชีนลูป　[machine loop]

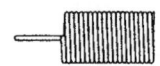

Fig M-1　machine loop

machining of end face
    machining processes such as milling and grinding performed on the end face of a workpiece.
    端面機械加工　[tanmen kikai kakou]
    工作物の端面に施す切削，研削などの機械加工．
    端面机械加工　[duān miàn jī xiè jiā gōng]
    pemesinan ujung　[pumesinan ujungu]
    끝면 기계 가공　[kkeunmyeon gigye gagong]
    การแต่งปลายด้วยจักรกล　[gaan taaeng plaai du-oi chak-kon]

macroscopic examination
    a test to visually inspect polished surface or cross section of an object after etching with hydrochloric acid, copper ammonium chloride, or nitrohydrochloric acid.
    マクロ組織試験　[makuro sosiki siken]
    研磨された断面又は表面を塩酸・塩化銅アンモニウム・王水などを用いて腐食し肉眼で判断する試験．
    微观组织试验　[wēi guān zǔ zhī shì yàn]
    pemerikasaan makroskopis　[pumerikasaan makurosukopisu]
    매크로 조직 검사　[maekeuro jojik geomsa]
    การทดสอบในระดับมหภาค　[gaan tod-sorb nai ra-dab ma-ha-paak]

magazine spring
    a spring used to push up bullets in weapon magazines. Rectangular compression springs are commonly used.
    マガジンスプリング　[magazin supuringu]
    銃器の弾倉に用いる弾の押上げ用のばね．一般に角型の圧縮コイルばねが多く用いられている．
    弹仓弹簧　[dàn cāng tán huáng]
    pegas kotak sediaan　[pugasu kotaku sediaan]
    탄창 스프링　[tanchang spring]
    สปริงยึดเข็มแทงชนวนปืน　[sa-pring yud khem taaeng sha-nuan pu-un]

magnetic field meter
    磁界強度計　[zikai kyoudokei]
    磁强计　[cí qiáng jì]
    pengukur medan magnet　[pungukuru medan maguneto]
    자계 강도계　[jagye gangdogye]
    เครื่องตรวจวัดสนามแม่เหล็ก　[kru-ang truat wad sa-nam maae-lek]

magnetic particle testing
    nondestructive testing which detects

the defects by means of adhesion of magnetic particles generated in the defective area by magnetizing steel springs or steel bars.
磁粉探傷試験　[zihun tansyou siken]
鉄鋼材料を用いたばねなどの強磁性体を磁化し，欠陥部に生じた磁極による磁粉の付着を利用して，欠陥を検出する非破壊試験方法．
磁粉探伤试验　[cí fěn tàn shāng shì yàn]
uji partikel magnet　[uji parutikeru maguneto]
자분 탐상 시험　[jabun tamsang siheom]
เครื่องตรวจหารอยร้าวด้วยผงแม่เหล็ก　[kru-ang truat haa roi-ra-o du-oi phong maae-lek]

magnetic separator
a machine that uses magnetism to remove grinding chips from coolant oil.
磁気分離機　[ziki bunriki]
磁石により研削油剤中の研削くずを取り除く装置．
磁场分离机　[cí chǎng fēn lí jī]
pemisah magnetik　[pumisahu magunetiku]
자기 분리기　[jagi bulligi]
เครื่องแยกด้วยแม่เหล็ก　[kru-ang yaaek du-oi maae-lek]

magnetic spring
spring using the repulsion force and attraction force by magnetism, as called by its type of function.
磁気ばね　[ziki bane]
ばねの機能の種類による名称で，磁気による反発力及び吸引力を利用するばね．
磁性弹簧　[cí xìng tán huáng]
pegas magnet　[pugasu maguneto]
자석 스프링　[jaseok spring]
สปริงแม่เหล็ก　[sa-pring maae-lek]

magnetic thickness tester
an instrument that uses magnetism to measure film thickness.
磁気膜厚計　[ziki makuatukei]
磁気を利用し膜の厚みを測定する計器．
磁场膜厚测试仪　[cí chǎng mó hòu cè shì yí]
penguji ketebalan magnetik　[punguji ketebaran magunetiku]
마그네틱 두께 측정기　[mageunetik dukke cheukjeonggi]
เครื่องทดสอบความหนาสนามแม่เหล็ก　[kru-ang tod-sorb kwaam-naa sa-naam maae-lek]

magnetic transformation
the transformation from ferromagnetic material to paramagnetic material or vice versa. Crystal structure remains unchanged.
磁気変態　[ziki hentai]
強磁性体から常磁性体へ，また常磁性体から強磁性体への変化．結晶構造の変化は伴わない．
磁场变态　[cí chǎng biàn tài]
tranformasi magnetik　[toranforumasi magunetiku]
자기 변태　[jagi byeontae]
การแปลงสนามแม่เหล็ก　[gaan plian-praaeng sa-naam maae-lek]

magnetostriction method
a measuring method of nondestructive stress measurement utilizing magnetostriction phenomenon.
磁気ひずみ法　[ziki hizumi hou]
磁気ひずみ原理を使った応力の非破壊測定法．
磁性应变法　[cí xìng yīng biàn fǎ]
metoda magneostriction　[metoda maguneosutorikution]
자기 변형율 법　[jagi byeonhyeongyul beop]

magnetostriction method — mandrel diameter

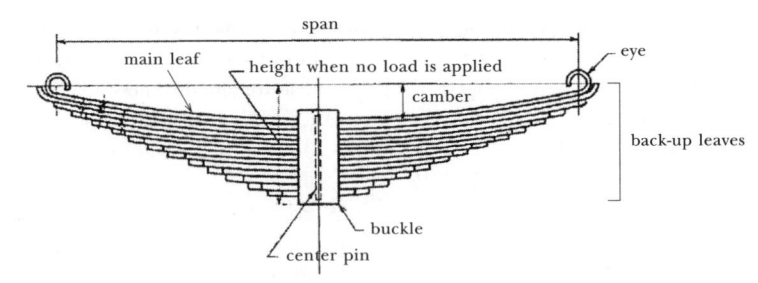

Fig M-2 main leaf

วิธีการสร้างความเครียดและความเค้นด้วยสนามแม่เหล็ก [wi-tii gaan sang kwaam kri-ead / ken du-oi sa-naam maae-lek]

main leaf
leaf with eyes on both ends. (see Fig M-2)
親板 [oyaita]
両端に荷重支持のための目玉又は取付け部をもつばね板．(Fig M-2 参照)
主簧片 [zhǔ huáng piàn]
daun utama [daun utama]
모판 [mopan]
แหนบแผ่นในการรับโหลด [naaep paaen lak nai gaan rab load]

main spring
spring always supporting the load in a progressive leaf spring. (see Fig M-3)
主ばね [syubane]
プログレッシブ重ね板ばねにおいて，常時荷重を受けるばね．(Fig M-3 参照)
主弾簧 [zhǔ tán huáng]

pegas utama [pugasu utama]
메인 스프링 [mein seupeuring]
แหนบหลัก [naaep lak]

malleability
easiness for metal to be formed into thin plate or foil.
展性 [tensei]
金属を打ち伸ばして薄い板や箔にできる性質．
可展性 [kě zhǎn xìng]
pemudahan dibentuk [pumudahan dibentuku]
전성 [jeonseong]
ความเหนียว [kwaam ni-eaw]

mandrel diameter
心金径 [singanekei]
心軸直径 [xīn zhóu zhí jìng]
dimeter poros pemegang [dimeteru porosu pumegangu]
맨드릴 지름 [maendeuril jireum]
เส้นผ่าศูนย์กลางของแกนม้วน [sen-pa-soon

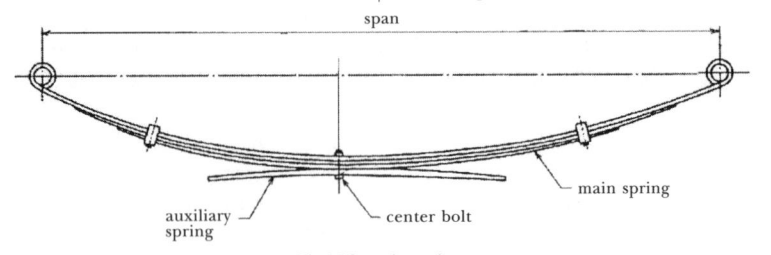

Fig M-3 main spring

glaang khong gaaen muan]

mandrel sliding
the lateral or vertical movement of a mandrel in a forming machine or torsion spring machine.
心金スライド [singane suraido]
フォーミングマシンやトーションスプリングマシンなどの心金が前後，上下などに移動すること．
心轴滑动 [xīn zhóu huá dòng]
geseran poros pemegang [geseran porosu pumegangu]
맨드릴 슬라이드 [maendeuril seullaideu]
การเลื่อนของแกนม้วน [gaan le-un khong gaaen muan]

mandrel-type coiling machine
a lathe-type coil forming machine.
心金式コイリングマシン [singanesiki koiringu masin]
旋盤式のコイリング成形機．
心轴式卷簧机 [xīn zhóu shì juǎn huáng jī]
mesin ulir jenis poros pemegang [mesin uriru jenisu porosu pumegangu]
맨드릴식 코일링기 [maendeurilsik koillinggi]
เครื่องม้วนชนิดมีแกนม้วน [kru-ang muan cha-nid mii gaaen]

man-machine system
a system in which humans and machines to be operated by humans compose a unified function.
人間機械系 [ningen kikai kei]
人間及び人間が操作する機械類を構成要素として機能する体系．
人机系统 [rén jī xì tǒng]
antar muka mesin dan orang [antaru muka mesin dan orangu]
인간기계 시스템 [ingangigye siseutem]
ระบบคน-เครื่องจักร [ra-bob kon kru-ang-chak]

maraging steel
tough and strong steel with Ni₃Mo and NiTi fine particle precipitated in martensite matrix.
マルエージング鋼 [marueezingu kou]
マルテンサイト母相にNi₃MoとNiTiが微細に析出し，高い強度とじん性を持つ鋼．
马氏体时效处理钢 [mǎ shì tǐ shí xiào chǔ lǐ gāng]
baja maraging [baja maragingu]
마르에이징강 [mareueijinggang]
เหล็กมาราจิ้ง [lek maraging]

maraging steel spring
springs made of maraging steel.
マルエージング鋼ばね [marueezingu kou bane]
マルエージング鋼で作られたばね．
马氏体时效处理钢弹簧 [mǎ shì tǐ shí xiào chǔ lǐ gāng tán huáng]
pegas baja marageing [pugasu baja marageingu]
마르에이징강 스프링 [mareueijinggang spring]
สปริงเหล็กมาร์เรจจิ้ง [sa-pring lek marageing]

marking
symbols, letters and/or numbers stamped on the goods or parts for identification.
マーキング [maakingu]
製品や部品の識別のためにスタンプされる記号，文字，あるいは数字．
记号 [jì hào]
penandaan, marking [punandaan, marukingu]
마킹 [making]
มาร์คกิ้ง [marking]

martempering
the process of quenching in cooling

liquid kept at a temperature in the upper region of the martensite formation temperature or slightly higher, ensuring that all parts are maintained at this temperature, then allowed to cool slowly. The purposes of this process are to prevent warping or cracking caused by quenching, and achieve an appropriate quenching structure.

マルテンパー　[marutenpaa]
マルテンサイト生成温度域の上部又はそれよりやや高い温度に保持した冷却剤中に焼入れして，各部が一様にその温度になるまで保持した後，徐冷する操作．その目的は焼入れなどによるひずみの発生や焼割れを防ぐとともに適当な焼入れ組織を得ることにある．

马氏体等温淬火　[mǎ shì tǐ děng wēn cuì huǒ]
martemper (kejut bertahap)
[marutenperu (kejuto berutahapu)]
마르템퍼　[mareutempeo]
การอบให้ได้ผลึกแบบมาร์เทนไซท์　[gaan ob hai dai pa-luk baaep martensite]

martensite
　the structure that forms due to transformation at temperatures below the Ms point without any diffusion when austenite is cooled rapidly. It is a solid solution formed of body-centered cubic crystals possessing the same chemical structure as the original austenite. Martensite may also form due to plastic deformation of austenite.

マルテンサイト　[marutensaito]
オーステナイトを急冷した場合にMs点以下の温度で拡散を伴わずに変態して生じる組織．元のオーステナイトと同じ化学組織を持つ体心立方晶の固溶体である．マルテンサイトは，オーステナイトの塑性変形によって生じることもある．

马氏体　[mǎ shì tǐ]
martensit　[marutensito]
마르텐사이트　[mareutensaiteu]
ผลึกแบบมาร์เทนไซท์　[pa-luk baaep martensite]

mass damper
　metal component attached to the large amplitude section of a leaf spring to suppress resonance by its mass effect. (see Fig M-4)

マスダンパー　[masu danpaa]
重ね板ばねの共振抑制のために，振幅の大きい部分に付けるおもり．（Fig M-4 参照）

质量阻尼　[zhì liàng zǔ ní]
peredam masa　[puredan masa]
매스 댐퍼　[maeseu daempeo]
ความหน่วงของมวล　[kwaam nu-ang khong muan]

mass effect
　differences in quench-hardened layer depth due to differences in mass and cross-sectional dimensions.

質量効果　[situryou kouka]
質量及び断面寸法の大小で，焼入硬化層深さの異なる度合い．

质量效应　[zhì liàng xiào yīng]

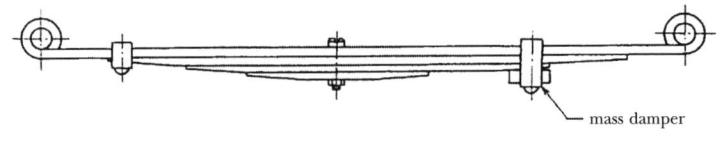

Fig M-4　mass damper

efek masa [efeku masa]
질량효과 [jillyanghyogwa]
ความแตกต่างของความลึกของชั้นชุบแข็งเนื่องจากมวลและขนาดที่ต่างกัน [kwaam taaek taang khong kwaam luk khong shan shub khaaeng ne-ung chaak mu-an laae kha-naad tii taang gan]

material
材料 [zairyou]
材料 [cái liào]
bahan [bahan]
재료 [jaeryo]
วัตถุดิบ [wat-tu-dip]

material flaw
flaws in materials which took place during manufacturing process.
材料欠陥 [zairyou kekkan]
材料の製造工程で発生した材料内部の欠陥.
材料缺陷 [cái liào quē xiàn]
cacat bahan [chachato bahan]
재료 결함 [jaeryo gyeolham]
รอยตำหนิของชิ้นงานจากกระบวนการผลิต [rooi tam-ni khong shin-ngaan chaak gra-buan gaan pa-lit]

material specification
documentation that stipulates such requirements as usage, chemical composition, and performance, and especially quality requirements.
材料仕様書 [zairyou siyousyo]
材料の使用目的, 化学成分, 性能などの要求事項, 特に品質の要求事項を記載し, 規定化した文書.
材料样本 [cái liào yàng běn]
spesifikasi bahan [supesifikasi bahan]
재료 시방서 [jaeryo sibangseo]
สเปคของวัตถุดิบ [spec khong wat-tu-dip]

material structure
micro structure of metal material.
材料組織 [zairyou sosiki]
金属材料の顕微鏡組織.
材料组织 [cái liào zǔ zhī]
susunan bahan [susunan bahan]
재료 조직 [jaeryo jojik]
โครงสร้างของวัตถุดิบ [krong-saang khong wat-tu-dip]

material testing
tests to measure the mechanical properties of materials. This includes tensile, impact, torsion, hardness, and bending tests.
材料試験 [zairyou siken]
材料の機械的な性質を測定するための試験. 引張試験, 衝撃試験, ねじり試験, 硬さ試験, 曲げ試験などがある.
材料试验 [cái liào shì yàn]
pengujian bahan [pungujian bahan]
재료 시험 [jaeryo siheom]
การทดสอบคุณสมบัติของวัตถุดิบ [gaan tod sorb ku-na-som-bat khong wat-tu-dip]

Maxibloc spring
an hourglass-shaped spring coiled from tapered wire. This gives a low compressed height, the absence of contact between coils and nonlinear characteristics. It also eliminates coil rattle noise.
マクシブロックスプリング [makusiburokku supuringu]
テーパ線を用いてコイリング成形されたつづみ形ばね. 密着高さが低くでき, 線間接触がなく非線形特性が得られるため, 線間たたき音を生じる心配がない.
Maxibloc 弹簧 [maxibloc tán huáng]
pegas maxiblok [pugasu makisiburoku]
맥시블록 스프링 [maeksibeullok spring]
สปริงแบบแมกซี่บล็อค [sa-pring baaep maxibloc]

maximum deviation
    maximum gap of distance, position or direction from designated value.
    最大偏差　[saidai hensa]
    定められた値からの隔たりで位置，距離，あるいは角度に対して用いられる．
    最大偏差　[zuì dà piān chā]
    penyimpangan maksimum　[puninpangan makusimun]
    최대 편차　[choedae pyeoncha]
    ค่าที่เบี่ยงเบนไปจากค่าที่ดีไซน์ไว้　[kaa tii biang-bain pai chaak kaa tii desig waai]

maximum load
    maximum load applied to a spring.
    最大荷重　[saidai kazyuu]
    ばねに加わる最大の荷重．
    最大负荷　[zuì dà fù hè]
    beban maksimum　[beban makusimun]
    최대 하중　[choedae hajung]
    ค่าโหลดสูงสุด　[kaa load soong-sud]

maximum size
    最大寸法　[saidai sunpou]
    最大尺寸　[zuì dà chǐ cùn]
    ukuran maksimum　[ukuran makusimun]
    최대 치수　[choedae chisu]
    ขนาดสูงสุด　[kha-naad soong-sud]

maximum stress
    maximum stress generated in a spring.
    最大応力　[saidai ouryoku]
    ばねに生じる最大の応力．
    最大应力　[zuì dà yīng lì]
    tegangan maksimum　[tegangan makusimun]
    최대 응력　[choedae eungnyeok]
    ค่าความเค้นในสปริง　[kaa kwaam ken nai sa-pring]

maximum torque
    the maximum rotational force that a rotating body takes around the axis of rotation.
    最大トルク　[saidai toruku]
    回転している物体がその回転軸の回りに受ける最大の回転力．
    最大扭矩　[zuì dà niǔ jǔ]
    momen puntir maksimum　[momen puntiru makusimun]
    최대 토-크　[choedae to-keu]
    ค่าแรงบิดทอร์กสูงสุด　[ka-load soong-sud]

mean center distance between coils
    the mean pitch between the centerline of adjacent coils on a coil spring.
    コイル平均ピッチ　[koiru heikin pitti]
    コイルばねの線断面中心ピッチの平均値．
    间距　[jiān jù]
    jarak pusat rata-rata antar ulir　[jaraku pusato rata-rata antaru uriru]
    코일 평균 피치　[koil pyeonggyun pichi]
    ระยะพิชท์เฉลี่ยระหว่างแกนกลางขด　[ra-ya pitch sha-lia ra-waang gaaen khod]

mean coil diameter
    mean value of the inner and the outer diameter of a coil, used for the calculating formula of the coil spring.
    コイル平均径　[koiru heikin kei]
    コイルばねの計算式に用いる，コイル内径と外径との平均値．
    弹簧中径　[tán huáng zhōng jìng]
    diameter ulir rata-rata　[diameteru uriru rata-rata]
    코일 평균지름　[koil pyeonggyunjireum]
    เส้นผ่าศูนย์กลางเฉลี่ยของขด　[sen-pa-soong-glaang sha-lia khong khod]

mean diameter by gravitational center of wire
    coil diameter measured by the gravitational center of the noncircular cross-section of the material.
    コイル重心径　[koiru zyuusin kei]
    異形断面コイルばねの，材料断面の図

心間の距離.
簧圈重心的平均直径 [huáng quān zhòng xīn de píng jūn zhí jìng]
diameter rata-rata di pusat grapitasi kawat [diameteru rata-rata di pusato gurapitasi kawato]
코일 중심지름 [koil jungsimjireum]
ค่าเฉลี่ยเส้นผ่าศูนย์กลางขดวัดโดยแกนกลางแนวดิ่งหน้าตัด [kaa sha-lia sen-pa-soon-glaang khod wat do-oi gaaen glaang naaew ding naa-tad]

Gravitational center of material section

mean diameter by gravitational center of wire

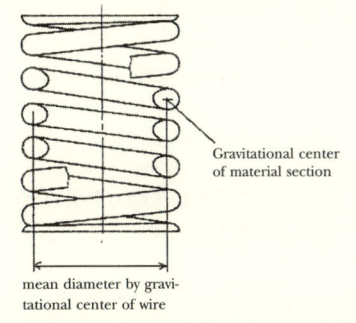

Fig M-5 mean diameter by gravitational center of wire

mean stress
　half of an algebraic sum of maximum and minimum stress generated in a spring under a repetitive load.
　平均応力 [heikin ouryoku]
　ばねに生じる繰返し応力の最大応力と最小応力の代数和の1/2.
　平均应力 [píng jūn yīng lì]
　tegangan rata-rata [tegangan rata-rata]
　평균응력 [pyeonggyuneungnyeok]
　ค่าความเค้นเฉลี่ย [kaa kwaam ken sha-lia]

mean time to failure
　a measure of reliability of equipment which is the average time before the first failure.
　故障までの平均時間 [kosyou made no heikin zikan]
最初の故障が生じるまでの平均時間で表す機器信頼性の指標.
　故障平均时间 [gù zhàng píng jūn shí jiān]
　waktu rata-rata kerusakan [wakutu rata-rata kerusakan]
　고장 날때까지의 평균시간 [gojang nalttaekkajiui pyeonggyunsigan]
　เวลาเฉลี่ยในการใช้งาน [wee-laa sha-lia nai gaan shai-gnaan]

mean time to repair
　a measure of reliability of repairable equipment which is the average time between repairs.
　平均修復時間 [heikin syuuhuku zikan]
　故障を修理してから次の故障が生じるまでの平均時間で表す修理可能な機器に対する信頼性指標.
　平均修复时间 [píng jūn xiū fù shí jiān]
　waktu rata-rata perbaikan [wakutu rata-rata perubaikan]
　평균 수리 시간 [pyeonggyun suri sigan]
　เวลาเฉลี่ยในการซ่อม [wee-laa sha-lia nai gaan sorm]

measured value
　the data value obtained from a measuring instrument at a given moment.
　測定値 [sokuteiti]
　ある瞬時において測定装置から得られる情報の値.
　测量值 [cè liáng zhí]
　nilai terukur [nirai terukuru]
　측정치 [cheukjeongchi]
　ค่าที่วัดได้ [kaa tii wad dai]

measuring accuracy
　the degree of consistency between measured and true values.
　測定精度 [sokutei seido]
　測定量の真の値との一致の度合い.
　测量精度 [cè liáng jīng dù]
　ketepatan pengukuran [ketepatan

pungukuran]
측정 정밀도 [cheukjeong jeongmildo]
ความแม่นยำในการวัด [kwaam maaen yam nai gaan wad]

measuring device
測定装置 [sokutei souti]
測量装置 [cè liáng zhuāng zhì]
sarana pengukuran [sarana pungukuran]
측정장치 [cheukjeongjangchi]
เครื่องมือวัด [kru-ang muu nai gaan wad]

measuring error
the difference between the measured value and true value.
測定誤差 [sokutei gosa]
測定値と測定量の真の値との違い．
測量误差 [cè liáng wù chā]
kesalahan pengukuran [kesarahan pungukuran]
측정오차 [cheukjeongocha]
ความผิดพลาดจากการวัด [kwaam pid plaad chaek gaan wad]

measuring method
測定方法 [sokutei houhou]
測量方法 [cè liáng fāng fǎ]
metoda pengukurang [metoda pungukurangu]
측정방법 [cheukjeongbangbeop]
วิธีการวัด [wi-tii gaan wad]

measuring range
the range between maximum and minimum amounts to be measured.
測定範囲 [sokutei han'i]
測定する量の最小値と最大値との範囲．
測量范围 [cè liáng fān wéi]
rentang pengukuran [rentangu pungukuran]
측정범위 [cheukjeongbeomwi]
ช่วงของการวัด [shu-ong khong gaan wad]

measuring spring
a spring used in measuring instruments. These include coil springs and spiral springs used in spring balances.
測定用ばね [sokuteiyou bane]
測定機器に用いるばね．ばね秤に用いるコイルばねやぜんまいなどがある．
測量用弾簧 [cè liáng yòng tán huáng]
pegas pengukur [pugasu pungukuru]
계측용 스프링 [gyecheungnyong spring]
สปริงที่เป็นตัววัดเทียบค่า [sa-pring tii pen to-a wad tieab kaa]

mechanical efficiency
the ratio of internal hose power (shaft hose power minus mechanical loss) and shaft hose power.
機械効率 [kikai kouritu]
軸動力から機械損失動力を引いた内部動力と軸動力との比．
机械效率 [jī xiè xiào lǜ]
efiensi mekanis [efiensi mekanisu]
기계효율 [gigyehyoyul]
ประสิทธิภาพเชิงกล [pra-sit-ti paab she-arng gon]

mechanical impedance analyzer
an equipment whereby a system characteristic can be displayed, by a parameter ratio of force and the speed of vibration. It is used for a vibration testing machine.
機械インピーダンス測定機 [kikai inpiidansu sokuteiki]
加振試験機において，加振力/振動速度などの比で系の特性を表示できる装置．
机械阻抗 [jī xiè zǔ kàng]
pengukuran tekanan secara mekanik [pungukuran tekanan sechara mekaniku]
기계적 임피던스 측정기 [gigyejeok impideonseu cheukjeonggi]

อุปกรณ์วัดที่แสดงค่าพารามิเตอร์ของการสั่น
[up-pa-gorn wad tii sa-daaeng kaa parameter khong gaan san]

mechanical loss
hose power that cannot be effectively extracted as shaft hose power, consisting of friction at bearings and gears, and power required to drive directly connected accessories.
機械損失　[kikai sonsitu]
軸受，歯車で発生する摩擦，直結補機を駆動する動力など，軸端出力として有効に取り出すことができない仕事量．
机械损耗　[jī xiè sǔn hào]
kehilangan mekanis　[kehirangan mekanisu]
기계손실　[gigyesonsil]
การสูญเสียเชิงกล　[gaan soon-siaa she-arng gon]

mechanical plating
a method of plating inside a tumbler using a mixture of glass beads and zinc powder as the impact medium. Suitable for high-stress springs, since there is less risk of hydrogen embrittlement.
機械衝撃めっき　[kikai syougeki mekki]
衝撃媒体として，ガラスビーズと亜鉛粉末を混ぜてタンブラー内で鍍金する方法．水素ぜい性の恐れが少ないため高応力ばねに適している．
机械冲击镀锌　[jī xiè chōng jī dù xīn]
penyepuhan mekanis　[punyepuhan mekanisu]
기계충격 도금　[gigyechunggyeok dogeum]
การชุบด้วยวิธีเชิงกล　[gaan shub du-oi wi-tii she-arng gon]

mechanical press
a press in which the slide moves mechanically.

機械プレス　[kikai puresu]
スライドを機械式機構によって駆動するプレス．
机械挤压　[jī xiè jī yā]
tekanan mekanis　[tekanan mekanisu]
기계 프레스　[gigye peureuseu]
การปั๊มอัดด้วยวิธีเชิงกล　[gaan pump ad du-oi wi-tii she-arng gon]

mechanical properties
properties related to mechanical deformation or fracture, including tensile strength, yield point, elongation, reduction of area, hardness, Charpy impact value, fatigue strength, and creep strength.
機械的性質　[kikaiteki seisitu]
引張強さ，降伏点，伸び，絞り，硬さ，衝撃値，疲れ強さ，クリープ強さなど機械的な変形及び破壊に関係する諸性質．
机械特性　[jī xiè tè xìng]
sifat mekanis　[sifato mekanisu]
기계적 성질　[gigyejeok seongjil]
คุณสมบัติเชิงกล　[kun-na-som-bat she-arng gon]

mechatronics
an engineering that incorporates the idea of mechanical and electric technology into a integrated system.
メカトロニクス　[mekatoronikusu]
機械工学技術と電子工学技術を統合して一つのまとまったシステムにする工学部門．
机电化　[jī diàn huà]
mekatronik　[mekatoroniku]
메커트로닉스　[mekeoteuronikseu]
ไฟฟ้าจักรกล　[fai-faa chak-kon]

merchant bar
hot rolled steel bar, usually rolled at high temperature of about 800°C. Also

called bar-in-coil.
マーチャントバー [maatyanto baa]
熱間圧延棒鋼のことで，通常800℃付近の高温で圧延された棒鋼．バーインコイルともいう．
热轧钢棒 [rè zhá gāng bàng]
batang ukur pedagang [batangu ukuru pudagangu]
열간 압연 봉강 [yeolganabyeon bonggang]
แท่งเหล็กม้วนร้อน [taaeng lek mu-an rorn]

metal deposition
the settling of microscopic metal particles contained in a liquid.
金属沈殿物 [kinzoku tindenbutu]
液体に混ざっている微小な金属が底部に沈んで溜まること．
金属沉淀物 [jīn shǔ chén diàn wù]
endapan logam [endapan rogam]
금속 침전물 [geumsok chimjeonmul]
การเซ็ทตัวของอนุภาคโลหะในของเหลว [gaan set tu-a khong a-nu-paak loo-ha nai khong leew]

metallic spring
a spring made of metal.
金属ばね [kinzoku bane]
金属材料を用いたばね．
金属弹簧 [jīn shǔ tán huáng]
pegas logam [pugasu rogam]
금속 스프링 [geumsok spring]
สปริงที่ทำจากโลหะ [sa-pring tii tam chaak loo-ha]

metallurgical microscope
an optical microscope used in the study of metal structure.
金属顕微鏡 [kinzoku kenbikyou]
金属組織の研究に用いられる光学顕微鏡．
金相显微镜 [jīn xiāng xiǎn wēi jìng]
mikroskop metalurgi [mikurosukopu metarurgi]

금속 현미경 [geumsok hyeonmigyeong]
กล้องจุลทรรศน์สำหรับเช็คโครงสร้างเหล็ก [glong-chun-la-tad sam-rap check krong-saang lek]

metric system
a system of units used in scientific work throughout the world. Its units of length, time, and mass are the meter, second, and kilogram respectively.
メートル系 [meetoru kei]
全世界で使用される科学分野に適用される単位系．長さ，時間，質量の単位はそれぞれメートル, 秒, キログラム．
公制 [gōng zhì]
sistem meter [sisutemu meteru]
메트릭 시스템 [meteurik siseutem]
ระบบเมตริก [ra-bob metric]

micrograph
a photograph of an inspection surface taken under a microscope. These include metallurgical micrographs and electron micrographs.
顕微鏡写真 [kenbikyou syasin]
被検査面を顕微鏡によって撮影された写真．金属顕微鏡写真，電子顕微鏡写真などがある．
显微镜图像 [xiǎn wēi jìng tú xiàng]
grapik mikro [gurapiku mikuro]
현미경 사진 [hyeonmigyeong sajin]
กราฟจุลภาค [gragh chun-la-paak]

micrograph analysis
the analysis and evaluation of an inspection surface using micrography to study conditions such as the shape of defects.
顕微鏡解析 [kenbikyou kaiseki]
被検査面を顕微鏡によって，その組織，欠点，状態，生態などを調べ解析・調査すること．
显微镜图像分析 [xiǎn wēi jìng tú xiàng fēn xī]

analisis grapik mikro [anarisisu gurapiku mikuro]
현미경 해석 [hyeonmigyeong haeseok]
การวิเคราะห์ผิวโดยกราฟจุลภาค [gaan wi kraorpiw do-oi gragh chun-la-paak]

micrometer
an instrument for measuring length precisely by using screw pitch. It is possible to measure with precision of 0.001 mm or 0.01 mm using a vernier. Analog or digital micrometers for measuring external and internal dimensions are available.
マイクロメータ [maikuro meeta]
ねじのピッチを応用して精密に測定できる長さの測定具．副尺を用いて0.001, 0.01 mmの精密測定ができる．外側用と内側用とがありデジタル式もある．
千分尺 [qiān fēn chǐ]
mikrometer [mikurometeru]
마이크로미터 [maikeuromiteo]
ไมโครมิเตอร์ [micrometer]

microscopic examination
examination of metal structure of the surface to be inspected using a microscope after surface etching.
ミクロ組織検査 [mikuro sosiki kensa]
材料の被検査面を腐食液で処理し顕微鏡によって金属組織などを調べる検査．
显微组织检查 [xiǎn wēi zǔ zhī jiǎn chá]
pemeriksaan mikroskopis [pumerikusaan mikurosukopisu]
현미경 조직검사 [hyeonmigyeong jojikgeomsa]
การตรวจสอบโครงสร้างผิวเหล็ก [gaan troat sorb kroong-saang piw lek]

microsensor
a very small (in the order of microns to millimeters) device that converts a nonelectrical phisical or chemical quantity into an electrical signal.
マイクロセンサ [maikuro sensa]
非電気的な物理量あるいは化学量を電気信号に変換する非常に小さな（数ミクロンから数ミリメートルの範囲）素子．
微型传感器 [wēi xíng chuán gǎn qì]
sensor mikro [sensoru mikuro]
마이크로 센서 [maikeuro senseo]
ไมโครเซ็นเซอร์ [micosensor]

military wrapper
3/4 turned second leaf wrapper.
ミリタリラッパ [miritari rappa]
板ばねの二番巻の一種で，巻いた部分が3/4 巻の形状．
军用包装材料 [jūn yòng bāo zhuāng cái liào]
wrapper militer [rapperu miriteru]
둘째판 3/4 감김 [duljjaepan 3/4 gamgim]
มิลลิทารี่ แร็พเพอร์ [millitary wrapper]

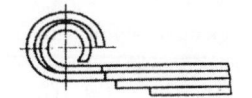

Fig M-6 military wrapper

mill edge
the edge of a steel sheet or strip formed naturally by rolling and remained uncut.
ミルエッジ [miru ezzi]
鋼板や鋼帯の端部が圧延によって自然にできたエッジそのままで，切断されないものをいう．
热轧缘边 [rè zhá yuán biān]
ujung tumbukan [ujungu tunbukan]
밀 에지 [mil eji]
ขอบกัด [khorb gad]

mini bloc spring
a barrel-shaped coil spring made from

tapered wire. The spring has a low compressed height, eliminates the contact between coils and provides nonlinear characteristics.
ミニブロックスプリング [mini burokku supuringu]
　テーパ線から作られたたる型コイルばね．密着高さが低くでき，線間接触なく非線形特性が得られる．
中凸形弾簧 [zhōng tū xíng tán huáng]
pegas blok mini [pugasu buroku mini]
미니블록 스프링 [minibeullok spring]
สปริงมินิบล็อค [sa-pring mini bloc]

mini bloc spring, conical
　a conical coil spring formed from tapered wire.
円すい形ミニブロックスプリング [ensuikei mini burokku supuringu]
　テーパ線を用いて成形された円すい形コイルばね．
锥形钢丝圆锥形弹簧 [zhuī xíng gāng sī yuán zhuī xíng tán huáng]
pegas blok mini, bentuk kerucut [pugasu buroku mini, bentuku kerukutu]
원추형 미니블록 스프링 [wonchuhyeong minibeullok spring]
สปริงมินิบล็อครูปกรวย [sa-pring mini bloc ruup gru-oi]

minimum distance (between coil)
　minimum distance between the inner diameter of the outer spring and the outer diameter of the inner spring of a combination spring. This term may signify the minimum pitch of coils.
コイル間最小寸法 [koirukan saisyou sunpou]
　組合せばねの外側のばねの内径と内側のばねの外径との最小距離．コイルピッチの最小寸法をいう場合もある．
组合弹簧簧圈间最小距离 [zǔ hé tán huáng

quān jiān zuì xiǎo jù lí]
jarak minimum (antara ulir) [jaraku minimumu (antara uriru)]
코일간 최소 간격 [koilgan choeso gangyeok]
ระยะต่ำสุดระหว่างขด [ra-ya tam sud ra-waang khod]

minimum preload
　application of minimum load to a spring before applying the specified load.
最小予圧 [saisyou yoatu]
　ばねに所定の荷重を加える前に最小限の荷重を加えること．
最小预压力 [zuì xiǎo yù yā lì]
beban awal minimum [beban awaru minimumu]
최소 예압 [choeso yeap]
แรงเริ่มต้นต่ำสุด [rang rerm ton tam-sud]

minimum stress
　minimum stress generated in a spring.
最小応力 [saisyou ouryoku]
　ばねに生じる最小の応力．
最小应力 [zuì xiǎo yīng lì]
tegangan minimum [tegangan minimumu]
최소 응력 [choeso eungnyeok]
ความเค้นต่ำสุด [kwaam ken tam sud]

minor and major axis of oval cross section
　the large (long) and small (short) diameters with respect to the axial direction of an ellipse cross section.
だ円断面線の長・短径 [daen danmensen no tyou-tankei]
　だ円素線断面の軸方向に対して小さい径と大きい径のこと．
椭圆截面的长、短径 [tuǒ yuán jié miàn de cháng、duǎn jìng]
sumber minor dan mayor penampang bujur telur [sunberu minoru dan mayoru

punanpangu bujuru teruru]
타원 단면 선재의 장단지름 [tawon danmyeon seonjaeui jangdanjireum]
แกนหลัก, รองของภาพตัดรูปไข่ [gaaen lak / rong khong paab tad ruup khai]

module

an assembly or a partial assembly which is prepared for a single maintenance.

モジュール [mozyuuru]
単位整備作業で扱えるようにした組立品又は部分組立品.
模量 [mó liàng]
modul [moduru]
모듈 [modyul]
โมดูล [module]

modulus of elasticity

the ratio of vertical stress acting on the bar cross section and the elongation or contraction per unit length (vertical distortion). The standard symbol is E. The unit of measure is MPa or N/mm^2.

弾性係数 [dansei keisuu]
棒の断面に働く垂直応力と，単位長さ当たりの伸び又は縮み（垂直ひずみ）との比．常用記号は E．単位は MPa 又は N/mm^2.
弾性模量 [tán xìng mó liàng]
modulus kekenyalan [modurusu kekenyaran]
탄성 계수 [tanseong gyesu]
โมดูลัสของความยืดหยุ่น [modulus khong kwaam yuud-yun]

moisture resistant

a property which does not absorb any humidity even under high humidity circumstance for long time.

耐湿性 [taisitusei]
長時間高湿度状態にさらされても湿気を吸収しない性質．
耐湿性 [nài shī xìng]

tahan lembab [tahan renbabu]
내습성 [naeseupseong]
การต้านทานความชื้น [gaan taan taan kwaam shu-un]

moment

the product of the force exerted and length of the perpendicular line to the line of action from the point where force is exerted.

モーメント [moomento]
力の作用線に下した垂線の長さとその力との積．
力矩 [lì jù]
momen [momen]
모멘트 [momenteu]
โมเมนต์ [moment]

moment curve

a graph which shows moment (torque) characteristics.

モーメント曲線 [moomento kyokusen]
モーメント（トルク）特性を表したグラフ．
力矩曲线 [lì jù qǔ xiàn]
kurva momen [kuruva momen]
모멘트 곡선 [momenteu gokseon]
เส้นโค้งโมเมนต์ [sen-kong moment]

moment measuring device

a testing machine which measure moment (torque).

モーメント測定装置 [moomento sokutei souti]
モーメント（トルク）を測定する試験装置．
力矩測量装置 [lì jù cè liáng zhuāng zhì]
sarana pengukur momen [sarana pungukuru momen]
모멘트 측정 장치 [momenteu cheukjeong jangchi]
อุปกรณ์วัดโมเมนต์ [up-pa-gorn wad moment]

moment of force
the product of the force exerted and the vertical distance from the rotation axis to the line of action.
力のモーメント　[tikara no moomento]
作用する力の大きさと回転の軸からの作用線までの垂直距離との積．
负荷力矩　[fù hè lì jǔ]
momen gaya　[momen gaya]
힘의 모멘트　[himui momenteu]
โมเมนต์ของแรง　[moment khong raaeng]

moment of inertia
magnitude of inertia around rotating axis. It is the sum of product of mass and its distance (from the axis) squared.
慣性モーメント　[kansei moomento]
物体のある回転軸に関する慣性の大きさを表す量で，その物体の微小部分の質量とその回転軸からの距離の2乗の積の総和．
惯性矩　[guàn xìng jǔ]
momen inersia　[momen inerusia]
관성 모멘트　[gwanseong momenteu]
โมเมนต์ของความเฉื่อยรอบแกนหมุน [moment khong kwaam shu-ai rorb gaaen mun]

mortor spring
springs formed from flat strip wound in the form of a spiral encased in a retainer to exert torque about an axis normal to the plane of the spiral.
接触形ぜんまい　[sessyokugata zenmai]
素線が互いに接触しているぜんまい．一端がばねケースに固定され，他端の軸周り回転トルクを利用するばね．
接触形渦卷弹簧　[jiē chù xíng wō juǎn tán huáng]
pegas mortor　[pugasu morutoru]
접촉형 태엽 스프링　[jeopchokhyeong taeyeop seupeuring]
สปริงมอเตอร์　[sa-pring motor]

Fig M-7 mortor spring

moving average
a series of arithmetical average of a certain number of time-series data. It is used for the analysis of time-series data.
移動平均　[idou heikin]
時系列データの各項に対し，それを中心とする前後一定項数の平均値．時系列データを解析するのに用いられる．
平均移动　[píng jūn yí dòng]
perpindahan rata-rata　[purupindahan rata-rata]
이동 평균　[idong pyeonggyun]
ค่าเฉลี่ยเลขคณิตที่ใช้ในการวิเคราะห์เวลา-ซีรีย์ [kaa sha-lia leek-kha-nid tii chai nai gaan wi kra-or wee-laa laae series]

muffle furnace
a furnace in which the workpiece is placed inside a heat-resistant chamber or container, then heated indirectly by combustion gas flowing around it.
マッフル炉　[mahhuru ro]
耐熱性の室又は容器に被加熱物を入れて，その周辺に燃焼ガスを流して間接的に加熱する炉．
马弗炉　[mǎ fú lú]
tunku muffle　[tunku mufufure]
머플로　[meopeullo]
เตาแบบปิด　[ta-o baaep pid]

multi peening
    shot peening performed several times under different conditions to effectively distribute residual stress.
    多段ピーニング　[tadan piiningu]
    効果的な残留応力分布を得ることを目的とし，条件を変えて数回行うショットピーニング．
    多段表面強化　[duō duàn biǎo miàn qiáng huà]
    peening ganda　[peeningu ganda]
    다단 피닝　[dadan pining]
    การทำช็อทพีนนิ่งหลายครั้ง　[gaan tam shot-peening laai krang]

multi-leaf spring
    leaf spring comprising laminated leaves having different lengths in incremental steps.
    マルチリーフスプリング　[maruti riihu supuringu]
    階段状に長さの異なるばね板を重ね合わせて構成された板ばね．
    复合板弹簧　[fù hé bǎn tán huáng]
    daun ganda　[daun ganda]
    겹판 스프링　[gyeoppan spring]
    แหนบชนิดหลายแผ่นความยาวต่าง ๆ กัน　[naaep cha-nid laai paaen kwam yaao taang taang gan]

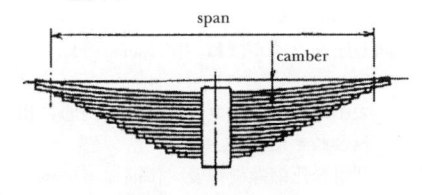

Fig M-8　multi-leaf spring

multi-slide press
    a press machine in which several slides placed around the piece of work operate to form a complicated spring shape.
    マルチスライドプレス　[maruti suraido puresu]
    ワークの周囲の複数のスライドによって複雑なばねの成形を行うプレス機械．
    多道冲压　[duō dào chōng yā]
    pegas daun ganda　[pugasu daun ganda]
    멀티 스라이드 프레스　[meolti seuraideu peureseu]
    เครื่องเพรสชนิดเลื่อนที่ละหลายแผ่น　[kru-ang press cha-nid liu-an tii la laai paaen]

multiple parallel stacking
    disk springs arranged in layers and in parallel.
    並列多重ばね　[heiretu tazyuu bane]
    皿ばねを並列に多段重ね合わせたものをいう．
    多重并列弹簧　[duō chóng bìng liè tán huáng]
    tumpukan sejajar ganda　[tunpukan sejajaru ganda]
    병렬 다중 접시 스프링　[byeongnyeol dajung jeopsi spring]
    สปริงที่วางซ้อนขนานกัน　[sa-pring tii waang sorn kha-naan gan]

music wire
    see "piano wire".
    ピアノ線　[pianosen]
    "piano wire" 参照．
    琴钢丝　[qín gāng sī]
    kawat musik, kawat piano　[kawato musiku, kawato piono]
    피아노 선　[piano seon]
    ลวดเปียโน　[lu-ad piano]

N

natural frequency
frequency of the free vibration of a vibrating system or of a single spring.
固有振動数　[koyuu sindousuu]
振動系の自由振動の振動数又はばね単体が自由振動するときの振動数．
固有振动频率　[gù yǒu zhèn dòng pín lǜ]
frekuensi alami　[furekuensi arami]
고유 진동수　[goyu jindongsu]
ความถี่ของการสั่นในสปริงเดี่ยว　[kwaam tii khong gaan san nai sa-pring dieaw]

natural high polymer spring
a spring made of natural high polymer like bamboo, hickory, etc.
天然高分子材ばね　[tennen koubunsizai bane]
竹，ヒッコリー材などの天然の高分子材を用いたばね．
天然高分子弹簧　[tiān rán gāo fēn zǐ tán huáng]
pegas polimer tinggi alami　[pugasu porimeru tingugi arami]
천연 고분자재료 스프링　[cheonnyeon gobunjajaeryo spring]
สปริงจากวัสดุธรรมชาติที่เป็นโพลีเมอร์สูง　[sa-pirng chaak was-sa-du tam-ma-chat tii pen polymer soong]

natural vibration
a vibration which keep vibrating without any force from outside. It is called a free vibration too.
固有振動　[koyuu sindou]
物体の振動において，特に外力を働かせなくても振動し続けるような振動．自由振動ともいう．
固有振动　[gù yǒu zhèn dòng]

getaran alami　[getaran arami]
고유 진동　[goyu jindong]
การสั่นโดยปราศจากแรงกระทำจากภายนอก [gaan san do-oi paad-sa chaak raaeng gra-tam chaak paai nork]

nest of spring
a set of springs whereby smaller piece is placed in the inside of larger piece coaxially.
ばねの入れ子組合せ　[bane no ireko kumiawase]
大小二つのばねのうち小さい方を大きい方の内側に入れた入れ子組合せのばね．
弹簧的组合　[tán huáng de zǔ hé]
susunan pegas　[susunan pugasu]
동심스프링의 병열 조합 [dongsimspringui byeongyeol johap]
ชุดของสปริงที่ประกอบแล้ว　[shud khong sa-pring tii pra-gorb laaew]

nested compression spring
a spring formed by combining large and small compression coil springs coaxially.
同心圧縮コイルばね　[dousin assyuku koiru bane]
大小二つの圧縮コイルばねを同心に組み合わせたばね．
并列组合压缩螺旋弹簧　[bìng liè zǔ hé yā suō luó xuán tán huáng]
pegas tekan tersusun　[pegasu tekan terususun]
동심 압축 코일 스프링　[dongsim apchuk koil spring]
สปริงขดอัดที่มีแกนกลางเดียวกัน　[sa-pring khod tii mii gaaen glaang dieaw gan]

nested ring spring

a spring formed by combining the outer ring with sloping friction surface inside, and the inner ring with sloping friction surface outside.

重ねた輪ばね [kasaneta wabane]
外輪は内側に，内輪は外側に傾斜がある摩擦面を持った輪を重ね合わせたばね．

环状弹簧 [huán zhuàng tán huáng]

pegas cincin tersusun [pugasu chinchin terususun]

조립 링 스프링 [jorip ring spring]

สปริงแหวนที่เรียงซ้อนกัน [sa-pring waaen tii riang sorn gan]

nib

small protruded spot of a leaf to prevent the displacement of the spring.

だぼ [dabo]
ばねのずれを防ぐために，ばね板の小部分を押し出したもの．

末端 [mò duān]

sela [sera]

배꼽 [baekkop]

จงอย [cha-ngoi]

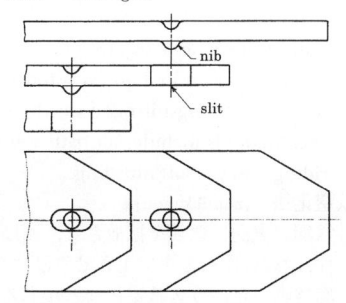

Fig N-1 nib

nickel alloy spring

a spring made of nickel alloy which has good corrosion resistance, non-magnetic property, and ability to be used at high temperature and low temperature below 0°C. Materials include Monel and Inconel.

ニッケル合金ばね [nikkeru goukin bane]
耐食性が良好で，非磁性で高温度あるいは0°C以下の低温で使用し得る特徴をもつニッケル合金を用いたばね．材料として，モネル，インコネルなどがある．

镍合金弹簧 [niè hé jīn tán huáng]

pegas paduan nikel [pugasu paduan nikeru]

니켈 합금 스프링 [nikel hapgeum spring]

สปริงเหล็กผสมนิคเกิล [sa-pring lek pa-som nickle]

nickel plating

ニッケルめっき [nikkeru mekki]

表面镀镍 [biǎo miàn dù niè]

pelapisan nikel [purapisan nikeru]

니켈 도금 [nikel dogeum]

การชุบนิคเกิล [gaan shub nickle]

nickel silver for spring

alloy consisting of copper, nickel, and zinc with an attractive silvery grey color. Nickel silver for springs normally contains 16.5~19.5% Ni and 54~58% Cu, with remaider consisting of Zn.

ばね用洋白 [baneyou youhaku]
Cu-Ni-Znの3元素合金で，美しい銀白色を有している．ばね用材ではNi 16.5〜19.5%，Cu 54〜58%，残部Znが普通である．

弹簧用银镍合金 [tán huáng yòng yín niè hé jīn]

nikel perak untuk pegas [nikeru puraku untuku pugasu]

스프링용 양백 [springyong yangbaek]

โลหะผสมทองแดง, นิคเกิล, สังกะสีสำหรับผลิตสปริง [lo-ha pa-som tong-daaeng, nickle, sang-ga-sii sam-rab pa-lit sa-pring]

nip

space between adjacent leaves in a leaf spring before cramping them by the center bolt.

ニップ [nippu]
重ね板ばねにおいて，センタボルトなどで締め付ける前の隣接するばね板とばね板とのすき間．
弾簧板间隙 [tán huáng bǎn jiān xì]
celah [cherahu]
닙 [nip]
ระยะระหว่างแผ่นแหนบก่อนยึดด้วยเซ็นเตอร์โบลท์ [ra-ya ra-waang paaen naaep gorn yud do-ui center bolt]

nip stress

bending stress generated in a leaf when the nip is tightened.

ニップ応力 [nippu ouryoku]
重ね板ばねのニップを締め付けたとき，ばね板に生じる曲げ応力．
弾簧板间隙应力 [tán huáng bǎn jiān xì yīng lì]
tegangan celah [tegangan cherahu]
닙 응력 [nip eungnyeok]
ความเค้นนิบ [kwaam ken nip]

nitrided layer

the layer of steel treated by nitriding to provide wear resistance and corrosion resistance.

窒化層 [tikkasou]
鉄鋼に窒化処理した処理層のことで，耐摩耗性，耐食性の特徴がある．
氮化层 [dàn huà céng]
lapisan nitrida [rapisan nitorida]
질화층 [jilhwacheung]
ชั้นของการชุบแข็งแบบไนไตร์ด [shan khong gaan shup-khaaeng baaeb nitride]

nitriding

operation to harden the surface layer of steel by diffusing nitrogen. This operation includes gas nitriding by re-solved ammonia gas and liquid nitriding by cyanide compounds.

窒化 [tikka]
鉄鋼の表面層に窒素を拡散させ，表面層を硬化する操作．方法には，アンモニア分解ガスによるガス窒化及び青酸塩による液体窒化がある．
氮化物 [dàn huà wù]
penitridaan [punitoridaan]
질화 [jilhwal]
การชุบแข็งด้วยไนโตรเจน [gaan shup khaaeng du-oi nitrogen]

nitriding steel

steel manufactured to be suitable for nitriding. It contains aluminum, chromium, molybdenum and other elements.

窒化鋼 [tikkakou]
表面を窒化するのに適するように造られた鋼．Al, Cr, Mo などを含む．
渗氮钢 [shèn dàn gāng]
baja nitrida [baja nitorida]
질화강 [jilhwagang]
เหล็กใช้สำหรับชุบแข็งด้วยไนโตรเจน [lek chai sam-rap shub khaaeng du-oi nitrogen]

nitrocarburizing

a method of heat treatment for improving wear and fatigue resistance of steel by diffusing nitrogen on the surface. Methods include salt bath soft nitriding and gas soft nitriding.

軟窒化法 [nantikkahou]
鉄鋼の表面に窒素を拡散させ，耐摩耗性，耐疲れ性などを向上させる熱処理．塩浴軟窒化，ガス軟窒化などがある．
软氮化法 [ruǎn dàn huà fǎ]
pelapisan karbon nitrida [purapisan karubon nitorida]
연질화법 [yeonjilhwabeop]
การทำไนโตรคาร์บิวไรซิ่ง [gaan tam

nitrocarburizing]

no-go gage
    a limit gage designed not to allow the test piece fit in place. It is usually employed with a go gage to establish the acceptable maximum and minimum limit.
    止りゲージ　[tomari geezi]
    テストピースが通らない設計された限界ゲージ．通常どおりゲージと一緒に使われて許容の最大値と最小値を確立する．
    止规　[zhǐ guī]
    gage pembatas　[gage penbatasu]
    정지 게이지　[jeongji geiji]
    โนโก เกจ　[no-go gauge]

noise regulation
    limit of noise level which is established to satisfy environmental standard.
    騒音規制　[souon kisei]
    騒音にかかわる環境基準を達成するために設けられた，騒音発生許容限界．
    噪声规范　[zào shēng guī fàn]
    batas kebisingan　[batasu kebisingan]
    소음 규제　[soeum gyuje]
    กฎข้อบังคับเกี่ยวกับเสียง　[god khor bang-kab gieaw gab si-ang]

nominal dimension
    the value or dimension indicating the standard value of a component.
    呼び寸法　[yobi sunpou]
    部品などの規格値を代表する値又は寸法．
    公称尺寸　[gōng chēng chǐ cùn]
    dimensi nominal　[dimensi nominaru]
    호칭 치수　[hoching chisu]
    ขนาดที่ระบุ　[kha-naad tii ra-bu]

nominal size
    size used for purposes of general identification. The actual size of a part will be approximately the same as the nominal size but need not be exactly the same.
    公称サイズ　[kousyou saizu]
    一般的な識別のために用いられるサイズ．実際の部品サイズは公称サイズに近いが，必ずしも正確に同一の値ではない．
    公称规格　[gōng chēng guī gé]
    ukuran nominal　[ukuran nominaru]
    공칭치수　[gongchingchisu]
    ขนาดที่ใช้ในการระบุทั่ว ๆ ไป　[kha-naard tii chai nai gaan ra-bu tu-a tu-a pai]

non-aging
    a property that does not show any aging deterioration on mechanical properties and machinability.
    非時効性　[hi zikousei]
    機械的性質及び加工性が実用上支障をきたすような経時変化をしない性質．
    非时效性　[fēi shí xiào xìng]
    tidak menua　[tidaku menua]
    비 시효성　[bi sihyoseong]
    ไม่แปรสภาพ　[mai praae sa-paap]

noncircular wire helical spring
    a coil spring made of the material the cross-section of which is noncircular.
    異形断面ばね　[ikei danmen bane]
    材料の断面形状が円形以外のコイルばね．
    异形截面钢丝螺旋弹簧　[yì xíng jié miàn gāng sī luó xuán tán huáng]
    pegas berpenampang tak serupa　[pugasu berupenanpangu taku serupa]
    이형 단면 선스프링　[ihyeong danmyeon seonspring]
    สปริงขดเกลียวชนิดลวดเหลี่ยม　[sa-pring khod gli-eaw cha-nid lu-ad liam]

noncontact measuring
    measuring objects without directly

touching them. Recently developed instruments can measure a wide range of properties using ultrasound or laser sensors.

非接触測定　[hisessyoku sokutei]
品物に直接接触させないでその品物を測定することで, 最近では超音波, レーザなどのセンサを応用していろいろな特性を測定できる機器が開発されている.

非接触測量　[fēi jiē chù cè liáng]
mengukur tanpa menyentuh [mengukuru tanpa menyentuhu]
비접촉 측정　[bijeopchok cheukjeong]
การวัดโดยไม่สัมผัส　[gaan wad du-oi mai sam-pad]

noncontact type spiral spring

spiral spring the coils of which do not contact each other.

非接触形ぜんまい　[hisessyokugata zenmai]
コイルが互いに接触していないぜんまい.

非接触形渦巻弾簧　[fēi jiē chù xíng wō juǎn tán huáng]
pegas mortor tanpa kontak　[pugasu morutoru tanpa kontaku]
비접촉형 태엽 스프링　[bijeopchokhyeong taeyeop seupeuring]
สปริงเกลียวชนิดไม่สัมผัส　[sa-pring glioew sha-nid nai sam-pad]

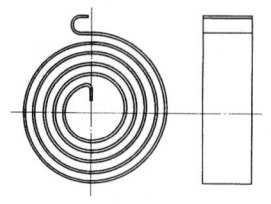

Fig N-2　noncontact type spiral spring

nondestructive test

test to investigate the existence, location, size, shape and distribution of defects in a material or a spring without destruction.

非破壊試験　[hihakai siken]
材料及びばねを破壊せずに, 欠陥の有無, その存在位置, 大きさ, 形状, 分布状態などを調べる試験.

非破坏性试验　[fēi pò huài xìng shì yàn]
uji tanpa merusak　[uji tanpa merusaku]
비파괴 시험　[bipagoe siheom]
การทดสอบแบบไม่เสียหาย　[gaan tod-sorb baaep mai siaa-haai]

nonferrous spring

spring made of nonferrous metal, as called by its material.

非鉄金属ばね　[hitetu kinzoku bane]
ばねの材料の種類による名称で, 非鉄金属材料を用いたばね.

非铁金属弹簧　[fēi tiě jīn shǔ tán huáng]
pegas bukan besi　[pugasu bukan besi]
비철 금속 스프링　[bicheol geumsok spring]
สปริงที่ไม่มีเหล็กเป็นส่วนผสม　[sa-pring tii mai mii lek pen suan pa-som]

nonlinear characteristics

spring characteristics where the relationship between the load and the deflection is not linear.

非線形特性　[hisenkei tokusei]
荷重とたわみとの関係が直線でないばね特性.

非线性特性　[fēi xiàn xìng tè xìng]
karakeristik tidak linear　[karakerisutiku tidaku rinearu]
비선형 특성　[biseonhyeong teukseong]
สปริงที่ความสัมพันธ์ระหว่างโหลดและการยุบตัว [sa-pring tii kwaam sam-pan ra-waang load gab gaan yub-to-a]

nonmagnetic steel
    alloy steel whose main alloy compositions are carbon, manganese, nickel, chromium, nitrogen, etc. and shows austenite structure.
    非磁性鋼　[hi ziseikou]
    C, Mn, Ni, Cr, N などを主な合金成分とし，オーステナイト組織を示す非磁性の合金鋼．
    非磁性钢　[fēi cí xìng gāng]
    baja tidak magnetis　[baja tidaku magunetisu]
    비자성강　[bijaseonggang]
    โลหะที่ไม่มีคุณสมบัติเป็นแม่เหล็ก　[lo-ha tii mai mee kun-na-som-bat pen maae-lek]

nonmetallic inclusion
    nonmetallic impurities precipitated or trapped into steel during solidification.
    非金属介在物　[hikinzoku kaizaibutu]
    鋼の凝固過程において鋼中に析出又は巻き込まれる非金属性の介在物．
    非金属介质　[fēi jīn shǔ jiè zhì]
    pemasukan bukan logam　[pumasukan bukan rogan]
    비금속 개재물　[bigeumsok gaejaemul]
    สิ่งเจือปนที่ไม่ใช่โลหะ　[sing chiu-a pon tii mai chai loo-ha]

nonmetallic spring
    a spring made of nonmetallic material. Includes resin springs, air springs, and ceramic springs.
    非金属ばね　[hikinzoku bane]
    ばねの種類からいう名称で非金属材料を用いたばね．樹脂ばね，空気ばね，セラミックばねなどがある．
    非金属弹簧　[fēi jīn shǔ tán huáng]
    pegas bukan logam　[pugasu bukan rogan]
    비금속 스프링　[bigeumsok spring]
    สปริงที่ไม่มีส่วนผสมของโลหะ　[sa-pring tii mai mii suan pa-som khong loo-ha]

normal distribution
    a most common random distribusion.
    正規分布　[seiki bunpu]
    最も一般的な確率分布．
    正态分布　[zhèng tài fēn bù]
    sebaran biasa　[sebaran biasa]
    정규 분포　[jeonggyu bunpo]
    การกระจายปกติ　[gaan gra-chaai pak-ka-ti]

normalizing
    the process of heating to appropriate temperature above the A_{C3} or A_{cm} point to form a uniform austenite structure before air cooling. The purposes are to eliminate any effects of previous processes, refine crystal grains, and improve mechanical properties.
    焼ならし　[yakinarasi]
    A_{C3} 点又は A_{cm} 点以上の適当な温度に加熱して一様なオーステナイト組織にした後空冷する操作．その目的は前加工の影響を除去し結晶粒を微細化して機械的性質を改善することである．
    正火, 常化　[zhèng huǒ, cháng huà]
    penormalan　[punorumaran]
    불림　[bullim]
    การอบคืนตัว　[gaan ob kuun to-a]

notch effect
    the effect of reduced strength due to concentrated stress caused by holes or grooves.
    ノッチ効果　[notti kouka]
    穴，溝などがある材料に力を加えると，応力集中効果によって強さが低下する効果．
    切口效应　[qiē kǒu xiào yīng]
    pengaruh goresan　[pungaruhu goresan]
    노치 효과　[nochi hyogwa]
    ผลกระทบจากความเค้นของรู, ร่อง　[pon gra-tob chaak kwaam ken khong ruu, rong]

notched-bar impact strength — number of leaves

notched-bar impact strength
  the energy required to break a test piece with concentrated stress caused by a groove or hole at single impact. The unit of measure is N·m or J.
  切欠き衝撃強さ　[kirikaki syougeki tuyosa]
  一回の衝撃により，溝や穴などの切欠きによる応力集中部を設けた試験片を破壊するのに必要なエネルギー．単位はN·m又はJ．
  切口冲击强度　[qiē kǒu chōng jī qiáng dù]
  kuat bentur batang tergores　[kuato benturo batango terugoresu]
  노치충격 강도　[nochichunggyeok gangdo]
  แรงกระแทกของแท่งบาก　[rang gra-taaek khong tang baak]

notched-bar impact test
  a test that breaks a test piece with concentrated stress caused by a groove or hole to evaluate the energy required, aspect of fractured surface, deformation behavior, and crack propagation.
  切欠き衝撃試験　[kirikaki syougeki siken]
  溝や穴などの切欠きによる応力集中部を設けた試験片に衝撃荷重を加えて破断し，要したエネルギーの大小，破断面の様相，変形挙動，き裂の進展挙動などによって，評価する試験．
  切口冲击试验　[qiē kǒu chōng jī shì yàn]
  uji bentur batang tergores　[uji benture batangu terugoresu]
  노치충격 시험　[nochichunggyeok siheom]
  การทดสอบแรงกระแทกของแท่งบาก　[gaan tod-sorb rang gra-taaek khong taaeng baak]

number of active coils
  number of turns used for the calculation of spring rate of a coil spring.
  有効巻数　[yuukou makisuu]
  コイルばねのばね定数の計算に用いる巻数．
  有効圏数　[yǒu xiào quān shù]
  jumlah ulir aktif　[junrahu uriru akutifu]
  유효 감김 수　[yuhyo gamgim su]
  จำนวนขดที่มีผลต่อค่าสปริงเรต　[cham-nuan khod tii mii pon to-a kaa sa-pring rate]

number of cycle to failure
  the number of repeated stress cycles until fatigue failure occurs.
  疲れ寿命　[tukare zyumyou]
  疲れ破壊を生じるまでの応力の繰返し回数．
  疲劳寿命　[pí láo shòu mìng]
  jumlah daur sampai rusak　[junrahu dauru sanpai rusaku]
  피로 수명　[piro sumyeong]
  จำนวนรอบที่ทำให้หัก　[cham-nuan rorb tii tam hai hak]

number of defects
  欠点数　[kettensuu]
  缺陷数　[quē xiàn shù]
  jumlah cacat　[junrahu kakato]
  결점수　[gyeoljeomsu]
  จำนวนของเสีย　[cham-nuan khong siaa]

number of free coils
  number of turns given by the calculation of total coils minus end coils.
  自由巻数　[ziyuu makisuu]
  コイルばねの総巻数から，両端の座巻数を引いた巻数．
  自由圏数　[zì yóu quān shù]
  jumlah ulir bebas　[junrahu uriru bebasu]
  자유 감김 수　[jayu gamgim su]
  จำนวนขดที่ไม่ทำงาน　[cham-nuan khod tii mai tam-ngaan]

number of leaves
  number of leaf spring of laminated spring.
  ばね板数　[baneitasuu]
  重ね板ばねのばね板の数．
  (板簧)片数　[(bǎn huáng) piàn shù]

jumlah daun [junrahu daun]
판 스프링 매수 [pan spring maesu]
จำนวนแผ่นของแหนบ [cham-nuan paaen khong naaep]

number of turns
the number of coils in a coil spring.
巻数 [makisuu]
コイルばねのコイルの数.
圈数 [quān shù]
jumlah putaran [junrahu putaran]
감김 수 [gamgim su]
จำนวนขด [cham-nuan khod]

number of twist
turned number at the breakage of test piece under twist test.
ねじり回数 [neziri kaisuu]
ねじり試験において，ねじり始めてから破断するまでの回数.
扭转圈数 [niǔ zhuǎn quān shù]
jumlah pintalan [junrahu pintaran]
비틀림 횟수 [biteullim hoetsu]
จำนวนครั้งของการบิดจนหักขณะทำการทดสอบการบิด [cham-nuan krang khong gaan bid chon hak ka-na tam gaan tod-sorb gaan bid]

O

offset hook
the hook on a helical extension spring formed from the coil outer circumference in the coil spring axial direction.
側面フック [sokumen hukku]
コイル外周部からコイルばね軸線方向に成形された，引張コイルばねのフック．
偏置钩环 [piān zhì gōu huán]
kait sisi [kaito sisi]
측면 고리 [cheungmyeon gori]
ตะขอด้านข้าง [ta cha-or daan khaang]

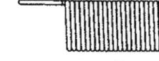

Fig O-1 offset hook

oil quenching
quenching in oil.
油焼入れ [abura yakiire]
冷却に油を用いて行う焼入れ．
油淬火 [yóu cuì huǒ]
pengejutan minyak [pungejutan minyaku]
유냉 [yunaeng]
การชุบแข็งด้วยน้ำมัน [gaan shub khaaeng]

oil seal ring spring
a sping to be used for tightening oil seal. It is a closely wound long slender coil spring with both ends connected to form a ring.
オイルシール用リングスプリング [oiru siiruyou ringu supuringu]
オイルシールの締付力を補助するために，密着巻きコイルばねの両端を連結して環状にして締付け用に用いるばね．
油水分离器 [yóu shuǐ fēn lí qì]
pegas cincin penyekat minyak [pugasu chinchin punyekato minyaku]
오일씰용 링 스프링 [oilssiryong ring seupeuring]
แหวนสปริงกันน้ำมันรั่ว [waaen sa-pring gan naam-man ru-a]

oil tempered wire for spring
steel wire manufactured through continuous oil quenching and tempering process in an uncoiled manner.
オイルテンパ線 [oiru tenpasen]
連続的にまっすぐな状態で油焼入れ・焼戻しを施して仕上げられた鋼線．
油淬火钢丝 [yóu cuì huǒ gāng sī]
kawat sepuh keras dikeraskan dengan minyak [kawato sepuhu kerasu dikerasukan dengan minyaku]
오일 템퍼선 [oil taempeoseon]
ลวดอบน้ำมันสำหรับสปริง [lu-ad ob nsam-man sam-rab sa-pring]

one end slides contact type taper leaf spring
a leaf spring with an eye formed only at one end.
片スライドタイプ重ね板ばね [katasuraido taipu kasaneita bane]
片側のみ目玉が成形された重ね板ばね．
单侧滑动复合板弹簧 [dān cè huá dòng fù hé bǎn tán huáng]
pegas daun mering bersentuh [pugasu daun meringu berusentuhu]
한쪽 끝 스라이드 접촉형 겹판 스프링

[hanjjok kkeut seuraideu jeopchokhyeong gyeoppan spring]
แหนบชนิดม้วนหูข้างเดียว　[naaep cha-nid mu-an huu khaang dieaw]

open end
　coil end which is not in contact with the adjacent coil.
　オープンエンド　[oopun endo]
　圧縮コイルばね端部の一種で，端末がコイル軸方向に隣りのコイルとすき間がある形状．
　端圏不并紧　[duān quān bù bìng jǐn]
　ujung terbuka　[ujungu terubuka]
　벌림 끝　[beollim kkeut]
　ปลายขดชนิดเปิด　[plaai khod cha-nid perd]

Fig O-2　open end

open flat end
　flat end of a helical compression spring whose end coil face is perpendicular to the spring axis, and end tip is not in contact with the adjacent coil.
　オープンフラットエンド　[oopun huratto endo]
　圧縮コイルばねの端部の一種で，端面が平らで，端末がコイル軸方向に隣接するコイルとすき間がある形状．
　端圏磨平不并紧　[duān quān mó píng bù bìng jǐn]
　ujung terbuka datar　[ujungu terubuka dataru]
　벌림 플랫형 끝　[beollim peullaethyeong kkeut]

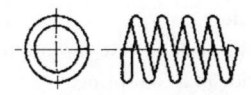

Fig O-3　open flat end

ปลายเปิดลักษณะแฟลต　[plaai-perd lak-sa-na flat]

open gap of leaf spring
　the gap between adjacent leaves in a leaf spring before tightening the center bolt.
　板間すきま　[bankan sukima]
　重ね板ばねにおいて，センタボルトなどで締め付ける前の隣接するばね板とばね板とのすきま．
　弹簧板间间隙　[tán huáng bǎn jiān jiān xì]
　pegas daun celah terbuka　[pugasu daun cherahu terubuka]
　판간 간격　[pangan gangyeok]
　ช่องว่างระหว่างแผ่นแหนบ　[shu-ong waang ra-wang paaen naaep]

open loop
　an open hook on a tension spring rather than an annular hook.
　開口フック　[kaikou hukku]
　引張ばねのフックが環状でなく口部が開けてあることをいう．
　开口环钩　[kāi kǒu huán gōu]
　lingkaran terbuka　[ringukaran terubuka]
　오픈 후크　[opeun hukeu]
　ตะขอแบบเปิดในสปริงดึง　[ta-khor baaep perd nai sa-pring dung]

operater
　a person who is in charge of operating a machine.
　オペレータ　[opereeta]
　機械を操作する担当者．
　操作工　[cāo zuò gōng]
　operator　[operatoru]
　조종자　[jojongja]
　เวิร์คเกอร์　[worker]

organic high polymer spring
　有機高分子材ばね　[yuuki koubunsizai bane]
　有机高分子弹簧　[yǒu jī gāo fēn zǐ tán huáng]

organic high polymer spring — outer guide

pegas organik polimer tinggi [pugasu oruganiku porimeru tingugi]
유기 고분자 재 스프링 [yugi gobunja jae spring]
สปริงที่ทำจากวัสดุอินทรีย์ที่เป็นโพลีเมอร์สูง [sa-pring tii tam chaak was-sa-du tii pen polymer soong]

orientation of bend axis

the orientation which governs the quality of bent parts in relation with the rolling direction of the steel strip. Transverse bend reveals good result whereas longitudinal bend tend to be poor.

曲げ方向 [mage houkou]
鋼帯の圧延方向との関連で曲げ成形品の品質を左右する方向．圧延と直角に曲げるのはよいが，圧延と平行に曲げるのはよくない傾向がある．

弯曲的方向 [wān qū de fāng xiàng]
arah bengkokan [arahu bengukokan]
굽힘방향 [gupimbanghyang]
ทิศทางของการโค้งงอ [tid-taang khong gaan koong-ngor]

oscillating stress

stress that fluctuates periodically.

振動応力 [sindou ouryoku]
周期的に変化する応力．

振动应力 [zhèn dòng yīng lì]
tegangan bolak balik [tegangan boraku bariku]
진동 응력 [jindong eungnyeok]
ความเค้นสั่นสะเทือนเป็นช่วง ๆ [kwaam ken san sa te-un pen shu-ang shu-ang]

oscillation damper

a device for attenuating the amplitude of oscillation by dissipating the energy.

ダンパ [danpa]
エネルギーを消散させる方法で，振動の振幅を軽減する装置．

减振器 [jiǎn zhèn qì]
peredam gerak bolak balik [puredan geraku boraku bariku]
진동진폭 감쇠 장치 [jindongjinpok gamsoe jangchi]
อุปกรณ์ลดการสั่น [up-pa-gorn lod gaan san]

oscillograph

a recording device of oscillation.

オッシログラフ [ossirogurahu]
振動現象の記録計．

示波器 [shì bō qì]
oscilograph [osuchurogurahu]
오실로그래프 [osillogeuraepeu]
กราฟแสดงการออสซิเลชั่น [graph sa-daaeng gaan oscillation]

outdoor exposure test

a test for investigating weather resistance, light resistance, and ozone resistance in a natural environment.

屋外暴露試験 [okugai bakuro siken]
自然環境状態で耐候性，耐光性，耐オゾン性を調べる試験．

露天试验 [lù tiān shì yàn]
uji diluar ruangan [uji diruaru ruangan]
옥외 폭로 시험 [okoe pongno siheom]
การทดสอบกลางแจ้ง [gaan tod-sorb glaang chaaeng]

outer coil diameter

outer diameter of a coil spring.

コイル外径 [koiru gaikei]
コイルばねの外径．

簧圈外径 [huáng quān wài jìng]
diameter ulir luar [diameteru uriru ruaru]
코일 바깥지름 [koil bakkatjireum]
เส้นผ่าศูนย์กลางนอกของขด [sen-pa-soon-glaang nork khong khod]

outer guide

a cylindrical guide mounted on the outside of a coil spring.

外側ガイド [sotogawa gaido]

コイルばねの外側に設ける筒形のガイド．
外側面导向　[wài cè miàn dǎo xiàng]
pemandu luar　[pumandu ruaru]
외측 가이드　[oecheuk gaideu]
ตัวนำด้านนอก　[tu-a nam daan nork]

outer ring
　a ring incorporated on the outside of a ring spring.
外輪　[sotowa]
輪ばねの外側に組み込まれる輪．
外轮　[wài lún]
cincin luar　[chinchin ruaru]
외륜　[oeryun]
แหวนด้านนอก　[waaen daan nork]

outer spring
　the outer coil spring of a double coil spring.
外側コイルばね　[sotogawa koiru bane]
２重コイルばねにおいて，外側に配置されるコイルばね．
外侧弹簧　[wài cè tán huáng]
pegas luar　[pugasu ruaru]
외측 코일 스프링　[oecheuk koil spring]
สปริงขดตัวนอก　[sa-pring khod to-a nork]

oval wire coil spring
　a noncircular wire helical spring made of wire with an oval cross-section.
だ円断面線コイルばね　[daen danmensen koiru bane]
異形断面ばねの一種で，材料の断面形状が卵形のコイルばね．
椭圆截面钢丝螺旋弹簧　[tuǒ yuán jié miàn gāng sī luó xuán tán huáng]
pegas ulir penampang bulat telur　[pugasu uriru punanpangu burato teruru]
타원 단면 선 코일 스프링　[tawon danmyeon seon koil spring]
สปริงขดหน้าตัดรูปไข่　[sa-pring khod naa tad ruup khai]

overaging
　aging that occurs at higher temperature or for longer time than the temperature or time necessary for maximum hardness and strength.
過時効　[kazikou]
硬さ，強さなどの性質が最高になる温度と時間よりも高い温度又は長い時間で起こる時効．
过时效　[guò shí xiào]
penuaan berlebih　[punuaan berurebihu]
과시효　[gwasihyo]
การแปรสภาพเกินเวลาที่กำหนด　[gaan praae sa-paap gern wee-laa tii gam-nod]

overall height
　in semi-elliptic spring, perpendicular distance, from the surface where compression stress is generated in use, of the smallest back-up leaf at the center pin or the center bolt, to the straight line connecting both eyes or connecting the load supporting points. In full-elliptic spring, distance between the most outer sides of the buckles. (see Fig O-4)
重ね板ばねの高さ　[kasaneita bane no takasa]
半だ円ばねでは，センタピン又はセンタボルトの位置の最小子板の使用時に，圧縮応力を生じる面から両目玉中心間又は荷重支持点間までを結んだ直線への垂直距離．だ円ばねでは，胴締めの最外側面間の距離．(Fig O-4参照)
重叠板弹簧的总高度　[chóng dié bǎn tán huáng de zǒng gāo dù]
tinggi keseluruhan　[tingugi keseruruhan]
판스프링의 높이　[panspringui nopi]
ความสูงแหนบทั้งชุด　[kwaam soong naaep tang shud]

overall length of torsion bar — overpeening

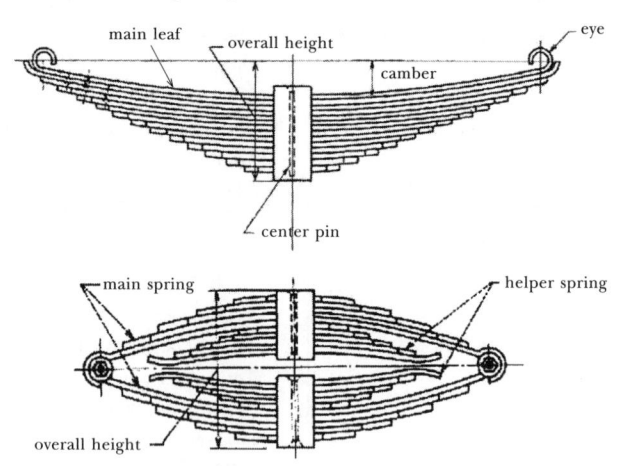

Fig O-4 overall height

overall length of torsion bar
トーションバーの全長　[toosyon baa no zentyou]
扭杆全长　[niǔ gǎn quán cháng]
panjang bantang puntir　[panjangu bantangu puntiru]
토션바의 전장　[tosyeonbaui jeonjang]
ความยาวรวมของทอร์ชั่นบาร์　[kwaam ya-o ro-um khong torsion bar]

overhaul
a maintenance procedure for machinery involving desassembly, replacing of damaged parts, and reassembly with functional inspection.
分解修理　[bunkai syuuri]
機械装置類を分解し，不良部品を交換し，再組立して機能テストを行うメンテナンス手順．
分解修理　[fēn jiě xiū lǐ]
overhaul　[overuhauru]
분해 수리　[bunhae suri]
โอเว่อร์ฮอลล์　[overhaul]

overheat
heating to the temperature high enough to impair metal properties.
過熱　[kanetu]
材料への諸性質が損傷されるほどの高温度まで加熱すること．
过热　[guò rè]
pemanasan berlebih　[pumanasan berurebihu]
과열　[gwayeol]
อุณหภูมิสูงเกินกำหนด　[un-na-ha-puum gern gam-nod]

overload
過負荷　[kahuka]
过载　[guò zǎi]
beban berlebih　[beban berurebihu]
과부하　[gwabuha]
โหลดเกินกำหนด　[load gern gam-nod]

overpeening
state processed by shot peening longer than the specified time. Shot peening has the optimal process time. Over time may decrease the improving effect of fatigue strength.
オーバーピーニング　[oobaa piiningu]
ショットピーニングの所定の投射時間

より長い時間をかけて，加工された状態．ショットピーニングには，最適加工条件がある．これより投射時間が長くなると，疲れ強さの向上効果が減少する可能性がある．
过度喷丸　[guò dù pēn wán]
peen berlebih　[peen berurebihu]
오버피닝　[obeopining]
การพ่นเม็ดโลหะมากเกินกำหนด　[gaan pon med loo-ha maak gern gam-nod]

oxidation
 a chemical reaction in which a substance is oxidized.
 酸化　[sanka]
 物質が酸素と化合する反応．
氧化　[yǎng huà]
oksidasi　[okusidasi]
산화　[sanhwa]
การทำปฏิกริยาของออกซิเจนกับธาตุอื่นๆ
[gaan tam pa-ti-ki-ri-ya khong taat u-un gab oxygen]

oxide film
 a layer of oxide formed on a surface of metal, such as iron and steel.
 酸化膜　[sankamaku]
 鉄鋼などの表皮部に形成された酸化物の膜をいう．
氧化膜　[yǎng huà mó]
lapisan tipis oksida　[rapisan tipisu okusida]
산화막　[sanhwamak]
แผ่นฟิล์มออกไซด์เคลือบผิว　[paaen film oxide klu-ab piw]

oxide inclusion
 metal oxide trapped in weld metal during solidification.
 酸化物の巻込み　[sankabutu no makikomi]
 凝固中の溶接金属の中に閉じ込められた金属酸化物．
氧化夹杂物　[yǎng huà jiā zá wù]
pemasukan oksida　[pumasukan okusida]
금속 산화 개재물　[geumsok sanhwa gaejaemul]
สารประกอบออกไซด์ของเหล็กในเหล็กเชื่อม
[saan pra-gorb oxide khong lek nai lek shi-um]

oxide layer
 see "oxide film".
 酸化物層　[sankabutu sou]
 "oxide film" 参照．
氧化层　[yǎng huà céng]
lapisan oksida　[lapisan oksida]
산화물 층　[sanhwamul cheung]
ชั้นของออกไซด์　[shan khong oxide]

P

parabolic cross section
    cross section of spring leaf with a half-moon shape.
    パラボリック断面 [paraborikku danmen] ばね板の断面で，半月形をしたばね板断面．
    抛物线截面 [pāo wù xiàn jié miàn]
    penampang parabola [punanpangu parabora]
    사다리꼴 단면 [sadarikkol danmyeon]
    หน้าตัดพาลาโบลิคในแหนบ [naa-tad palabolic nai naaep]

parabolic leaf spring
    leaf spring tapered for approximately entire length except for the clamping area to other components. Also called parabolic leaf spring because parabolic tapering is often employed in order to obtain more uniform stress distribution. (see Fig P-1)
    パラボリックリーフスプリング [paraborikku riihu supuringu]
    他部品との取付部分を除き，ほぼ全体にわたってテーパが施してある板ばね．より均一な応力分布を得るため，放物線状にテーパを施す場合もよくあるので，パラボリックリーフスプリングともいう．(Fig P-1 参照)
    抛物线板弹簧 [pāo wù xiàn bǎn tán huáng]
    pegas daun parabolik [pugasu daun paraboriku]
    파라보릭 판 스프링 [paraborik pan spring]
    แหบบชนิดพาลาโบลิค [naaep cha-nid palabolic]

parallel arrangement
    parallel arrangement of disc springs (stacked with concave and convex sides facing each other).
    ばねの並列配列 [bane no heiretu hairetu] 皿ばねを並列（凹面側と凸面側とを合わせて積み重ねたもの）に配置すること．
    弹簧的并列配合 [tán huáng de bìng liè zǔ hé]
    pengaturan sejajar [pungaturan sejajaru]
    스프링의 병렬 배열 [springui byeongnyeol baeyeol]
    การเรียงซ้อนกันของสปริงแผ่น [gaan riang sorn gan khong sa-pring paaen]

parallel combination of spring
    a combining way or springs whereby the force of the combined spring is multiplied by the number of springs.
    ばねの並列組合せ [bane no heiretu kumiawase]
    ばねの力がばねの個数倍になるように組み合わせる方法．
    并列组合压缩弹簧 [bìng liè zǔ hé yā suō tán huáng]
    pegas kombinasi paralel [pugasu konbinasi parareru]
    스프링의 병렬조합 [springui byeongnyeoljohap]
    สปริงขดเรียงอัดซ้อนกัน [sa-ring khod riang

Fig P-1 parabolic leaf spring

ad sorn gan]

parallel cut
　　the untreated cut end (cut perpendicularly) of back-up leaves of a leaf spring.
　　平かいさき　[hira kaisaki]
　　重ね板ばねの子板の端部を切断したままの形状（端面に直角）のこと．
　　平行切削　[píng xíng qiē xuē]
　　potongan sejajar　[potongan sejajaru]
　　평탄끝　[pyeongtankkeut]
　　ตัดขนาน　[tad kha-naan]

parallel stacking
　　see "parallel arrangement".
　　ばねの並列重ね合せ　[bane no heiretu kasaneawase]
　　"parallel arrangement" 参照．
　　并列组合弹簧　[bìng liè zǔ hé tán huáng]
　　susunan sejajar　[susunan sejajaru]
　　병렬 적층　[byeongnyeol jeokcheung]
　　สปริงวางเรียงซ้อนขนานกัน　[sa-pring waang riang sorn kha-naan gan]

parallel type coiled wave spring
　　a coiled wave spring made from two overlaid flat wire.
　　並列タイプコイルドウェーブスプリング　[heiretu taipu koirudo ueebu supuringu]
　　重ね合わせた 2 本の平線から作られた波形コイルスプリング．
　　并列螺旋波弹簧　[bìng liè luó xuán bō tán huáng]
　　pegas ulir gelombang pararel　[pugasu uriru gerobangu parareru]
　　병렬형 코일드 웨이브 스프링　[byeongnyeolhyeong koildeu weibeu spring]
　　สปริงรูปคลื่นขดชนิดวางขนานกัน　[sa-pring ruup klu-un cha-nid waang kha-naan gan]

parallelism
　　deviation from parallel position for a straight portion of a machine against reference straight line.
　　平行度　[heikoudo]
　　機械の直線部分が参照直線に対して平行から逸脱している程度．
　　平行度　[píng xíng dù]
　　kesejajaran　[kesejajaran]
　　평행도　[pyeonghaengdo]
　　ระดับการขนาน　[ra-dab gaan kha-naan]

Pareto diagram
　　a diagram showing grouped data arranged in descending order of occurrence frequency together with cumulative total.
　　パレート図　[pareeto zu]
　　項目別に層別して，出現度数の大きさの順に並べるとともに，累積和を示した図．
　　帕累托图　[pà lèi tuō tú]
　　diagram pareto　[diaguran pareto]
　　파레토 도　[pareto do]
　　แผนภูมิพาเรโต　[paaen puum pareto]

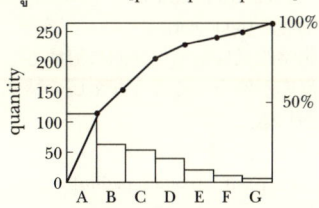

Fig P-2　Pareto diagram

partial decarburisation
　　a phenomenon in which carbon is partially or locally lost from an iron or steel surface when heated in an atmosphere that reacts with carbon.
　　局部脱炭　[kyokubu dattan]
　　炭素と反応する雰囲気の中で鉄鋼を加熱するとき，表面から部分的又は局部的に炭素が失われた現象をいう．
　　局部脱碳　[jú bù tuō tàn]
　　dekarburisasi sebagian　[dekaruburisasi

partial decarburisation — pearlite

sebagian]
국부 탈탄 [gukbu taltan]
การสูญเสียคาร์บอนเป็นส่วนๆ [gaan soon siaa carbon pen suan saun]

passive corrosion protection

a corrosion protection technology to create passive state layer on metal surface.
不動態防食 [hudoutai bousyoku]
金属表面に不動体を生成させて行う防食技術.
钝化防蚀 [dùn huà fáng shí]
perlindungan korosi pasif [pururindungan korosi pasihu]
부동태 부식방지 [budongtae busikbangji]
การป้องกันการกัดกร่อนทางอ้อม [gaan ponggan gaan gad-gron tang orm]

passive state

a special surface condition of metal under which chemical or electrochemical reaction ceases.
不動態 [hudoutai]
化学的又は電気化学的に溶解もしくは反応が停止するような金属の特殊な表面状態.
不动态 [bù dòng tài]
kondisi pasif [kondisi pasihu]
부동태 [budongtae]
ผิวเหล็กชนิดพิเศษไม่เคลื่อนไหว [piw lek cha-nid pi-ses mai klu-an wai]

patenting

a heat treatment for wire whereby fine pearlite structure is obtained. The wire is heated to up above A_{c3} transformation temperature and held at the temperature for a certain time, then quenched to temperature just below A_{r1} transformation temperature then goes through isothermal transformation.
パテンチング [patentingu]
線材をA_{c3}変態点以上に加熱保持した後, A_{r1}変態点以下で恒温変態させて微細なパーライト組織を得る熱処理.
正火 [zhèng huǒ]
mematenkan [mematenkan]
파텐팅 [patenting]
พาเทนติ้ง [patenting]

pay-off speed

a speed at which the wire is fed from a reel stand to a coiling machine.
繰出し速度 [kuridasi sokudo]
コイリングマシンに送られるときの線材の送り込まれる速度.
送料速度 [sòng liào sù dù]
kecepatan pengulangan [kesepatan pungurangan]
연속 선재 공급 속도 [yeonsok seonjae gonggeup sokdo]
ความเร็วส่ง [kwaam rew song]

peak stress

maximum stress caused by stress concentration of notch.
ピーク応力 [piiku ouryoku]
ノッチ部への応力集中の結果生じる最大応力.
峰值应力 [fēng zhí yīng lì]
tegangan puncak [tegangan punkaku]
피크응력 [pikeueungnyeok]
ความเค้นสูงสุด [kwaam ken soong sud]

pearlite

lamellar structure of ferrite and cementite formed in eutectoid transformation when austenite is cooled.
パーライト [paaraito]
オーステナイトの冷却に際し, 共析変態に生じたフェライトとセメンタイトの層状組織.
最大负荷指示 [zuì dà fù hè zhǐ shì]
perlit [perurito]

퍼얼라이트 [peoeollaiteu]
โครงสร้างเหล็กกระยะเพียร์ไลต์ [krong-sang lek ra-ya pearlite]

peel test
a test to peel a weld zone and check the broken state. This is applied for spot welding, plug welding, brazing, etc.
はくり試験 [hakuri siken]
溶接部を引きはがしその破断状態を調べる試験．スポット溶接，プラグ溶接，ろう溶接などに適用される．
剥离试验 [bāo lí shì yàn]
uji pengupasan lapisan [uji pungupasan rapisan]
박리 시험 [bangni siheom]
การทดสอบสภาพการแตกโดยการลอกผิวบริเวณเชื่อม [gaan tod-sorb sa-paap gaan taaek dooi gaan lork piw bo-ri-wen shi-um]

peening barrel
rotating drum for shot peening.
ピーニングドラム [piiningu doramu]
ショットピーニングを施すときの，回転式のドラム形の躯体．
喷丸滚筒 [pēn wán gǔn tǒng]
tong peen [tongu peen]
피닝 드럼 [pining deureom]
ถังบรรจุเม็ดเหล็ก [tang ban-chu med lek]

peening effect
the effects of improved machining characteristics generated by shot peening. These effects include improvements of residual stress in the machined layer, fatigue strength due to work-hardened layer, resistance to abrasion and so on.
ピーニング効果 [piiningu kouka]
ショットピーニングにより発生する被加工の特性を向上させる効果のこと．効果は，加工層の圧縮残留応力や加工硬化層に起因する疲れ強さ，耐摩耗性の向上などがある．
表面强化效应 [biǎo miàn qiáng huà xiào yīng]
efek peen [efeku peen]
피닝 효과 [pining hyogwa]
ผลจากการทำพีนนิ่ง [pon chaak gaan tam peening]

peening nozzle
blasting nozzle of shot for shot peening.
ピーニングノズル [piiningu nozuru]
ショットピーニングを施すときの，ショットを投射する噴射口．
喷丸嘴 [pēn wán zuǐ]
penyembur peen [punyenburu peen]
피닝 노즐 [pining nojeul]
หัวพ่นเม็ดเหล็ก [ho-a pon med lek]

peening speed
the speed at which shot is ejected in shot peening. This is a major factor in determining peening strength. Speeds from 30 m/s to 80 m/s are normally used. High-speed airflow system and high-speed rotary vane system are used.
投射速度 [tousya sokudo]
ショットピーニングのショットを投射する速度で，ピーニング強度に大きく影響する因子であり，通常30〜80 m/sで使用されている．加速方式に高速空気流式，高速回転翼車などがある．
喷射速度 [pēn shè sù dù]
kecepatan peen [kechepatan peen]
투사 속도 [tusa sokdo]
ความเร็วในการพ่น [kwaam rew nai gaan pon]

peening time
ピーニング時間 [piiningu zikan]
喷射时间 [pēn shè shí jiān]

waktu peen   [wakutu peen]
피닝 시간   [pining sigan]
ระยะเวลาในการทำพีนนิ่ง   [ra-ya wee-laa nai gaan tam peening]

perforation erosion
dents created on material surface by physical or chemical attack.
穴状浸食   [anazyou sinsyoku]
材料の表面が物理的又は化学的現象によって腐食・浸食してできた穴状のへこみ.
穴状腐蚀   [xué zhuàng fǔ shí]
erosi karena pelubangan   [erosi karena purubangan]
공상 침식   [gongsang chimsik]
การกัดเซาะเป็นรู   [gaan gad sa-or pen ruu]

period of vibration
the minimum interval of a state to reveal itself in the same level within the same period under a vibrating condition.
振動の周期   [sindou no syuuki]
同一時間ごとに同一の状態が繰り返される場合，同じ状態が繰り返される時刻の間の最小の間隔をいう.
振动周期   [zhèn dòng zhōu qī]
selang waktu getaran   [serangu wakutu getaran]
진동 주기   [jindong jugi]
รอบของการสั่นสะเทือน   [rorb khong gaan san sa te-un]

periodic check
a check on an equipment in a designated interval for preventive maintenance.
定期点検   [teiki tenken]
予防保全のために，一定の時間間隔で行う設備点検.
定期检查   [dìng qī jiǎn chá]
pengecekan berkala   [pungechekan berukara]
정기 점검   [jeonggi jeomgeom]
การตรวจเช็คเป็นระยะตามกำหนด   [gaan truat check pen re ya tam gam-nod]

permanent deformation
the state in which a spring, rubber material, or synthetic resin remains deformed and does not return to its original shape after removing the load applied.
永久変形   [eikyuu henkei]
ばね，ゴム材料，合成樹脂などにある変形を与え，次いで負荷を取り去り放置しても，完全には原形に戻らないで残る状態.
永久変形   [yǒng jiǔ biàn xíng]
peruahan bentuk tetap   [puruahan bentuku tetapu]
영구변형   [yeonggubyeonhyeong]
การเปลี่ยนรูปอย่างถาวร   [gaan pli-an ruup yaang taa-worn]

permanent load
a constant load that does not change its magnitude and direction during a service life.
不変荷重   [huhen kazyuu]
時間的に大きさ，方向ともに変化しない荷重.
静載荷   [jìng zǎi hè]
beban tetap   [beban tetapu]
불변 하중   [bulbyeon hajung]
โหลดคงที่   [load kong tii]

permanent set in fatigue
a permanent deformation of a spring caused by repeated load.
疲労変形   [hirou henkei]
ばねが繰り返し荷重を受けたときに生じる永久変形.
疲劳変形   [pí láo biàn xíng]
keadaan tetap dalam kelelahan   [keadaan

tetapu daran kererahan]
피로 변형 [piro byeonhyeong]
การเปลี่ยนรูปถาวรจากการโหลดซ้ำ [gaan plian ruup taa-worn chaak gaan load sam]

permanent strain
    residual strain caused by permanent deformation.
    永久ひずみ [eikyuu hizumi]
    永久変形によって生じる残留ひずみ.
    永久応変 [yǒng jiǔ yīng biàn]
    regangan tetap [regangan tetapu]
    영구 변형 [yeonggu byeonhyeong]
    ความเครียดตกค้างถาวร [kwaam kri-ead tok-kaang taa-worn]

permeability
    flux density (B) and magnetic field (H) ratio (B/H). This is quantity to indicate magnetic characteristic of a magnetic material, and normally initial permeability and maximum permeability are used.
    透磁率 [touziritu]
    磁束密度(B)と磁界(H)との比(B/H). 磁性体の磁気特性を示す量で, 一般に初透磁率と最大透磁率が用いられる.
    导磁性 [dǎo cí xìng]
    permeabilitas [purumeabiritasu]
    투자율 [tujayul]
    คุณสมบัติการเป็นแม่เหล็ก [kun-na-som-bat gaan pen maae-lek]

perpendicularity
    the degree of deviation of a line from a perpendicular position against reference line.
    直角度 [tyokkakudo]
    直線と直線, 直線と平面, あるいは平面と平面の組合せにおいて, どちらか一方を基準として直角な直線, あるいは平面の狂いの大きさをいう.
    垂直度 [chuí zhí dù]

kesikuan [kesikuan]
직각도 [jikgakdo]
ตั้งฉาก [tang shaak]

pH value
    an abbreviation of potential hydrogen. It indicates the degree of acidity or alkalinity of water. Normally it is called pH value, and pH 7 is neutrality.
    ピーエッチ値（ペーハー値） [piietti ti (peehaa ti)]
    Potencial Hydlogen の略で, 水などの酸性やアルカリ性の度合いを示す数値. 水素イオン濃度の逆数の常用対数. 一般にペーハー値といい pH 7 を中性といっている.
    pH 值 [pH zhí]
    nilai ph [nirai pH]
    페하값 [pehagap]
    ค่าพีเอช [kaa PH]

phosphate coating
    the process of treating metal with a solution, usually of phosphoric acid and soluble phosphate to form an insoluble phosphate film on its surface. Phosphate coating is used for surface preparation for painting, lubrication for forging, or temporary rust prevention, and zinc-based, manganese-based and iron-based phosphate coatings are available.
    りん酸塩皮膜処理 [rinsan'en himaku syori]
    りん酸及び可溶性りん酸塩を主体とする水溶性液で金属を処理し, その表面に不溶性りん酸塩皮膜をつくる処理. 塗装の下地, 鍛造の潤滑, 一時的防せいなどに用いられ, 亜鉛系, マンガン系, 鉄系がある.
    磷化处理 [lín huà chǔ lǐ]
    pelapisan pospat [purapisan posupato]

phosphate coating — pickling inhibitor

인산염 피막 처리 [insannyeom pimak cheori]
การชุบฟอสเฟต [gaan shup phosphate]

phosphor bronze for spring

a typical Cu-Sn-based spring material that has long been used. Phosphorus is added for deoxidizing, and a content of 0.03% to 0.35% is thought to increase hardness and wear resistance. It provides better spring characteristics than brass, and is a spring material treated through low-temperature annealing. It is also used for switches and relays.

ばね用りん青銅 [baneyou rinseidou]
りん青銅は，Cu-Sn 系の代表的なばね材料で古くから用いられている．りんは脱酸のために添加されるものであるが 0.03 ～ 0.35% の含有は硬さや耐摩耗性を高めるといわれている．ばね性は黄銅より優れ，低温焼なまし型ばね材料である．用途はスイッチ，リレーなどである．

弹簧用青铜 [tán huáng yòng qīng tóng]
perunggu pospor untuk pegas [purungugu posuporu untuku pugasu]
스프링용 인청동 [springyong incheongdong]
โลหะผสมฟอสฟอรัส-ทองเหลืองสำหรับทำสปริง [lao-ha pa-som phosporus - tong le-arng sam-rab tam sa-pring]

physical vapor deposition coating (PVD)

the formation of a thin film on the surface of an object by depositing physically vaporized materials.

物理蒸着法（PVD） [buturi zyoutyakuhou (piibuidjii)]
物理的方法で物質を蒸発させ，対象部品表面に薄膜を形成すること．

物理蒸镀法 [wù lǐ zhēng dù fǎ]
pelapisan deposisi uap fisika [purapisan deposisi uapu fisika]
물리 증착법 (PVD) [mulli jeungchakbeop (PVD)]
การเคลือบไอน้ำ [gaan klu-ab ai-naam]

piano wire

steel wire drawn from cold-working piano wire material after normal patenting. It is used in high quality springs and tire cord. Also called music wire.

ピアノ線 [pianosen]
ピアノ線材を用い，通常パテンチング後伸線など冷間加工し仕上げられた鋼線．高級ばねやタイヤコードなどの製品に用いられる．ミュージックワイヤともいう．

琴钢丝 [qín gāng sī]
kawat piano [kawato piano]
피아노선 [pianoseon]
สายเปียโน [saai piano]

pickling

cleaning by immersing in acidic solution to remove mill scale and corrosion on metal parts. Hydrochloric acid and sulfuric acid are usually used.

酸洗い [san'arai]
金属製品のミルスケールやさびの層を除くために酸性の溶液に浸して生地を清浄にすること．一般的には塩酸，硫酸などが用いられている．

酸洗 [suān xǐ]
pengasaman [pungasaman]
산세척 [sansecheok]
การล้างด้วยกรด [gaan laang du-oi grod]

pickling inhibitor

an additive to prevent rapid or localized chemical or electrochemical reaction during pickling.

酸洗抑制剤 [sansen yokuseizai]
酸洗時の化学反応又は電気化学反応の

急激もしくは局部的な進行を妨げる添加剤.
酸洗抑制剤　[suān xǐ yì zhì jì]
penghalang pengasaman　[punguharangu pungasaman]
산세척 억제제　[sansecheok eokjeje]
สารเติมเพื่อป้องกันสารเคมีอย่างรวดเร็ว　[saan term pe-ua pong-gan saan ke-mii yaang ru-ad rew]

pickling process
　the process to clean and remove mill scale or corrosion on materials (before rolling) and products (before plating).
酸洗工程　[sansen koutei]
　材料（圧延前など）や製品（めっき前など）などのスケール又はさびなどを除去し，清掃するために行う工程．
酸洗过程　[suān xǐ guò chéng]
proses pengasaman　[purosesu pungasaman]
산세척 공정　[sansecheok gongjeong]
ขบวนการล้างด้วยกรด　[kha-buan gaan laang du-oi grod]

pickling test
　a test to check for surface defects such as flaws, cracks, and wrinkles using an acidic solution such as hydrochloric acid with an oxidation inhibitor added.
酸洗い検査　[san'arai kensa]
　酸化抑制剤を添加した塩酸などの酸液を使用して，主として，表面欠陥のきず，割れ，しわなどを調べる試験．
酸洗检查　[suān xǐ jiǎn chá]
uji pengasaman　[uji pungasaman]
산세척 검사　[sansecheok geomsa]
การทดสอบรอยที่ผิวด้วยกรด　[gaan tod-sorb roi tii piw du-oi grod]

piercing punch
　a punch used to pierce part of a workpiece (using a press).

孔あけポンチ　[anaake ponti]
　材料などの一部に孔をあけるためのパンチ（プレスなどで行う）．
冲孔　[chōng kǒng]
punch pelubang　[punti purubangu]
천공 펀치　[cheongong peonchi]
การเจาะรู　[gaan cho-a ruu]

pigtail end
　coil end which is wound with reduced radius.
ピッグテールエンド　[piggu teeru endo]
　圧縮コイルばね端部の一種で，端末がコイル径方向内側に巻き込まれた形状．
猪尾弹簧　[zhū wěi tán huáng]
ujung kuncir　[ujungu kunsiru]
피그테일 끝 (돼지 꼬리 끝)　[pigeuteil kkeut (dwaeji kkori kkeut)]
ปลายขดแบบหางหมู　[plaai khod baaep haang muu]

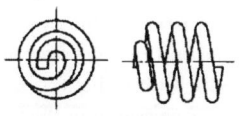

Fig P-3　pigtail end

piston ring
　a notched metal ring which has elasticity in radial direction. It fits in a piston groove.
ピストンリング　[pisuton ringu]
　外周方向へばね性を持ち，合い口を持つ金属製のリング．ピストンの溝にはめて使用する．
活塞环　[huó sāi huán]
cincin torak　[chinchin toraku]
피스톤 링 스프링　[piseuton ring spring]
สปริงแหวนลูกสูบ　[sa-pring waaen luuk-suub]

pit
　a dent on the surface of metal plate.
くぼみ　[kubomi]

pit — pitch correction

金属表面にあるくぼみ．
凹坑　[āo kēng]
lubang　[rubangu]
피트　[piteu]
หลุมบนผิวแผ่นเหล็ก　[lum bon piw paaen lek]

pit depth
くぼみ深さ　[kubomi hukasa]
凹坑深度　[āo kēng shēn dù]
kedalaman lubang　[kedaraman rubangu]
피트깊이　[piteu gipi]
ความลึกหลุม　[kwaam luk lum]

pit diameter
くぼみ径　[kubomi kei]
凹坑直径　[āo kēng zhí jìng]
diameter lubang　[diameteru rubangu]
피트지름　[piteu jireum]
เส้นผ่าศูนย์กลางหลุม　[sen-pa-soon-glaang lum]

pitch
distance between the centers of the cross-section of adjacent coils in the plane including the axis of the coil spring, as measured in the axial direction.

ピッチ　[pitti]
コイルばねの中心線を含む断面で，互いに隣り合うコイルの中心線に平行な

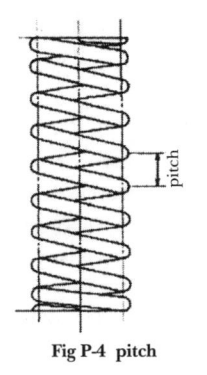

材料断面の中心間距離．
节距, 齿距　[jié jù, chǐ jù]
jarak antara　[jaraku antara]
피치　[pichi]
พิทช์　[pitch]

pitch angle
angle made by the centerline of the material of a coil spring in relation to the plane perpendicular to the centerline of the coil spring.

ピッチ角　[pitti kaku]
コイルばねの材料の中心線がばねの中心線に直角な平面となす角．
螺旋角　[luó xuán jiǎo]
sudut jarak antara　[suduto jaraku antara]
피치각　[pichigak]
มุมพิทช์　[mum pitch]

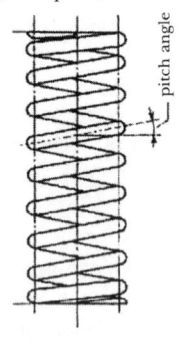

Fig P-5　pitch angle

pitch correction
the correction of coil pitch after coiling operation.

ピッチ修正　[pitti syuusei]
コイルばねなどのピッチをコイリング時に修正又はコイリング後に所定のピッチに修正すること．
节距修正　[jié jù xiū zhèng]
koreksi jarak antara　[korekusi jaraku antara]
피치 수정　[pichi sujeong]

Fig P-4　pitch

การแก้ไขพิทช์ [gaan gaae khai pitch]
pitch error
difference between actual pitch and specified pitch of a coil spring. It is positive if larger than the specification or negative if smaller. Normally it is for one pitch but may be for two or more pitches.
ピッチ誤差 [pitti gosa]
製品の実際のピッチと規定上のピッチとの差をいい，規定より大きい場合は正，小さい場合は負とする．一般に1ピッチについていうが，2ピッチ以上についていうこともある．
节距误差 [jié jù wù chā]
kesalahan jarak antara [kesarahan jaraku antara]
피치 오차 [pichi ocha]
ความผิดปกติของพิทช์ [kwaam pid-pak-ka-ti khong pitch]

pitch tool
tool which provides the pitch when forming the coil spring. There are two types of the tool. One is the finger type which provides the pitch by extruding the coil, and the other is the wedge type which provides the pitch by pushing in the wedge-shaped tool.
ピッチツール [pitti tuuru]
コイルばねをコイリングマシンで成形するとき，ピッチを与えるために使用する工具．コイルを押し出してピッチ付けを行うフィンガ式と，くさび形工具を押し込んでピッチ付けを行うくさび式とがある．
节距刀 [jié jù dāo]
alat jarak antara [arato jaraku antara]
피치 툴 [pichi tul]
อุปกรณ์ทำพิทช์ [up-pa-gorn tam pitch]

pitch wedge
a tool with wedge-shaped tip, used for coiling a piched spring. Different types of tools (e.g., right-handed, left-handed winding, fixed and adjustable types) are available, each with different end profiles.
くさび形ピッチツール [kusabigata pitti tuuru]
ピッチあきのばねを加工するときに使用する先端がくさび形の工具．右巻用と左巻用，固定式と調整式などがありそれぞれ先端の形状が異なっている．
楔形块节距刀 [xiē xíng kuài jié jù dāo]
ropemecah jarak antara [ropemekahu jaraku antara]
쐐기형 피치 툴 [sswaegihyeong pichi tul]
อุปกรณ์พิทช์ชนิดลิ่ม [up-pa-gorn pitch cha-nid lim]

pitting corrosion
hole-shaped corrosion progressing to the inside of the metal from the local corrosion.
孔食 [kousyoku]
局部腐食が金属内部に向かって，孔状に進行する腐食．
点蚀 [diǎn shí]
berlubang karena korosi [berurubangu karena korosi]
공식 [gongsik]
การกัดกร่อนเป็นหลุม [gaan gad gron pen lum]

planetary gear
遊星歯車 [yuusei haguruma]
行星齿轮 [xíng xīng chǐ lún]
roda gigi planet [roda gigi puraneto]
유성 치차 [yuseong chicha]
เกียร์ที่ใช้เฟืองเล็กหมุนรอบเฟืองใหญ่ [gear tii chai fe-ung lek mun rorb fe-ung yai]

plastic analysis

an analysis to study loads to cause stress, strain, deformation, and rupture in consideration of plastic characteristics including strain hardening of material.

塑性解析　[sosei kaiseki]
材料のひずみ硬化性を含む塑性挙動を考慮して，応力，ひずみ，変形及び破壊を生じる荷重などを求める解析をいう．
塑性解析　[sù xìng jiě xī]
analisa plastis　[anarisa purasutisu]
소성 해석　[soseong haeseok]
การวิเคราะห์คุณสมบัติการเป็นพลาสติค　[gaan wi kra-or kun-na-som-bat gaan pen plastic]

plastic coating

coating with synthetic resin. For wire harness clamp etc., the parts may be partially covered with plastic tubes for preventing damage.

プラスチックコーティング　[purasutikku kootjingu]
品物などに合成樹脂を被せること．ワイヤハーネスクランプなどでは損傷防止のため部品の一部分にプラスチックチューブを被せる場合もある．
涂塑工艺　[tú sù gōng yì]
pelapisan plastis　[purapisan purasutisu]
플라스틱 코팅　[peullaseutik koting]
การชุบพลาสติค　[gaan shub plastic]

plastic deformation

a type of deformation that does not allow the shape of an object return to its original form after the force is removed.

塑性変形　[sosei henkei]
物体に力を加えて変形させた後，力を取り除いても原形に戻らないこと．
塑性变形　[sù xìng biàn xíng]
bentukan plastis　[bentukan purasutisu]
소성 변형　[soseong byeonhyeong]
การผิดรูปชนิดไม่คืนตัว　[gaan pid ruup cha-nid mai kuun tu-a]

plastic range

the range of deformation under the stress exceeding the elastic limit. In plastic range, the shape of an object does not return to its original form even after the force applied is removed.

塑性域　[soseiiki]
弾性限度以上の応力における変形の範囲のことで，応力を取り除いても原形に戻らなくなる領域をいう．
塑性范围　[sù xìng fàn wéi]
rentang plastis　[rentangu purasutisu]
소성역　[soseongyeok]
ช่วงการเปลี่ยนรูปอย่างถาวร　[shu-ong gaan plian ruup yaang taa-worn]

plastic shot

a hard plastic particle to be used in shot peening process.

樹脂系ショット　[zyusikei syotto]
ショットピーニングに用いられる硬い樹脂の粒子．
树脂丸粒　[shù zhī wán lì]
tembakan dengan plastik　[tenbakan dengan purasutiku]
수지계 쇼트　[sujigye syoteu]
การช็อตด้วยเม็ดพลาสติค　[gaan shot du-oi med plastic]

plastic spring

プラスチックばね　[purasutikku bane]
塑料弹簧　[sù liào tán huáng]
pegas plastis　[pugasu purasutisu]
플라스틱 스프링　[peullaseutik spring]
สปริงพลาสติค　[sa-pring plastic]

plate cam

a type of plane cams. The motion of

the follower is determined by the outer periphery of the plate. From the shape of cam, it is also called heart cam or disc cam.

板カム　[ita kamu]
　平面カムの一種，従節の運動が板の外周で決められるカム．カムの形からハートカム又はディスクカムとも呼ばれている．
平板凸轮　[píng bǎn tū lún]
nok pelat　[noku perato]
판 캠　[pan kaem]
แคมแผ่น　[cam paaen]

Poisson distribution
　a distribution of provability for the number of occurence when independent events are repeated.
ポアソン分布　[poason bunpu]
　独立の事象を繰り返したとき，ある事象の出現する確率が表される分布．
泊松分布　[bó sōng fēn bù]
penyebaran poison　[punyebaran poison]
프아송 분포　[peuasong bunpo]
การกระจายของพอยซ์ซัน　[gaan gra-chaai khong Poisson]

Poisson's ratio
　the ratio of longitudinal strain and lateral strain of a material when loaded in the longitudinal direction within the elastic limit.
ポアソン比　[poason hi]
　材料の弾性限界内で縦方向に荷重を加えたとき，縦ひずみと横ひずみとの比．
泊松比　[bó sōng bǐ]
rasio poisson　[rasio poisuson]
프아송비　[peuasongbi]
อัตราส่วนของพอยซ์ซัน　[at-tra suan khong Poisson]

population
　the group consisting of all specimens that possess the characteristics being investigated.
母集団　[bosyuudan]
　考察の対象となる特性を持つすべてのものの集団．
总体　[zǒng tǐ]
populasi　[popurasi]
모집단　[mojipdan]
ประชากร　[pra-shaa-gorn]

population mean
　the average of all items in a population.
母平均　[boheikin]
　母集団の平均．
总体平均　[zǒng tǐ píng jūn]
rata-rata populasi　[rata-rata popurasi]
모집단 평균　[mojipdan pyeonggyun]
ค่าเฉลี่ยของประชากร　[kaa sha-liaa khong pra-sha-gorn]

positioning
位置決め　[itigime]
定位　[dìng wèi]
penempatan　[punenpatan]
위치 결정　[wichi gyeoljeong]
การกำหนดตำแหน่ง　[gaan gam nod tam naaeng]

post-annealing
　annealing to remove residual stress after welding or a heavy plastic deformation.
後焼なまし　[ato yakinamasi]
　溶接や過酷な成形の後に，残留応力を除去するために行う後熱処理．
二次退火　[èr cì tuì huǒ]
pasca anneal　[pasuka annearu]
응력 제거 풀림　[eungnyeok jegeo pullim]
การอบเพื่อลดความเค้นตกค้างหลังเชื่อมหรือการเปลี่ยนรูปถาวร　[gaan ob pe-ua lod kwaam ken tok-kaang lang shi-um]

post-treatment
　a secondary process which is necessary

post-treatment — precision measurement

after the primary technological treatment.
後処理　[atosyori]
主要な工程の後で必要となる2次的な工程．
后处理　[hòu chǔ lǐ]
pasca perlakuan　[pasuka pururakuan]
후처리　[hucheori]
กระบวนการหลัง　[gra-buan gaan lang]

powder coating
a method of coating that uses synthetic resin-based powder paint containing neither water nor solvent, which are applied to surfaces electrostatically or using heat, and then hot cured to form a coating film.
粉体塗装　[huntai tosou]
表面に，水及び溶剤を含まない合成樹脂系の粉末状の塗料を静電気又は熱によって付着させ，加熱硬化して塗膜を形成させる方式．
粉末涂层　[fěn mò tú céng]
pelapisan serbuk　[purapisan serubuku]
분체 도장　[bunche dojang]
การเคลือบสีฝุ่น　[gaan klu-ab sii fun]

power spring
see "mortor spring".
ぜんまい　[zenmai]
"mortor spring"の項参照．
发条　[fā tiáo]
pegas tenaga　[pugasu tenaga]
태엽　[taeyeop]
สปริงเพาเวอร์　[sa-pring power]

power train
a vehicle mechanism to connect the engine to driven axle. It include drive shaft, clutch, transmission, and differential gear.
伝導機構　[dendou kikou]
自動車のエンジンと駆動車軸を連結する機構．ドライブシャフト，クラッチ，トランスミッション，差動ギヤからなる．
传动机构　[chuán dòng jī gòu]
penghantaran tenaga　[punguhantaran tenaga]
동력전달 장치　[dongnyeokjeondal jangchi]
กลไกที่เชื่อมเครื่องจักรกับเพลาขับ　[kon-kai tii shi-um kru-ang chak gap plao-khab]

precipitation
the phenomenon whereby crystals in a different phase separate and grow from solid solution.
析出　[sekisyutu]
固溶体から異相の結晶が分離成長する現象．
析出　[xī chū]
pengendapan　[pungendapan]
석출　[seokchul]
การตกตะกอน　[gaan tok ta-gorn]

precipitation hardening
hardening caused by the precipitation of particles like carbide and intermetallic compounds from supersaturated solid solution.
析出硬化　[sekisyutu kouka]
過飽和固溶体から炭化物，金属間化合物などの異相が析出するために起こる硬化．
析出硬化　[xī chū yìng huà]
pengerasan melalui pengendapan　[pungerasan merarui pungendapan]
석출 경화　[seokchul gyeonghwa]
ทำให้แข็งโดยการตกตะกอน　[tam-hai khaaeng dooi gaan tok-ta-gorn]

precision measurement
精密測定　[seimitu sokutei]
精密測量　[jīng mì cè liáng]
pengukurran tepat　[pungukururan

tepato]
정밀 측정 [jeongmil cheukjeong]
การวัดละเอียด [gaan wad la-iad]

preheating
preor heating in preparation for welding or thermal cutting.
予熱 [yonetu]
溶接又は熱切断の操作に先立って母材に熱を加えること．
预热 [yù rè]
pemanasan awal [pumanasan awaru]
예열 [yeyeol]
การอุ่นให้ความร้อนก่อนปฏิบัติการ [gaan un hai rorn gorn pa-ti-bad-gaan]

preheating furnace
a furnace to operate with lower elevated temperature to be placed in front of high temperature furnace.
予熱炉 [yoneturo]
材料などを高温加熱する必要がある場合，はじめ比較的低温で加熱する炉．
预热炉 [yù rè lú]
tungku pemanasan awal [tunguku pumanasan awaru]
예열로 [yeyeollo]
เตาเผาอุ่นให้ความร้อนต่ำ [ta-o pa-o un hai kwaam rorn tam]

press
a machine by which pressure is applied to a work piece.
プレス [puresu]
被加工物に圧力を加える機械．
冲压 [chōng yā]
pres [puresu]
프레스 [peureseu]
เครื่องเพรส [kru-ang press]

press quenching
quenching carried out in the pressed condition to prevent quenching distortion.
プレスクエンチング [puresu kuentingu]
焼入変形を規制するために，プレスした状態で行う焼入れ．
加圧淬火 [jiā yā cuì huǒ]
kejut kempa [kejuto kenpa]
프레스 퀜칭 [peureseu kwenching]
การชุบแข็งภายใต้การกดอัด [gaan shub-khaaeng paai taai gaan god ad]

press tempering
tempering carried out in the pressed condition.
プレステンパ [puresu tenpa]
プレスした状態で行う焼戻し．
加圧回火 [jiā yā huí huǒ]
sepuh keras kempa [sepuhu kerasu kenpa]
프레스 템파 [peureseu tempa]
การอบภายใต้การกดอัด [gaan ob paai taai gaan god ad]

pressing bend method
a method of bending test in which the sample is placed on two supports and gradually pressed into a specified shape by applying a pushing metal tool at its center.
押曲げ法 [osimage hou]
曲げ試験の一種で，試験片を2個の支えに載せ，その中央に押金具を当てて，徐々に荷重を加えて規定の形に曲げる試験方法．
压弯试验法 [yā wān shì yàn fǎ]
metoda benkok tekan [metoda benkoku tekan]
프레스 굽힘시험법 [peureseu gupimsiheombeop]
วิธีการทดสอบการโค้งงอด้วยการกด [wi-tii gaan tod-sorb gann koong ngor du-oi gaan god]

pressure control system
a system to maintain, raise, or lower

the pressure in a hydraulic machine.
圧力制御系 [aturyoku seigyokei]
油圧機械の圧力を上下させたり保持するシステム．
压力控制系统 [yā lì kòng zhì xì tǒng]
rangkaian kontrol tekanan [rangukaian kontroru tekanan]
압력제어 회로 [amnyeokjeeo hoero]
ระบบการควบคุมแรงดัน [ra-bob gaan ku-ab kum raaeng dan]

pressure plate
    a piece of plate to press a spring or a piece of plate to secure the material in press die setting.
押さえ板 [osaeita]
ばねを押さえる板やプレス金型の素材押さえ板など．
压板 [yā bǎn]
pelat tekanan [purato tekanan]
누름 판 [nureum pan]
แผ่นรับแรงดัน [paaen rab rang dan]

pretension
    the internal force of a helical extension spring with no load exerted.
初張力 [syotyouryoku]
無荷重時に密着している引張コイルばねの内力．
初拉力 [chū lā lì]
tarikan awal [tarikan awaru]
초장력 [chojangnyeok]
แรงดึงตั้งต้นขณะไม่มีโหลด [rang dung tang ton kha-n amai-mii load]

pretreatment
    a prior treatment to be made on the surface of work in preparation of a primary surface treatment.
前処理 [maesyori]
表面処理の主工程の前に施す準備的な処理．
前处理 [qián chǔ lǐ]

olahan awal [orahan awaru]
전처리 [jeoncheori]
การเตรียมผิวของชิ้นงาน [gaan triam piw khong shin-ngaan]

prevention of pollution
    an activity to prevent air and water from being contaminated with particular substances.
汚染の予防 [osen no yobou]
空気や水などが特定の物質により汚染されることを防ぐための活動．
预防污染 [yù fáng wū rǎn]
pencegahan polusi [punchegahan porusi]
오염 예방 [oyeom yebang]
การป้องกันมลพิษ [gaan pong gan mon-la-pid]

preventive maintenance
    a well planned maintenance to prevent machinery and equipment from getting into trouble.
予防保全 [yobou hozen]
設備，機器などの使用中の故障を未然に防止するために計画的に行う保守．
维护保养 [wéi hù bǎo yǎng]
perawatan pencegahan [purawatan punchegahan]
예방 보전 [yebang bojeon]
การบำรุงรักษาเชิงป้องกัน [gaan bam-rung lak-sa she-arng pong-gan]

primary stress
    a normal stress or shear stress necessary to satisfy the law of balance against an imposed load.
主応力 [syouryoku]
負荷に対して釣合いの法則を満足するのに必要な垂直応力又はせん断応力．
一阶应力 [yī jiē yīng lì]
tegangan utama [tegangan utama]
주 응력 [ju eungnyeok]
ความเค้นปฐมภูมิ [kwam ken pa-tom-puum]

primary winding
> process to form a power spring which provides the material with a curvature.
> 一次巻　[itizimaki]
> ぜんまいの成形法で，材料に一定方向の曲率を与える加工．
> 初巻　[chū juǎn]
> gulungan utama　[gurungan utama]
> 일차 감기 (태엽 스프링)　[ilcha gamgi (taeyeop spring)]
> การม้วนเริ่มต้นในการผลิตเพาเวอร์สปริง　[gaan muan rerm-ton nai gaan pa-lit power sa-pring]

priming coat
> the first wating to be applied on the surface of a work. It increases the adhesion strength of secondary coating.
> 下地塗装　[sitazi tosou]
> 対象物の表面に最初に施す塗装．次工程の塗装の付着強さを高める．
> 底层涂漆　[dǐ céng tú qī]
> lapisan dasar　[rapisan dasaru]
> 하부 도장　[habu dojang]
> การเคลือบสีพื้น　[gaan klu-ab sii peuun]

printed circuit
> a conductive pattern printed on the surface of an insulating base.
> 印刷回路　[insatu kairo]
> 絶縁基板上に印刷された導体のパターン回路．
> 印刷回路　[yìn shuā huí lù]
> rangkaian tercetak　[rangukaian teruchetaku]
> 인쇄 회로　[inswae hoero]
> รูปแบบวงจรไฟฟ้า　[ruup-baaep wong-chorn fai-faa]

printed circuit board
> a boad with slots for integrated circuit chips and connectors to other component. The necessary circuits are printed on the board.
> 印刷回路基板　[insatu kairo kiban]
> IC用のスロットや他のコンポーネント用のコネクタを搭載した基板．必要な回路は基板上に印刷されている．
> 印刷回路板　[yìn shuā huí lù bǎn]
> papan rankaian tercetak　[papan rankaian teruchetaku]
> 인쇄 회로 기판　[inswae hoero gipan]
> บอร์ดวงจรไฟฟ้า　[board wong-chorn fai-faa]

process annealing
> an annealing performed at an appropriate temperature above the recrystallization temperature and below the A_1 point to facilitate cold working after softening steel hardened by cold working.
> 中間焼なまし　[tyuukan yakinamasi]
> 冷間加工で硬化した鋼を軟化し，引き続いて行う冷間加工を容易にする目的で，再結晶温度以上A_1点以下の適当な温度で行う焼なまし．
> 中间退火　[zhōng jiān tuì huǒ]
> anneal melalui proses　[annearu merarui purosesu]
> 중간 풀림　[junggan pullim]
> กระบวนการอบ　[gra-buan op]

process capability
> an achievement limit which may be accomplished reasonably in a stable industrial process.
> 工程能力　[koutei nouryoku]
> 安定した工程において合理的に達成可能な能力の限界．
> 工程能力　[gōng chéng néng lì]
> indeks kapabilitas proses　[indekusu kapabiritasu purosesu]
> 공정능력　[gongjeongneungnyeok]
> โพรเซสคาพาบิลิตี้　[process capability]

process capability index
> an index of tolerance divided by 6 σ.

process capability index — progressive leaf spring

工程能力指数　[koutei nouryoku sisuu]
公差を工程能力の６σで割った値．
工程能力指数　[gōng chéng néng lì zhǐ shù]
kapabilitas proses　[kapabiritasu purosesu]
공정능력 지수　[gongjeongneungnyeok jisu]
ดัชนีโพรเซสคาพาบิลิตี้　[dad-sha-nii process capability]

process specification
 a chart showing the work title, details, and machinery used in accordance with work procedures.
工程表　[kouteihyou]
作業の手順に従って，作業の名称，内容及び使用機械を示した図表．
工程表　[gōng chéng biǎo]
spesifikasi proses　[supesifikasi purosesu]
공정표　[gongjeongpyo]
โพรเซสชาร์ท　[process chart]

profile of dent
 the shape of a dent produced by shot peening. It may have a raised perimeter or be a simple sunken dent.
こん形状　[kon keizyou]
ショットピーニング加工のショットの圧こん形状のことでこんの周囲が盛り上がる場合と単に凹部を示す場合とがある．
凹痕形状　[āo hén xíng zhuàng]
bentuk cekungan　[bentuku chekungan]
압흔 형상　[apheun hyeongsang]

ลักษณะของรอยบุ๋ม　[lak-sa-na khong roi bum]

profile projector
 an apparatus that measures the profile and dimensions of an object by projecting an optically enlarged image onto a screen at a specified magnification.
投影検査器　[touei kensaki]
物体を定められた倍率で光学的に拡大投影し投影面上でその形状及び寸法などを測定する機器．
投影仪　[tóu yǐng yí]
proyeksi profil　[puroyekusi purofiru]
투영 검사기　[tuyeong geomsagi]
อุปกรณ์วัดลักษณะ, ขนาดของวัตถุ　[up-pa-gorn wad lak-sa-na, kha-naad khong wad-tu]

progressive leaf spring
 a leaf spring comprising a main spring which always supports the load and an auxiliary spring which supports the load supplementary as the load increases. (see Fig P-6)
プログレッシブ重ね板ばね　[puroguressibu kasaneita bane]
常時荷重を受けるばね（主ばね）と，荷重の増加とともに補助的に働くばね（補助ばね）とで構成された重ね板ばね．(Fig P-6 参照)
重叠板弹簧　[chóng dié bǎn tán huáng]
pegas daun progresif　[pugasu daun

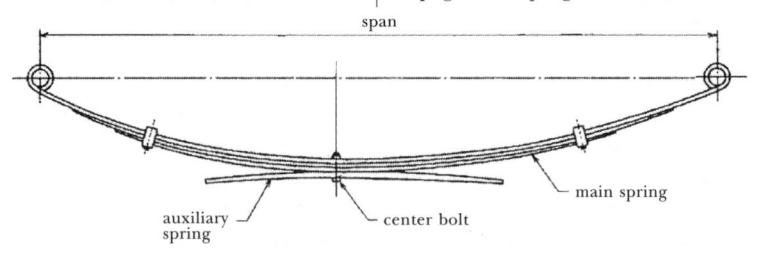

Fig P-6　progressive leaf spring

puroguresifu]
프로그레시브 겹판 스프링
[peurogeuresibeu gyeoppan spring]
แหนบแผ่นชนิดโพรเกรสซีฟ [naaep paaen cha-nid progressive]

progressive rate spring
    a spring the deflection of which is not linear to the load applied.
    非線形特性ばね [hisenkei tokusei bane]
    荷重とたわみとの関係が直線でないばね.
    非线性弹簧 [fēi xiàn xìng tán huáng]
    pegas progresif [pugasu puroguresifu]
    비선형 특성 스프링 [biseonhyeong teukseong spring]
    สปริงที่ความสัมพันธ์ของการยุบตัวและโหลดไม่เป็นเส้นตรง [sa-pring tii kwaam sam-pan khong gaan yup tu-a laae load mai pen sen trong]

proof load
    the maximum permissible axial force exerted on threaded fastners such as nuts and bolts when used to clamp objects.
    保証荷重 [hosyou kazyuu]
    ボルト，ナットなどのねじ部品で品物を締め付けて使用するとき，そのねじ部品に許される最大の軸力.
    安全负荷 [ān quán fù hè]
    beban coba [beban choba]
    보증 하중 [bojeung hajung]
    ค่าโหลดสูงสุดบนอุปกรณ์ขัน [kaa load soong-sud bon up-pa-gorn khan]

protective atmospheric furnace
    a heating furnace in which the interior atmosphere can be controlled for such operations as oxidizing, carburizing, or nitriding steel during heating.
    雰囲気調整炉 [hun'iki tyousei ro]
    鋼の加熱酸化あるいは浸炭又は浸室などの操作のために炉内の雰囲気状態が制御できる加熱炉.
    可控气热处理炉 [kě kòng qì rè chǔ lǐ lú]
    tungku dengan atmosfer pelindung [tunguku dengan atomosuferu purindungu]
    분위기 조정노 [bunwigi jojeongno]
    เตาเผาชนิดที่ควบคุมบรรยากาศภายในได้ [ta-o pa-o cha-nid tii ku-ab-kum ban-ya-gaad paai nai daai]

protective coating
    保護塗装 [hogo tosou]
    保护膜 [bǎo hù mó]
    lapisan pelindung [rapisan purindungu]
    보호 도장 [boho dojang]
    การเคลือบป้องกัน [gaan klu-ab pong gan]

protective device
    a device that senses faults at startup or during operation, and performs specific controls or emits signals to prevent major damage due to these faults.
    保安装置 [hoan souti]
    起動時もしくは運転中の異常を感知して，その異常によって大きな損傷や被害が生じることを防ぐように所定の制御をしたり又は信号を発する装置.
    安全装置 [ān quán zhuāng zhì]
    sarana pelindung [sarana purindungu]
    보안 장치 [boan jangchi]
    อุปกรณ์ป้องกัน [up-pa-gorn pong gan]

proximity detector
    a sensing device that produces an electrical signal when approached by an object.
    近接検出器 [kinsetu kensyutuki]
    物体が接近したときに電気信号を発生する感知装置.
    接近検測器 [jiē jìn jiǎn cè qì]
    deteksi sentuhan [detekusi sentuhan]
    접근 검출기 [jeopgeun geomchulgi]
    อุปกรณ์ตรวจจับการเข้าใกล้ [up-pa-gorn truat-chab gaan khao-klai]

pseudoelasticity

the phenomenon whereby deformation caused by stress-induced martensitic transformation returns to its original state due to reverse transformation when the load is removed.

超弾性 [tyoudansei]
応力誘起マルテンサイト変態によって生じた変形が，除荷時に逆変態によって元に戻る性質．
超弾性 [chāo tán xìng]
psedoelastis [pusedoerasutisu]
초탄성 [chotanseong]
คุณสมบัติยืดหยุ่นเทียม [kun-na-som-bat yuud-yun tiam]

pull out load

a load neccessary to pull out a bushing from assembled state.

離脱荷重 [ridatu kazyuu]
圧入した重ね板ばねのブシュを逆に取り去るときの荷重をいう．
卸荷 [xiè hè]
beban penarik [beban punariku]
이탈 하중 [ital hajung]
แรงดึงออก [rang dung ork]

pulsating stress

stress alternating between zero and a maximum value or minimum value.

片振応力 [kataburi ouryoku]
ゼロと最大値又はゼロと最小値との間を繰り返す応力．
脉动应力 [mài dòng yīng lì]
tegangan berdenyut [tegangan berudenyuto]
편진응력 [pyeonjineungnyeok]
ความเค้นสลับเป็นจังหวะ [kwaam ken sa-lab pen chang-wa]

punching force

the force required to punch holes in a material during stamping process.

打抜力 [utinukiryoku]
材料に打抜加工で穴をあけるときに要する力．
冲孔力 [chōng kǒng lì]
gaya punch [gaya puntchi]
편칭 휘스 [peonching hwoseu]
แรงเจาะ [rang cha-or]

punching property

the shearing characteristic of a material.

打抜加工性 [utinuki kakousei]
材料のせん断加工性．
冲孔加工性 [chōng kǒng jiā gōng xìng]
kualitas punch [kuaritasu punchi]
편칭 가공성 [peonching gagongseong]
คุณสมบัติของการเจาะ [kun-na-som-bat khong gaan cha-or]

pusher type furnace

a type of continuous furnace in which trays and workpieces are intermittently fed in by an insertion pusher.

プッシャー型炉 [pussyaa gata ro]
連続挿入炉の一つで挿入用プッシャーでトレイや品物そのものを断続的に送り込む方式の炉．
连续推进式炉 [lián xù tuī jìn shì lú]
tungku jenis dorong [tunguku jenisu dorongu]
푸샤형 로 [pusyahyeong ro]
เตาเผาชนิดผลักชิ้นงานเข้าอย่างต่อเนื่อง [ta-o pa-o cha-nid plak shin nhaan kha-o yaang to-a ne-ung]

push-on spring nut

thin leaf spring for fastening, with a hole and tabs at its center for inserting a stud. By the spring action and the locking action of the tabs, this nut holds the component to be held by holding the stud and fastening the component to the panel.

押込みばね板ナット　[osikomi bane ita natto]
　薄板ばねの中心部に，スタッドを押し込むつめをもつ締結部品．ばね作用及びつめ部の抜け止め作用によってスタッドを保持し，被保持部品とパネルとを締結，保持する．

嵌入式弾簧螺帽　[qiàn rù shì tán huáng luó mào]

mur pegas tekan　[muru pugasu tekan]

누름 스프링판 너트　[nureum springpan neoteu]

นัทเกลียวสปริงแบบกด　[nat gli-aew sa-pring baaep god]

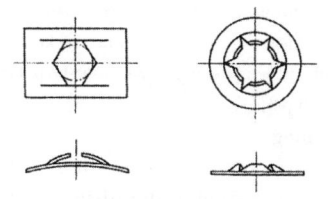

Fig P-7　push-on spring nut

Q

quality assurance
　　a series of systematic actions to provide adequate confidence in the quality of products or service.
品質保証　[hinsitu hosyou]
　　製品やサービスの品質に対する適切な保証を行う系統的な活動.
质量保证　[zhì liàng bǎo zhèng]
jaminan mutu　[jaminan mutu]
품질보증　[pumjilbojeung]
การประกันคุณภาพ　[gaan pra-gan kun-na-paab]

quality control
　　operational techniques to keep the quality level of a product or service.
品質管理　[hinsitu kanri]
　　製品やサービスの水準を維持するための業務手法.
质量管理　[zhì liàng guǎn lǐ]
kendali mutu　[kendari mutu]
품질관리　[pumjilgwalli]
การควบคุมคุณภาพ　[gaan ku-ab kum kun-na-paab]

quality control chart
　　a chart to indicate the quality index of a product.
品質管理図　[hinsitu kanrizu]
　　製品の品質指標を表示する管理図.
质量管理图　[zhì liàng guǎn lǐ tú]
diagram kendali mutu　[diaguramu kendari mutu]
품질관리도　[pumjilgwallido]
แผนผังการควบคุมคุณภาพ　[paaen-pang gaan ku-ab kum kun-na-paab]

quench aging
　　aging of steel material that occurs when steel quenched from high temperature is held at room temperature or slightly higher temperature.
急冷時効　[kyuurei zikou]
　　高温から急冷した鉄鋼を室温又はそれより少し高い温度に保持したときに起こる時効.
淬火时效　[cuì huǒ shí xiào]
penuaan kejut　[punuaan kejuto]
담금질 시효　[damgeumjil sihyo]
การเปลี่ยนแปลงของเหล็กชุบแข็งเนื่องจากอุณหภูมิเปลี่ยนไป　[gaan plian-praaeng khong lek shub khaaeng ne-ung chaak un-na-ha-poom pliean pai]

quenching
　　an operation of hardening by rapid cooling from austenitizing temperature. This may also signify the operation of simple rapid cooling not necessarily aiming at hardening. This term also includes the case where the rolling is carried out in the austenitized state immediately followed by the quenching, and this may be called direct quenching after rolling.
焼入れ　[yakiire]
　　オーステナイト化温度から急冷して硬化させる操作. 必ずしも硬化を目的とせず, 単に急速に冷却する操作をいうこともある. なお, オーステナイト状態で圧延を行い, その後, 圧延ライン上で直ちに行う焼入れもこれに含み, これを圧延後直接焼入れということがある.
淬火　[cuì huǒ]
kejut　[kejutu]

담금질 [damgeumjil]
การชุบแข็งโดยทำให้วัตถุเย็นลงอย่างเร็ว [gaan shub khaaeng dooi tam hai wad-tu yen tu-a long yaang rew]

quenching and sizing machine
machine which quenches the heated helical compression spring while pressing it by the free height adjusting equipment and the distortion prevention equipment.
コイルばね焼入機 [koiru bane yakiireki]
加熱した圧縮コイルばねを，自由高さ調節装置及び変形防止装置によってプレスしたまま焼入れする機械．
分級淬火 [fēn jí cuì huǒ]
mesin pengejut dan pembentuk [mesin pungejutu dan punbentuku]
코일 스프링 담금질기 [koil spring damgeumjilgi]
เครื่องม้วนสปริงขดที่ชุบแข็งพร้อมกับเพรส [kru-ang muan sa-pring khod tii shub khaaeng prorm gab press]

quenching and tempering
the process of reheating steel to an appropriate temperature below the A_1 point and then cooling, after quenching and hardening from the austenitizing temperature. This process is carried out to promote transformation or precipitation for obtaining specific characteristics and states.
焼入焼戻し [yakiire yakimodosi]
オーステナイト化温度から急冷硬化させた後，変態又は析出を進行させて所定の性質及び状態を与えるために，A_1点以下の適当な温度に再加熱，冷却する操作．
调质 [tiáo zhì]
pengejutan dan sepuh keras [pungejutan dan sepuhu kerasu]

담금질, 뜨임 [damgeumjil, tteuim]
การชุบแข็งและการอบในกระบวนการรีฮีทดิ้ง [gaan shub kaaeng laae gaan ob nai gra-buan gaan reheating]

quenching crack
a crack caused by quenching stress.
焼割れ [yakiware]
焼入応力によって生じる割れ．
淬裂 [cuì liè]
retak karena kejutan [retaku karena kejutan]
담금질 균열 [damgeumjil gyunnyeol]
การแตกจากความเค้นชุบแข็ง [gaan taaek chaak kwaam ken shub kaaeng]

quenching distortion
a change in shape or size caused by quenching.
焼入変形 [yakiire henkei]
焼入れによって生じる形状又は寸法の変化．
淬火变形 [cuì huǒ biàn xíng]
distorsi karena kejutan [disutorusi karena kejutan]
담금질 변형 [damgeumjil byeonhyeong]
การเปลี่ยนรูปจากการชุบแข็ง [gaan plian-ruup chaak gaan shub khaaeng]

quenching stress
the residual stress caused by quenching. Quenching stress is normally a combination of thermal stress caused by difference of cooling times between exterior and interior sections and transformation stress accompanied by transformation.
焼入応力 [yakiire ouryoku]
焼入れで生じる残留応力．焼入応力には，内外部の冷却時間的なずれに起因する熱応力と変態に伴う変態応力とがあり，一般に両者が組み合わされて生じる．

淬火应力　[cuì huǒ yīng lì]
tegangan kejut　[tegangan kejuto]
담금질 응력　[damgeumjil eungnyeok]
ความเค้นตกค้างจากการชุบแข็ง　[kwaam ken tok-kaang chaak gaan shub khaaeng]

quenching temperature
 the temperature of an object when immersed in quenching solution.
焼入温度　[yakiire ondo]
 鋼などを焼入れするとき，焼入液に投入するときの品物の温度．
淬火温度　[cuì huǒ wēn dù]
suhu kejutan　[suhu kejutan]
담금질 온도　[damgeumjil ondo]
อุณหภูมิขณะชุบแข็ง　[un-na-ha-puum kha-na shub-khaaeng]

R

radial load
    the load perpendicular to the central axis of a bearing.
    ラジアル荷重　［raziaru kazyuu］
    軸受中心軸に垂直な方向に働く荷重．
    径向负荷　［jìng xiàng fù hè］
    beban jari-jari　［beban jari-jari］
    레이디얼 하중　［reidieol hajung］
    เรเดียลโหลด　［ras-sa-mii khong su-an koong］

radius of curvature
    the radius of a small section of a curve that can be regarded as an arc.
    曲率半径　［kyokuritu hankei］
    曲線の微小部分を考えると円弧と見なすことができ，この円弧の半径をいう．
    曲率半径　［qū lǜ bàn jìng］
    jari-jari lengkung　［jari-jari rengukungu］
    곡률 반경　［gongnyul bangyeong］
    รัศมีของส่วนโค้ง　［ras-sa-mii khong at-tra kong］

rail fastening spring
    a spring used to secure rails to a sleeper.
    レール締結用ばね　［reeru teiketuyou bane］
    鉄道のレールと枕木とを締結させるばね．
    导轨紧固弹簧　［dǎo guǐ jǐn gù tán huáng］
    pegas daun pengencang rel　［pugasu daun pungencangu reru］
    레일 클립　［reil keullip］
    สปริงรัดรางรถไฟ　［sa-pring rad raang rod-fai］

raised hook
    U-shaped hook of an extension spring.
    U フック　［yuu hukku］
    引張コイルばね端末のU字形フック．
    U 形钩　［U xíng gōu］
    kait bentuk U　［kaito bentuku U］
    유 (U) 형 고리　［you (U) hyeong gori］
    ตะขอรูปตัวยู　［ta-khor ruup tu-a yuu］

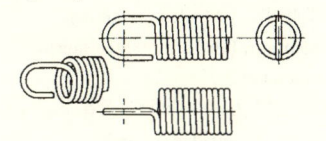

Fig R-1　raised hook

rate of operation
    the ratio of a certain production demand to the full production capacity.
    稼働率　［kadouritu］
    生産量を加工するのに，その設備能力でフル操業したときの定時能力に対する需要の割合をいう．
    开机率　［kāi jī lǜ］
    operasi rata-rata　［operasi rata-rata］
    가동율　［gadongyul］
    อัตราการผลิตสูงสุด　［at-tra gaan pa-lit soong-sud］

ratio of yield point to tensile strength
    the ratio of yield stress or proof stress to tensile strength.
    降伏点の引張強さ比　［kouhukuten no hippari tuyosa hi］
    引張強さに対する降伏点又は耐力との割合．
    屈强比　［qū qiáng bǐ］
    perbandingan titik mulur terhadap kuat tarik　［purubandingan titiku murure teruhadapu kuato tariku］
    항복점의 인장강도 비　［hangbokjeomui injanggangdo bi］
    อัตราส่วนของจุดล้าต่อกำลังการดึง　［at-tra suan

khong chud laa to-a gam-lang gaan duung]

rattle noise
    sound caused by the sudden contact of springs or the spring and the adjacent component.
    たたき音　[tatakion]
    ばね同士又はばねと周辺部品とが，衝撃的に接触したときに生じる音．
    振动噪声　[zhèn dòng zào shēng]
    kebisingan berderak　[kebisingan beruderaku]
    래틀 노이즈 (따닥따닥 소음)　[laeteul no-ijeu (ttadagttadag so-eum)]
    เสียงกระทบกัน　[si-ang gra-top gan]

reaction
    a reversed force against an applied force on a body.
    反作用　[hansayou]
    物体が他の物体に力を及ぼすとき，他の物体から受ける反対方向の力．
    反作用　[fǎn zuò yòng]
    reaksi　[reakusi]
    반작용　[banjangnyong]
    ปฏิกิริยา　[pa-ki-ki-ri-ya]

rebound force
    a force exerted on a leaf spring when a vehicle goes through a rebound moment.
    リバウンド力　[ribaundo ryoku]
    車両のリバウンド時に板ばねに加わる力．
    反冲力　[fǎn chōng lì]
    gaya gerak balik　[gaya geraku bariku]
    반발력　[banballyeok]
    แรงสะท้อนกลับ　[rang sa-torn glab]

rebound leaf
    a leaf to protect the main leaf against the load in reverse direction. (see Fig R-2)
    押さえばね板　[osae baneita]
    親板を逆方向の荷重に対して保護するためのばね板．（Fig R-2 参照）
    回弹板　[huí tán bǎn]
    daun gerak balik　[daun geraku bariku]
    누름 스프링판　[nureum springpan]
    แหนบช่วย　[naaep shu-oi]

rebound stroke
    stroke from the position under the commonly used load to the position of the rebound stopper in the suspension spring.
    リバウンドストローク　[ribaundo sutorooku]
    懸架ばねにおける，常用荷重時からリバウンドストッパで止まるまでの行程．
    回弹冲击　[huí tán chōng jī]
    langkah gerak balik　[rangukahu geraku bariku]
    리바운드 스트로크　[ribaundeu seuteurokeu]
    สโตรคการสะท้อนกลับ　[stroke gaan sa-torn glab]

recarburization
    carburizing treatment to restore the carbon content in the decarburized layer of steel.
    復炭　[hukutan]
    鋼の脱炭層を元の炭素量に回復させるために行う浸炭処理．
    再碳化　[zài tàn huà]

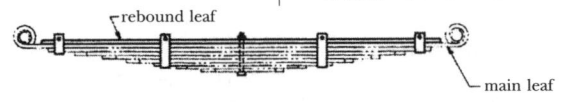

Fig R-2　rebound leaf

dekarburisasi [dekaruburisasi]
복탄 [boktan]
กระบวนการนำคาร์บอนกลับคืนสู่ผิวเหล็ก [grabuan gaan nam carbon glab ku-un suu piw lek]

recrystallization annealing
annealing performed to cause recrystallization. Recrystallization is the phenomenon whereby new crystal nuclei without internal strain replace the original crystal grains.
再結晶焼なまし [saikessyou yakinamasi]
再結晶を起こさせるために行う焼なまし。再結晶とは元の結晶粒から内部ひずみのない新しい結晶の核が発生し、元の結晶粒と置き換わっていく現象。
再结晶退火 [zài jié jīng tuì huǒ]
anneal rekristalisasi [annearu rekurisutarisasi]
재결정 풀림 [jaegyeoljeong pullim]
การอบให้ตกผลึกใหม่ [gaan ob hai tok pa-luk mai]

recrystallization temperature
the temperature that triggers recrystallization.
再結晶温度 [saikessyou ondo]
再結晶を起こす温度。
再结晶温度 [zài jié jīng wēn dù]
suhu rekristalisasi [suhu rekurisutarisasi]
재결정 온도 [jaegyeoljeong ondo]
อุณหภูมิตกผลึกใหม่ [un-na-ha-puum tok pa-luk mai]

rectangular cross section torsion bar spring
a torsion bar spring that has a rectangular cross section.
四角断面タイプトーションバー [sikaku danmen taipu toosyon baa]
長方形断面を持つトーションバー。
方形截面扭杆 [fāng xíng jié miàn niǔ gǎn]
pegas puntir berpenampang kotak
[pugasu puntiru burupenanpangu kotaku]

사각단면 토션바 [sagakdanmyeon tosyeonba]
ทอร์ชั่นบาร์ชนิดหน้าตัดสี่เหลี่ยม [torsion bar cha-nid naa-tad sii-liam]

rectangular hook
square hook at the end of a helical extension spring.
角フック [kaku hukku]
引張コイルばねの端部の一種で、角形のフック。
方形钩环 [fāng xíng gōu huán]
kait tegak lurus [kaito tegaku rurusu]
각형 고리 [gakhyeong gori]
ตะขอเหลี่ยมที่ปลายสปริงดึง [ta-khor lieam tii plaai sa-pring dung]

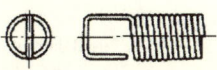

Fig R-3 rectangular hook

rectangular section
rectangular cross-section of a leaf.
長方形断面 [tyouhoukei danmen]
断面形状が長方形のばね板の断面。
方形截面 [fāng xíng jié miàn]
penampang empat persegi panjang
[punanpangu enpato purusegi panjangu]
각형 단면 [gakhyeong danmyeon]
หน้าตัดแนวยาวของแหนบ [naa-tad naaew-ya-o khong naaep]

rectangular wire coil spring
a coil spring made from wire with a rectangular or square cross section. One of the noncircular wire coil spring.
角ばね [kaku bane]
異形断面ばねの一種で、材料の断面形状が長方形又は正方形のコイルばね。
矩形钢丝弹簧 [jǔ xíng gāng sī tán huáng]
pegas empat persegi panjang [pugasu enpato purusegi panjangu]

rectangular wire coil spring — reduced hook

각형 스프링 [gakhyeong spring]
สปริงขดลวดหน้าตัดรูปเหลี่ยม [sa-pring khod lu-ad naa tad ruup lieam]

rectangular wire formed boots band
 a retaining ring made from rectangular cross section wire to be used mainly as a dust cover of the driving unit.
角断面ブーツバンド [kaku danmen buutu bando]
 止め輪の一種で，主に駆動部位などの防じん用カバー止めに用いられている角断面線を成形したバンド．
防尘罩固定用长方形截面板条 [fáng chén zhào gù dìng yòng cháng fāng xíng jié miàn bǎn tiáo]
pita bernampang kotak [pita berunanpangu kotaku]
각단면 브츠 밴드 [gakdanmyeon beucheu baendeu]
บูทส์แบนด์จากลวดหน้าตัดรูปเหลี่ยม [boots band chaak lu-ad naa tad roop liam]

rectangular wire formed hose clip
 a hose clamping clip made from rectangular cross section wire.
角断面ホースクリップ [kaku danmen hoosu kurippu]
 ホース接続部用のクリップの一種で，角断面線を用いたホースクリップ．
接管用长方形截面线夹 [jiē guǎn yòng cháng fāng xíng jié miàn xiàn jiā]
penjepit selang berpenampang kotak [punjepito serangu berupenanpangu kotaku]
각단면 호스 클램프 [gakdanmyeon hoseu keullaempeu]
โฮสคลิปจากลวดหน้าตัดรูปเหลี่ยม [hose clip chaak lu-ad naa tad roop liam]

rectangularly wound compression spring
 a compression coil spring with a rectangular coil shape.
圧縮角形コイルばね [assyuku kakugata koiru bane]

コイル形状が四角形をした圧縮コイルばね．
方形圧縮弾簧 [fāng xíng yā suō tán huáng]
pegas tekan penampang kotak [pugasu tekan punanpangu kotaku]
각형 압축코일 스프링 [gakhyeong apchukkoil spring]
สปริงขดรับแรงกดรูปเหลี่ยม [sa-pring khod rab rang god ruup liam]

red shortness
 the characteristic of steel that becomes brittle in the temperature range for hot-working.
赤熱ぜい性 [sekinetu zeisei]
 熱間加工の温度範囲で鋼がもろくなる性質．
热脆性 [rè cuì xìng]
kerapuhan panas [kerapuhan panasu]
적열 취성 [jeongnyeol chwiseong]
คุณสมบัติเหล็กที่เปราะเนื่องจากกระบวนการแบบร้อน [kun-na-som-bat lek tii pra-or ne-ung chaak gra-buan gaan baaep rorn]

reduced hook
 a hook with its diameter smaller than the coil diameter. There are two cases: center reduced hook and side reduced hook.
絞りフック [sibori hukku]
 フック部の直径がコイル直径より小さいフックをいう．絞りフックにはコイル中心に位置させたものとコイル径上に立ち上げたものとがある．
缩式钩环 [suō shì gōu huán]
lingkaran (kait) yang diperkecil [ringukaran (kaito) yangu diperukechiru]
축소 고리 [chukso gori]
ตะขอที่มีขนาดเส้นผ่าศูนย์กลางเล็กกว่าเส้นผ่าศูนย์กลางของขด [ta-khor tii mii sen-pa-soon-glaang lek gwaa sen-pa-soon-glaang khong khod]

redundancy
　　a safety measure to keep the function of the system by providing more than one device of the operation.
冗長性　[zyoutyousei]
　　ある機能システムに二つ以上の操作手段を与えることによって機能維持を図る安全手段．
冗余性　[rǒng yú xìng]
tidak diperlukan lagi　[tidaku diperurukan ragi]
여유성　[yeoyuseong]
การวัดอย่างปลอดภัยโดยใช้อุปกรณ์มากกว่า 1 ชนิด　[gaan wad yaang plord-pai]

reel
　　a rotating frame to wind wire.
巻取枠　[makitoriwaku]
　　線材を巻き取るための枠をいう．
卷绕筒　[juǎn rào tǒng]
gulungan　[gurungan]
릴　[ril]
เฟรมหมุนสำหรับม้วนลวด　[frame mun sam-rab mu-an luad]

reference line
　　a line used as the datum for design, manufacturing, positioning, and assembly.
基準線　[kizyunsen]
　　設計・製作・配置・組立などの基準とする線．
基准线　[jī zhǔn xiàn]
garis acuan　[garisu achuan]
기준선　[gijunseon]
เส้นอ้างอิง　[sen a-ang-ing]

regression analysis
　　the presumption of an equation and coefficient to be used in the equation when the mean of probability varaiables can be expressed by a polinomial.
回帰分析　[kaiki bunseki]
　　確率変数の期待値が多項式で表される場合に係数や式について推定，検定を行うこと．
回归分析　[huí guī fēn xī]
analisa regresi　[anarisa reguresi]
회기 분석　[hoegi bunseok]
การวิเคราะห์การถดถอย　[gaan wi-kra-or gaan tod-tooi]

relative frequency
　　the ratio of the frequency of a particular measured value against the total number of measurements.
相対度数　[soutai dosuu]
　　ある測定値の出現度数を測定値の総数で割った値．
相对频率　[xiāng duì pín lǜ]
frekuensi relatif　[furekuensi reratifu]
상대 도수　[sangdae dosu]
ความถี่สัมพัทธ์　[kwaam tii sam-pat]

relative permeability
　　a ratio between magnetic permeability of the material and magnetic permeability of vacuum.
比透磁率　[hi touziritu]
　　物質の絶対透磁率と真空透磁率の比．
相对导磁率　[xiāng duì dǎo cí lǜ]
permeabilitas relatif　[purumeabiritasu reratifu]
비 투자율　[bi tujayul]
คุณสมบัติแม่เหล็กสัมพัทธ์　[kun-na-son-bat maae-lek sam-pat]

relaxation
　　phenomenon where the stress decreases gradually when a spring is held in the extended or compressed state to a certain length.
応力緩和　[ouryoku kanwa]
　　ばねを一定の長さに伸長又は圧縮した状態に保持したとき，応力が徐々に低

relaxation — reliability

下する現象.
松弛 [sōng chí]
pengenduran [pungenduran]
응력 완화 (이완) [eungnyeok wanhwa (iwan)]
การคลายตัวของความเค้น [gaan klaai to-a khong kwaam ken]

relaxation measurement
to measure load decrease as an index of relaxation.
リラクゼーション測定 [rirakuzeesyon sokutei]
応力緩和の指標として荷重の低下を測定すること.
松弛測定法 [sōng chí cè dìng fǎ]
pengukuran pengenduran [pungukuran pungenduran]
릴랙세이션 측정 [rillaekseisyeon cheukjeong]
การวัดการคลายตัวของความเค้น [gaan wad gaan klaai to-a khong kwaam ken]

relaxation rate
the percentage of decrease in load compared with the original load.
荷重低下率 [kazyuu teika ritu]
初荷重に対する減少した荷重の百分率.
松弛率 [sōng chí lǜ]
laju pengenduran [raju pungenduran]
하중저하율 [hajungjeohayul]
อัตราการคลายตัวของความเค้น [at-traa gaan gaan klaai to-a khong kwaam ken]

relaxation test
a test to measure load decrease of time under the condition that the total strain is constant.
リラクゼーション試験 [rirakuzeesyon siken]
全ひずみ一定の条件の下で, 荷重（応力）の時間的低下を測定する試験.

松弛試験 [sōng chí shì yàn]
uji pengenduran [uji pungenduran]
이완 시험 [iwan siheom]
การทดสอบการวัดโหลดที่ลดลงภายใต้ความเครียดคงที่ [gaan tod-sorb gaan wad load tii lod-long paat-taai kwaam kriead kong-tii]

relaxation value
the percentage of decrease in load compared with the original load.
リラクゼーション値 [rirakuzeesyon ti]
初期荷重に対する減少した荷重の百分率.
松弛值 [sōng chí zhí]
nilai pengenduran [nirai pungenduran]
이완 값 [iwan gap]
ค่าการลดลงของโหลดเปรียบเทียบกับโหลดเดิม [kaa gaan lod-long khong load prieab-tieab gab load derm]

relay spring
a spring used in relays in which a terminal opens and closes the contact directly or indirectly by cutting off current to a solenoid coil.
リレー用ばね [rireeyou bane]
電磁コイルへの通電と停止によって接触子が直接又は間接的に接点の開閉を行うリレーに用いるばね.
继电器弹簧 [jì diàn qì tán huáng]
pegas penerus [pugasu punerusu]
릴레이용 스프링 [rilleiyong spring]
สปริงสำหรับรีเลย์ [sa-pring sam-rab relay]

reliability
the ability of a machine to keep its function within a specified term and condition.
信頼性 [sinraisei]
定められた期間及び定められた条件のもとで装置が求められた機能を果たす能力.
可靠性 [kě kào xìng]

keandalan　[keandaran]
신뢰성　[silloeseong]
ความเชื่อถือได้　[kwaam shi-ua tu-u daai]

repeated intermittent test
　　a repeated test performed intermittently under constant conditions.
繰返し間欠試験　[kurikaesi kanketu siken]
　　ある一定条件で間欠的に繰り返し行う試験．
反复间歇试验　[fǎn fù jiān xiē shì yàn]
uji berulang　[uji berurangu]
간헐 반복 시험　[ganheol banbok siheom]
การทดสอบซ้ำเป็นช่วง ๆ ภายใต้เงื่อนไขคงที่
[gaan tod sorb sam pen she-ong paai-taai nguan khai kong tii]

repeated test
　　a test repeated within a relatively short time using the same method, same sample, and same conditions.
繰返し試験　[kurikaesi siken]
　　同一の方法で同一の測定対象を同じ条件で比較的短い時間に繰り返し行う試験．
重复试验　[chóng fù shì yàn]
uji ulang　[uji urangu]
반복 시험　[banbok siheom]
การทวนสอบ　[gaan tu-an sorb]

reproducibility
　　the degree of consistency of individual measurements repeated within a relatively short time using the same method, same sample, and same conditions.
再現性　[saigensei]
　　同一の方法で同一の測定対象を同じ条件で比較的短い時間に繰り返し測定した場合，個々の測定値が一致する性質又は度合い．
再现性　[zài xiàn xìng]
mampu dibuat ulang　[manpu dibuato urangu]

재현성　[jaehyeonseong]
คุณสมบัติการผลิตซ้ำ　[kun-na-som-bat gaan pa-lit sam]

residual compressive stresses
　　compressive stress remaining inside a metal under no external force or thermal gradient.
残留圧縮応力　[zanryuu assyuku ouryoku]
　　外力又は熱勾配がない状態で，金属内部に残っている圧縮応力．
残余压应力　[cán yú yā yīng lì]
tegangan tekan tersisa　[tegangan tekan terusisa]
압축 잔류 응력　[apchuk jallyu eungnyeok]
ความเค้นอัดตกค้าง　[kwaam ken ad tok kaang]

residual hydrogen content
　　density of residual hydrogen in metal after dehydrogen process.
残存水素量　[zanzon suisoryou]
　　金属から水素を放出させた後に，残った水素の濃度．
残余氢含量　[cán yú qīng hán liàng]
isi hidrogen tersisa　[isi hidorogen terusisa]
잔존 수소량　[janjon susoryang]
ปริมาณไฮโดรเจนตกค้างในเหล็ก　[pa-ri-man hydrogen tok-kaang nai lek]

residual magnetism
　　magnetism which stays in material after removing the magnetic field.
残留磁気　[zanryuu ziki]
　　磁界を取り去っても材料に残存する磁気．
残留磁性　[cán liú cí xìng]
magnet tersisa　[maguneto terusisa]
잔류자기　[jallyujagi]
อำนาจแม่เหล็กตกค้างในวัตถุ　[am-naat maae-lek tok-kaang ani wad-tu]

residual shearing strain
　　shearing strain left in a spring after re-

residual shearing strain — resistance to thermal shocks

moving the applied load or moment.
残留せん断ひずみ [zanryuu sendan hizumi]
ばねに荷重又はトルクを加えて変形させ，次に荷重又はトルクを除いたときに残るせん断ひずみ．
残余剪切应变 [cán yú jiǎn qiē yīng biàn]
regang risan tersisa [regangu risan terusisa]
잔류 전단 변형 [jallyu jeondan byeonhyeong]
ความเครียดเฉือนตกค้าง [kwaam kriead shi-un tok-kaang]

residual stress

stress remaining inside a metal under no external force or thermal gradient.
残留応力 [zanryuu ouryoku]
外力又は熱勾配がない状態で，金属内部に残っている応力．
残余应力 [cán yú yīng lì]
tegangan tersisa [tegangan terusisa]
잔류 응력 [jallyu eungnyeok]
ความเค้นตกค้าง [kwaam ken tok-kaang]

residual stress distribution

distribution of residual stress in a spring. Generally, residual stress is not distributed uniformly but unevenly in particular portions.
残留応力分布 [zanryuu ouryoku bunpu]
ばねの内部に残留している応力の分布状態．一般には，一様な分布をせず，特定部分に偏って分布している．
残余应力分布 [cán yú yīng lì fēn bù]
sebaran tegangan tersisa [sebaran tegangan terusisa]
잔류 응력 분포 [jallyu eungnyeok bunpo]
การกระจายของความเค้นตกค้าง [gaan grachaai khong kwaam ken tok-kaang]

residual tensile stresses

tensile stress remaining inside a metal under no external force or thermal gradient.
残留引張応力 [zanryuu hippari ouryoku]
外力又は熱勾配がない状態で，金属内部に残っている引張応力．
残余拉伸应力 [cán yú lā shēn yīng lì]
tegangan tarik tersisa [tegangan tariku terusisa]
인장 잔류 응력 [injang jallyu eungnyeok]
ความเค้นดึงตกค้าง [kwaam ken duung tok-kaang]

resistance to buckling

the strength whereby a structure or component can endure buckling due to external forces.
座屈強度 [zakutu kyoudo]
外力により構造物や部材が座屈に耐える力．
屈服强度 [qū fú qiáng dù]
tahanan terhadap tekuk [tahanan teruhadapu tekuku]
좌굴 강도 [jwagul gangdo]
ความต้านทานการยุบตัว [kwaam taan taan gaan yub to-a]

resistance to corrosion

the property whereby a material can endure corrosion in certain environments.
耐食性 [taisyokusei]
ある環境における腐食作用に耐える性質．
耐腐蚀性 [nài fǔ shí xìng]
tahanan terhadap korosi [tahanan teruhadapu korosi]
내식성 [naesikseong]
ความต้านทานการกัดกร่อน [kwaam taan taan gaan gad gron]

resistance to thermal shocks

the resistance to thermal shock failure caused by thermal stress accompanied

by rapid changes in temperature.
熱衝撃抵抗　[netusyougeki teikou]
急激な温度変化に伴う熱応力によって破壊される熱衝撃現象に対する抵抗強さ.
热冲击阻力　[rè chōng jī zǔ lì]
tahanan terhadap kejutan termal [tahanan teruhadapu kejutan terumaru]
열충격 저항　[yeolchunggyeok jeohang]
ความต้านทานเทอร์มอลช็อค　[kwaam taan taan thermal shock]

resistance welding
to join metal parts by pressing them together and applying electric current to obtain high temperature.
抵抗溶接　[teikou yousetu]
接合させた部材の接触部を通して電流を流し, 抵抗加熱するとともに加圧を加えて行う溶接.
电阻焊接　[diàn zǔ hàn jiē]
las tahanan　[rasu tahanan]
저항 용접　[jeohang yongjeop]
การเชื่อมต้านทาน　[gaan shi-um taan taan]

resonance
phenomenon where the amplitude increases significantly when the frequency of the external force approaches the inherent frequency of the spring or of the system including it.
共振　[kyousin]
外力の振動数がばねを含む系又はばね自身の固有振動数に近づくと, 振幅が非常に大きくなる現象.
共振　[gòng zhèn]
resonansi　[resonansi]
공진　[gongjin]
การสั่นอย่างรุนแรงของสปริง　[gaan san yaang run-raaeng khong sa-pring]

resonance frequency
frequency to cause resonance.

Resonance frequency may differ according to the nature of the amount measured. For example, velocity resonance may occur at the different frequency from displacement resonance. If confusing, it shall be clearly stated as "velocity resonance frequency".
共振振動数　[kyousin sindousuu]
共振しているときの振動数. 共振振動数は, 測定される量の性質によって異なることがある. 例えば, 速度共振は, 変位共振と異なる振動数のときに起き得る. まぎらわしいときは, 速度共振振動数と明記する.
共振频率　[gòng zhèn pín lǜ]
frekuensi resonansi　[frekuensi resonansi]
공진 진동수　[gongjin jindongsu]
ความถี่ที่ทำให้เกิดรีโซแนนซ์　[kwaam tii tii tam hai gerd resonance]

retained austenite
the residual austenite remained due to incomplete transformation from austenite to martensite in quenching.
残留オーステナイト　[zanryuu oosutenaito]
焼入れによってオーステナイトからマルテンサイトへの変態が完全に行われない場合に一部のオーステナイトが残留することをいう.
残留奥氏体　[cán liú ào shì tǐ]
austenit yang ditahan　[ausutenito yangu ditahan]
잔류 오스테나이트　[jallyu oseutenaiteu]
ออสเทนไนท์ตกค้าง　[austenite tok kaang]

retaining ring
a ring to retain parts on shafts or in cylinders.
止め輪　[tomewa]
軸あるいは円筒内側で部品を保持する

retaining ring — reverse torsion test of steel wire

リング.
挡圈　[dǎng quān]
cincin penahan　[chinchin penahan]
멈춤링　[meomchumring]
แหวนกัน　[waaen gan]

retaining washer
　a spring washer used for fastening bolts.
締付座金　[simetuke zagane]
ボルトの締付けに用いるばね座金.
固定垫圈　[gù dìng diàn quān]
ring pipih penahan　[ringu pipihi punahan]
체결 와셔　[chegyeol wasyeo]
แหวนรองยึดโบลท์　[waaen rong yud bolt]

retort furnace
　a furnace incorporating a flame-resistant container to prevent the direct heating of workpieces.
レトルト炉　[retoruto ro]
直接工作物が加熱されないように耐火性容器を設けた炉.
蒸馏炉　[zhēng liú lú]
tungku retor　[tunguku retoru]
레토르트 로　[retoreuteu ro]
เตาชนิดมีอุปกรณ์ด้านเปลวไฟไม่ให้สัมผัสชิ้นงานโดยตรง　[ta-o cha-nid mii up-pa-gorn taan pleew-fai mai-hai sam-pad ngaan do-oi trong]

retractable mandrel
　a coiling machine mandrel with a mechanism that moves in and out after each coiling process.
引込式心金　[hikikomisiki singane]
コイリングマシンの巻付け用心金がコイリング後，毎回出入運動する機構の心金.
嵌入式心轴　[qiàn rù shì xīn zhóu]
poros pemegang yang dapat ditarik [porosu pumegangu yangu dapato ditariku]
인입식 맨드릴　[inipsik maendeuril]

แกนม้วนชนิดหดเข้าออกได้　[gaaen mu-an cha-nid hod khao-oork dai]

reverse bend test
　a test in which a test piece is fixed at one end and clamped with a chuck at the other end, and is bent repeatedly 90 degrees backwards and forwards. Also called reverse bending test.
反復曲げ試験　[hanpuku mage siken]
試験片の一端を固定し，自由端をつかみ金具によって左右に90°ずつ反復して曲げる試験．繰返し曲げ試験という場合もある.
反复弯曲试验　[fǎn fù wān qū shì yàn]
uji bengkok terbalik　[uji bengukoku terubariku]
반복굽힘 시험　[banbokgupim siheom]
การทดสอบการดัดกลับ　[gaan tod-sorb gaan tad glab]

reverse torsion test of steel wire
　a test in which the test specimen is twisted at a certain number of times in one direction at the first stage, and in the opposite direction in the second stage until it breaks.
鋼線の逆ねじり試験　[kousen no gyakuneziri siken]
試験片の両端を規定のつかみ間隔で硬くつかみ，たわまない程度に緊張しながら規定回数ねじった後，破断するまで逆方向にねじり，そのときのねじれの状況及び破断面の状況を調べる試験.
钢丝的反复扭转试验　[gāng sī de fǎn fù niǔ zhuǎn shì yàn]
uji puntir terbalik kawat baja　[uji puntiru terubariku kawato baja]
강선 역 비틀림 시험　[gangseon yeok biteullim siheom]
การทดสอบการบิดกลับ　[gaan tod-sorb gaan

bid glab]

reverse wind
 process to form a power spring where the power spring, annealed in low-temperature after the primary wind, is rewound in the reverse direction to its primarily wound curvature. This is to enhance the repulsive force of the power spring and to obtain high torque.
 二次巻 [nizi maki]
 ぜんまいの成形法で，一次巻後に低温焼なましを施したぜんまいを，その曲率とは逆の方向に再度巻き付ける加工．この目的は，ぜんまいの反発力を高め，高トルクを得るためである．
 二次卷曲 [èr cì juǎn qū]
 penggulungan terbalik [pungugurungan terubariku]
 이차 감기 [icha gamgi]
 กระบวนการม้วนกลับในการผลิตเพาเวอร์สปริง [gra bu-an gaan muan glab nai gaan pa-lit power sa-pring]

reversed stress
 repeated stress generated by an external force with the same magnitude and opposite directions.
 両振応力 [ryouburi ouryoku]
 大きさが同じで符号が正負逆方向の外力によって生じる繰返し応力．
 交変応力 [jiāo biàn yīng lì]
 tegangan dibalik [tegangan dibariku]
 양진 응력 [yangjin eungnyeok]
 ความเค้นย้อนกลับ [kwaam ken yorn glab]

revolution per minute
 毎分回転数（RPM） [maihun kaitensuu (aarupiiemu)]
 转速（RPM） [zhuǎn sù (RPM)]
 putaran per menit [putaran peru menito]
 매분 회전수 [maebun hoejeonsu]

 รอบหมุนต่อนาที [rorp mun to-a naa-tii]

ribbed and grooved section
 rectangular cross-section of a leaf with a groove on one side and a rib on the opposite side at the center of the width.
 リブ付き断面 [ributuki danmen]
 長方形断面の幅方向の中央の，片面に溝を，片面にリブを付けたばね板の断面．
 加强筋弹簧钢 [jiā qiáng jīn tán huáng gāng]
 penampang dengan rib [punanpangu dengan ribu]
 마루골 스프링강 [marukol springgang]
 หน้าตัดที่เป็นร่องและขอบยกขึ้น [naa-tad tii pen rong laae khob yok khun]

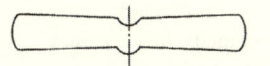

Fig R-4 ribbed and grooved section

right hand coil
 turning direction of coiling like a right-hand thread.
 右巻 [migimaki]
 右ねじと同じようなコイルばねの，コイルの巻き方向．
 右旋 [yòu xuán]
 ulir kanan [uriru kanan]
 오른쪽 감기 [oreunjjok gamgi]
 สปริงขดชนิดม้วนด้านขวา [sa-pring khod cha-nid muan daan khwaa]

rigidity
 the deformation resistance against external force. This is expressed in a number of ways depending on the shape and structure of the object.
 剛性 [gousei]
 外力に対する変形抵抗．物体の形状・構造などによっていろいろな表し方がある．

刚性（度） [gāng xìng (dù)]
kekekaran [kekekaran]
강성 [gangseong]
ความแข็งแกร่งต้านการเปลี่ยนรูป [kwaam khaaeng-graaeng daan gaan pliean-roop]

rimmed steel
    steel allowed to solidify with the characteristic rimming action of molten steel in which oxygen and carbon in the steel inside the casting mold react to generate carbon monoxide.

リムド鋼 [rimudo kou]
鋳型内で溶鋼中の酸素と炭素が作用して一酸化炭素を発生し，溶鋼が特有の沸騰攪拌運動（リミングアクション）をしながら凝固した鋼．
沸腾钢 [fèi téng gāng]
baja rimmed [baja rinmedo]
림드 강 [rimdeu gang]
เหล็กริม [lek rim]

ring gauge
    a ring-shaped measuring instrument with a precision-finished inner diameter used for determining the dimensions of a circle.

リングゲージ [ringu geezi]
内径を精密に仕上げた環状の測定工具のことで，円の寸法を決定するために用いられる．
环规 [huán guī]
pembanding cincin [punbandingu chinchin]
링 게이지 [ring geiji]
เกจรูปแหวน [guage ruup waaen]

ring spring
    a combination of rings having a compression spring function where the outer ring has an inner slanted friction surface and the inner ring has an outer slanted friction surface.

輪ばね [wabane]
外輪は内側に，内輪は外側に傾斜がある摩擦面をもった輪状のばねを重ね合わせたばね．
环状弹簧 [huán zhuàng tán huáng]
pegas cincin [pugasu chinchin]
링 스프링 [ring spring]
สปริงแหวน [sa-pring waaen]

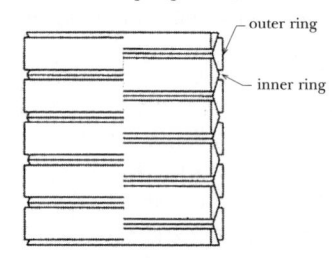

Fig R-5　ring spring

rinse
    the process of washing off penetrant or emulsifier adhering to the test piece surface with water. Also performed before surface treatment.

洗浄処理 [senzyou syori]
試験体表面に付着している浸透液及び乳化剤などを水で洗い流す操作．表面処理などの前処理にも行われる．
洗涤处理 [xǐ dí chǔ lǐ]
membilas [menbirasu]
세정처리 [sejeongcheori]
กระบวนการล้างผิว [gaan laang laang-piw]

ripple
    undesirable fluctuations with periodic but non-sinusoidal patterns around the mean value of indicated values, readings or supplied values.

リップル [rippuru]
指示値・表示値又は供給値の平均付近における周期的であるが非正弦波状の好ましくない変化．
皱纹 [zhòu wén]

riak [riaku]
리플 [ripeul]
ริปเปิ้ล [ripple]

Rockwell hardness
    a measure of hardness as determined by the Rockwell hardness tester.
ロックウエル硬さ [rokkuueru katasa]
ロックウエル硬さ試験機を用いて測定される硬さ.
洛氏硬度 [luò shì yìng dù]
kekerasan Rockwell [kekerasan Rokkuweru]
로크웰 경도 [rokeuwel gyeongdo]
ความแข็งร็อคเวล [kwaam khaaeng Rockwell]

rod diameter
ロッド径 [roddo kei]
杆径 [gǎn jìng]
diameter batang [diameteru batangu]
봉 직경 [bong jikgyeong]
เส้นผ่าศูนย์กลางของแท่ง [sen-pa-soon-glaang khong taaeng]

rolled steel
    steel plates or strips rolled on a rolling mill. Hot or cold rolling processes may be used.
圧延鋼材 [atuen kouzai]
圧延機によって伸延された鋼板又は帯鋼.熱間又は冷間で圧延されたものをいう.
轧制钢材 [zhá zhì gāng cái]
baja dirol [baja diroru]
압연 강재 [abyeon gangjae]
เหล็กแผ่นรีด [lek paaen riid]

rolled wire
    wires or wire rods rolled on a rolling mill. Hot or cold rolling processes may be used.
圧延線 [atuen sen]
圧延機によって伸線された線又は線材.熱間又は冷間で伸線されたものをいう.
轧制线材 [zhá zhì xiàn cái]
kawat dirol [kawato diroru]
압연 선재 [abyeon seonjae]
เหล็กเส้นรีด [lek sen riid]

roller bearing
    a shaft bearing which consists of parrallel or tapered steel rollers confined between outer and inner rings.
ころ軸受 [koro zikuuke]
外リングと内リングの間に配置された並行ローラ又はテーパローラからなる軸受.
滚动轴承 [gǔn dòng zhōu chéng]
bantalan rol [bantaran roru]
롤러 베어링 [rolleo beeoring]
แบริ่งลูกกลิ้ง [baering luuk gling]

roller straightening unit
    a device for correcting the distortion in materials or forgings by using rollers.
ローラ式矯正装置 [roorasiki kyousei souti]
材料や鍛造品などに生じているひずみをローラで矯正する装置.
辊式矫直装置 [gǔn shì jiǎo zhí zhuāng zhì]
unit pelurus rol [unito pururusu roru]
롤러식 교정 장치 [rolleosik gyojeong jangchi]
ตัวดัดตรงชนิดลูกกลิ้ง [to-a dad trong cha-nid luuk gling]

rolling defect
    a defect that occurs during rolling or roller straightening processes.
ロールきず [rooru kizu]
圧延又はロール矯正のときに生じるきず.
轧制缺陷 [zhá zhì quē xiàn]
kerusakan karena penggerollan [kerusakan karena pungugeroruran]
롤 흠 [rol heum]

rolling defect — rough roll

งานเสียจากการรีด　[ngaan sia-a chaak gaan riid]

rolling rate
　the ratio of steel thickness before and after rolling when steel strip reduces its thickness between pressure rollers.
　圧下率　[akka ritu]
　鋼片がロールの間で圧縮されて厚さを減少させるとき，ロールの通過前と通過後の厚さの比．
　圧下率　[yā xià lǜ]
　laju mengerol　[raju mengeroru]
　압하율　[apayul]
　อัตราการรีด (ความหนาก่อนและหลังการรีด)　[at-traa gaan riid]

rotary hearth furnace
　a furnace with a rotating hearth and the inlet and outlet positioned next to each other.
　炉床回転式炉　[rosyou kaitensiki ro]
　炉床が回転して挿入側と排出側が隣合せになった炉．
　旋转炉底炉　[xuán zhuǎn lú dǐ lú]
　tungku pengapian berputar　[tunguku pungapian beruputaru]
　로상 회전로　[rosang hoejeonno]
　เตาชนิดโรตารี่　[ta-o cha-nid rotary]

rotating bending fatigue test
　one of the methods to test fatigue strength of the material by applying repetitive bending to a test bar or wire in all directions. The repetitive bending moment is exerted by rotating the test piece around its axis with chucking arbors placed at an angle.
　回転曲げ疲労試験　[kaiten mage hirou siken]
　材料の疲れ強さを試験する方法の一つで，丸棒試験片又は線材試験片に全方向繰返し曲げを与える疲労試験．繰返し曲げモーメントは，一定の角度を保った試験片の両端把持部を回転させることによって得られる．
　回转弯曲疲劳试验　[huí zhuǎn wān qǔ pí láo shì yàn]
　uji kelelahan putar benkok　[uji kererahan putaru benkoku]
　회전 굽힘 피로시험　[hoejeon gupim pirosiheom]
　การทดสอบความล้าหมุนงอ　[gaan tod-sorb kwaam laa mun-ngor]

rotating speed
　the number of rotations per unit of time.
　回転速度　[kaiten sokudo]
　単位時間当たりの回転数．
　转速　[zhuǎn sù]
　kecepatan beputar　[kechepatan beputaru]
　회전 속도　[hoejeon sokdo]
　ความเร็วการหมุน　[kwaam rew gaan mun]

rough grinding
　a grinding process that leaves margin on a workpiece for finishing.
　荒削り研削　[arakezuri kensaku]
　工作物の仕上げに要する取代を残して研削あるいは研磨加工する作業．
　粗切削　[cū qiē xuē]
　penggerindaan kasar　[pungugerindaan kasaru]
　거친연삭　[geochinnyeonsak]
　การเจียรหยาบ　[gaan chiaa yaab]

rough roll
　the initial stage of rolling process of steel.
　粗圧延　[soatuen]
　鋼塊から製品の最終的工程のうち，最初の段階の圧延．
　粗轧　[cū zhá]
　rol kasar　[roru kasaru]
　조압연　[joabyeon]

การรีดหยาบ [gaan riid yaab]
rough surface
generic term of surface defects to be found on the spring surface such as scale attachment, roughness caused by overheating, rust and corrosion.
はだ荒れ [hadaare]
ばね表面のスケールの付着，過熱による表面荒れ，さび，腐食などの表面欠陥の総称.
粗糙表面 [cū cāo biǎo miàn]
permukaan kasar [purumukaan kasaru]
표면 조도 [pyomyeon jodo]
ผิวหยาบเป็นรอย [piw yaab pen rooi]

roughness
the degree of unevenness on a material surface. Surfaces with no unevenness are classed as smooth surface; surfaces with less unevenness as semismooth surface; and surfaces with much unevenness as rough surface.
粗さ [arasa]
表面の凹凸の程度をいい，その区分は，滑らかで凹凸などがない状態を平滑面，凹凸などが少ない状態を準平滑面,凹凸などが多い状態を粗面という．
粗糙度 [cū cāo dù]
kekasaran [kekasaran]
거칠기 [geochilgi]
ความหยาบ [kwaam yaab]

roughness measurement
a measurement of the surface roughness (smooth, semismooth, or rough). Roughness is measured using optical interference, contact probe, electrical capacitance, air leakage, or optical roughness gauges.
粗さ測定 [arasa sokutei]
表面の凹凸の程度（平滑面，準平滑面，粗面）の程度を測定すること．光干渉式，触針式，電気容量式，空気漏洩式，光学式などの粗さ計を用いて測定する．
粗糙度測量 [cū cāo dù cè liáng]
pengukuran kekerasan [pungukuran kekerasan]
조도 측정 [jodo cheukjeong]
การวัดความหยาบ [gaan wad kwaam yaab]

round edge of flat bar
the side edge of flat steel bar with the radius of about half the thickness.
丸こば [marukoba]
板ばねの平鋼の側部の曲率半径を板厚のほぼ半分にとった場合を，丸こばという．
圆端 [yuán duān]
batang flat bersisi bulat [batangu furato berusisi burato]
판의 둥근 모서리 [panui dunggeun moseori]
ขอบมนของเหล็กแผ่น [khorb mon khong lek paaen]

round steel
steel rolled or forged into bars with a circular cross section.
丸鋼 [marukou]
棒状に圧延又は鍛造された鋼で断面の形状が円形の鋼材.
圆钢 [yuán gāng]
baja bulat [baja burato]
환강 [hwangang]
เหล็กรีดแผ่นชนิดหน้าตัดกลม [lek riid paaen cha-nid naa-tad glom]

round wire
wire with a circular cross section.
丸線 [marusen]
断面が円形の線.
圆形金属丝 [yuán xíng jīn shǔ sī]
kawat bulat [kawato burato]
원형 선재 [wonhyeong seonjae]

ลวดกลม　　[lu-ad glom]

rounded edge
  the rounded edge of steel plate for leaf springs which has a curvature radius of about half the thickness.
丸端面（丸こば）　[maru tanmen (marukoba)]
板厚のほぼ半分の曲率半径で丸みをつけた，板ばね用平鋼の側部．
圓形端面　[yuán xíng duān miàn]
tepi yang dibulatkan　[tepi yangu diburatokan]
둥근 끝부 모서리　[dunggeun kkeutbu moseori]
ขอบมน　　[khorb mon]

rounded shot
  the generic term for spherical shot. May be steel, cast iron, or glass.
丸形ショット　[marugata syotto]
球形をなしているショットの総称で，スチールショット，鋳鋼ショットガラス系ショットなどがある．
圓形鋼丝切丸　[yuán xíng gāng sī qiē wán]
kawat dipotong bulat　[kawato dipotongu burato]
라운드 컷트 와이어　[raundeu keotteu waieo]
เม็ดช็อตรูปกลม　[med shot ruup glom]

rubber covered compression spring
  a helical compression spring covered with rubber. Used for railway cars, supports of vibrating equipment, or as resonance springs and antivibration springs. Commonly known as Eligo spring.
圧縮ゴム被覆コイルばね　[assyuku gomu hihuku koiru bane]
ゴムで被覆した圧縮コイルばね．鉄道車両用，振動機器の支持用，あるいは共振ばね，防振用ばねとして用いられ

ている．通称，エリゴばねともいう．
橡胶包层压缩弹簧　[xiàng jiāo bāo céng yā suō tán huáng]
pegas ulir berlapis karet tekan　[pugasu uriru berurapisu kareto tekan]
고무 피복 압축 코일 스프링　[gomu pibok apchuk koil spring]
สปริงขดรับแรงกดชนิดติดยาง　[sa-pring khod rab rang god cha-nid tid yaang]

rubber mount type
  eye of a bar stabilizer with the positioning rubber bush.
ゴムマウントタイプ　[gomu maunto taipu]
バースタビライザ目玉部形状の一種で，ゴムブシュにより目玉を挟み固定する形式．
橡胶骨架　[xiàng jiāo gǔ jià]
karet jenis mouunt　[kareto jenisu mouunto]
고무 마운트 형　[gomu maunteu hyeong]
หูสตาบิไลเซอร์ชนิดใส่บุชยาง　[huu stabilizer cha-nid sai bush yaang]

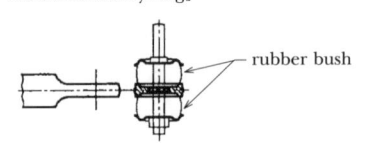

Fig R-6　rubber mount type

rubber pad
ゴムパッド　[gomu paddo]
橡胶垫　[xiàng jiāo diàn]
bantalan karet　[bantaran kareto]
고무 패드　[gomu paedeu]
แผ่นยาง　[paaen yaang]

rubber spring
  spring using the elasticity of rubbers, as called by its material.
ゴムばね　[gomu bane]
ばねの材料の種類からいう名称で，ゴ

ムの弾性を利用するばね．
橡胶弹簧　[xiàng jiāo tán huáng]
pegas karet　[pugasu kareto]
고무 스프링　[gomu spring]
สปริงยาง　[sa-pring yaang]

rubbing corrosion
 a damage or corrosion due to friction occurred on the interface between metals or a metal and another material due to repeated micro-skid in the air.
擦過腐食　[sakka husyoku]
空気中で金属とそれに接触する金属又は他の物質との接触面で相対的な繰返し微小滑りが起きたとき金属表面に生じる損傷又は摩擦腐食現象．
摩擦腐蚀　[mó cā fǔ shí]
korosi karena gosokan　[korosi karena gosokan]
마찰 부식　[machal busik]
การกัดกร่อนจากการเสียดสี　[gaan gad-gron chaak gaan seiad-sii]

rust
 the compound primarily made up of hydroxides or oxides formed on a ferrous metal surface. More commonly refers to corrosion formed on a metal surface.
さび　[sabi]
普通には，鉄表面に生成する水酸化物又は酸化物を主体とする化合物．広義には，金属表面にできる腐食成分をいう．
锈　[xiù]
karat　[karato]
녹　[nok]
สนิม　[sa-nim]

rust protection
 surface protection against rust on metals. This is normally achieved using plating, coating, or rust preventive oil.

防せい　[bousei]
金属にさびが発生するのを防ぐこと．一般にめっき，塗装，防せい油などの処理を施している．
防锈　[fáng xiù]
perlindungan terhadap karat　[pururindungan teruhadapu karato]
방청　[bangcheong]
การป้องกันสนิม　[gaan pong-gan sa-nim]

rust resisting steel
 alloy steel containing chromium or chromium and nickel to increase corrosion resistance. Alloys containing more than about 11% chromium are generally called stainless steel, which can be subdivided into martensitic stainless steel, ferritic stainless steel, austenitic stainless steel, austenitic/ferritic stainless steel, and precipitation hardening stainless steel depending on the structure.
不銹鋼　[husyuukou]
耐食性を向上させる目的でクロム又はクロムとニッケルを含有させた合金鋼．一般には，クロム含有量が約11%以上の鋼をステンレス鋼といい，主として，その組織によって，マルテンサイト系，フェライト系，オーステナイト系，オーステナイト・フェライト系及び析出硬化系の五つに分類される．
不锈钢　[bù xiù gāng]
baja pencegah karat　[baja punchegahu karato]
내식강　[naesikgang]
เหล็กโลหะผสมทนสนิม　[lek loo-ha pa-som ton sa-nim]

rust resistivity
 metal characteristics to show toughness against rust or oxidation.

rust resistivity

耐さび性　[taisabisei]
　金属がさびに侵され（酸化され）にくい性質.
耐锈　[nài xiù]
mencegah karat　[menchegahu karato]
내식성　[naesikseong]
การทนทานต่อสนิม　[gaan ton-taan to-a sa-nim]

S

safety belt spring
 the generic term for springs used in car safety belts. These include power springs to wind the belt and flat springs used in buckles.
 安全ベルト用ばね　[anzen berutoyou bane]
 自動車などの安全ベルトに用いられているばねの総称で，ベルト巻取り用のぜんまいばねやバックル部に用いられている薄板ばねなどがある．
 安全带弹簧　[ān quán dài tán huáng]
 pegas sabuk keselamatan　[pugasu sabuku keseramatan]
 안전 벨트용 스프링　[anjeon belteuyong spring]
 สปริงเข็มขัดนิรภัย　[sa-pring khem-khad ni-ra-pai]

safety device
 a hazard prevention device or alarm system of machines for an operator.
 安全装置　[anzen souti]
 機械操作者のための危険防止あるいは警報装置．
 安全装置　[ān quán zhuāng zhì]
 sarana penyelamatan　[sarana punyeramatan]
 안전 장치　[anjeon jangchi]
 อุปกรณ์นิรภัย　[up-pa-gorn ni-ra-pai]

safety factor on margin
 rate of the load (stress) causing spring's breakdown (yield or fatigue) to the load (stress) under the normal use condition.
 破壊安全率　[hakai anzenritu]
 ばねが破壊（降伏，疲れ）を起こす荷重（応力）と通常の使用状況下における荷重（応力）との比．
 安全系数　[ān quán xì shù]
 faktor keselamatan pada batas　[fakutoru keseramatan pada batasu]
 파괴 안전률　[pagoe anjeollyul]
 อัตราความปลอดภัย　[at-traa kwaam plord-pai]

safety load
 a load that does not produce stress beyond the limits of a structure or body.
 安全荷重　[anzen kazyuu]
 構造物，物体などに制限以上の応力を生じさせない荷重．
 安全负荷　[ān quán fù hè]
 beban keselamatan　[beban keseramatan]
 안전 하중　[anjeon hajung]
 โหลดในระดับที่ปลอดภัย　[load nai ra-dap tii plord-pai]

safety sign
 a sign which gives a general safety message obtained by a combination of color and geometric shape.
 安全標識　[anzen hyousiki]
 色と形状の組合せで得られる一般的な安全上の伝達内容を表す標識．
 安全标识　[ān quán biāo shí]
 tanda keamanan　[tanda keamanan]
 안전표지　[anjeonpyoji]
 สัญลักษณ์ความปลอดภัย　[san-ya-lak kwaam plord-pai]

salt bath heat treatment
 heat treatment performed in a salt bath.
 塩浴熱処理　[en'yoku netusyori]
 塩浴中で行う熱処理．

salt bath heat treatment — scale

盐浴　[yán yù]
olah panas bak garam　[orahu panasu baku garan]
염욕 열처리　[yeomnyok yeolcheori]
ชุบแข็งด้วยน้ำเกลือ　[shub khaaeng du-oi naam-gluua]

salt spray test
test which examines rust and blister after installing the test piece or the spring in the test chamber sprayed with 5% aqueous solution of natrium chloride kept at 35°C. This test is applied to surface treated test pieces such as plating and coating, and stainless steel.
塩水噴霧試験　[ensui hunmu siken]
5％塩化ナトリウム水溶液を35℃に保って噴霧させた試験装置内へ試験片又はばねを静置して，さび，膨れなどの発生状態を調べる試験．めっき，塗覆装などの表面処理を施したもの，ステンレス鋼などに用いられる．
盐雾试验　[yán wù shì yàn]
uji semprot garam　[uji senpuroto garan]
염수 분무 시험　[yeomsu bunmu siheom]
การทดสอบการกัดกร่อนโดยพ่นน้ำเกลือ　[gaan tod-sorb gaan gad gorn dooi pon naam gluua]

salt water immersion test
a test that examine corrosion or deterioration of metal by dipping samples into a solution of sodium chloride.
塩水浸せき試験　[ensui sinseki siken]
試料を塩化ナトリウム溶液中に浸して，腐食，さび，劣化などの状態を調べる試験．
盐水浸渍试验　[yán shuǐ jìn zì shì yàn]
uji rendam air garam　[uji rendan airu garan]
염수 침적 시험　[yeomsu chimjeok siheom]
การทดสอบการแช่น้ำเกลือ　[gaan tod-sorb gaan shiaa naam gluua]

sample size
the number of items inspected in a sample.
サンプルの大きさ　[sanpuru no ookisa]
サンプルに含まれる検査単位の数．
样品尺寸　[yàng pǐn chǐ cùn]
ukuran contoh　[ukuran chontohu]
시료 크기　[siryo keugi]
ขนาดของกลุ่มตัวอย่าง　[kha-naad khong glum to-a yaang]

sand blasting
process similar to shot peening, where the grained polishing agents are blasted on the surface of the spring by compressed air, centrifugal force or other method. Also used for the surface cleaning of the spring.
サンドブラスト　[sando burasuto]
粒状の研磨材を圧縮空気，遠心力又はその他の方法によって，ばねの表面に打ち付けて行うショットピーニングに類似した加工．ばねの表面清浄目的にも用いられる．
喷丸机　[pēn wán jī]
semprot pasir　[senpuroto pasiru]
샌드 블라스트　[saendeu beullaseuteu]
การยิงขัดผิวด้วยเม็ดทราย　[gaan ying khad piw do-ui med yaab]

scale
scalelike oxide film on the surface of steel when heated.
スケール　[sukeeru]
鋼の加熱の際，酸化によって生じた表面のうろこ状の皮膜．
氧化皮　[yǎng huà pí]
skala, sisik kerak, timbangan　[sukara, sisiku keraku, tinbangan]
스케일　[seukeil]
ผิวสเกลของเหล็ก　[piw scale khong lek]

scattergram

a chart designed for visual grasp of the nature of data by plotting x and y data on x-y plane.

散布図　[sanpuzu]

xとyの関係をx-y平面上にプロットして視覚的に把握できるようにした図.

分布图　[fēn bù tú]

began tebaran　[began tebaran]

산포도　[sanpodo]

กราฟแสดงข้อมูล　[graph sa-daaeng ko-a moon]

scrap reel

an equipment used to coil punched-out metal scrap ejected from a press.

スクラップリール　[sukurappu riiru]

プレスから送り出される抜きかすを巻き取る装置.

落料清除装置　[luò liào qīng chú zhuāng zhì]

penggulung sampah besi　[pungugurungu sanpahu besi]

스크랩 릴　[seukeuraep ril]

ตัวดันเอาสแครปออกจากเครื่องเพรส　[tu-a dan a-o scarp org chaak kre-ung press]

scratch

a continuous or intermittent linear mark formed on the surface of material during handling.

すりきず　[surikizu]

取扱い中に表面にできた連続又は断続した線状のきず.

划痕　[huá hén]

goresan　[goresan]

스크래치　[seukeuraechi]

รอยขีดข่วน　[roi chiid khu-an]

seal

密封　[mippuu]

密封　[mì fēng]

penyekat　[punyekato]

실　[sil]

ผนึก　[pa-nuk]

seasoning crack

a crack occurring in steel during storage after quenching or tempering. Also called natural cracking.

置割れ　[okiware]

焼入れ又は焼入焼戻した鉄鋼が放置中に生じる割れ．自然割れともいう.

放置开裂　[fàng zhì kāi liè]

retak musim　[retaku musimu]

자연 균열　[jayeon gyunnyeol]

การแตกด้วยตัวเอง　[gaan taaek du-oi tu-a e-eng]

seat

generic term of the component where the spring is seated.

座　[za]

ばねを据える部位，相手部品などの総称.

基座　[jī zuò]

dudukan　[dudukan]

시트　[siteu]

แท่นรอง　[taaen rong]

seat spring

generic term for springs for cushions used in the seats, chairs and beds of vehicles. This includes zigzag springs, helical compression springs, barrel-shaped springs and air springs.

シートばね　[siito bane]

車両用の座席，いす，寝具などに用いるクッション用のばねの総称．ジグザグばね，圧縮コイルばね，たる形コイルばね，空気ばねなどが用いられる.

座椅弹簧　[zuò yǐ tán huáng]

pegas tempat duduk　[pugasu tenpato duduku]

시트 스프링　[siteu spring]

สปริงเบาะรอง　[sa-pring ba-or rorng]

seat spring — self hardening

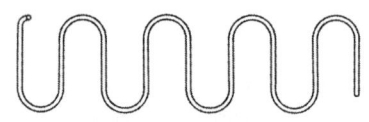

Fig S-1 seat spring

second leaf wrapper
   turned section at the end of second leaf next to the main leaf wrapping around the eye of main leaf.
   二番巻　[niban maki]
   親板の次のばね板の端部を，親板の目玉に沿って巻いた部分．
   次級簧片套　[cì jí huáng piàn tào]
   lilitan kedua　[riritan kedua]
   둘째판 감김 (1/4)　[duljjaepan gamgim (1/4)]
   ขดของปลายแผ่นที่ 2　[khod khong plaai paaen tii song]

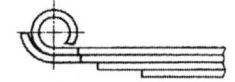

Fig S-2 second leaf wrapper

second stage leaf
   a spring that works complementarily under increased load in a progressive leaf spring.
   補助ばね　[hozyo bane]
   プログレッシブ重ね板ばねにおいて，荷重の増加とともに補助的に働くばね．
   辅助弹簧片　[fǔ zhù tán huáng piàn]
   daun tahap kedua　[daun tahapu kedua]
   보조 스프링　[bojo spring]
   แหนบช่วย　[naaep shu-oi]

secondary stress
   self-limited vertical stress and shear stress that are generated by self restriction or restriction from a neghiboring structure.
   二次応力　[nizi ouryoku]
   構造物の隣接部分の拘束又は自己拘束によって生じ，自己制限的性質を持つ垂直応力又はせん断応力．
   二次应力　[èr cì yīng lì]
   tegangan kedua　[tegangan kedua]
   이차 응력　[icha eungnyeok]
   ความเค้นทุติยภูมิ　[kwaam ken tut-ti-ya-puum]

section modulus
   the value obtained by dividing geometrical moment of inertia around the neutral axis of cross-section of a beam by the distance from the neutral axis to outer surface.
   断面係数　[danmen keisuu]
   はりの断面の中立軸に関する断面二次モーメントの値を中立軸から外表面までの長さで除したもの．
   截面系数　[jié miàn xì shù]
   batang berpenampang　[batangu berupunanpangu]
   단면 계수　[danmyeon gyesu]
   โมดูลัสหน้าตัด　[modulus naa-tad]

segregation
   the nonuniform distribution of alloying elements and impurities.
   偏析　[henseki]
   合金元素や不純物が不均一に偏在している状態．
   偏析　[piān xī]
   pemisahan　[pumisahan]
   편석　[pyeonseok]
   การแยกออกไป　[gaan yaaek ork pai]

self hardening
   the property of a certain steel whereby austenite transforms into martensite readily by air cooling from quenching temperature accompanied by hardness increments.
   自硬性　[zikousei]
   焼入温度から空気中で冷却する程度で

self hardening — separated hook extension spring

も，容易にオーステナイトからマルテンサイトを生じて硬化する鋼の性質．
自硬性　[zì yìng xìng]
pengerasan diri　[pungerasan diri]
자경성　[jagyeongseong]
การเกิดความแข็งเอง　[gaan gerd kwaam khaaeng e-eng]

semi-elliptic spring
   leaf spring shaped like a semi-ellipse. (see Fig S-3)
半だ円ばね　[handaen bane]
半だ円のような形状をした，重ね板ばね．（Fig S-3 参照）
半椭圆弹簧　[bàn tuǒ yuán tán huáng]
pegas semi eliptik　[pugasu semi eriputiku]
반타원 스프링　[bantawon spring]
สปริงกึ่งรูปไข่　[sa-pring gung roop khai]

semi-round mandrel
   a mandrel made of a semiround bar or rectangular bar with one face rounded.
半丸心金　[hanmaru singane]
半丸棒状，又は一つの面に丸みの付いた角柱状の心金．
半圆心轴　[bàn yuán xīn zhóu]
poros pemegang setengah bulat　[porosu pumegangu setengahu burato]
반원 맨드릴　[banwon maendeuril]
แกนม้วนครึ่งวงกลม　[gaaen muan krung wong-glom]

sensor
   a device that senses a phisical quantity, and converts it to an electronic signal for a system control unit.
センサ　[sensa]
物理量を感知して電気信号に変換しシステムの制御に用いる素子．
传感器　[chuán gǎn qì]
sensor　[sensoru]
센서　[senseo]
ตัวเซ็นเซอร์　[tu-a sensor]

sensory test
   a test method for screening the characteristics of a test piece using the five senses of human beings.
官能検査　[kannou kensa]
人間の五官によって，試験体の特性を選別する試験方法．
感度试验　[gǎn dù shì yàn]
uji kepekaan　[uji kepekaan]
관능 검사　[gwanneung geomsa]
การทดสอบการตอบสนอง　[gaan tod-sorb gaan torb-sa-nong]

separated hook extension spring
   a helical extension spring with independent hooks inserted at both ends. In some cases, the hooks are screwed in.
自在フック式引張コイルばね　[zizai hukkusiki hippari koiru bane]
独立したフックを両端に挿入した引張コイルばね．ねじ込む場合もある．
分体钩环拉簧　[fēn tǐ gōu huán lā huáng]
pegas tarik kait terpisah　[pugasu tariku kaito terupisahu]

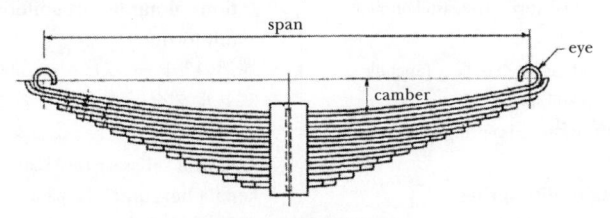

Fig S-3 semi-elliptic spring

separated hook extension spring — serrated head

고리분리형 인장 코일 스프링
[goribullihyeong injang koil spring]
สปริงขดรับแรงดึงชนิดตะขอแยก　　[sa-pring khod rab raaeng dung cha-nid ta-khor yaaek]

Fig S-4　separated hook extension spring

sequential control
　　a process control method where the process is controled by a pre-determined program.
シーケンス制御　[siikensu seigyo]
あらかじめ定められた順序に従って，制御の各段階を逐次進めていく制御．
序列控制　　[xù liè kòng zhì]
pengendalian berurutan　　[pungendarian berurutan]
시퀀스 제어　　[sikwonseu jeeo]
การควบคุมซีเควนเชียล　　[gaan ku-ob kum sequential]

series combination coil spring
　　coil springs combined in series. The spring constant is reduced in halves if two identical springs are combined.
圧縮直列組合せコイルばね　　[assyuku tyokuretu kumiawase koiru bane]
直列式に組み合わせたコイルばね．同じばねを2個組み合わせるとばね定数が1/2になる．
直列组合压缩螺旋弹簧　　[zhí liè zǔ hé yā suō luó xuán tán huáng]
pegas kombinasi seri　　[pugasu konbinasi seri]
압축 직열조합 코일 스프링　　[apchuk jingnyeoljohap koil spring]
สปริงขดรวมเป็นซีรี่ย์　　[sa-pring khod ru-am pen series]

series stacking of disc spring
　　disc springs stacked in series with adjacent tops or bottoms facing each other.
皿ばねの直列積重ね　　[sara bane no tyokuretu tumikasane]
皿ばねを直列に重ね合わせたことをいう．底面同士が向かい合うように重ねる．
对合组合碟形弹簧　　[duì hé zǔ hé dié xíng tán huáng]
tumpukan berangkai (dari cangkram tunggal)　　[tunpukan berangukai (dari changukuran tungugaru)]
접시 스프링 직열 적층　　[jeopsi spring jingnyeol jeokcheung]
การเรียงกันเป็นตั้งของสปริงแผ่น　　[gaan riang gan pen tang khong sa-pring paaen]

series type coiled wave spring
　　a coiled wave spring wound so that adjacent peaks contact.
直列式コイルドウェーブスプリング
[tyokuretusiki koirudo ueebu supuringu]
凸部同士が接触するようにコイリングされたコイルドウェーブスプリング．
直列式波片弹簧　　[zhí liè shì bō piàn tán huáng]
pegas ulir gelombang seri　　[pugasu uriru geronbangu seri]
직열식 코일드 웨이브 스프링
[jingnyeolsik koildeu weibeu spring]
สปริงขดชนิดเป็นซีรี่　　[sa-pring khod cha-nid pen series]

serrated head
　　the head of a torsion bar with serrations along its circumference to transmit torque.
端部のセレーション　　[tanbu no sereesyon]
トルクを伝達するために周囲に溝を切った形状のトーションバーの頭部．
端部花键　　[duān bù huā jiàn]
kepala bergurat　　[kepara berugurato]
끝부 세레이션　　[kkeutbu sereisyeon]

ส่วนหัวที่เจียร์ร่องฟัน [suan hu-a tii chiaa rong-fan]

serration profile
 a profile curve of serration on a shaft or corresponding hole.
 セレーション形状 [sereesyon keizyou]
 軸側及び嵌合する穴側に設けたセレーションの断面形状.
 花键形状 [huā jiàn xíng zhuàng]
 profil bergurat [purofiru berugurato]
 세레이션 형상 [sereisyeon hyeongsang]
 ลักษณะการเจียร์ร่องฟัน [luk-sa-na gaan chiaa rong-fan]

service life
 the length of time during which a machine or a tool can be operated before breakdown.
 耐用年数 [taiyou nensuu]
 機械や工具が壊れる前に使用できる時間の長さ.
 使用寿命 [shǐ yòng shòu mìng]
 umur layanan [umuru rayanan]
 내용 연수 [naeyong yeonsu]
 อายุการใช้งาน [aa-yu gan shai-ngaan]

set
 generic term for the permanent distortion of a spring. This includes static, dynamic and warm set.
 へたり [hetari]
 ばねの永久変形の総称. 静的へたり, 動的へたり及び温間へたりなどがある.
 变形 [biàn xíng]
 pasang [pasangu]
 변형 [byeonhyeong]
 การเปลี่ยนรูปอย่างถาวรของสปริง [gaan plian ruup yaang ta-worn khong sa-pring]

set angle
 the angle at which a torsion coil spring is mounted.

取付角度 [torituke kakudo]
ねじりコイルばねの取付角度.
安装角度 [ān zhuāng jiǎo dù]
sudut pemasangan [suduto pumasangan]
장착 각도 [jangchak gakdo]
มุมในการประกอบ [mum nai gaan pra-gorb]

set height
 height of a spring when mounted. Set height is used for a leaf spring and a helical compression spring, and set length for helical extension spring.
 ばねの取付高さ [bane no torituke takasa]
 取付時のばねの高さ. 重ね板ばね, 圧縮コイルばねのときは, 取付高さといい, 引張コイルばねのときは, 取付長さという.
 安装高度 [ān zhuāng gāo dù]
 tinggi yang ditetapkan [tingugi yangu ditetapukan]
 스프링 장착 높이 [spring jangchak nopi]
 ความสูงประกอบ [kwaam soong pra-gorb]

set length
 set length for helical extension spring.
 ばねの取付長さ [bane no torituke nagasa]
 引張コイルばねの取付長さ.
 安装长度 [ān zhuāng cháng dù]
 panjang pegas yang ditetapkan [panjangu pugasu yangu ditetapukan]
 스프링 장착 길이 [spring jangchak giri]
 ระยะประกอบ [ra-ya pra-gorb]

setting
 operation to increase stress relaxation resistance and durability, by applying a load or torque in advance which exceeds the maximum value in service generating a certain degree of permanent distortion.
 セッチング [settingu]
 ばねにあらかじめ使用される最大値を超える荷重又はトルクを加えて, ある

程度の永久変形を生じさせ，ばねの弾
性限を高め，耐へたり性，耐久性を向
上させる操作．
立定処理　[lì dìng chǔ lǐ]
setting　[settingu]
세팅　[seting]
การเซ็ทเพื่อเพิ่มความคงทน　[gaan set pe-uu perm kwaam kong-ton]

setting height
height of a spring when setting load is applied.
セッチング高さ　[settingu takasa]
ばねをセッチングするときの高さ．
立定処理時的高度　[lì dìng chǔ lǐ shí de gāo dù]
tinggi penentu　[tingugi penentu]
세팅 높이　[seting nopi]
ความสูงเซ็ท　[kwaam soong set]

setting load
load required to pre-relax a spring in order to improve stress relaxation resistance in service.
セッチング荷重　[settingu kazyuu]
ばねをセッチングするときの荷重．
立定処理時的负荷　[lì dìng chǔ lǐ shí de fù hè]
beban penentu　[beban punentu]
세팅 하중　[seting hajung]
เซ็ทดิ้งโหลด　[setting load]

setting stress
stress generated in a spring when setting load is applied.
セッチング応力　[settingu ouryoku]
ばねをセッチングするときの応力．
立定処理時的应力　[lì dìng chǔ lǐ shí de yīng lì]
pegas penentu　[pugasu punentu]
세팅 응력　[seting eungnyeok]
ความเค้นจากการเซ็ท　[kwaam ken chaak gaan set]

shackle
an U-shaped metal fitting to connect an eye of a leaf spring and a body chasis. It compensates the change of span when the spring is deflected.
シャックル　[syakkuru]
重ね板ばねの板のスパンの変化を逃すために目玉とシャシを連結するU字形の金具．
带销U形钩　[dài xiāo U xíng gōu]
penyangga, anting anting　[punyanguga, antingu antingu]
샤클　[syakeul]
แช็คเคิล　[shackle]

shackle eye
holes for the connection pin of a shackle.
シャックルアイ　[syakkuru ai]
シャックルの目玉．ピンを差し込む穴．
U形钩的孔　[U xíng gōu de kǒng]
gelang penyangga　[gerangu punyanguga]
샤클 아이　[syakeul ai]
รูสำหรับใส่พิน　[ruu sam-rab sai pin]

shake-proof washer
a washer used to prevent bolts or nuts from loosening.
ゆるみどめ座金　[yurumidome zagane]
ボルトやナットなどのねじがゆるまないように締結部にはめ込む座金．
缓冲垫圈　[huǎn chōng diàn quān]
ring pipih tahan guncangan　[ringu pipihu tahan gunchangan]
이완 방지 와셔　[iwan bangji wasyeo]
แหวนสปริงป้องกันการหลวม　[waaen sa-pring pong-gan gaan lu-am]

shape memory alloys
an alloy that possesses a shape-memory effect. These include Ti-Ni and Ca-Zn-Al alloys, which can recover their original shape due to temperature change.

形状記憶合金　[keizyou kioku goukin]
　形状記憶効果を有する合金．Ti-Ni 合金，Ca-Zu-Al 合金などがあり，温度変化によって形状回復する性質を持つ．
形状记忆合金　[xíng zhuàng jì yì hé jīn]
paduan memori bentuk　[paduan memori bentuku]
형상 기억 합금　[hyeongsang gieok hapgeum]
โลหะผสมเซฟเมมโมรี่　[loo-ha pa-som shape memory]

shape memory alloy spring
　a spring made of shape memory alloy. It returns to its original shape by means of temperature change.
形状記憶合金ばね　[keizyou kioku goukin bane]
　形状記憶合金で作られたばね．このばねは温度変化によって形状回復する．
形状记忆合金弹簧　[xíng zhuàng jì yì hé jīn tán huáng]
pegas logam paduan mengingat bentuk　[pegasu rogan paduan mengingato bentuku]
형상 기억 합금 스프링　[hyeongsang gieok hapgeum spring]
สปริงโลหะผสมเซฟเมมโมรี่　[sa-pring loo-ha pa-som shape memory]

shape of bar end
　the shape of the end of an object that has a bar shape.
バー端末形状　[baa tanmatu keizyou]
　棒状に加工されたものの端部の形状のこと．
杆端形状　[gǎn duān xíng zhuàng]
bentuk ujung batang　[bentuku ujungu batangu]
봉 끝단면 형상　[bong kkeutdanmyeon hyeongsang]
รูปร่างของปลายแท่ง　[ruup-rang khong plaai taaeng]

shaving
　process to finish the outer or inner circumference of the spring by shaving press.
シェービング　[syeebingu]
　ばねなどの外周，内周などを，所定の寸法に削り取って仕上げる加工方法．
修边　[xiū biān]
menyerut　[menyeruto]
셰이빙　[syeibing]
ลอกออก　[lork-ork]

shear
せん断　[sendan]
剪断　[jiǎn duàn]
irisan, mengiris　[irisan, mengirisu]
전단　[jeondan]
การตัดออก　[gaan tad ork]

shear angle
　the inclination of the upper blade to the lower blade when cutting sheet metal.
せん断角度　[sendan kakudo]
　板金を切断するとき下刃に対して上刃の傾きをいう．
剪断角度　[jiǎn duàn jiǎo dù]
sudut irisan　[suduto irisan]
전단 각도　[jeondan gakdo]
มุมในการตัด　[mum nai gaan tad]

shear modulus
　ratio of shearing stress to shearing strain within the elastic limit. Commonly designated by G. Unit is MPa or N/mm^2.
横弾性係数　[yoko dansei keisuu]
　弾性限度内におけるせん断応力とせん断ひずみとの比．量記号：G，単位記号：MPa 又は N/mm^2．
剪切模量　[jiǎn qiē mó liàng]
modulus irisan　[modurusu irisan]
횡탄성 계수　[hoengtanseong gyesu]

โมดุลัสในการตัด   [modulus nai gaan tad]

shear strength
the maximum shear stress that a material can withstand without shear fracture. The unit of measure is MPa or N/mm^2.
せん断強さ   [sendan tuyosa]
せん断力を加えたとき材料がせん断破壊を生じるまでの最大応力をいう．単位はMPa又はN/mm^{2}で表す．
剪切強度   [jiǎn qiē qiáng dù]
kuat iris   [kuato irisu]
전단 강도   [jeondan gangdo]
กำลังในการตัด   [gam-lang nai gaan tad]

shear stress
stress defined by the quotient of force acting in parallel direction to the cross section divided by the area of the cross section.
せん断応力   [sendan ouryoku]
物体に形状変化をもたらすせん断方向の応力．
剪切応力   [jiǎn qiē yīng lì]
tegangan iris   [tegangan irisu]
전단 응력   [jeondan eungnyeok]
ความเค้นในการตัด   [kwaam ken nai gaan tad]

shearing load
a load which is applied to neighboring parallel planes with equal magnitude and opposite direction to each other. Shearing stress is induced by shearing load.
せん断荷重   [sendan kazyuu]
せん断力を加えたとき材料がせん断破壊を生じるまでの最大荷重．
剪切负荷   [jiǎn qiē fù hè]
beban irisan   [beban irisan]
전단 하중   [jeondan hajung]
โหลดในการตัด   [load nai gaan tad]

shearing strain
strain in the shearing direction indicating the distortion of an object.
せん断ひずみ   [sendan hizumi]
物体の形状変化を表すせん断方向のひずみ．
剪切応変   [jiǎn qiē yīng biàn]
regang irisan   [regangu irisan]
전단 변형   [jeondan byeonhyeong]
ความเครียดในการตัด   [kwaam kriead nai gaan tad]

sheet gauge
nominal thickness of sheet products.
板厚標準寸法   [itaatu hyouzyun sunpou]
板状製品の厚さの標準寸法．
片規   [piàn guī]
tebal lembaran   [tebaru renbaran]
판 두께 표준 치수   [pan dukke pyojun chisu]
เกจสำหรับวัดความหนาแผ่น   [guage sam-rab wad kwaam naa paaen]

sheet metal gauge
a gauge used to measure gap size by combining metal pieces with predetermined thickness.
板ゲージ   [ita geezi]
あらかじめ厚さを測定済みの板小片を組み合わせたゲージ．すきまの測定に使用する．
塞尺   [sāi chǐ]
tebal logam lembaran   [tebaru rogan renbaran]
판 게이지   [pan geiji]
เกจสำหรับวัดช่องว่างระหว่างแผ่น   [guage sam-rab wad shu-ong ra-waang paaen]

sheet steel
rolled steel plate that has thickness of about 1 mm and is cut into specified size. Uncut steel strip is called band steel.

薄板鋼板　[usuita kouhan]
　一般に厚さ1mm程度の圧延鋼板を所定の大きさに切断されたものをいう．切断されずに帯状のものを帯鋼という．
薄钢板　[báo gāng bǎn]
baja lembaran　[baja renbaran]
박판 강판　[bakpan gangpan]
โลหะแผ่นบาง　[loo-ha paaen baang]

sherardizing
　the process that heats steel in zinc powder or a mixed powder containing zinc and diffuses zinc into the steel surface to form a corrosion resistant film.
シェラダイジング　[syeradaizingu]
　粉末亜鉛又はこれを含む混合粉末中で鉄鋼を加熱してその表面に亜鉛を拡散させて耐食性皮膜をつくる操作．
粉末镀锌　[fěn mò dù xīn]
pelapisan pada permukaan seng tambahan　[purapisan pada purumukaan sengu tanbahan]
세라다이징　[seradaijing]
การชุบด้วยดีบุก / ตะกั่ว　[gaan shub du-oi dii-buk / ta-gu-a]

shim
　a thin piece of metal inserted between objects to adjust gap size.
シム　[simu]
　物と物との間に挿入して間隙調整に用いる薄板状のもの．
调整垫　[tiáo zhěng diàn]
penyisip　[punyisipu]
심　[sim]
แผ่นชิมปรับขนาดช่อง　[paaen shim prab khanaad shu-ong]

shock absorber
　an equipment to alleviate mechanical shock. Springs, rubber, air pressure or hydraulic pressure is used to absorb kinetic energy.
緩衝器　[kansyouki]
　機械的衝撃を緩和する装置で，ばね，ゴム，空気圧，油圧などを利用して運動エネルギーを吸収する．
缓冲器　[huǎn chōng qì]
peredan kejut　[puredan kejuto]
완충기　[wanchunggi]
ตัวกันสะเทือน　[tu-a gan sa-te-un]

shock absorber spring
　a spring used to alleviate shock.
ショックアブソーバ用ばね　[syokku abusoobayou bane]
　衝撃を緩和するために用いられるばね．
缓冲弹簧　[huǎn chōng tán huáng]
pegas peredan kejut　[pugasu puredan kejuto]
완충 스프링　[wanchung spring]
สปริงกันสะเทือน　[sa-pring gan sa-te-un]

shock wave
衝撃波　[syougekiha]
冲击波　[chōng jī bō]
gelombang kejut　[geronbangu kejuto]
충격 파　[chunggyeok pa]
คลื่นกระแทก　[klu-un gra-taaek]

Shore hardness
　an index of hardness calculated using the rebound height of a hammer dropped on the surface of a test-piece from a fixed height. The symbol is HS.
ショアー硬さ　[syoaa katasa]
　一定の高さから試料の試験面上に落下させたハンマの跳ね上がり高さで算出される硬さ指数．記号はHSで示す．
肖氏硬度　[xiāo shì yìng dù]
kekerasan shore　[kekerasan suhore]
쇼아 경도　[syoa gyeongdo]
ความแข็งชอร์　[kwaam khaaeng Shore]

short hook ends torsion spring
a solid coiling torsion spring with short hooks on the outside.
ショートフック型密着ねじりコイルばね [syooto hukkugata mittyaku neziri koiru bane]
密着巻きのねじりコイルばねで、外側に短いフックを有するもの．
短钩环扭转弹簧 [duǎn gōu huán niǔ zhuǎn tán huáng]
pegas puntir berujung kait pendek [pugasu puntiru berujungu kaito pundeku]
짧은고리 밀착 토션 스프링 [jjalbeungorimilchak tosyeon spging]
สปริงขดหมุนชนิดปลายติดตะขอ [sa-pring khod mun cha-nid plaai tid ta-khor]

Fig S-5 short hook ends torsion spring

shortness
a hard, weak and less deformable characteristic, as determined from the impact value or fractured surface in impact tests. Also called embrittlement.
ぜい性 [zeisei]
硬くてもろく、変形能の小さい性質．衝撃試験における衝撃値の大小又は破面の状況によって評価される．ぜい化ともいう．
脆性 [cuì xìng]
hal pendeknya [haru pundekunya]
취성 [chwiseong]
ความเปราะ [kwaam pra-or]

shot
small steel grains used for shot peening. Steel wire shot (cut wire) and cast steel shot are available.
ショット [syotto]
ショットピーニング加工に用いる鋼粒のことで、鋼線ショット（カットワイヤ）と鋳鋼ショットとがある．
弹丸 [dàn wán]
peluru [peruru]
쇼트 [syoteu]
เม็ดช็อต [med shot]

shot blast
process to remove scales and rusts and generate compressive residual stress on the surface layer, by blasting shots (steel balls) on the metal surface using centrifugal force or air pressure.
ショットブラスト [syotto burasuto]
ショット（鋼球など）を遠心力、空気圧などを利用して金属表面に投射して、スケール、さびなどを除去するとともに表面層に圧縮残留応力を生じさせる加工．
喷丸 [pēn wán]
semburan peluru [senburan pururu]
쇼트 블라스트 [syoteu beullaseuteu]
การขัดผิวโดยพ่นเม็ดเหล็ก [gaan khad piw do-oi pon med lek]

shot blasting time
ショットブラスト投射時間 [syotto burasuto tousya zikan]
喷丸处理时间 [pēn wán chǔ lǐ shí jiān]
waktu semburan peluru [wakutu senburan peruru]
쇼트블라스트 투사 시간 [syoteubeullaseuteu tusa sigan]
เวลาในการขัดผิว [we-laa nai gaan khad piw]

shot blasting wheel
a rotor with curved vanes to be used in a shot-blasting machine.
投射用羽根車 [tousyayou haneguruma]
ショットブラスト装置に使われる回転体で曲率を持った羽根が取り付けられている．
喷丸叶轮 [pēn wán yè lún]
roda penyembur peluru [roda

shot blasting wheel — shot peening

punyenburu peruru]
쇼트블라스트 임펠라 [syoteubeullaseuteu impella]
ล้อปัดเม็ดเหล็ก [loor pad med lek]

shot density
the density of shot defined as the mass of shots divided by the volume of shots. The volume of shots is obtained by means of pouring 50 ml of water into 100 ml container, adding 100 g of shots to it, and measuring the rise of water level.
ショット密度 [syotto mitudo]
100 ml の容器中に 50 ml まで水を入れ，次いで 100 g のショットを入れて水位の上昇を測り，ショットの容積でショットの質量を割って定義される値．
喷丸密度 [pēn wán mì dù]
rapat peluru [rapato pururu]
쇼트 밀도 [syoteu mildo]
ความหนาแน่นของการช็อต [kwaam naa-naaen khong gaan shot]

shot distribution
distribution of the shots blasted to the spring from the spray nozzle or the wheel. This depends on blasting type and the nozzle shape. Projection density and processed result vary between the center and the outer circumference of wheel.
ショット投射分布 [syotto tousya bunpu]
噴射ノズル又は翼車からばねなどに投射されるショットの散布状態．投射機又はノズルの形状によって異なり，中央部と外周部とでは投射密度が異なり，加工状態も変化する．
丸粒分布 [wán lì fēn bù]
distribusi butiran peluru [disutoribusi butiran pururu]

쇼트 투사 분포 [syoteu tusa bunpo]
การกระจายของเม็ดช็อต [gaan gra-chaai khong med-shot]

shot hardness
optimum hardness of steel ball shot is between 450 and 520 H_V. For the cut wire shot, the hardness is defined by wire itself.
ショットの硬さ [syotto no katasa]
鋼球ショットの適正硬さは 450～520 H_V である．カットワイヤショットは元の鋼線の硬さで規定される．
喷丸硬度 [pēn wán yìng dù]
sembur peluru [senburu pururu]
쇼트 경도 [syoteu gyeongdo]
ความแข็งของเม็ดช็อต [kwaam khaaeng khong med shot]

shot intensity
an measure indicating the degree of shot peening. It is defined by the arc height value of an almen strip and kinetic energy of the shot.
ショットの強さ [syotto no tuyosa]
ショットピーニング加工の程度を表す尺度のことで，アルメンストリップのアークハイト値及びショットの運動のエネルギーが用いられる．
喷丸強度 [pēn wán qiáng dù]
intensitas peluru [intensitasu pururu]
쇼트 강도 [syoteu gangdo]
ความรุนแรงการช็อต [kwaam run-rang gaan shot]

shot peening
process to improve fatigue strength by blasting the surface of the spring with steel pellets generating residual compressive stress in the surface layer.
ショットピーニング [syotto piiningu]
ショット（鋼球など）をばね表面に高速で打ち付け，主として表面層に圧縮

273

残留応力を生じさせ，疲れ強さを向上させるために行う加工．
喷丸　[pēn wán]
peen peluru　[peen pururu]
쇼트 피닝　[syoteu pining]
การช็อตพีนนิ่ง　[gaan shotpeening]

shot peening time
time period to process shot peening. It depends on part size, discharging condition, shot types, aim of arc height value, and coverage, etc.
ショットピーニング投射時間　[syotto piiningu tousya zikan]
ショットピーニング加工を施す時間をいう．実際の時間は，品物の大きさ投射条件，ショットの種類，目的とするアークハイト値，カバレージなどによる．
喷丸时间　[pēn wán shí jiān]
waktu peen peluru　[wakutu peen pururu]
쇼트 피닝 시간　[syoteu pining sigan]
เวลาในการช็อตพีนนิ่ง　[we-laa nai gaan shotpeening]

shot profile
the shape of shot used in shot peening process.
ショット形状　[syotto keizyou]
ショットピーニングに使用されるショットの形状．
丸粒形状　[wán lì xíng zhuàng]
profil peluru　[purofiru pururu]
쇼트 형상　[syoteu hyeongsang]
ลักษณะการช็อต　[luk-sa-na gaan shot]

side trimming
process to cut off the side area of the full length leaf other than the main leaf to prevent interference with nearby components.
幅落し　[habaotosi]
重ね板ばねにおいて，親板以外の全長板などの端部を周辺部品との干渉を避けるために，ばね板の側面を切り落とす加工．
去飞边　[qù fēi biān]
buang sisi　[buangu sisi]
사이드 트림　[saideu teurim]
การตัดขอบด้านข้าง　[gaan tad khorb daan khaang]

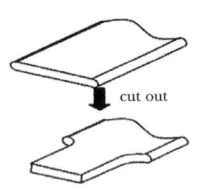

Fig S-6 side trimming

silencer
rubber or plastics tip attached at the end of the leaves to suppress squeak noise and rattle noise. In the case of coil springs, a piece of ringed material is inserted between end coil and spring seat.
サイレンサ　[sairensa]
きしみ音，たたき音などの抑制のために，重ね板ばねの先端部に付ける，又はコイルばねとばね座との間に入れるゴム製，合成樹脂製などの部品．
插头　[chā tóu]
peredam (sisipan ujung)　[puredan (sisipan ujungu)]
사이렌서　[sairenseo]
ไซเลนเซอร์　[silencer]

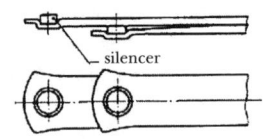

Fig S-7 silencer

silent bush
a bush that is inserted in the eye of leaf springs to muffle noise.

サイレントブシュ [sairento busyu]
重ね板ばねなどの目玉部に挿入して消音などの目的で用いられているブシュ．
消音軸套 [xiāo yīn zhóu tào]
gelang bantalan sunyi [gerangu bantaran sunyi]
사일런트 부시 [sailleonteu busi]
ไซเลนท์บุช [silent bush]

single disc deflection
the deflection of a single disc spring.
単一皿ばねのたわみ [tan'itu sarabane no tawami]
皿ばね単体のたわみ．
单片碟簧挠度 [dān piàn dié huáng náo dù]
lenturan cakram tunggal [renturan chakuran tungugaru]
단일 접시 스프링 휨 [danil jeopsi spring hwim]
การยุบตัวของสปริงแผ่นเดี่ยว [gaan yup to-a khong sa-pring paaen di-eaw]

single leaf
a leaf spring formed of a single leaf.
シングルリーフ [singuru riihu]
ばね板が１枚の板ばね．
单板弹簧 [dān bǎn tán huáng]
daun tunggal [daun tungugaru]
단매물 스프링 [danmaemul spring]
แหนบแผ่นเดี่ยว [naaep paaen di-eaw]

single point coiling system
a method of winding coil springs using one coiling pin.
一本ピンコイリング方式 [ippon pin koiringu housiki]
コイルばねを巻くとき，１本のコイリングピンで巻く方式をいう．
单卷簧销机构 [dān juǎn huáng xiāo jī gòu]
sistem pengulir titik tunggal [sisutemu punguriru titiku tungugaru]
원편 코일링 시스템 [wonpin koilling

siseutem]
รูปแบบการม้วนทีละ 1 อัน [ruup baaep gaan mu-an tii-la an]

single wire series extension spring
an extension spring made of single wire having two coils in series. Hooks are located on both ends.
ダブル引張コイルばね [daburu hippari koiru bane]
同一線材で２か所のコイルを成形し，両端にフックを有する引張ばね．
单丝连接双拉簧 [dān sī lián jiē shuāng lā huáng]
pegas ulir dengan penarik ganda [pugasu uriru dengan penariku ganda]
더블 인장 코일 스프링 [deobeul injang koil sring]
สปริงดึงชนิดเส้นลวดเดี่ยว [sa-pring duung cha-nid sen lu-ad di-eaw]

sintered iron alloy
iron material made by sintering process. The mechanical properties are improved by added alloy elements.
焼結鉄合金 [syouketu tetu goukin]
焼結による鉄系材料で，合金成分により優れた機械的性質を示す．
烧结合金 [shāo jié hé jīn]
besi paduan sintered [besi paduan sinteredo]
소결 철 합금 [sogyeol cheol hapgeum]
โลหะผสมเหล็กเส้นใยแก้ว [loo-ha pa-som lek sen yai kaaew]

slant angle of wire cross section
angle made by the lateral axis of the noncircular cross-section wire in relation to the cross-sectional plane of the coil when no load is applied.
断面の倒れ角 [danmen no taorekaku]
無荷重時に異形断面コイルばねの材料断面の横方向軸線が，コイル横断面と

なす角度.
断面倒角 [duàn miàn dǎo jiǎo]
sudut kemiringan penambang [suduto kemiringan punanbangu]
단면의 기울음 각 [danmyeonui giureum gak]
มุมเอียงของลวดหน้าตัด [mum-iang khong luad naa tad]

sleeve
    a tubular parts to be inserted to a shaft.
    スリーブ [suriibu]
    軸にはめ込む管状の部品.
    轴套 [zhóu tào]
    bungkus [bungukusu]
    슬리브 [seullibeu]
    ปลอกหุ้มเพลา [plork hum pla-o]

slenderness ratio of coil spring
    ratio of the free height of a coil spring to the mean diameter of the coil.
    縦横比 [tateyoko hi]
    コイルばねの自由高さとコイル平均径との比.
    高径比 [gāo jìng bǐ]
    perbandingan kerampingan pegas ulir [purubandingan keranpingan pugasu uriru]
    종횡비 (코일 스프링) [jonghoengbi (koil spring)]
    อัตราส่วนของความสูงอิสระและเส้นผ่าศูนย์กลางเฉลี่ยในสปริงขด [at-traa su-an kong kwaam soong is-sa-ra laae sen-pa-soon-glaang sha-lia nai sa-pring khod]

slide
    displacement between the bush and the stabilizer at the fixed position to the body in the direction of the stabilizer axis, occurring when the working stroke of the stabilizer increases to cause the larger load input.
    横ずれ [yokozure]
    スタビライザの作動ストロークが増え，荷重入力が大きくなった場合，車両への取付部でブシュとスタビライザとの間でスタビライザの軸方向に生じるずれ.
    横向偏移 [héng xiàng piān yí]
    geseran samping [geseran sanpingu]
    밀림 [millim]
    การเลื่อนตัวออกด้านข้าง [gaan le-uan to-a ork daan khaang]

slide pad
    a metal pad to bear the load from an eyeless leaf spring.
    スライドパッド [suraido paddo]
    目玉のない重ね板ばねの荷重を受ける金属パッド.
    滑板 [huá bǎn]
    bantalan luncur [bantaran runkuru]
    슬라이드 받침쇠 [batchimsoe]
    แผ่นเหล็กรับโหลด [paaen lek rap load]

sliding end
    the flat end of a leaf spring without eyes, allowing the fulcrum to shift when the spring is deflected.
    滑面端末 [katumen tanmatu]
    重ね板ばねの端末が目玉ではなく，たわんだときに支点が移動できる平らな端末.
    滑动末端 [huá dòng mò duān]
    ujung luncur [ujungu runkuru]
    미끄럼 끝면 [mikkeureom kkeunmyeon]
    ปลายแผ่นแหนบที่ไม่มีหู [plaai paaen naaep tii mai mii huu]

sliding mandrel
    the mandrel moving in and out of the processing surface.
    しゅう動心金 [syuudou singane]
    加工面へ出入りする機構の心金.
    滑动芯轴 [huá dòng xīn zhóu]
    poros pemegang luncur [porosu pumegangu runkuru]

미끄럼 심봉 [mikkeureom simbong]
แกนม้วนชนิดเคลื่อนที่เข้า-ออก [gaaen mu-an cha-nid klu-an tii kha-o org]

Smith diagram
see "fatigue limit diagram".
スミスの疲れ限度線図 [sumisu no tukare gendo senzu]
"fatigue limit diagram" 参照.
史密斯疲劳极限图 [shǐ mì sī pí láo jí xiàn tú]
bagan Smith [bagan Sumisu]
스미스 내구 한도 선도 [seumiseu naegu hando seondo]
แผนภูมิของสมิธ [paaen poom khong Smith]

smoothed edge
the edge of a material or work piece smoothed by removing burrs and corners.
仕上端面 [siage tanmen]
ばりや角部を取って滑らかにした材料や工作物の端面.
光滑端面 [guāng huá duān miàn]
tepi yang dihaluskan [tepi yangu diharusukan]
끝면 다듬질 [kkeunmyeon dadeumjil]
ขอบเรียบ [khorb riab]

S–N curve
diagram drawn by taking the stress amplitude on the ordinate and the number of repetition times to the breakdown (including the number of repetition times in case the breakdown does not occur) on the abscissa.
S–N 線図 [esu-enu senzu]
縦軸に応力振幅, 横軸に破壊までの繰返数（破壊せずに試験を終了した場合の繰返数を含む）をとって描いた線図.
S-N(疲劳)曲线 [S-N (pí láo) qǔ xiàn]
kurva S-N [kuruva S-N]
에스엔 (S-N) 곡선 [S-N gokseon]
เส้นโค้งเอส-เอน [sen kong S-N]

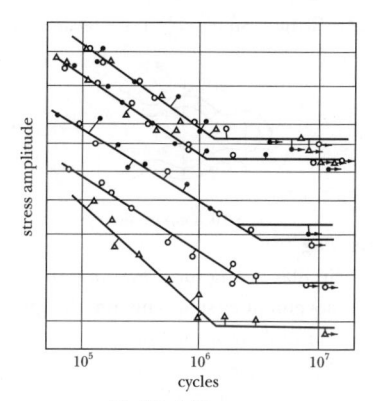

Fig S-8 S–N curve

snap pin
a pin inserted into a shaft hole in radial direction to prevent relative displacement.
スナップピン [sunappu pin]
軸の径方向孔に差し込み, 軸の相対移動を防ぐピン.
弹簧销 [tán huáng xiāo]
pin pengunci [pin pungunsi]
스냅 핀 [seunaep pin]
สลักกันเลื่อน [sa-luk gan leu-un]

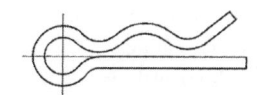

Fig S-9 snap pin

snap retainer
retainer inserted into the groove of the shaft preventing relative displacement of the shaft.
スナップリテーナ [sunappu riteena]
軸の溝に差し込み, 軸の相対移動を防ぐリテーナ.
弹簧挡圈 [tán huáng dǎng quān]
pemegang kunci [pumegangu kunsi]
스냅 리테이너 [seunaep riteineo]

สแนปรึเทนเนอร์กันเลื่อน [snap retainer gan leu-un]

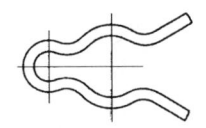

Fig S-10 snap retainer

snap ring
circular spring for preventing the axial movement by snapping into the groove on the shaft or the hole.
止め輪 [tomewa]
軸又は穴に付けた溝にはめて，軸方向の移動を防ぐ輪状のばね．
弹簧挡圈 [tán huáng dǎng quān]
cincin pengunci [chinchin pungunchi]
스냅 링 [seunaep ring]
แหวนกันเลื่อน [waaen gan leu-un]

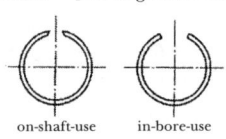

on-shaft-use   in-bore-use
Fig S-11 snap ring

snubber
a device connecting between axle and frame to slow the recoil of the spring and reduce jolting.
緩衝器 [kansyouki]
懸架ばねの戻りを減速し揺れを少なくするために車軸とフレームの間を連結する装置．
缓冲器 [huǎn chōng qì]
peredam tumbukan [puredan tunbukan]
완충기 [wanchunggi]
ตัวเชื่อมต่อระหว่างเพลากับเฟรม [tu-a shi-um ra-waang pla-o gab frame]

soaking time
the duration of heating time for a workpiece at a certain temperature in order to minimize uneven temperature distribution in the workpiece.
均熱時間 [kinnetu zikan]
材料の内外の温度差が少なくなるようにする目的で一定の温度に保持する時間．
均热时间 [jūn rè shí jiān]
waktu rendam [wakutu rendan]
균열 시간 [gyunnyeol sigan]
ระยะเวลาในการฮีทติ้ง [ra-ya wee-laa nai gaan heating]

soft annealing of end
annealing performed to soften only the ends of a spring.
ばね端末焼なまし [bane tanmatu yakinamasi]
ばねの端部のみを軟化させるために行う焼なまし．
端部退火 [duān bù tuì huǒ]
ujung anneal lunak [ujungu annearu runaku]
끝단부 풀림 [kkeutdanbu pullim]
การอบปลาย [gan ob plaai]

soldering
a metal joinning method with solder.
はんだ付け [handazuke]
はんだを用いて行う金属の接合法．
焊接 [hàn jiē]
solder [soruderu]
납땜 [napttaem]
การบัดกรี [gaan bad-grii]

solid coiling
process to wind coils tightly so that the coils make contact each other.
密着巻 [mittyaku maki]
コイルばねにおいて，互いに隣り合うコイルが密着した状態に成形する加工．
密圈弹簧 [mì quān tán huáng]
ulir padat [uriru padato]
밀착감김 [milchagamgim]

solid coiling — solid stabilizer bar with screw ends

การม้วนที่ทำให้ขดที่ติดกัน　[khod tii tam hai khod tid gan]

solid height
    height of a helical compression spring when all adjacent coils close tightly.
    密着高さ　[mittyaku takasa]
    圧縮コイルばねの互いに隣り合うコイルが密着しているときの高さ.
    压并高度　[yā bìng gāo dù]
    tinggi rapat　[tingugi rapato]
    밀착 높이　[milchak nopi]
    ความสูงโซลิด　[kwaam soong solid]

solid length
    the length of an extension spring with all adjacent coils in contact.
    密着長さ　[mittyaku nagasa]
    引張ばねの互いに隣り合うコイルが密着しているときの長さ.
    压并长度　[yā bìng cháng dù]
    panjang rapat　[panjangu rapato]
    밀착 길이　[milchak giri]
    ความยาวโซลิด　[kwaam ya-o solid]

solid position
    the solid height of a friction coil spring.
    密着位置　[mittyaku iti]
    摩擦ばねの密着高さ.
    并紧位置　[bìng jǐn wèi zhì]
    letak rapat　[retaku rapato]
    밀착 위치　[milchak wichi]
    ตำแหน่งโซลิด　[tam-naaeng solid]

solid solution
    a uniform crystalline phase composed of two or more elements.
    固溶体　[koyoutai]
    2種類以上の元素によって形成される均一な固体の結晶質の相.
    固溶体　[gù róng tǐ]
    larutan rapat　[rarutan rapato]
    고용체　[goyongche]
    โซลิดโซลูชั่น　[solid solution]

solid stabilizer bar
    a solid stabilizer to reduce rolling of a vehicle body subjected to centrifugal force.
    中実スタビライザ　[tyuuzitu sutabiraiza]
    車体に遠心力が作用した場合の車体の横揺れを少なくするスタビライザで，中実材から作られる.
    实心稳定杆　[shí xīn wěn dìng gǎn]
    batang stabiliser padat　[batangu sutabiriseru padato]
    중실 스테비라이저　[jungsil seutebiraijeo]
    เหล็กกันโคลงแบบตัน　[lek gan klong baaep tan]

solid stabilizer bar with ball joint ends
    a solid stabilizer with ball joints at both ends for connection.
    ボールジョイント式中実スタビライザ　[booru zyointosiki tyuuzitu sutabiraiza]
    両端末がボールジョントで組み付けられるように成形されたスタビライザで，中実材から作られる.
    两端铰接实心稳定杆　[liǎng duān jiǎo jiē shí xīn wěn dìng gǎn]
    batang stabilisator padat berujung balljoint　[batangu sutabirisatoru padato berujungu baurujointo]
    볼 조인트식 중실 스테비라이저　[bol jointeusik jungsil seutebiraijeo]
    เหล็กกันโคลงแบบตันชนิดปลายเป็นบอลจอยท์　[lek gan klong baaep tan cha-nid plaai pen ball-joint]

solid stabilizer bar with screw ends
    a solid stabilizer with screw joints at both ends.
    ねじ式中実スタビライザ　[nezisiki tyuuzitu sutabiraiza]
    両端末がねじ継手に成形されたスタビライザで，中実材から作られる.

solid stabilizer bar with screw ends — spaced coil garter spring

两端螺纹连接实心稳定杆 [liǎng duān luó wén lián jiē shí xīn wěn dìng gǎn]
batang stabilisator padat berujung baut [batangu sutabirisatoru padato berujungu bauto]
나사형 중실 스테비라이저 [nasahyeong jungsil seutebiraijeo]
เหล็กกันโคลงแบบต้นชนิดปลายเป็นเกลียว [lek gan klong baaep tan cha-nid plaai pen glieaw]

solution heat treatment
 the process of heating and holding a steel alloy at or above the temperature at which a solid solution is formed, then cooling the alloy rapidly to prevent the alloying elements from precipitating.
固溶化熱処理 [koyouka netusyori]
鋼の合金成分が固溶体化する温度以上に加熱保持した後，急冷してその合金成分が析出するのを阻止する操作．
固溶化热处理 [gù róng huà rè chǔ lǐ]
olah kalor dalam larutan [orahu karoru daran rarutan]
고용화 열처리 [goyonghwa yeolcheori]
การทำให้เป็นโซลิดโซลูชั่น [gaan tam hai pen solid solution]

sorbite
 an intermixed structure of ferrite and granulated carbide precipitation obtained by tempering martensite at about 600°C.
ソルバイト [sorubaito]
マルテンサイトを約600℃で焼戻しをして，粒状に析出成長した炭化物とフェライトの混合組織．
索氏体 [suǒ shì tǐ]
sorbit [sorubito]
소르바이트 [soreubaiteu]
เหล็กโครงสร้างซอร์ไบท์ [lek krong saang sorbite]

space between coils
 space between the adjacent coils in the plane including the axis of the coil spring, as measured in the axial direction.
線間すきま [senkan sukima]
コイルばねの中心線を含む断面で，互いに隣り合うコイルの中心線に平行な材料断面間のすきま．
弹簧的螺旋间距 [tán huáng de luó xuán jiān jù]
jarak antar lilitan [jaraku antaru riritan]
선간 틈새 [seongan teumsae]
ระยะห่างระหว่างขด [ra-ya haang ra-wang khod]

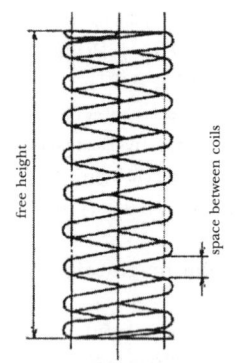

Fig S-12 space between coils

spaced coil garter spring
 a garter spring with space between wires.
ピッチ巻きガータスプリング [pitti maki gaata supuringu]
ピッチを付けたガータスプリング．
非接触形环状螺旋弹簧 [fēi jiē chù xíng huán zhuàng luó xuán tán huáng]
pegas ulir berjarak garter [pugasu uriru berujaraku garuteru]
피치있는 가터 스프링 [pichiinneun gateo spring]

spaced coil garter spring — spare turns

สปริงการ์เตอร์ชนิดมีระยะระหว่างขด [sa-pring garter cha-nid mii ra-ya ra-waang khod]

spaced coil torsion spring
    a torsion spring with space between wires.
ピッチ巻きねじりコイルばね [pitti maki neziri koiru bane]
ピッチを付けたねじりコイルばね.
非接触形环状扭转弹簧 [fēi jiē chù xíng huán zhuàng niǔ zhuǎn tán huáng]
pegas puntir ulir berspasi [pugasu puntiru uriru berusupasi]
피치 있는 비틀림 코일 스프링 [pichiinneun biteullim koil spring]
สปริงทอร์ชั่นชนิดมีระยะระหว่างขด [sa-pring torsion cha-nid mii ra-ya ra-waang khod]

spacer
    a component to secure the necessary space in a leaf spring.
スペーサ [supeesa]
重ね板ばねで，必要な間隔を保持するために装着する板状の部品.
调整垫片 [tiáo zhěng diàn piàn]
pembuat jarak [punbuato jaraku]
스페이서 [seupeiseo]
สเปเซอร์ [spacer]

spacer bush
    a tubular parts to provide a certain space between objects.
スペーサブシュ [supeesa busyu]
物と物の間にスペースを与えるのに用いられる円筒状の部品.
间隙调整套 [jiān xì tiáo zhěng tào]
gelang pembuat jarak [gerangu punbuato jaraku]
스페이서 부시 [seupeiseo busi]
สเปเซอร์บุช [spacer bush]

span
    distance between the load supporting points of a leaf spring. (see Fig S-13)
スパン [supan]
板ばねの荷重支持点間の距離. (Fig S-13 参照)
跨度 [kuà dù]
span [supan]
스팬 [seupaen]
ระยะ [ra-ya]

spare turns
    number of turns from free state to the lower limit of active turns when winding up the power spring.
予巻 [yomaki]
ぜんまいの自由状態から，使用される範囲の巻数の下限値まで巻き付ける巻数.
预卷 [yù juǎn]
ulir cadangan [uriru kadangan]

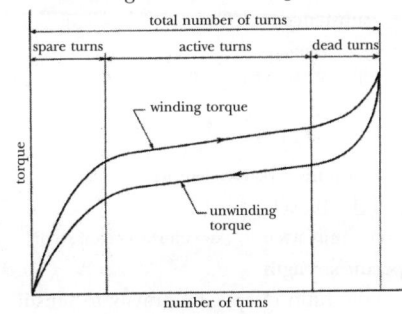

Fig S-14 spare turns

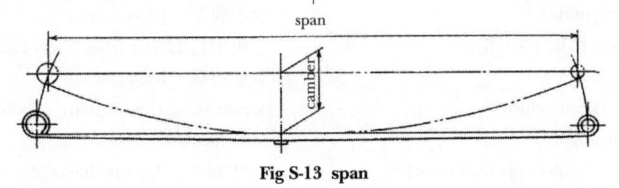

Fig S-13 span

여유 감김 (스파이럴 스프링) [yeoyu gamgim (seupaireol spring)]
จำนวนขดในภาวะอิสระ [cham nu-an khod nai paa-wa is-sa-ra]

spark test
> a test to estimate the type of steel by grinding a steel product (or a semi-finished steel workpiece such as steep strip or steel material) and characterizing the sparks produced.

火花試験 [hibana siken]
鋼製品(鋼片，鋼材等の半製品を含む)をグラインダで研削し，発生する火花の特徴を観察することによって，鋼種の推定を行う試験．
放電試験 [fàng diàn shì yàn]
uji percik [uji peruchiku]
불꽃 시험 [bulkkot siheom]
การทดสอบการจุดไฟสตาร์ท [gaan tod-sorb gaan chud fai start]

specific heat
> the amount of heat required to raise the temperature of a unit mass of a substance by 1°C.

比熱 [hinetu]
単位質量の物質の温度を1℃高めるのに要する熱量．
比热 [bǐ rè]
kalor jenis [karoru jenisu]
비열 [biyeol]
ความร้อนจำเพาะ [kwaam rorn cham pa-or]

specific strength
> the ratio of specific gravity to tensile strength.

比強度 [hikyoudo]
引張強さと比重との比．
比强度 [bǐ qiáng dù]
kuat jenis [kuato jenisu]
비강도 [bigangdo]
กำลังจำเพาะ [kam lang cham pa-or]

specific weight
> the weight per unit volume of a substance.

比重量 [hizyuuryou]
物質の単位体積当たりの重量．
比重 [bǐ zhòng]
berat jenis [berato jenisu]
비중량 [bijungnyang]
น้ำหนักจำเพาะ [naam nak cham pa-or]

specification requirement
> requirements specified in specification document.

規格要求事項 [kikaku youkyuu zikou]
仕様書に記載されている要求事項．
技术要求 [jì shù yāo qiú]
permintaan spesifikasi [purumintaan supesifikasi]
규격 요구사항 [gyugyeok yogusahang]
หัวข้อกำหนดเฉพาะ [ho-a khor gam nod sha-pa-or]

specifications
> a quantitative description of the required characteristics of an industrial product.

仕様書 [siyousyo]
工業製品に要求される特性値を数値で記述した文書．
技术规范书 [jì shù guī fàn shū]
spesifikasi [supesifikasi]
시방서 [sibangseo]
สเปค [spec]

specified load
> specified spring load according to the purpose of use.

指定荷重 [sitei kazyuu]
使用目的から指定するばね荷重．
额定负荷 [é dìng fù hè]
beban yang ditentukan [beban yangu ditentukan]
지정 하중 [jijeong hajung]

โหลดที่กำหนด [load tii gam nod]

specimen
試供体 [sikyoutai]
试件 [shì jiàn]
contoh [chontohu]
시료 [siryo]
ตัวอย่างทดสอบ [to-a yang tod-sorb]

spherical washer
   a plain washer with a spherical surface on one side. Concave and convex spherical washers with the same diameter are combined.
球面座金 [kyuumen zagane]
平座金の片面が球面状になっている座金．同一半径の凸球面と凹球面とを組にして用いる．
球面垫圈 [qiú miàn diàn quān]
ring pipih bulat [ringu pipihu burato]
구면 와셔 [gumyeon wasyeo]
แหวนรองทรงกลม [waaen rong song glom]

spheroidzing
   annealing process of steel to transform precipitated carbide from a flaky form to a stable spherical form.
球状化焼なまし [kyuuzyouka yakinamasi]
鉄鋼中の析出した炭化物を片状の形態から安定した球状の形態に発達させる焼なまし．
球状化退火 [qiú zhuàng huà tuì huǒ]
anneal sferoidal [annearu suferoidaru]
구상화 풀림 [gusanghwa pullim]
การอบให้เป็นรูปทรงกลม [gaan ob hai pen ruup song glom]

spindle ends extension spring
   a helical extension spring with both ends shaped into cones.
両絞り引張コイルばね [ryousibori hippari koiru bane]
両端末を円すい形に形成した引張コイルばね．

両端呈圆锥状拉簧 [liǎng duān chéng yuán zhuī zhuàng lā huáng]
pegas tarik berujung spindel [pugasu tariku berujungu supinderu]
양단 수축형 인장 코일 스프링 [yangdan suchukhyeong injang koil spring]
สปริงขดดึงชนิดปลายรูปกรวย [sa-pring khod duung plaai roop glu-i]

spiral coiling machine
ぜんまい成形機 [zenmai seikeiki]
涡卷弹簧成形机 [wō juǎn tán huáng chéng xíng jī]
mesin ulir spiral [mesin uriru supiraru]
태엽 성형기 [taeyeop seonghyeonggi]
เครื่องม้วนขดก้นหอย [kru-ang mu-an khod gon hoi]

spiral diameter
   inside or outside diameter of a spiral.
ら旋径 [rasen kei]
ら旋形状の内径あるいは外径のこと．
螺旋直径 [luó xuán zhí jìng]
diameter spiral [diameteru supiraru]
나선경 [naseongyeong]
เส้นผ่าศูนย์กลางของขดก้นหอย [sen-pa-soon-glaang khong khod gon hoi]

spiral spring
   a spring with spiral shape in a plane.
渦巻ばね [uzumaki bane]
平面内で渦巻形をしているばね．
涡卷弹簧 [wō juǎn tán huáng]
pegas spiral [pugasu supiraru]

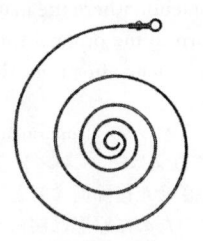

Fig S-15 spiral spring

spiral spring — spring characteristic

스파이럴 스프링 [seupaireol spring]
สปริงขดก้นหอย [sa-pring khod gon hoi]

spray painting
  吹付塗装 [hukituke tosou]
  喷涂料 [pēn tú liào]
  pengecatan semprot [pungechatan senpuroto]
  스프레이 도장 [seupeurei dojang]
  การพ่นสี [gaan pon sii]

spring
  mechanical element designed to utilize accumulated energy of elastic deflection.
  ばね [bane]
  物体の弾性変形によって蓄積されたエネルギーを利用することを主目的とする機械要素.
  弹簧 [tán huáng]
  pegas [pugasu]
  스프링 [spring]
  สปริง [sa-pring]

spring axis
  the centerline of a coil spring.
  ばね軸線 [bane zikusen]
  コイルばねのコイル中心を結んだ線.
  弹簧轴线 [tán huáng zhóu xiàn]
  sumbu pegas [sunbu pugasu]
  스프링 축선 [spring chukseon]
  แนวแกนของสปริง [naaew gaaen khong sa-pring]

spring back
  phenomenon where the material tends to return to the original form because of its elasticity after removing load or moment.
  スプリングバック [supuringu bakku]
  材料に力又はモーメントを加えて塑性変形させた後除荷すると，材料の持つ弾性のために原形に戻ろうとする現象.
  回弹 [huí tán]
  melenting kembali [merentingu kenbali]
  스프링 백 [spring baek]
  สปริงแบ็ค [sa-pring back]

spring balance
  a device to measure weight of a substance by balancing the weight of the substance against spring force, then measuring spring deflection.
  ばね秤 [banebakari]
  品物の重量とばねの力をバランスさせ，ばねのたわみから重量を読み取る装置.
  弹簧秤 [tán huáng chèng]
  neraca pegas [neracha pugasu]
  스프링 저울 [spring jeoul]
  ตาชั่งสปริง [taa-shang sa-pring]

spring barrel of clock
  the casing for a contact-type spiral spring, and container for a clock spring.
  時計のぜんまいケース [tokei no zenmai keesu]
  接触形渦巻ばねのケース（こう箱ともいう）のことで，時計のぜんまいを格納する容器をいう.
  发条 [fā tiáo]
  kotak pegas [kotaku pugasu]
  태엽 스프링 케이스 (시계용) [taeyeop spring keiseu (sigyeyong)]
  กล่องสปริงนาฬิกา [klong sa-pring naa-ri-kaa]

spring case
  ばねケース [bane keesu]
  弹簧盒 [tán huáng hé]
  tong pegas [tongu pugasu]
  스프링 케이스 [spring keiseu]
  หีบบรรจุสปริง [hiip ban-chu sa-pring]

spring characteristic
  relationship between the load applied to a spring and the deflection caused

by the load.
ばね特性　[bane tokusei]
　ばねに加わる荷重とそれによって生じるばねのたわみとの関係．
弾簧特性　[tán huáng tè xìng]
karakteristik pegas　[karakuterisutiku pugasu]
스프링 특성　[spring teukseong]
ความสัมพันธ์ระหว่างค่าโหลดและการยุบตัวของสปริง　[kwaam sam-pan ra-waang kaa load laae gaan yup tu-a khong sa-pring]

spring characteristic test
　generic term of tests to determine the spring characteristics.
ばね特性試験　[bane tokusei siken]
　ばね特性を測定する試験の総称．
弾簧特性試験　[tán huáng tè xìng shì yàn]
uji sifat pegas　[uji sifato pugasu]
스프링 특성 시험　[spring teukseong siheom]
การทดสอบลักษณะเฉพาะของสปริง　[gaan tod-sorb lak-sa-na sha-paor khong sa-pring]

spring clip
　a clip to prevent adjacent leaves of a leaf spring from separating or shifting laterally.
ばねクリップ　[bane kurippu]
　重ね板ばねのばね板が相互に離れること及び横ずれすることを防ぐ金具．
弾簧压板　[tán huáng yā bǎn]
klip pegas　[kuripu pugasu]
스프링 클립　[spring keullip]
คลิปรัดแหนบสปริง　[clip rad naaep sa-pring]

spring clip bolt
　a bolt used in a clip.
ばね取付けねじ　[bane torituke nezi]
　クリップに用いるボルト．
弾簧夹紧螺栓　[tán huáng jiā jǐn luó shuān]
baut klip pegas　[bauto kuripu pugasu]
스프링 클립 볼트　[spring keullip bolteu]
คลิปโบลท์สำหรับแหนบสปริง　[clop bolt sam-rab naaep sa-pring]

spring constant
　force or moment required to cause the unit distortion (deflection or deflection angle) in a spring. This term generally indicates the static spring rate against the static force.
ばね定数　[bane zyousuu]
　ばねに単位の変形（たわみ又はたわみ角）を与えるのに必要な力又はモーメント．一般には，静的荷重に対する静ばね定数のことをいう．
弾簧刚度（常数）　[tán huáng gāng dù (cháng shù)]
tetapan pegas　[tetapan pugasu]
스프링 상수　[spring sangsu]
ค่าคงที่สปริง　[kaa kong tii sa-pring]

spring deflection
ばねのたわみ　[bane no tawami]
弾簧挠度　[tán huáng náo dù]
defleksi pegas　[defurekusi pugasu]
스프링 휨　[spring hwim]
การโก่ง/ยุบตัวของสปริง　[gaan kong / yup to-a khong sa-pring]

spring deflection limit
　characteristic value to predict and evaluate the long-time creeping distortion of thin plates in a short time.
ばね限界値　[bane genkaiti]
　薄板ばね材料の長期クリープ変形量を，短時間に推定・評価するための特性値．
弾性挠度极限　[tán xìng náo dù jí xiàn]
batas defleksi pegas　[batasu defurekusi pugasu]
스프링 한계치　[spring hangyechi]
ค่าจำกัดของการโก่ง/ยุบตัวของสปริง　[kaa cham gad khong gaan kong to-a / yup to-a khong sa-pring]

spring end — spring hook

spring end

the configuration of coil spring ends. A coil spring with open ends has end tips which do not touch the adjacent coils. A coil spring with closed ends has end tips which touch the adjacent coils. The latter provides the better squareness of the spring axis.

ばね端末　[bane tanmatu]
コイルばね終端の形状．オープン端末の場合は隣のコイルと接触していない．クローズ端末の場合は隣のコイルと接触する．後者はばね軸の直角度がよい．
弾簧端部　[tán huáng duān bù]
ujung pegas　[ujungu pugasu]
스프링 끝　[spring kkeut]
ปลายสปริง　[plaai sa-pring]

spring end grinding machine

a machine used for grinding the spring end of a helical compression spring.

ばね端面研削盤　[bane tanmen kensakuban]
圧縮コイルばねの座を研削する機械．
端面磨簧机　[duān miàn mó huáng jī]
mesing gerinda ujung pegas　[mesingu gerinda ujungu pugasu]
스프링 좌면 연삭기　[spring jwamyeon yeonsakgi]
เครื่องเจียร์ปลายสปริง　[kru-ang chiaa plaai sa-pring]

spring eye

the rounded end of a spring leaf.

ばねの目玉　[bane no medama]
ばね板の端部を丸く巻いた部分．
吊耳　[diào ěr]
gelang pegas　[gerangu pugasu]
스프링 귀　[spring gwi]
หูสปริง　[hoo sa-primg]

spring eye center

the center of a leaf eye of laminated leaf spring.

目玉の中心　[medama no tyuusin]
板ばねの親板の目玉の中心位置．
吊耳中心　[diào ěr zhōng xīn]
pusat gelang pegas　[pusato gerangu pugasu]
스프링 귀 중심　[spring gwi jungsim]
ศูนย์กลางหูสปริง　[soon glang hoo sa-pring]

spring force

ばね力　[baneryoku]
弾簧力　[tán huáng lì]
gaya pegas (beban pegas)　[gaya pugasu (beban pugasu)]
스프링 힘　[spring him]
แรงโหลดของสปริง　[rang load khong sa-pring]

spring hardness

ばね硬さ　[bane katasa]
弾簧硬度　[tán huáng yìng dù]
sekeras pegas　[sekerasu pugasu]
스프링 경도　[spring gyeongdo]
ความแข็งของสปริง　[kwaam khaaeng khong sa-pring]

spring holder

a piece of metal to keep the stability of spring.

ばね受け具　[bane ukegu]
ばねを安定させて支える金属部品．
弾簧座　[tán huáng zuò]
pemegang pegas　[pumegangu pugasu]
스프링 홀더　[spring holdeo]
ตัวยึดสปริง　[to-a yud sa-pring]

spring hook

a hook of an extension coil spring with a variety of shape like a round loop, square, V-shaped, etc.

ばねフック　[bane hukku]
引張コイルばねのフックのことで, 丸,

角，V字型等いろいろな形状がある．
弾簧钩环 [tán huáng gōu huán]
kait pegas [kaito pugasu]
스프링 고리 [spring gori]
ตะขอของสปริง [ta-khor khong sa-pring]
spring index
 ratio of the mean diameter of a coil to the material's diameter or the width in the direction of the coil diameter.
ばね指数 [bane sisuu]
 コイル平均径と，材料の直径又はコイル径方向の幅との比．
弹性指数（旋绕比） [tán xìng zhǐ shù (xuán rào bǐ)]
indeks pegas [indekusu pugasu]
스프링 지수 [spring jisuu]
ค่าสัมประสิทธิ์สปริง [kaa sam-pra-sit sa-pring]
spring leaf
 a plate composing a leaf spring.
ばね板 [baneita]
 重ね板ばねを構成する板．
弹簧板 [tán huáng bǎn]
daun pegas [daun pugasu]
스프링 판 [spring pan]
แผ่นสปริง [paaen sa-pring]
spring length
ばね長さ [bane nagasa]
弹簧长度 [tán huáng cháng dù]
panjang pegas [panjangu pugasu]
스프링 길이 [spring giri]
ความยาวของสปริง [kwaam ya-o khong sa-pring]
spring load
 force exerted on or by a spring.
ばね荷重 [bane kazyuu]
 ばねに加わる又はばねから生じる荷重．
弹簧负荷 [tán huáng fù hè]
beban pegas [beban pugasu]
스프링 하중 [spring hajung]

โหลดของสปริง [load khong sa-pring]
spring looping machine
 a machine used for forming a hook on an extension spring.
フック成形機 [hukku seikeiki]
 引張ばねのフック部を成形する機械．
钩环成形机 [gōu huán chéng xíng jī]
mesin pelingkar pegas [mesin puringukaru pugasu]
스프링 고리 성형기 [spring gori seonghyeonggi]
เครื่องทำสปริงลูป [kru-ang tam luup sa-pring]
spring measuring device
 equipments to measure spring characteristics. Includes load tester, hardness tester, and durability tester.
ばね測定装置 [bane sokutei souti]
 ばね特性を測定する機器の総称．荷重試験機，硬さ試験機，耐久試験機などがある．
弹簧测定装置 [tán huáng cè dìng zhuāng zhì]
sarana pengukur pegas [sarana pungukuru pugasu]
스프링 측정 장치 [spring cheukjeong jangchi]
อุปกรณ์วัดค่าของสปริง [up-pa-gorn wad kaa khong sa-pring]
spring mounting system
 the method of spring mounting. A pivot type allows the rotation of the spring ends. A fixed type allows only the axial deflection.
ばね取付方式 [bane torituke housiki]
 ばねの取付方法．ピボット方式はばね端の回転を許容する．固定方式は軸方向たわみだけを許容する．
弹簧装配系统 [tán huáng zhuāng pèi xì tǒng]
sistem pemasangan pegas [sisutemu pumasangan pugasu]

spring mounting system — spring steel

스프링 장착 방식　[spring jangchak bangsik]
ระบบสปริงเม้าท์ติ้ง　[ra-bob sa-pring mounting]

spring nut
　　thin leaf spring for fastening, with a nut section and an insertion section to be inserted into the panel. After inserting the insertion section into the panel using its spring action, place the component to be held and fasten by the screw.
　　ばね板ナット　[baneita natto]
　　薄板ばねで，パネルに差し込む挿入部及びナット部のある締結部品．挿入部のばね作用を利用して，あらかじめパネルに差し込んだ後，被保持部品を重ねてねじ部品で締結する．
　　弾簧螺母　[tán huáng luó mǔ]
　　mur pegas　[muru pugasu]
　　스프링 너트　[spring neoteu]
　　แผ่นสปริงยึดน๊าท　[sa-pring yud nat]

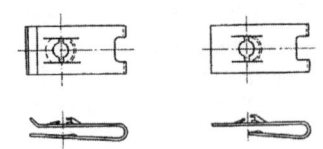

Fig S-16　spring nut

spring pin
　　a kind of spring pin made by rolling elastic plate cylindrically and used for connecting adjacent parts utilising its spring action in radial direction when inserted into a hole.
　　スプリングピン　[supuringu pin]
　　弾性がある板を円筒状に丸め，その半径方向のばね作用を利用し，穴に打ち込んで隣接部片を連接するピン．
　　弾簧销　[tán huáng xiāo]
　　pin pegas　[pin pugasu]
　　스프링 핀 (롤 핀)　[spring pin (rol pin)]
　　สลักกุญแจสปริง　[sa-lak gun chaae sa-pring]

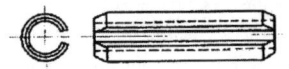

Fig S-17　spring pin

spring separator
　　a machine to untwist entwined springs, primarily using vibration or rotary motion.
　　ばね分離機　[bane bunriki]
　　ばねとばねが絡み合ったものをほぐす機器をいう．一般に振動や回転運動などを用いた機構のものが多い．
　　弹簧分离机　[tán huáng fēn lí jī]
　　pemisah pegas　[pumisahu pugasu]
　　스프링 분리기　[spring bulligi]
　　เครื่องคลายสปริง　[kre-ung klaai sa-pring]

spring sorter
　　a machine or euipment to select springs compliant with specified standards.
　　ばね選別機　[bane senbetuki]
　　定められた規格に対するばねの良否を選別する機器又は装置．
　　弹簧分选机　[tán huáng fēn xuǎn jī]
　　pemilah pegas　[pumirahu pugasu]
　　스프링 선별기　[spring seonbyeolgi]
　　เครื่องแยกจัดเรียงสปริง　[kru-ang chad riang sa-pring]

spring steel
　　steel that is suitable for manufacturing all kinds of spring products due to its excellent elasticity when quenched and tempered. It can provide a certain load without causing any permanent deformation when the load is removed. It includes piano wire, hard steel wire, stainless steel wire, oil tempered steel wire, and cold-rolled steel strip. These

are cold-rolled and heat-treated to improve elastic properties, and formed into wire or flat springs.
ばね鋼　［banekou］
焼入焼戻し状態でのその優れた弾力性によって，あらゆる種類の"ばね"部品製造に特に適した鋼．この鋼には負荷が除去されたとき，鋼が一切の永久変形を示すことなしに，ある一定の範囲内で負荷を与えることができる．ピアノ線，硬鋼線，ステンレス鋼線，オイルテンパ線，冷間圧延鋼帯などのように冷間加工及び熱処理によってばね性能を高め，そのまま線ばね，薄板ばねなど小物ばねに成形する鋼も含む．
弹簧钢　［tán huáng gāng］
baja pegas　［baja pugasu］
스프링 강　［spring gang］
เหล็กสปริง　［lek sa-pring］

spring steel strip
steel material for spring that has been rolled and coiled.
ばね用帯鋼　［baneyou obikou］
圧延され，コイル状に巻かれたばね鋼鋼材．
弹簧带钢　［tán huáng dài gāng］
kepinan baja pegas　［kepinan baja pugasu］
스프링용 강대　［springyong gangdae］
แถบเหล็กแผ่นสปริง　［taaep lek paaen sa-pring］

spring steel wire for spring
the generic term for wires or wire rods used for springs. Includes hard steel wire, piano wire, stainless steel wire, and oil tempered steel wire.
ばね用線　［baneyou sen］
硬鋼線，ピアノ線，ステンレス鋼線，オイルテンパ線などがある．
弹簧线材　［tán huáng xiàn cái］
kawat baja pegas　［kawato baja pugasu］

스프링용 선재　［springyong seonjae］
ลวดเหล็กสำหรับสปริง　［lu-ad lek sam-rab sa-pring］

spring testing machine
spring load tester or fatigue tester.
ばね試験機　［bane sikenki］
ばね荷重測定機又は疲労試験機．
弹簧试验机　［tán huáng shì yàn jī］
mesin penguji pegas　［mesin punguji pugasu］
스프링 시험기　［spring siheomgi］
เครื่องทดสอบโหลดและความทนทานของสปริง　［kru-ang tod-sorb load laae kwaam ton-taan khong sa-pring］

springs in series
springs arranged in series so that the spring deflection is added sequentially.
直列のばね　［tyokuretu no bane］
ばねのたわみが加算されるように組み合わせたばね．
直列弹簧　［zhí liè tán huáng］
pegas dalam sen　［pugasu dalan sen］
직열 스프링　［jingnyeol spring］
แผ่นสปริงที่เรียงเป็นลำดับตามการยุบตัว　［paaen sa-pring tii rieang pen lam-dap］

square wire
wire or wire rod with a square cross section.
角線　［kakusen］
断面の形状が正方形をした線又は線材．
方形截面线材　［fāng xíng jié miàn xiàn cái］
kawat persegi　［kawato purusegi］
각 선　［gak seon］
ลวดเหลี่ยม　［lu-ad liam］

squareness
the degree of perpendicularity of the coil spring axis to the fitting surface.
直角度　［tyokkakudo］
コイルばねの軸線と取付面がなす角の

垂直度.
垂直度　[chuí zhí dù]
sudut siku　[suduto siku]
직각도　[jikgakdo]
ความตรง　[kwaam trong]

squeak noise
sound caused by the frictional contact of springs or the spring and the adjacent component.
きしみ音　[kisimion]
ばね同士又はばねと周辺部品とがこすれ合うことによって生じる音.
噪音　[zào yīn]
suara berdecit　[suara burudechito]
마찰음　[machareum]
เสียงดังเอี๊ยดๆ　[siang dang iead iead]

stabilizer bar
springs to minimize rolling of a vehicle body subjected to centrifugal force.
スタビライザ　[sutabiraiza]
車体に遠心力が作用した場合の車体の横揺れを少なくするために取り付けられているばね.
稳定杆　[wěn dìng gǎn]
batang pemantap　[batangu pumantapu]
스테빌라이저　[seutebillaijeo]
เหล็กกันโคลง　[lek gan klong]

stabilizer eye
end of a bar stabilizer to be fastened to the chassis.
スタビライザ目玉　[sutabiraiza medama]
バースタビライザ端部の相手部品への

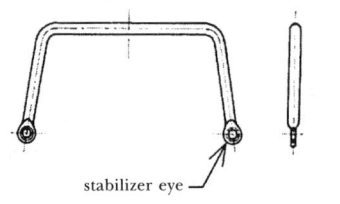

Fig S-18　stabilizer eye

取付部分.
稳定器耳孔　[wěn dìng qì ěr kǒng]
gelang pemantap　[gerangu pumantapu]
스테비라이저 아이　[seutebiraijeo ai]
หูเหล็กกันโคลง　[hoo lek gan klong]

stabilizing treatment
heat treatment to prevent a steel product from dimensional or structural changes with time.
安定化熱処理　[anteika netusyori]
時間経過による寸法又は組織の変化を防ぐために行う鉄鋼製品の熱処理.
去应力热处理　[qù yīng lì rè chǔ lǐ]
olahan pemantap　[orahan pumantapu]
안정화 열처리　[anjeonghwa yeolcheori]
การทำให้คงตัว　[gaan tam hai kong to-a]

stacked in parallel
combination of springs so that the spring force is added.
並列積重ね　[heiretu tumikasane]
同一たわみに対し，ばねの力が加算されるように組み合わせること．
并列组合　[bìng liè zǔ hé]
ditumpuk sejajar　[ditunpuku sejajaru]
병렬 조합　[byeongyeol johap]
เรียงซ้อนขนานกัน　[riang sorn kha-naan gan]

stacked in series
combination of springs so that the spring deflection is added.
直列積重ね　[tyokuretu tumikasane]
ばねのたわみが加算されるように組み合わせること．
直列组合　[zhí liè zǔ hé]
ditumpuk seri　[ditunpuku seri]
직열 적중　[jingnyeol jeokjung]
เรียงซ้อนเป็นลำดับตามการยุบตัว　[gaan riang sorn pem lam-dab]

stain
dirty color change caused by incomplete combustion of surface oil.

油焼け [aburayake]
　表面に付着した油の不完全燃焼によって生じる変色．
油污 [yóu wū]
kejut oli [kejuto ori]
스테이닝 [seuteining]
เป็นรอยเปื้อน [pen roi pe-uun]

stainless steel
　an alloyed steel containing chromium element or nickel and chromium element to improve corrosion resistance. Stainless steel is defined as an alloyed steel that contains at least 11% chromium. It includes martensitic, ferritic, austenitic, austenitic/ferritic and precipitation-hardened stainless steels mainly depending on their structures.
ステンレス鋼 [sutenresu kou]
　耐食性を向上させる目的でクロム又はニッケルとクロムを含有させた合金鋼．一般にクロム含有量が約11％以上の鋼をステンレス鋼といい，主としてその組織によって，マルテンサイト系，フェライト系，オーステナイト系，オーステナイト・フェライト系及び析出硬化系の五つに分類される．
不锈钢 [bù xiù gāng]
baja tahan karat [baja tahan karato]
스테인레스강 [seuteilleseugang]
เหล็กกันสนิม [lek gan sa-nim]

stainless steel shot
　steel shot for shot peening which is made of stainless steel.
ステンレス鋼製ショット [sutenresukousei syotto]
　ステンレス鋼製のショットピーニング用鋼粒．
不锈钢喷丸 [bù xiù gāng pēn wán]
peluru baja tahan karat [pururu baja tahan karato]
스테인레스강재 쇼트 [seuteilleseugangjae syoteu]
เม็ดช็อทชนิดสเตนเลส [med shot cha-nid stainless]

stainless steel spring
　springs made of stainless steel.
ステンレス鋼ばね [sutenresukou bane]
　ステンレス鋼を素材にしたばね．
不锈钢弹簧 [bù xiù gāng tán huáng]
pegas baja antikarat [pugasu baja antikarato]
스테인레스강재 스프링 [seuteilleseugangjae spring]
สปริงเหล็กกันสนิม [sa-pring lek gan sa-nim]

standard deviation
　a quantity which appears in statistics denoted by the square root of variance.
標準偏差 [hyouzyun hensa]
　分散の平方根で表される統計量．
标准差 [biāo zhǔn chā]
penyimpangan standar [punyinpangan sutandaru]
표준 편차 [pyojun pyeoncha]
การเบี่ยงเบนมาตรฐาน [gaan biang bain maat-tra-taan]

standard spring
　a standardized spring with established selection charts.
規格ばね [kikaku bane]
　標準化され，選定表が確立しているばね．
标准弹簧 [biāo zhǔn tán huáng]
pegas standar [pugasu sutandaru]
규격 스프링 [gyugyeok spring]
สปริงมาตรฐาน [sa-pring maat-tra-taan]

standard value
標準値 [hyouzyunti]
标准值 [biāo zhǔn zhí]
nilai standar [nirai sutandaru]
표준치 [pyojunchi]

standard value — statistical process control

ค่ามาตรฐาน [kaa maat-tra-taan]
standardization
 the preparation of generally accepted procedures, dimensions, materials, or parts in order to achieve reasonably economical and effective manufacturing of products.
 標準化 [hyouzyunka]
 製品を合理的な経済性と効率で生産するために部品，材料，各部寸法，広く受容されている生産手順などを事前に整えておくこと．
 标准化 [biāo zhǔn huà]
 standarisasi [sutandarisasi]
 표준화 [pyojunhwa]
 การทำให้เป็นมาตรฐาน [gaan tam hai pen maat-tra-taan]

starter spring
 a double-coil torsion spring to provide return action of starter pedal.
 始動機ばね [sidouki bane]
 二輪車の始動機のペダル戻り用二重巻ねじりコイルばね．
 起动弹簧 [qǐ dòng tán huáng]
 pegas setater [pugasu setateru]
 시동기 스프링 [sidonggi spring]
 สปริงจุดสตาร์ท [sa-pring chud sa-tart]

static friction
 see "friction force".
 静止摩擦 [seisi masatu]
 "friction force" 参照．
 静摩擦 [jìng mó cā]
 gesekan statis [gesekan sutatisu]
 정 마찰 [jeong machal]
 การเสียดทานสถิต [gaan siead-taan sa-tid]

static load
 a load that does not fluctuate.
 静的荷重 [seiteki kazyuu]
 時間的にほぼ変動しない荷重．
 静负荷 [jìng fù hè]

beban statis [beban sutatisu]
정하중 [jeonghajung]
โหลดคงที่ [load kong tii]

static strain
 the strain caused by a static load.
 静ひずみ [seihizumi]
 静的荷重を受けたときのひずみ．
 静变形 [jìng biàn xíng]
 regang statis [regangu sutatisu]
 정변형 [jeongbyeonhyeong]
 ความเครียดคงที่ [kwaam kriead kong tii]

static stress
 the stress caused by a static load.
 静応力 [seiouryoku]
 静的荷重を受けたときの応力．
 静应力 [jìng yīng lì]
 tegangan statis [tegangan sutatisu]
 정응력 [jeongeungnyeok]
 ความเค้นคงที่ [kwaam ken kong tii]

static test
 a testing condition under low strain velocity. Tensile, bending, and compression test are conducted in static test condition.
 静的試験 [seiteki siken]
 荷重又は変形速度がゆるやかな試験．引張試験，曲げ試験，圧縮試験などがある．
 静态试验 [jìng tài shì yàn]
 uji statis [uji sutatisu]
 정적 시험 [jeongjeok siheom]
 การทดสอบการคงที่ [gaan tod-sorb gaan kong tii]

statistical process control
 a process control technique by way of statistical handling of data.
 統計的工程管理 [toukeiteki koutei kanri]
 統計手法を用いて工程の管理を行うこと．
 统计的过程管理 [tǒng jì de guò chéng guǎn lǐ]

kendali proses secara statistik　[kendari purosesu sechara sutatisutiku]
통계적 공정관리　[tonggejeok gongjeonggwalli]
การควบคุมกระบวนการโดยใช้สถิติ　[gaan kuab kum gra-buan-gaan do-oi shai sa-ti-ti]

statistical quality control
 the use of statistical techniques as a means of controlling the quality of a product or process.
 統計的品質管理　[toukeiteki hinsitu kanri]
 製品あるいは製造工程の品質管理の手段として統計的手法を用いること．
 统计质量管理　[tǒng jì zhì liàng guǎn lǐ]
 statistik kendali mutu　[sutatisutiku kendari mutu]
 통계적 품질관리　[tonggyejeok pumjilgwalli]
 การควบคุมคุณภาพโดยใช้สถิติ　[gaan ku-ab kum kun-na-pab do-oi shai sa-ti-ti]

steel
 鋼　[hagane]
 钢　[gāng]
 baja　[baja]
 강　[gang]
 เหล็กกล้า　[lek-glaa]

steel bars
 steel which is rolled or forged into rod and cut to a specified length.
 棒鋼　[boukou]
 棒状に圧延又は鍛造された鋼で所定の長さに切断された鋼材．
 棒钢　[bàng gāng]
 batang baja　[batangu baja]
 강봉　[gangbong]
 แท่งเหล็กกล้า　[taaeng lek-glaa]

steel grit
 angular cast iron or cast steel particles with angled edges.
 鋼製グリット　[kousei guritto]

りょう角を持つ角張った形状の鋳鉄又は鋳鋼の粒子．
钢砂　[gāng shā]
butiran baja　[butiran baja]
강제 그릿　[gangje geurit]
ผงเหล็กกล้า　[pong lek-glaa]

steel shot
 鋼製ショット　[kousei syotto]
 钢丸　[gāng wán]
 peluru baja　[pururu baja]
 강제 쇼트　[gangje syoteu]
 เหล็กช็อต　[lek shot]

steel spring
 spring made of steel materials, as called by its material.
 鋼製ばね　[kousei bane]
 ばねの材料の種類による名称で，鋼材料を用いたばね．
 钢制弹簧　[gāng zhì tán huáng]
 pegas baja　[pugasu baja]
 강재 스프링　[gangjae spring]
 สปริงเหล็กกล้า　[sa-pring lek glaa]

steel strip
 rolled and coiled steel.
 鋼帯　[koutai]
 圧延され，コイル状に巻かれた鋼材．
 钢带　[gāng dài]
 kepingan baja　[kepingan baja]
 강대　[gangdae]
 แผ่นเหล็กกล้า　[paaen lek glaa]

steel wire
 鋼線　[kousen]
 钢丝　[gāng sī]
 kawat baja　[kawato baja]
 강선　[gangseon]
 เส้นลวดเหล็กกล้า　[sen lu-ad lek glaa]

step
 step caused by the difference of the lengths of adjacent leaves of a leaf spring.

step — straight span

ステップ [suteppu]
重ね板ばねの隣接するばね板の長さの相違からできる段.
分阶 [fēn jiē]
langkah [rangukahu]
스텝 [seutep]
ขั้นตอน [khan-torn]

stepped quenching
secondary quenching after initial quenching for carburized steel. Heating temperature shall be above the A_{C1} point for the carburized layer in order to harden the carburized layer.
二次焼入れ [nizi yakiire]
浸炭した鋼の浸炭層を硬化する目的で，一次焼入れののち，浸炭層の A_{C1} 点以上の適当な温度に加熱して行う焼入れ.
二次淬火 [èr cì cuì huǒ]
kejut bertahap [kejuto berutahapu]
이차 담금질 [icha damgeumjil]
การชุบขั้นที่ 2 [gaan shub khan tii song]

stock reel
a reel on which wire rod is wound.
ストックリール [sutokku riiru]
線材巻取り用の巻枠.
料架 [liào jià]
kumpaan bahan [kunpaan bahan]
스토크 릴 (선대) [seutokeu ril (seondae)]
แท่นสำหรับวางกองเหล็กลวด [taaen sam-rab waang kong lek lu-ad]

straight arm ends torsion spring
a torsion coil spring in which the arm for loading at the ends of a spring is formed linearly toward the tangential direction of the coil diameter.
密着ストレートフック形ねじりコイルばね [mittyaku sutoreeto hukkugata neziri koiru bane]
端末の負荷用の腕がコイル径から接線方向に直線的に成形されたねじりコイルばね.
直臂端密圈扭转弹簧 [zhí bì duān mì quān niǔ zhuǎn tán huáng]
pegas puntir berujung lurus [pugasu puntiru berujungu rurusu]
밀착 스트레이트 고리형 비틀림코일스프링 [milchak seuteureiteu gorihyeong biteullimkoil spring]
สปริงทอร์ชั่นปลายแขนตรง [sa-pring torsion plaai khaaen trong]

straight offset ends torsion spring
a torsion coil spring in which the arm for loading at the ends of a spring is formed perpendicularly to the coil circle.
密着直線起こし形フックねじりコイルばね [mittyaku tyokusen okosigata hukku neziri koiru bane]
端末の負荷用の腕がコイル円に対して垂直に成形されたねじりコイルばね.
偏置直臂端扭转弹簧 [piān zhì zhí bì duān niǔ zhuǎn tán huáng]
pegas puntir berujung lurus menepi [pugasu puntiru berujungu rurusu menepi]
밀착 직선 수직 고리형 비틀림 코일 스프링 [milchak jikseon sujik gorihyeong biteullim koil spring]
สปริงทอร์ชั่นปลายแขนเยื้อง [sa-pring torsion plaai khaaen ye-ung]

straight span
distance between the load supporting points of a leaf spring measured in flattened position. (see Fig S-19)
ストレートスパン [sutoreeto supan]
板ばねの荷重支持点間を水平に展開した長さ．(Fig S-19 参照)
直线跨度 [zhí xiàn kuà dù]
span lurus [supan rurusu]
스트레이트 스팬 [seuteureiteu seupaen]

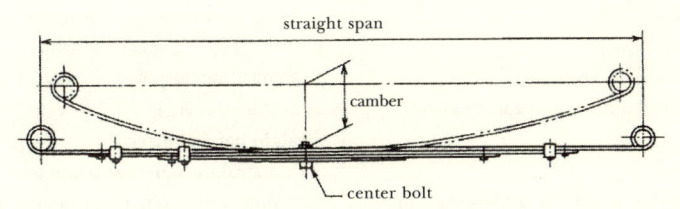

Fig S-19 straight span

ระยะตรง　　[ra-ya trong]

straightener
    equipment to straighten the wire material by eliminating bending, twisting, or residual strain.
    ストレートナー　[sutoreetonaa]
    材料がもつ曲がり，ねじれなどのくせ及び残留ひずみを矯正又は除去する装置．
    校直机　[jiào zhí jī]
    pelurus　[pururus]
    교정기　[gyojeonggi]
    อุปกรณ์ทำให้ตรง　[up-pa-gron tam hai trong]

straightening annealing
    annealing process performed by heating to and maintaining below the transformation point while applying a load in order to remove strain in steel material or castings.
    ひずみ取り焼なまし　[hizumi tori yakinamasi]
    鋼材又は鋳物に生じたひずみを除去するために，荷重を加えながら変態点以下の温度に加熱保持して行う焼なまし．
    校正退火　[jiào zhèng tuì huǒ]
    anneal untuk meluruskan　[annearu untuku merurusukan]
    응력제거 풀림　[eungnyeokjegeo pullim]
    การอบเพื่อไม่ให้เปลี่ยนรูป　[gaan ob pe-uu mai hai pliean-ruup]

straightening roller
    a machine to straighten or flatten wire and sheet by passing them between a number of rollers staggered above and below.
    矯正ローラ　[kyousei roora]
    交互に配列した多数のロールの間に線や板材などを通して，まっすぐ又は平らにする装置．
    校直滚轮　[jiào zhí gǔn lún]
    rol untuk meluruskan　[roru untuku merurusukan]
    교정 롤러　[gyojeong rolleo]
    ตัวรีดให้ตรง　[to-a riid hai trong]

straightness
    the magnitude of deviation from being straight in geometrical meaning.
    真直度　[sintyokudo]
    直線部分の幾何学的直線からの狂いの大きさ．
    直线度　[zhí xiàn dù]
    kelurusan　[kerurusan]
    진직도　[jinjikdo]
    ความตรง　[kwaam trong]

strain
    rate of shape or volume variation caused by force. Example: extension strain, compression strain, shear strain.
    ひずみ　[hizumi]
    力によって生じる形及び体積の変化率．引張ひずみ，圧縮ひずみ及びせん断ひずみがある．

应变 [yīng biàn]
regang [regangu]
변형 [byeonhyeong]
ความเครียดบิดเบี้ยว [kwaam kriead bid bieaw]

strain aging
aging that occurs in cold-worked material.
ひずみ時効 [hizumi zikou]
冷間加工した材料に起こる時効.
应变时效 [yīng biàn shí xiào]
penuaan karena regang [punuaan karena regangu]
변형 시효 [byeonhyeong sihyo]
การเปลี่ยนรูปของวัตถุในกระบวนการแบบเย็น [gaan plian-ruup khong wad-tu nai gra bu-an gaan baaeb yen]

strain gauge
an apparatus to measure strain using the physical phenomenon by which resistance of an electrical resistor changes when subjected to strain.
ストレインゲージ [sutorein geezi]
電気抵抗体にひずみを与えたときの抵抗値が変化する物理現象を応用してひずみの測定を行うもの.
应变仪 [yīng biàn yí]
pembanding regang [punbandingu regangu]
스트레인게이지 [seuteureingeiji]
เกจวัดความเครียด [gauge wad kwaam kriead]

strain hardening
the phenomenon whereby a cold-worked material is hardened as the processing strain increases.
加工硬化 [kakou kouka]
加工を冷間で行ったとき、加工ひずみの増大に伴って材料が硬化する現象.
应变硬化 [yīng biàn yìng huà]
pengerasan regang [pungerasan regangu]
가공 경화 [gagong gyeonghwa]
ความเครียดขณะชุบแข็ง [kwaam kriead khana shub-khaaeng]

straight bar stabilizer
a straight bar that bears torsion loads only, with a separate moment arm.
直棒式バースタビライザ [tyokubousiki baa sutabiraiza]
ねじり部から連続したモーメントアーム部の代わりにアームを別部品とし、ねじり荷重を受け持つだけの機能を持つ真直棒.
直臂端稳定杆 [zhí bì duān wěn dìng gǎn]
batang stabiliser lurus [batangu sutabiriseru rurusu]
스트레이트 바 스테비라이저 [seuteureiteu ba seutebiraijeo]
เหล็กกันโคลงแบบตรง [lek gan klong trong baaep trong]

stranded wire spring
coil spring made of a stranded wire.
より線ばね [yorisen bane]
より線状の材料を用いたコイルばね.
多股螺旋弹簧 [duō gǔ luó xuán tán huáng]
pegas dari kawat berhelai helai [pugasu dari kawato beruherai herai]
꼬임선 스프링 [kkoimseon spring]
สปริงขดทำจากลวดขด [sa-pring khod tam chaak lu-ad khod]

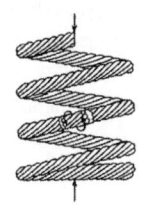

Fig S-20 stranded wire spring

stress
the value obtained by dividing the ex-

ternal axial force exerted on a test piece by its initial cross-sectional area.
応力　[ouryoku]
試験片に加わる軸方向外力を試験片の初期断面積で除した値.
应力　[yīng lì]
tegangan　[tegangan]
응력　[eungnyeok]
ความเค้น　[kwaam ken]

stress amplitude
half of an algebraic difference of maximum and minimum stress generated in a spring under a repetitive load.
応力振幅　[ouryoku sinpuku]
ばねに生じる繰返し応力の最大応力と最小応力との代数差の1/2.
应力幅　[yīng lì fú]
amplitude tegangan　[anpuritude tegangan]
응력 진폭　[eungnyeok jinpok]
ขนาดของความเค้น　[kha-nad khong kwaam ken]

stress concentration factor
stress at material defects like a crack or stress at a local area where a sharp change of cross section exists is larger than ordinary stress. The ratio of increased stress and ordinary stress is refered to as the stress concentration factor.
応力集中係数　[ouryoku syuutyuu keisuu]
材料中のき裂などの欠陥部分や, 断面が急激に変化する箇所に, 周辺部に比較して大きな応力が生ずる. この大きさの増加倍率を応力集中係数という.
应力集中系数　[yīng lì jí zhōng xì shù]
faktor konsentrasi tegangan　[fakutoru konsentorasi tegangan]
응력 집중 계수　[eungnyeok jipjung gyesu]
องค์ประกอบการเกิดความเค้น　[ong-pra-gorb gaan gerd kwaam ken]

stress correction factor
factor to be multiplied to the simple torsion stress to calculate the maximum shearing stress. Maximum shearing stress in the cross-section of the circular material of a coil spring is generated at the inner side of the coil, owing to the bending of the wire and the direct shearing action.
コイルばねの応力修正係数　[koiru bane no ouryoku syuusei keisuu]
コイルばねの円形材料断面に生じるせん断応力は, 線のわん曲や直接せん断作用のため, コイル内側で最大となる. この最大せん断応力を算出するために, 単純なねじり応力に乗じる係数.
应力修正系数　[yīng lì xiū zhèng xì shù]
faktor koreksi tegangan　[fakutoru korekusi tegangan]
응력 수정 계수　[eungnyeok sujeong gyesu]
องค์ประกอบการแก้ไขความเค้น　[ong-pra-gorb gaan gaae-khai kwaam ken]

stress corrosion cracking
phenomenon where the material subjected to extension stress breaks at the lower stress than the normal breakdown stress level, under the corrosive environment.
応力腐食割れ　[ouryoku husyoku ware]
引張応力を受ける材料が, 腐食環境下で通常の破壊応力水準より低い応力で割れを生じる現象.
应力腐蚀裂纹　[yīng lì fǔ shí liè wén]
retakan koreksi tegangan　[retakan korekusi tegangan]
응력 부식 균열　[eungnyeok busik gyunnyeol]
การแตกร้าวแบบกัดกร่อนจากความเค้น　[gaan taaek ra-o baaep gad-gron chaak kwaam ken]

stress cracking
 an external or internal crack in a solid body caused by a stress below the breaking strength.
 応力き裂　[ouryoku kiretu]
 破壊強さよりも小さい応力により材料の表面あるいは内部に生じるき裂.
 应力开裂　[yīng lì kāi liè]
 retakan tegangan　[retakan tegangan]
 응력 균열　[eungnyeok gyunnyeol]
 การแตกร้าวที่เกิดจากความเค้น　[gaan taek ra-o tii gerd chaak kwaam ken]

stress cycle
 a cycle in which stress progresses from initial value to next initial value going through maximum and minimum value.
 応力サイクル　[ouryoku saikuru]
 応力がある初期値から始まり代数的最大値及び最小値を通り初期値に戻るサイクル.
 应力循环　[yīng lì xún huán]
 daur tegangan　[dauru tegangan]
 응력 사이클　[eungnyeok saikeul]
 รอบของความเค้น　[rorb khong kwaam ken]

stress cycle frequency
 occurrence count of stress cycle per unit time.
 応力サイクル周波数　[ouryoku saikuru syuuhasuu]
 単位時間当たりの応力サイクル数.
 应力循环周期　[yīng lì xún huán zhōu qī]
 frekuensi daur tegangan　[furekuensi dauru tegangan]
 응력 사이클 주파수　[eungnyeok saikeul jupasu]
 ความถี่ของการเกิดรอบความเค้น　[kwaan tii khong gaan gerd rorb kwaam ken]

stress endurance diagram
 diagram drawn by taking the stress amplitude on the ordinate and the number of repetition times to the breakdown (including the number of repetition times in case the breakdown does not occur) on the abscissa. (see Fig S-8)
 S-N 線図　[esu-enu senzu]
 縦軸に応力振幅，横軸に破壊までの繰返数（破壊せずに試験を終了した場合の繰返数を含む）をとって描いた線図．(Fig S-8 参照)
 疲劳 (S-N) 曲线　[pí láo (S-N) qǔ xiàn]
 bagan ketahanan tegangan　[bagan ketahanan tegangan]
 응력 반복수 선도 (S-N 선도)　[eungnyeok banboksu seondo (S-N seondo)]
 แผนภาพความทนทานต่อความเค้น　[paaen paab kwaam ton taan to-u kwaam ken]

stress gradient
 inclination of stress from center to outer area in stressed member.
 応力勾配　[ouryoku koubai]
 応力を受けている部材において中心部から表面への応力の傾き．
 应力梯度　[yīng lì tī dù]
 lereng tegangan　[rerengu tegangan]
 응력 구배　[eungnyeok gubae]
 แนวของความเค้น　[naaew khong kwaam ken]

stress induced martensitec transformation
 the transformation to martensite induced by applied stress.
 応力誘起マルテンサイト変態　[ouryoku yuuki marutensaito hentai]
 加えた応力によってマルテンサイトが誘起される変態．
 应力诱发马氏体变形　[yīng lì yòu fā mǎ shì tǐ biàn xíng]
 tranformasi martensit dengan induksi tegangan　[toranforumasi marutensite dengan induksi tegangan]

응력 유기 마르텐사이트 변태
[eungnyeok yugi mareutensaiteu byeontae]
การเปลี่ยนเป็นภาวะมาร์เทนไซด์โดยความเค้น
[gaan pli-an pen paa-wa martensite dooi kwaam ken]

stress range
the difference between the maximum and minimum stress of repeated stress. The stress range is twice the stress amplitude.
応力範囲　[ouryoku han'i]
繰返し応力の最大応力と最小応力の差．なお，応力範囲は応力振幅の2倍である．
应力范围　[yīng lì fàn wéi]
rentang tegangan　[rentangu tegangan]
응력 범위　[eungnyeok beomwi]
ช่วงของความเค้น　[shu-ong khong kwaam ken]

stress ratio
ratio of the minimum to maximum value of the repetitive stress generated in a spring. R=$\sigma_{min}/\sigma_{max}$
応力比　[ouryokuhi]
ばねに繰返し応力が加わっているとき，最小応力の最大応力に対する代数比．R=$\sigma_{min}/\sigma_{max}$
最小最大应力比　[zuì xiǎo zuì dà yīng lì bǐ]
perbandingan tegangan　[purubandingan tegangan]
응력비　[eungnyeokbi]
อัตราส่วนของความเค้น　[at-traa su-an khong kwaam ken]

stress relaxation
phenomenon where the stress decreases gradually when a spring is held in the extended or compressed state to a certain length.
応力緩和　[ouryoku kanwa]
ばねを一定の長さに伸長又は圧縮した状態に保持したとき，応力が徐々に低下する現象．
应力松弛　[yīng lì sōng chí]
pengenduran tegangan　[pungenduran tegangan]
응력 완화　[eungnyeok wanhwa]
การคายความเค้น　[gaan kaai kwaam ken]

stress relaxation curve
a diagram showing the relationship between stress or load and time in relaxation test.
リラクゼーション曲線　[rirakuzeesyon kyokusen]
リラクゼーション試験における応力と時間又は荷重と時間の関係を表す曲線．
应力松弛曲线　[yīng lì sōng chí qū xiàn]
kurva pengenduran tegangan　[kuruva pungenduran tegangan]
응력 완화 곡선　[eungnyeok wanhwa gokseon]
เส้นโค้งการคายความเค้น　[sen koong gaan kaai kwaam ken]

stress relief
an operation whereby a workpiece is heated to proper temperature below the transformation point to remove residual stress charged to the workpiece during manufacturing process.
応力除去　[ouryoku zyokyo]
成形その他の工程中に生じた望ましくない残留応力を変態点以下の適当な温度に加熱して除去する操作．
消除应力　[xiāo chú yīng lì]
pembebasan tegangan　[punbebasan tegangan]
응력 제거　[eungnyeok jegeo]
การคลายความเค้น　[gaan kla-ai kwaam ken]

stress relief annealing
heat treatment to reduce internal

stress relief annealing — stress-strain-temperature diagram

stress without changing the metallurgical structure by heating to an appropriate temperature, and then cooling at an appropriate rate.

応力除去焼なまし　[ouryoku zyokyo yakinamasi]
本質的に組織を変えることなく，内部応力を減らすために適切な温度へ加熱し，適切な速度で冷却する熱処理.

去应力退火　[qù yīng lì tuì huǒ]
anneal bebas tegangan　[anneru bebasu tegangan]
응력 제거 풀림　[eungnyeok jegeo pullim]
การอบคลายความเค้น　[gaan ob klaai kwaam ken]

stress relief heat treatment

a heat treatment for a workpiece which is welded or cut by heat to remove residual stress.

後熱処理　[ato netusyori]
溶接部又は熱切断部の残留応力を除去するために行う熱処理.

去应力后热处理　[qù yīng lì hòu rè chǔ lǐ]
olah panas pelepas tegangan　[orahu panasu purepasu tegangan]
응력제거 열처리　[eungnyeokjegeo yeolcheori]
การคลายความเค้นโดยให้ความร้อน　[gaan klaai kwaam ken dooi hai kwaam rorn]

stress-peening

shot peening carried out under the static stress in the direction the spring is used. By this process, higher residual stress is obtained than ordinary shot peening.

ストレスピーニング　[sutoresu piiningu]
ばねにその使用される方向の静的応力を与えた状態で行うショットピーニング. ショットピーニングより高い残留応力を得るために行う加工.

应力喷丸强化　[yīng lì pēn wán qiáng huà]
peen tegangan　[peen tegangan]
스트레스 피닝　[seuteureseu pining]
การทำสเตรสพีนนิ่ง　[gaan tam stresspeening]

stress-strain diagram

curve showing the relationship between the nominal stress and the extension strain of the parallel part of the test piece in whole extension test process.

応力-ひずみ線図　[ouryoku-hizumi senzu]
引張試験の全過程における試験片平行部の公称応力と伸びとの関係を表す曲線.

应力-应变曲线　[yīng lì- yīng biàn qū xiàn]
bagan tegangan-regang　[bagan tegangan-regangu]
응력-변위 선도　[eungnyeok-byeonwi seondo]
แผนภูมิความเค้น - ความเครียด　[paaen poom kwaam ken-kwaam krieat]

stress-strain-temperature diagram

a diagram showing temperature dependence of the stress-strain curve, strain dependence of the stress-temperature curve, and stress dependence of the strain-temperature curve in triaxial coordinate system comprising of stress, strain, and temperature.

応力-ひずみ-温度線図　[ouryoku-hizumi-ondo senzu]
応力-ひずみ曲線の温度依存性，応力-温度曲線のひずみ依存性，及びひずみ-温度曲線の応力依存性を応力-ひずみ-温度の3軸座標で表した線図.

应力-应变-温度曲线　[yīng lì- yīng biàn-wēn dù qū xiàn]
bagan tegangan regang suhu　[bagan tegangan regangu suhu]
응력-변위-온도선도　[eungnyeok-

byeonwi-ondoseondo]
แผนภูมิความเค้น - ความเครียด - อุณหภูมิ
[paaen poom kwaam ken-kwaam krieat-un-na-ha-puum]

stress vs. creep rate curve
    a graph indicating the relationship between stress (logarithmic scale) and the minimum creep rate (logarithmic scale) or steady-state creep rate (logarithmic scale) measured under varying temperature conditions.
    応力対クリープ速度線図　[ouryoku tai kuriipu sokudo senzu]
    応力（対数目盛）と最小クリープ速度（対数目盛）又は定常クリープ速度（対数目盛）との関係を，異なる温度条件で実測した線図．
    应力蠕变图　[yīng lì rú biàn tú]
    kurva laju tegangan creep　[kuruva raju tegangan kureepu]
    응력 대 크리프 속도 선도　[eungnyeok dae keuripeu sokdo seondo]
    กราฟแสดงความสัมพันธ์ระหว่างความเค้นและอัตราการเกิดครีป　[grahp sa-daaeng kwaam sampan ra-waang kwaam ken laae at-traa kwaam gaan gerd creep]

stress vs. creep rupture time curve
    a curve indicating the relationship between stress (logarithmic scale) and creep rupture time (logarithmic scale) under varying temperature conditions.
    応力クリープ破断時間曲線　[ouryoku kuriipu hadan zikan kyokusen]
    異なる温度条件下における応力（対数目盛）とクリープ破断時間（対数目盛）との関係を表す曲線．
    应力蠕变断裂时间曲线　[yīng lì rú biàn duàn liè shí jiān qū xiàn]
    kurva pecah karena tegangan　[kuruva pekahu karena tegangan]
    응력 크리프 파단속도 곡선　[eungnyeok keuripeu padansokdo gokseon]
    เส้นโค้งความเค้น, ครีปและระยะเวลาการแตก　[sen koong kwaam ken, creep laae raya we-laa gaan taaek]

strip
    belt-like material with uniform width sheared from thin plate.
    ストリップ　[sutorippu]
    薄板を一定の幅に裁断した帯状の材料．
    带，条　[dài, tiáo]
    strip　[sutorippu]
    강대　[gangdae]
    แถบ　[taaeb]

stroke
    a distance from the nominal loaded position to the bump stopper or rebound stopper position in a suspension leaf spring.
    ストローク　[sutorooku]
    懸架ばねにおける常用荷重時からバンプストッパ又はリバウンドストッパにあたるまでの行程．
    行程　[xíng chéng]
    langkah　[rangukahu]
    스트로크　[seuteurokeu]
    ระยะสโตรค　[ra-ya stroke]

structural transformation
    the phenomenon whereby one crystalline structure transforms into another when temperture rises or falls.
    組織変態　[sosiki hentai]
    温度を上昇又は下降させたとき，ある結晶構造から他の結晶構造に変化する現象．
    组织变化　[zǔ zhī biàn huà]
    perubahan struktural　[purubahan suturukuturaru]
    조직 변태　[jojik byeontae]

structural transformation — sulphurizing

การเปลี่ยนรูปของโครงสร้าง　　[gaan plian ruup khong krong-saang]

stud
 a bar with a threaded end on both sides to be screwed into a machine base.
 スタッド　[sutaddo]
 棒の両端にねじがあり一方のねじを機械本体などに植え込んで用いられる植込みボルトをいう．
 双头螺栓　[shuāng tóu luó shuān]
 baut tak berkepala　[bauto taku berukepara]
 스터드　[seuteodeu]
 สลักเกลียว　[sa-lak gli-aw]

sub-zero treatment
 the heat treatment of steel by cooling an object below 0°C immediately after quenching in order to improve wear resistance and prevent deformation due to aging of steel products.
 サブゼロ処理　[sabu zero syori]
 鉄鋼製品の耐摩耗性の向上，経年変形の防止を目的として，焼入れ後直ちに0°C以下の低温度に冷却する処理．
 低温处理　[dī wēn chǔ lǐ]
 olahan dibawah nol　[orahan dibawahu noru]
 서브제로 처리　[seobeujero cheori]
 การชุบแข็งภายใต้ศูนย์องศา　[gaan shub khaaeng paai tai soon ong-sa]

sulfur dioxide corrosion test
 a test to investigate the propagation of corrosion by leaving a test piece in a test chamber containing 0.5 to 2 vol. % of sulfur dioxide at 40°C and 95% RH.
 亜硫酸ガス試験　[aryuusan gasu siken]
 二酸化硫黄 0.5～2%（容量）を含む温度40°C，相対湿度 95% に保った試験装置内へ試験片を静置し，さびの発生状況などを調べる試験．
 亚硫酸气体试验　[yà liú suān qì tǐ shì yàn]
 uji korosi gas sulfur　[uji korosi gasu surufuru]
 아황산 가스 시험　[ahwangsan gaseu siheom]
 การทดสอบการกัดกร่อนโดยก๊าซซัลเฟอร์ไดออกไซด์　[gaan tod-sord gaan gad gron do-oi gas sulfur dioxide]

sulphide
 the general term of the compound combined with sulphur.
 硫化物　[ryuukabutu]
 硫黄と化合してできた化合物の総称．
 硫化物　[liú huà wù]
 sulfida　[surufida]
 유화물　[yuhwamul]
 สารประกอบของซัลเฟอร์　[saan pra-gorb khong sulphur]

sulphur print examination
 a test to investigate the sulphur distribution in steel using a sulphur print, which is obtained by pressing the cross-section of a test piece onto photographic printing paper moistened with sulphuric acid.
 サルファプリント試験　[sarufa purinto siken]
 鋼の断面に硫酸でぬらした写真用印画紙を密着させることによって，サルファプリントを得て鋼における硫黄の分布状況を調べる試験．
 硫印试验　[liú yìn shì yàn]
 uji cetak sulfur　[uji chetaku surufuru]
 설퍼 프린트 시험　[seolpeo peurinteu siheom]
 การทดสอบซัลเฟอร์พรินท์　[gaan tod-sorb sulphur print]

sulphurizing
 an operation whereby sulphur is dif-

fused on steel surface.
浸硫法　[sinryuuhou]
硫黄を鉄鋼の表面に拡散させる操作．
浸硫法　[jìn liú fǎ]
pemasukan sulfur　[pumasukan surufuru]
침황처리　[chimhwangcheori]
กรรมวิธีซัลเฟอร์ไรซิ่ง　[gaam-ma-wi-tii sulphurizing]

supercarburizing
　a case in which a workpiece is carburized excessively.
過剰浸炭　[kazyou sintan]
浸炭層が目標値以上になる現象．
过量渗碳　[guò liàng shèn tàn]
superkarburisasi　[superukaruburisasi]
과잉 침탄　[gwaing chimtan]
การเพิ่มคาร์บอนพิเศษ　[gaan perm carbon pen pi-ses]

superelasticity
　the property whereby deformation caused by stress-induced martensitic transformation returns to its initial state in reverse transformation when load is removed. Strain of up to 10 times and more of the elastic strain which is not accompanied by martensite deformation is recovered.
超弾性　[tyoudansei]
応力誘起マルテンサイト変態によって生じた変形が除荷時に逆変態によって元に戻る性質．変態を伴わない弾性ひずみの数倍～数十倍のひずみが除荷時に回復する．
超弹性　[chāo tán xìng]
super elastisitas　[superu erasutisitasu]
초탄성　[chotanseong]
ความยืดหยุ่นพิเศษ　[kwaam yuud-yun pi-ses]

surface condition
表面状態　[hyoumen zyoutai]
表面状态　[biǎo miàn zhuàng tài]

keadaan permukaan　[keadaan purumukaan]
표면 상태　[pyomyeon sangtae]
สภาพผิว　[sa-paab piw]

surface corrosion
表面腐食　[hyoumen husyoku]
表面腐蚀　[biǎo miàn fǔ shí]
korosi permukaan　[korosi purumukaan]
표면 부식　[pyomyeon busik]
การกัดกร่อนของผิว　[gaan gad gron khong piw]

surface crack
　a crack which initiates from the surface of metal parts under stress corrosion environment or fatigue loading.
表面割れ　[hyoumen ware]
応力腐食環境あるいは疲労荷重のもとで金属製品表面を起点として発生する割れ．
表面裂纹　[biǎo miàn liè wén]
retak permukaan　[retaku purumukaan]
표면 균열　[pyomyeon gyunnyeol]
การแตกของผิว　[gaan taaek khong piw]

surface decarburization
　reduction of carbon content in the surface layer of steel due to overheating.
表面脱炭　[hyoumen dattan]
過熱によって鋼の表層部の炭素量が減少すること．
表面脱碳　[biǎo miàn tuō tàn]
dekarburisasi permukaan　[dekaruburisasi purumukaan]
표면 탈탄　[pyomyeon taltan]
การสูญเสียคาร์บอนที่ผิว　[gaan soon siaa carbon tii piw]

surface defect
　a surface condition that damages the function and durability of a spring. A crack, rust, and decarburization are listed under the term.

表面欠陥　[hyoumen kekkan]
材料や品物などの表面状態が要求事項を満たさないこと．一般に，割れ，さび，被覆むらなどをいう．
表面缺陷　[biǎo miàn quē xiàn]
cacat permukaan　[chachato purumukaan]
표면 결함　[pyomyeon gyeolham]
งานเสียบริเวณผิว　[ngaan siaa bo-ri-wen piw]

surface finish
a process to add exterior beauty to the product by deburring, grinding, painting, or polishing.
表面仕上げ　[hyoumen siage]
製品表面のばり取り，研磨，塗装，つや出しなど外観を向上させる工程．
表面抛光　[biǎo miàn pāo guāng]
penyelesaian permukaan　[punyeresaian purumukaan]
표면 다듬질　[pyomyeon dadeumjil]
ผิวสำเร็จ　[piw sam-ret]

surface hardening
a process performed to harden the surface layer of metal products.
表面硬化処理　[hyoumen kouka syori]
金属製品の表面層を硬化させるために行う処理．
表面硬度　[biǎo miàn yìng dù]
pengerasan permukaan　[pungerasan purumukaan]
표면 경화 처리　[pyomyeon gyeonghwa cheori]
การชุบแข็งที่ผิว　[gaan shub khaeng tii piw]

surface hardness
hardness of the surface of a material or a product.
表面硬さ　[hyoumen katasa]
素材及び製品の表面の硬さ．
表面硬化処理　[biǎo miàn yìng huà chǔ lǐ]
kekerasan permukaan　[kekerasan purumukaan]

표면경도　[pyomyeon-gyeongdo]
ความแข็งที่ผิว　[kwaam khaaeng tii piw]

surface oxidation
a phenomenon in which metal starts making oxide on the surface of a workpiece.
表面酸化　[hyoumen sanka]
金属が空気中の酸素と化合して表層部より酸化物を作る現象．
表面氧化　[biǎo miàn yǎng huà]
oksidasi permukaan　[okusidasi purumukaan]
표면 산화　[pyomyeon sanhwa]
การเกิดออกซิเดชั่นที่ผิว　[gaan gerd oxidation tii piw]

surface protection
protection for surface oxidation.
表面保護　[hyoumen hogo]
表面酸化から保護すること．
表面防护　[biǎo miàn fáng hù]
perlindungan permukaan [pururindungan purumukaan]
표면 보호　[pyomyeon boho]
การป้องกันผิว　[gaan pong gan piw]

surface roughness
the degree of microscopic unevenness of a metal surface.
表面粗さ　[hyoumen arasa]
金属表面の微小な凹凸の程度．
表面粗糙度　[biǎo miàn cū cāo dù]
kekasaran permukaan　[kekasaran purumukaan]
표면 조도　[pyomyeon jodo]
ความขรุขระของผิว　[kwaam khru-khra khong piw]

surface strain hardening
a process to harden the surface of metal by utilizing work hardening effect.
表面加工硬化　[hyoumen kakou kouka]
加工硬化現象を利用して金属表面を硬

化させる工程.
表面加工硬化　[biǎo miàn jiā gōng yìng huà]
pengerasan regang permukaan
[pungerasan regangu purumukaan]
표면 가공 경화　[pyomyeon gagong gyeonghwa]
การชุบแข็งความเครียดที่ผิว　[gaan shub khaaeng kwaam kriead tii piw]

surface tension
 a tension force that acts on the surface of liquid to make the surface area as small as possible.
表面張力　[hyoumen tyouryoku]
液体の表面に作用する表面積をできるだけ小さくしようとする張力.
表面张力　[biǎo miàn zhāng lì]
tegangan permukaan　[tegangan purumukaan]
표면 장력　[pyomyeon jangnyeok]
แรงตึงผิว　[rang tung piw]

surface treatment
 a method of altering surface properties or adding new characteristics by modifying the surface state.
表面処理　[hyoumen syori]
表面の状態を変えることによって, 表面の性質を変えたり, 新しい機能を付加すること.
表面处理　[biǎo miàn chǔ lǐ]
olahan permukaan　[orahan purumukaan]
표면 처리　[pyomyeon cheori]
การเตรียมผิว　[gaan triam piw]

surging
 resonance which occurs where the inherent frequency of the spring is equivalent to the frequency of the external force. When a coil spring is subjected to vibration, torsional shock waves run along the coil wire reciprocally.
サージング　[saazingu]

コイルばねを加振するとねじりが衝撃波となり, コイル素線に沿ってばね両端間を往復する. 往復に要する時間が, 外力による加振の周期に等しいときに生じる共振現象.
共振　[gòng zhèn]
penyentak　[punyentaku]
공진　[gongjin]
ความถี่ที่สอดคล้องกัน　[kwaam tii tii sord-klong gan]

suspension leaf spring
 a generic term for leaf springs used to support the bodies of motor vehicles and train cars.
懸架用重ね板ばね　[kengayou kasaneita bane]
一般に自動車・鉄道車両などの車体を支える重ね板ばね.
悬架重叠板弹簧　[xuán jià chóng dié bǎn tán huáng]
pegas daun penumpu　[pugasu daun punumnpu]
현가용 겹판 스프링　[hyeongayong gyeoppan spring]
แหนบแผ่นรับแรงสั่นสะเทือน　[naaep paaen rab-rang san sa-te-un]

suspension spring
 spring which generally supports the bodies of automobiles, railway cars or the like.
懸架ばね　[kenga bane]
一般に自動車, 鉄道車両などの車体を支えるばね.
悬架弹簧　[xuán jià tán huáng]
pegas penumpu　[pugasu punumupu]
현가 스프링　[hyeonga spring]
สปริงรับแรงสั่นสะเทือน　[sa-pring rab rang san sa-te-un]

suspension system
 a suspension unit which combines

suspension system — system

springs, shock absorbers, and sway braces to be used for an automobil.
懸架装置　[kenga souti]
ばね，緩衝器，スウェイブレースを組み合わせた自動車用の懸架装置．
悬架装置　[xuán jià zhuāng zhì]
sistem penumpu　[sisutemu punumupu]
현가장치　[hyeongajangchi]
ระบบรับแรงสั่นสะเทือนในรถยนต์　[ra-bop rab rang san sa-te-un nai rod-yon]

sweep

a horizontal or a circular movement of electron beam on the screen of cathode-ray tube.
掃引　[souin]
ブラウン管のスクリーン上で電子ビームが水平又は円を描いて動くこと．
扫描　[sǎo miáo]
menyapu　[menyapu]
스위프　[seuwipeu]
การเคลื่อนที่ของลำแสงอิเลคตรอน　[gaan klu-ion tii khong lam saaeg electron]

swing axle spring

a leaf spring used for independent suspension system on the rear axle of a truck.
スイングアクスル式ばね　[suingu akusurusiki bane]
トラック後輪側の独立懸架装置に用いる重ね板ばねのこと．
摆轴弹簧　[bǎi zhóu tán huáng]
pegas gandar berayun　[pugasu gandaru berayun]
스윙 액슬 스프링　[swing aeksseul spring]
สปริงสำหรับการเหวี่ยงของเพลา　[sa-pring sam-rab gaan wi-ang chong pla-o]

swivel hook

a rotational hook of an extension spring.
スイベルフック　[suiberu hukku]

引張ばねの回転自由なフック．
回转钩环　[huí zhuǎn gōu huán]
kait pemutar　[kaito pumutaru]
회전 고리　[hoejeon gori]
ตะขอหมุนที่ติดกับสปริงดึง　[ta-khor mun tii tid gab sa-pring dung]

Fig S-21　swivel hook

symmetric leaf spring

a leaf spring the center bolt or the center pin of which is located at the center of the span.
対称ばね　[taisyou bane]
センタボルト又はセンタピンの位置が，スパンの中央にある重ね板ばね．
对称板弹簧　[duì chèn bǎn tán huáng]
pegas daun simetris　[pugasu daun simetorisu]
대칭 스프링　[daeching spring]
สปริงแหนบแผ่นชนิดสมดุล　[sa-pring naaep paaen cha-nid som-dun]

synthetic resin spring

the general term of springs made of synthetic resin.
合成樹脂ばね　[gouseizyusi bane]
合成樹脂材料を用いたばねの総称．
合成树脂弹簧　[hé chéng shù zhī tán huáng]
pegas resin sintetis　[pugasu resin sintetisu]
합성수지 스프링　[hapseongsuji spring]
สปริงเรซินสังเคราะห์　[sa-pring resin sang-kra-or]

system

a combination of several pieces of equipment integrated to perform a specific function.
システム　[sisutemu]
いくつかの装置を統合して，特定の機能を実行するようにした要素の集合

体.
系统 [xì tǒng]
sistim [sisutimu]
시스템 [siseutem]
ระบบ [ra-bop]

systems engineering
 a technology to design and analize the interrelation of many elements and their controlling devices to maximize the performance of a system.
システム工学 [sisutemu kougaku]
 システムの目的を最もよく達成するためにその構成要素，制御装置などの相互関係を設計，分析する技術．
系统工程学 [xì tǒng gōng chéng xué]
teknik sistem [tekuniku sisutemu]
시스템 공학 [siseutem gonghak]
วิศวกรรมระบบ [wis-sa-wa-gam ra-bop]

T

tandem axle type suspension
a suspension type that has two axles for the purpose of reducing load to ground shock and increasing towing capacity.
タンデムアクスル式懸架　[tandemu akusurusiki kenga]
路面への圧力及び車体への反力を軽減し又は牽引力を増加するために車軸を二本取り付け懸架する方式.
串联式悬架装置　[chuàn lián shì xuán jià zhuāng zhì]
suspensi jenis sumbu tandem　[susupensi jenisu sunbu tanden]
탠덤 액슬형 현가　[taendeom aekseulhyeong hyeonga]
ซัสเพนชั่นชนิดแทนเด็ม　[suspension cha-nid tan-dem]

tangent tail ends
coil end which is straight and tangent to the coiling circle.
タンジェントテールエンド　[tanzyento teeru endo]
圧縮コイルばね端部の一種で，端末がコイル接線方向に伸ばされた形状.
切向直尾端压缩弹簧　[qiē xiàng zhí wěi duān yā suō tán huáng]
ujung ekor tangen　[ujungu ekoru tangen]
접선 꼬리 끝　[jeopseon kkori kkeut]
ปลายขดตรงและสัมผัสกับขดกลม　[plaai khod trong laae sam-pas gab khod glom]

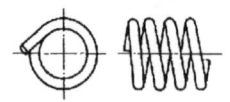

Fig T-1 tangent tail ends

tapar grinding
process to form the leaf of a leaf spring or the wire of a coil spring into the tapered shape by rolling, forging or grinding.
テーパ加工　[teepa kakou]
重ね板ばねのばね板，コイルばねなどの線を，圧延，鍛造，研削などによってテーパ状にする加工.
锥度加工（磨削）　[zhuī dù jiā gōng (mó xuē)]
gerinda tirus　[gerinda tirusu]
테이퍼 가공 (연삭)　[teipeo gagong (yeonsak)]
การเจียร์ให้เรียว　[gaan chiaa hai rieaw]

taper
a gradual thin shape of spring leaf tip or wire tip.
テーパ　[teepa]
ばね板端部あるいは線の先端部の先細り形状.
锥度　[zhuī dù]
tirus　[tirusu]
테이퍼　[teipeo]
ลักษณะปลายแคบเรียวลง　[lak-sa-na kaaep rieaw long]

taper forging
a forging process to produce tapered leaf or tapered wire.
テーパ鍛造　[teepa tanzou]
ばね板や線をテーパ形状にする鍛造.
锥度加工（锻造）　[zhuī dù jiā gōng (duàn zào)]
tumbuk tirus　[tunbuku tirusu]
테이퍼 가공 (단조)　[teipeo gagong (danjo)]
การทุบขึ้นรูปให้ปลายเรียว　[gaan tup khun ruup hai plaai rieaw]

taper key
テーパキー [teepa kii]
楔形键 [xiē xíng jiàn]
kunci tirus [kunsi tirusu]
테이퍼 키 [teipeo ki]
เทเปอร์คีย์ [taper key]

taper roll bearing
a bearing which sustains axial loads and prevents axial movement of a loaded shaft by using tapered rolling piece.
テーパロールベアリング [teepa rooru bearingu]
転動体として円すいころを用いたスラスト軸受.
锥形滚动轴承 [zhuī xíng gǔn dòng zhóu chéng]
bantalan rol tirus [bantaran roru tirusu]
테이퍼 롤링 베어링 [teipeo rolling beeoring]
แบริ่งโรลเลอร์สำหรับขึ้นรูปปลายเทเปอร์ [bearing roller sam-rab khun ruup plaai taper]

taper rolling
a rolling to produce tapered leaf or tapered wire.
テーパ圧延 [teepa atuen]
ばね板や線をテーパ形状にする圧延加工.
锥度加工（压延） [zhuī dù jiā gōng (yā yán)]
rol tirus [roru tirusu]
테이퍼 가공 (압연) [teipeo gagong (abyeon)]
การรีดขึ้นรูปปลายเทเปอร์ [gaan riid khun ruup plaai taper]

taper rolling machine
rolling machine for taper forming.
テーパロールマシン [teepa rooru masin]
テーパ加工するためのロール機.
锥端辊压机 [zhuī duān gǔn yā jī]
mesin rol tirus [mesin roru tirusu]
테이퍼 롤링기 [teipeo rollinggi]

เครื่องรีดขึ้นรูปปลายเทเปอร์ [kru-ang riid khun ruup plaai taper]

tapered end
tapered end of a helical compression spring or a leaf spring, worked by rolling or forging.
テーパエンド [teepa endo]
圧縮コイルばね及び板ばねのばね板端部の一種で，圧延，鍛造，研削などによってテーパ加工を施した形状.
锥端 [zhuī duān]
ujung tirus [ujungu tirusu]
테이퍼 끝 [teipeo kkeut]
ปลายเทเปอร์ [plaai taper]

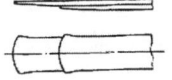

Fig T-2 tapered end

tapered end with circuler hook
a bechive shape of an extension spring with circular hook.
絞り丸フック [sibori maruhukku]
引張コイルばねの端末を絞り込んだ形に成形し円形のフックを成形したもの.
带回转钩环的锥形端 [dài huí zhuǎn gōu huán de zhuī xíng duān]
ujung tirus dengan kait melingkar [ujungu tirusu dengan kaito meringukaru]
오므린 원형 고리 [omeurin wonhyeong gori]
ปลายเทเปอร์ที่มีตะขอกลม [plaai taper tii mii ta-khor glom]

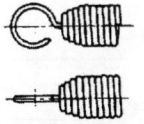

Fig T-3 tapered end with circuler hook

tapered eye type — technical maintenance

tapered eye type
> eye of a bar stabilizer with the tapered hole for tapered arbor.
> テーパアイタイプ　[teepa ai taipu]
> バースタビライザ目玉部形状の一種で，テーパ状の孔に固定する形式．
> 锥形吊耳　[zhuī xíng diào ěr]
> jenis lubang tirus　[jenisu rubangu tirusu]
> 테이퍼 아이 형　[teipeo ai hyeong]
> แหนบชนิดหูเทเปอร์　[naaep cha-nid huu taper]

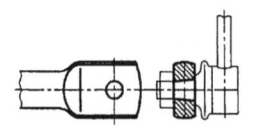

Fig T-4 tapered eye type

tapered leaf spring
> leaf spring tapered for approximately entire length except for the clamping area to other components. Also called parabolic leaf spring because parabolic tapering is often employed in order to obtain more uniform stress distribution. (see Fig T-5)
> テーパリーフスプリング　[teepa riihu supuringu]
> 他部品との取付部分を除き，ほぼ全体にわたってテーパが施してある板ばね．より均一な応力分布を得るため，放射線状にテーパを施す場合もよくあるので，パラボリックリーフスプリングともいう．(Fig T-5 参照)
> 变截面单板弹簧　[biàn jié miàn dān bǎn tán huáng]
> pegas daun tirus　[pugasu daun tirusu]
> 테이퍼리프 스프링　[teipeoripeu spring]
> สปริงแหนบแผ่นชนิดปลายเทเปอร์　[sa-pring naaep paaen cha-nid plaai taper]

tapered length
> the length of the tapered section of a spring leaf.
> テーパ長さ　[teepa nagasa]
> ばね板のテーパに加工された部分の長さをいう．
> 锥形长度　[zhuī xíng cháng dù]
> panjang tirus　[panjangu tirusu]
> 테이퍼 길이　[teipeo giri]
> ความยาวเทเปอร์　[kwaam ya-o taper]

technical inspection
> inspection of equipment to determine whether it is serviceable for continued use or needs repairs.
> 技術検査　[gizyutu kensa]
> 機器が引き続き使用可能か，又は修理が必要かを調べる検査．
> 技术检查　[jì shù jiǎn chá]
> inspeksi teknis　[insupekusi tekunisu]
> 기술검사　[gisulgeomsa]
> การตรวจเช็คทางเทคนิค　[gaan truat check taang technic]

technical maintenance
> a category of maintenance that includes the replacement of unserviceable major parts, and precision adjustment, testing, and alignment of internal components.
> 技術メンテナンス　[gizyutu mentenansu]

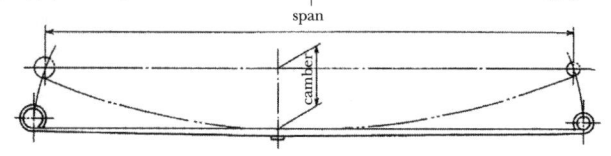

Fig T-5 tapered leaf spring

継続使用困難な主要部品の交換，精密調整，テスト，内部部品の調整等を行うメンテナンス．
技术维护　[jì shù wéi hù]
perawatan teknis　[purawatan tekunisu]
테크니컬 정비　[tekeunikeol jeongbi]
การบำรุงรักษาทางเทคนิค　[gaan bam-rung rak-saa taang tec-nic]

technical manual
a document containing detailed information on operational instructions, handling, maintenance, and repair of the equipment.
技術マニュアル　[gizyutu manyuaru]
機器の操作方法，取扱方法，メンテナンス要領，修理に関する事項を記述した文書．
技术手册　[jì shù shǒu cè]
manual teknis　[manuaru tekunisu]
테크니컬 매뉴얼　[tekeunikeol maenyueol]
คู่มือด้านทางเทคนิค　[kuu meuu daan tec-nic]

technical specifications
a detailed description of technical requirements to form the basis for the actual design of equipment.
技術仕様書　[gizyutu siyousyo]
機器の設計の基本項目となる技術要求事項を詳細に記述した文書．
技术规范　[jì shù guī fàn]
spesifikasi teknis　[supesifikasi tekunisu]
기술 시방서　[gisul sibangseo]
สเปคด้านเทคนิค　[spec daan tec-nic]

Teflon coating
テフロン皮膜　[tehuron himaku]
聚四氟乙烯膜　[jù sì fú yǐ xī mó]
lapisan teflon　[rapisan tefuron]
테프론 코팅　[tepeuron koting]
การเคลือบด้วยเทฟล่อน　[gaan klu-ab du-oi teflon]

technical maintenance — temper brittlenness

telemetering
transmitting an electronic signal of an instrument reading to a remote location by means of wires, radio waves, or other means.
遠隔測定　[enkaku sokutei]
計器の読取値の電気信号を有線，無線，その他の手段で離れた場所に伝達すること．
远距离测定　[yuǎn jù lí cè dìng]
telemetering　[teremeteringu]
원격측정　[won-gyeokcheukjeong]
การวัดในระยะไกล　[gaan wad nai ra-ya glai]

telethermometer
a temperature measuring system in which the temperature sensor is located at a distance from the temperature indicator.
遠隔温度計　[enkaku ondokei]
温度表示器から離れた場所に温度センサを設置した温度測定システム．
远距离温度计　[yuǎn jù lí wēn dù jì]
telemeter　[teremeteru]
원격 온도계　[wongyeok ondogye]
ตัววัดอุณหภูมิในระยะไกล　[to-a wad un-na-ha-puum ra-ya glai]

temper brittlenness
a property of steel whereby a quenched workpiece becomes susceptible to brittle fracture when it is kept at a certain tempering temperature or it is cooled down gradually from that temperature.
焼戻ぜい性　[yakimodosi zeisei]
焼入れした鉄鋼をある焼戻温度に保持した場合又は焼戻温度から徐冷した場合，ぜい性破壊が生じやすくなる現象．
回火脆性　[huí huǒ cuì xìng]
kerapuhan karena sepuh keras [kerapuhan karena sepuhu kerasu]
뜨임 취성　[tteuim chwiseong]

temper brittlenness — tempered martensite

การเปราะจากการอบ [gaan pra-or chak gaan ob]

temper colour
the color of oxide film formed on a steel surface when tempered.
テンパーカラー [tenpaa karaa]
焼戻しの際に鉄鋼の表面に現れる酸化膜の色.
回火颜色 [huí huǒ yán sè]
warna sepuh keras [waruna sepuhu kerasu]
템퍼 칼라 [tempeo kalla]
สีของแผ่นออกไซด์บนผิวงานเกิดขึ้นหลังจากอบ [sii khong paaen oxide bon piw ngaan gerd khun lang chaak ob]

temper furnace
a furnace used for tempering.
焼戻炉 [yakimodosi ro]
焼戻加工のために用いる炉.
回火炉 [huí huǒ lú]
tungku sepuh keras [tunguku sepuhu kerasu]
뜨임로 [tteuimno]
เตาอบ [ta-o ob]

temperature control
温度制御 [ondo seigyo]
温度控制 [wēn dù kòng zhì]
pengendalian suhu [pungendarian suhu]
온도 제어 [ondo jeeo]
การควบคุมอุณหภูมิ [gaan ku-ab kum un-na-ha puum]

temperature profile
a temperature distribution in a furnace along its long direction.
温度分布 [ondo bunpu]
炉の長手方向への温度分布.
温度分布 [wēn dù fēn bù]
distribusi suhu [disutoribusi suhu]
온도 분포 [ondo bunpo]
การกระจายของอุณหภูมิ [gaam gra-chaai khong un-na-ha-poom]

temperature sensor
a sensing device to respond to temperature stimulation.
温度センサ [ondo sensa]
温度に感応する感知器.
温度传感器 [wēn dù chuán gǎn qì]
sensor suhu [sensoru suhu]
온도 센서 [ondo senseo]
เซ็นเซอร์จับอุณหภูมิ [sensor chap un-na-ha-puum]

temperature time curve
the diagram that shows the relation between the temperature and the time.
温度−時間曲線 [ondo-zikan kyokusen]
温度−时间曲线 [wēn dù- shí jiān qǔ xiàn]
kurva suhu waktu [kuruva suhu wakutu]
온도-시간 곡선 [ondo-sigan gokseon]
เส้นโค้งอุณหภูมิ - เวลา [sen kong khong un-na-ha puum-wee laa]

temperature transducer
a device in an automatic temperature-control system that converts the temperature into some other quantity such as mechanical movement, pressure, or electric voltage to be applied to control the system.
温度変換器 [ondo henkanki]
自動温度制御システムに使われる素子で，温度を他の機械的な動き，圧力，電圧などの物理量に変換して系の制御を行う.
温度交换器 [wēn dù jiāo huàn qì]
pengubah suhu [pungubahu suhu]
온도 변환기 [ondo byeonhwan-gi]
อุปกรณ์เปลี่ยนอุณหภูมิ [up-pa-gorn plian un-na-ha-puum]

tempered martensite
a microstructure derived from temper-

ing martensite.
焼戻マルテンサイト　[yakimodosi marutensaito]
　マルテンサイトを焼き戻して得られる組織．
回火马氏体　[huí huǒ mǎ shì tǐ]
martensit disepuh keras　[marutensito disepuhu kerasu]
템퍼드 마르텐사이트　[tempeodeu mareutensaiteu]
เหล็กโครงสร้างมาร์เทนไซท์ที่ผ่านการอบ　[lek krong sang martensite tii paan gaan ob]

tempering
　the processing that the material is heated and kept with a proper temperature under A_1 point to proceed the transformation or precipitation to become a stable structure, and cooled to room temperature to get the specific property. There is a case that it is practiced after normalizing.
焼戻し　[yakimodosi]
　焼入れで生じた組織を，変態又は析出を進行させて安定な組織に近づけ，所要の性質及び状態を与えるために，A_1点以下の適当な温度に加熱，冷却する操作．焼ならしの後に用いることもある．
回火　[huí huǒ]
sepuh kerah　[sepuhu kerahu]
뜨임　[tteuim]
การอบให้คงรูป　[gaan ob hai kong ruup]

tempering crack
　a crack that occurs due to rapid heating and cooling, or structural changes when quenched steel is annealed.
焼戻割れ　[yakimodosi ware]
　焼入れした鉄鋼を焼戻しする際，急熱・急冷又は組織変化のために生じる割れ．

回火裂缝　[huí huǒ liè fèng]
retak karena sepuh keras　[retaku karena sepuhu kerasu]
뜨임 균열　[tteuim gyunnyeol]
การแตกจากการอบ　[gaan taaek chaak gaan ob]

tempering temperature
　a temperature at which tempering is achieved.
焼戻温度　[yakimodosi ondo]
　焼戻しを行う温度．
回火温度　[huí huǒ wēn dù]
suhu disepuh keras　[suhu disepuhu kerasu]
뜨임 온도　[tteuim ondo]
อุณหภูมิในการอบ　[un-na-ha puum nai gaan ob]

tempering time
　a holding period for tempering in a heated furnace.
焼戻時間　[yakimodosi zikan]
　焼戻しに必要な加熱した炉中での保持時間．
回火时间　[huí huǒ shí jiān]
waktu disepuh keras　[wakutu disepuhu kerasu]
뜨임 시간　[tteuim sigan]
เวลาในการอบ　[wee-laa nai gaan ob]

template
　a thin plate with punched out figures and letters for a trace writing.
テンプレート　[tenpureeto]
　図形や文字をならい書きするときに用いる薄い板．
型板　[xíng bǎn]
pola　[pora]
형판　[hyeongpan]
แผ่นแบบตัวอักษร　[paaen baaep tu-a ak-sorn]

tensile load
　a load to extend a material toward the

tensile load — tension ring

axis line.
引張荷重　[hippari kazyuu]
軸線方向に引き伸ばすように働く荷重.
拉伸載荷　[lā shēn zǎi hé]
beban tarik　[beban tariku]
인장 하중　[injang hajung]
โหลดของการดึง　[load khong gaan dung]

tensile strength
the value obtained by dividing maximum tensile load applied to a test piece by the original cross sectional area of the parallel section of the test piece.
引張強さ　[hippari tuyosa]
最大引張荷重を平行部の原断面積で除した値.
拉伸強度　[lā shēn qiáng dù]
kuat tarik　[kuato tariku]
인장 강도　[injang gangdo]
ความแข็งแรงต่อการดึง　[kwaam khaaeng rang to-u gaan duung]

tensile stress
the value obtained by dividing the tensile load applied to a test piece by the original cross sectional area of the parallel section of the test piece.
引張応力　[hippari ouryoku]
試験片に加えられた引張荷重を, 試験片の平行部の原断面積で除した値.
拉伸応力　[lā shēn yīng lì]
tegangan tarik　[tegangan tariku]
인장 응력　[injang eungnyeok]
ความเค้นแรงดึง　[kwaam ken rang duung]

tensile test
a test to measure the yield point, proof stress, tensile strength, yield elongation, fracture elongation, and reduction of area by gradually pulling a test piece or product using a tensile tester.

引張試験　[hippari siken]
引張試験機を用いて試験片又は製品を徐々に引っ張り, 降伏点, 耐力, 引張強さ, 降伏伸び, 破断伸び, 絞りなどを測定する試験.
拉伸試験　[lā shēn shì yàn]
uji tarik　[uji tariku]
인장 시험　[injang siheom]
การทดสอบแรงดึง　[gaan tod-sorb rang duung]

tensile test at elevated temperature
a test whereby a test piece is kept at a certain elevated temperature during a tensile test.
高温引張試験　[kouon hippari siken]
試験片を高温度に保って行う引張試験.
高温拉伸試験　[gāo wēn lā shēn shì yàn]
uji tarik suhu tinggi　[uji tariku suhu tingugi]
고온 인장시험　[goon injangsiheom]
การทดสอบแรงดึงที่อุณหภูมิสูง　[gaan tod-sorb rang duung tii un-na-ha-puum]

tension annealing
low-temperature heat treatment performed while applying constant tension to a wire rod.
テンションアニーリング　[tensyon aniiringu]
線材料に一定張力を与えながら行う低温熱処理.
応力退火　[yīng lì tuì huǒ]
anneal tarik　[annearu tariku]
인장 풀림　[injang pullim]
การอบอ่อนคลายความดึง　[gaan ob-orn klaai kwaam tung]

tension ring
a circular spring that gives tension force to a gasket ring.
テンションリング　[tensyon ringu]

ガスケットリングに張力を与える円形
ばね．
拉伸环　[lā shēn huán]
cincin tarik　[chinchin tariku]
텐션 링　[tensyeon ring]
สปริงกลมในแกสเกตริง　[sa-pring glom nai gasket ring]

tension set
    the difference between the original gauge length of a test piece and the gauge length after applying and subsequently removing the load in a tensile test.
永久伸び　[eikyuu nobi]
引張試験においてある荷重を与え，次にこれを除去した後の標点間の長さと元の標点距離との差．
永久拉伸变形　[yǒng jiǔ lā shēn biàn xíng]
perpanjangan tetap　[purupanjangan tetapu]
영구 신연　[yeonggu sinnyeon]
การเซ็ทถาวร　[gaan set taa-worn]

tension spring coiling machine
    a machine that forms the coil and hook of a helical extension spring.
引張ばね成形機　[hippari bane seikeiki]
主として引張コイルばねのコイル部及びフック部を成形する機械．
拉伸弹簧成形机　[lā shēn tán huáng chéng xíng jī]
mesin ulir pegas tarik　[mesin uriru pugasu tariku]
인장 스프링 성형기　[injang spring seonghyeonggi]
เครื่องม้วนสปริงขดชนิดดึงและตะขอ　[kru-ang muan sa-pring khod cha-nid dung laae ta-khor]

tension spring testing machine
    the generic term for machines that test tension springs. Includes tensile load testers and endurance testers.
引張ばね試験機　[hippari bane sikenki]
引張ばねを試験する機械の総称で，引張荷重試験機や耐久試験機などがある．
拉伸弹簧试验机　[lā shēn tán huáng shì yàn jī]
mesin pengujian pegas tarik　[mesin pungujian pugasu tariku]
인장 스프링 시험기　[injang spring siheomgi]
เครื่องทดสอบสปริงขดชนิดดึง　[kru-ang tod-sorb sa-pring khod cha-nid dung]

terminal curve of spiral spring
    an end tip bend of spiral spring at inner or outer end.
渦巻ばねの端末曲げ　[uzumaki bane no tanmatu mage]
渦巻ばねの外径側あるいは内径側の端末部の曲げ．
涡卷弹簧的末端弯曲　[wō juǎn tán huáng de mò duān wān qǔ]
kurva akhir pegas spriral　[kuruva akuhiru pugasu supuriraru]
스파이럴 스프링 단부 굽힘　[seupaireol spring danbu gupim]
ปลายโค้งของสปริงก้นหอย　[plaai kong khong sa-pring gon hoi]

test condition
    necessary conditions to practice a test.
試験条件　[siken zyouken]
試験を実施するために必要な条件．
试验条件　[shì yàn tiáo jiàn]
keadaan uji　[keadaan uji]
시험 조건　[siheom jogeon]
เงื่อนไขในการทดสอบ　[nge-un khai nai gaan tod-sorb]

test load
    static load applied to the spring to determine the spring characteristics. Usually the load is applied once.
試験荷重　[siken kazyuu]

test load — thermal plasticity

  ばね特性を測定するために，ばねに加える静的荷重．通常1回負荷する．
  試験負荷　[shì yàn fù hé]
  beban uji　[beban uji]
  시험 하중　[siheom hajung]
  โหลดที่ใช้ในการทดสอบ　[load tii chai nai gaan tod-sorb]

test piece
  a piece of material mechined to the suitable dimensions for the mechanical property test.
  試験片　[sikenhen]
  機械的試験の目的に応じた寸法に加工した試料片．
  试验片　[shì yàn piàn]
  benda pengujian　[benda pungujian]
  시험편　[siheompyeon]
  ชิ้นงานทดสอบ　[shin-ngaan tod-sorb]

test result
  試験結果　[siken kekka]
  试验结果　[shì yàn jié guǒ]
  hasil uji　[hasiru uji]
  시험 결과　[siheom gyeolgwa]
  ผลการทดสอบ　[pon gaan tod-sorb]

test temperature
  a temperature at which the test is conducted.
  試験温度　[siken ondo]
  試験を実施したときの温度．
  试验温度　[shì yàn wēn dù]
  suhu uji　[suhu uji]
  시험 온도　[siheom ondo]
  อุณหภูมิที่ใช้ในการทดสอบ　[un-na-ha puum tii chai nai gaan tod-sorb]

testing method
  a procedure to achieve the purpose of a test.
  試験方法　[siken houhou]
  試験の目的を達成させるための方法．
  试验方法　[shì yàn fāng fǎ]
  metode penngujian　[metode punngujian]
  시험 방법　[siheom bangbeop]
  วิธีการทดสอบ　[wi-tii gaan tod-sorb]

testing time
  試験時間　[siken zikan]
  试验时间　[shì yàn shí jiān]
  waktu pengujian　[wakutu pungujian]
  시험 시간　[siheom sigan]
  ระยะเวลาในการทดสอบ　[ra-ya wee-laa nai gaan tod-sorb]

thermal conductivity
  the ratio of amount of heat flowing perpendicularly through the unit area of an isothermal plane taken perpendicular to the direction of heat conduction per unit time to the thermal gradient in the same direction.
  熱伝導率　[netu dendouritu]
  熱の伝わる方向に垂直にとった等温面の単位面積を通って単位時間に垂直に流れる熱量とこの方向の温度勾配との比．
  导热率　[dǎo rè lǜ]
  hantaran termal　[hantaran terumaru]
  열 전도율　[yeol jeondoyul]
  อัตราการนำความร้อน　[at-traa gaan nam kwaam rorn]

thermal expansion
  an increase of volume of a material resulting from a rise of temperature.
  熱膨張　[netuboutyou]
  物質の体積が温度の上昇に伴って増加する現象．
  热膨胀　[rè péng zhàng]
  muai termal　[muai terumaru]
  열팽창　[yeolpaengchang]
  การขยายตัวจากความร้อน　[gaan kha-yaai to-a chaak kwaam rorn]

thermal plasticity
  a repeatable characteristic whereby a

material becomes soft by heating and hard by cooling.
熱可塑性　[netu kasosei]
　熱を加えれば軟らかくなり，冷却すれば硬くなることを繰り返す性質．
热可塑性　[rè kě sù xìng]
plastisitas panas　[purasutisitasu panasu]
열 가소성　[yeol gasoseong]
คุณสมบัติความเป็นพลาสติกจากความร้อน
[kun-na som-bat kwaam pen plastic chaak kwaam rorn]

thermal stress
　stress induced in a body when the body are not free to expand in response to changes in temperature. It may be caused by external support conditions, uneven temperature distribution inside a material, or the difference in thermal expansion.
熱応力　[netu ouryoku]
　自由熱膨張を拘束するために生じる応力で，外部支持条件で生じることもあり，一物体内の温度の不均一分布又は熱膨張差によって生じることもある．
热应力　[rè yīng lì]
regangan termal　[regangan terumaru]
열응력　[yeoreungnyeok]
ความเค้นที่เกิดจากความร้อน　[kwaam ken tii gerd chaak kwaam rorn]

thermocouple
　a temperature sensor consisting of two dissimilar conductors joined together at their ends to generate electricity at the joint due to thermoelectric effect.
熱電対　[netudentui]
　熱電効果によって接合部で起電力を生じる一対の異種材料からなる導体を用いた温度センサ．
热电偶　[rè diàn ǒu]
termocouple　[terumochoupure]

열전대　[yeoljeondae]
เทอร์โมคอปเปิล　[thermcouple]

thermograph
　an instrument that senses, and records the temperature.
自記温度計　[ziki ondokei]
　温度を感知して記録する機器．
温度记录器　[wēn dù jì lù qì]
termograp　[terumogurapu]
기록 온도계　[girok ondogye]
ตัววัดบันทึกอุณหภูมิ　[to-a wad ban-tuk un-na-ha-puum]

thermomechanical treatment
　a process to provide a material with specific properties which cannot be achieved by heat treatment alone, by performing final plastic forming process within a certain temperature range.
加工熱処理　[kakou netusyori]
　最終の塑性加工がある温度範囲で行われ，熱処理だけでは得られない特定の性質を持つ材料を生じさせる加工工程．
热处理加工　[rè chǔ lǐ jiā gōng]
olahan termomekanik　[orahan terumomekaniku]
가공 열처리　[gagong yeolcheori]
กรรมวิธีความร้อนเชิงกล　[gam-ma-wi-tii kwaam rorn she-ong kon]

thermostat
　a device which measures changes in temperature, and controls the heat source to maintain a desired temperature of the system.
サーモスタット　[saamosutatto]
　温度の変化を感知し，熱源を制御して系全体を望ましい温度に維持する素子．
自动湿度，温度调节装置　[zì dòng shī dù,

thermostat — threaded plug engagement of extension coil spring

wēn dù tiáo jié zhuāng zhì]
termostat　[terumosutato]
서모스탯　[seomoseutaet]
เทอร์โมสแตด　[thermostat]

thickness gauge
    an instrument for measuring the thickness of a sheet of metal, or the thickness of a coating. The simplest way for the former is given by comparing the object with a set of thin metal sheets with guaranteed accuracy of the thickness. Other examples include backscattering radioactive thickness gauges and ultrasonic thickness gauges etc.
    シクネスゲージ　[sikunesu geezi]
    金属の薄板や塗膜の厚さの測定器．前者の最も簡単な方法はあらかじめ測定した板厚を持つ一組の金属片と測定物を比較することで得られる．ほかには放射線板厚計，超音波板厚計などがある．
    塞尺　[sāi chǐ]
    penaksir tebal　[punakusiru tebaru]
    두께 게이지　[dukke geiji]
    เกจวัดความหนา　[gauge wad kwaam nan]

thickness of wire cross section
    thickness of a noncircular cross-section of the material.
    短径　[tankei]
    異形断面コイルばねに用いる材料の横断面の厚さ．
    短轴　[duǎn zhóu]
    ketebalan penampang kawat　[ketebaran punanpangu kawato]

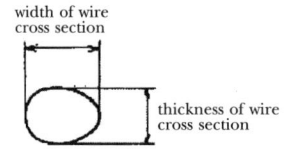

Fig T-6 thickness of wire cross section

단경　[dangyeong]
ความหนาของหน้าตัดเส้นลวด　[kwaam naa khong naa tad sen lu-ad]

thin sheet
薄板　[usuita]
薄板　[báo bǎn]
lembaran tipis　[renbaran tipisu]
박판　[bakpan]
แผ่นบาง　[paaen baang]

threaded mandrel
    a mandrel with a spiral groove. Material is wound onto this spiral groove to form a coil spring.
    溝付心金　[mizotuki singane]
    ら旋状の溝を持つ心金のことで，このら旋形状の谷部に材料を挿入させてコイルばねを成形する．
    螺纹心轴　[luó wén xīn zhóu]
    poros pemotong ulir　[porosu pumotongu uriru]
    나선형 맨드릴　[naseonhyeong maendeuril]
    แกนม้วนชนิดมีร่องเกลียว　[gaaen mu-an chanid mii rong glieaw]

threaded plug engagement of extension coil spring
栓形ねじ込みフック　[sengata nezikomi hukku]
圆形螺纹拧入接头（钩环）　[yuán xíng luó wén níng rù jiē tóu (gōu huán)]
pegas tarik berkait ulir　[pegasu tariku burukaito uriru]
플러그형 나사 박음 고리　[peulleogeuhyeong nasa bageum gori]
ตะขอชนิดมีร่องเกลียวแบบกด　[ta-khor cha-nid

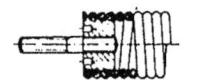

Fig T-7 threaded plug engagement of extension coil spring

mii rong glieaw baaep god]

threaded-on plate engagement of extension coil spring

板形ねじ込みフック [itagata nezikomi hukku]

片形螺纹拧入接头（钩环） [piàn xíng luó wén nǐng rù jiē tóu (gōu huán)]

pegas tarik berkait pipih [pegasu tariku burukaito pipihu]

판형 나사 박음 고리 [panhyeong nasa bageum gori]

ตะขอชนิดมีร่องเกลียวแบบแผ่น [ta-khor chanid mii rong glieaw baaep paaen]

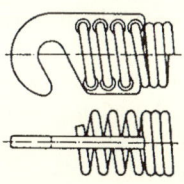

Fig T-8 thread-on plate engagement of extension coil spring

through feed grinding

a grinding method whereby a work is transferred from an entrance to the opposite side such as a center-less grinding.

通し送り式研削 [toosi okurisiki kensaku]

無心研削盤（センタレス）のように被研削物が一方方向に送られる研削方法。

连续进给磨削 [lián xù jìn gěi mó xuē]

penggerindaan umpan terus [pungugerindaan umupan terusu]

관통 이송 연삭 [gwantong isong yeonsak]

การเจียรชนิดป้อนต่อเนื่อง [gaan chiaa chanid porn to-a ne-ung]

throughput

the capacity of processing in a specified time.

処理能力 [syori nouryoku]

与えられた時間内に遂行される仕事量.

处理能力 [chǔ lǐ néng lì]

kemampuan olah [kemanpuan orahu]

처리능력 [cheorineungnyeok]

ความสามารถในการทำงานตามกำหนด [kwaam sa-maad nai gaan tam ngaan taam gam-nod]

tightening torque

the torque required to tighten nuts or screws in order to generate initial clamping force.

締付トルク [simetuke toruku]

初期締付力を生じさせるために，ナット又はおねじ部品を締め付けるのに必要なトルク．

紧固扭矩 [jǐn gù niǔ jǔ]

momen puntir pengencang [momen puntiru pungenkangu]

조임 토-크 [joim to-keu]

แรงบิดตึงแน่น [rang bid tung naaen]

time and motion study

observation, analysis, and measuring of the steps in the performance of a job to determine a standard time.

作業研究（TMS） [sagyou kenkyuu (tjiiemues)]

実行されている作業の各段階を観察，分析，測定して標準作業時間を決めること．

作业研究（TMS） [zuò yè yán jiū (TMS)]

studi waktu dan gerak kerja [sutudi wakutu dan geraku keruja]

작업 연구 [jakeop yeongu]

การศึกษาด้านเวลาและการเคลื่อนไหว [gaan suk-saa daan wee-laa laae gaan klu-an wai]

time switch

a clock-controlled switch to open or close a circuit at one or more predetermined times.

時限スイッチ　[zigen suitti]
　1回あるいはそれ以上のあらかじめ決めた時刻に回路を閉じたり開放する，時計で制御されたスイッチ．
限时开关　[xiàn shí kāi guān]
switch waktu　[suwitti wakutu]
타임 스위치　[time seuwichi]
สวิตช์เวลา　[switch wee-laa]

time-temperature-transformation curve
　a diagram indicating the start and end lines of transformation by drawing lines connecting the transformation start and end times when eutectoid carbon steel is cooled to and maintained at a number of different temperatures from the austenite region.
T.T.T 線図　[tjii-tjii-tjii senzu]
　共析炭素鋼をオーステナイト域から種々の温度に急冷し，その温度に保持して変態の開始時間と終了時間を測定し，これらの点を結んで変態開始線と終了線を表した図．
时间-温度-变形（TTT）曲线　[shí jiān-wēn dù- biàn xíng (TTT) qǔ xiàn]
kurva waktu-suhu-tranformasi　[kuruva wakutu-suhu-toranforumasi]
티티티 (T.T.T) 선도　[tititi (T.T.T) seondo]
เส้นโค้งเวลา, อุณหภูมิ, การแปรรูป　[sen we-laa, un-na-ha-puum, gaan praae ruup]

tip-contact method
　design method of a leaf spring based on the assumption that the force on a leaf is transferred to adjacent leaf only at the end of the leaf.
板端法　[bantanhou]
　ばね板から隣接するばね板への力の伝達が，板の先端だけで行われるという仮定を基礎とする重ね板ばねの設計方法．
板端法　[bǎn duān fǎ]
metoda sentuh ujung　[metoda sentuhu ujungu]
판단법　[pandanbeop]
วิธีการติดปลาย　[wi-tii gaan tid plaai]

titanium alloy spring
　a spring made of pure titanium or titanium alloys such as α-alloy, α–β-alloy, or β-alloy.
チタン合金ばね　[titan goukin bane]
　純チタン，α合金，α–β合金，β合金等のチタン合金を用いたばね．
钛合金弹簧　[tài hé jīn tán huáng]
pegas paduan nikel　[pugasu paduan nikeru]
티탄 합금 스프링　[titan hapgeum spring]
สปริงโลหะผสมไททาเนียม　[sa-pring loo-ha pa-som titanium]

toggle switch
　a switch that is operated by flipping a projecting lever that is combined with a spring to provide a snap action of swiching.
トグルスイッチ　[toguru suitti]
　ばねと係止されているレバーを操作して，開閉のスナップアクションを行うスイッチ．
触点开关　[chù diǎn kāi guān]
switch togle　[suwitti togure]
토글 스위치　[togeul seuwichi]
สวิตช์ปิด-เปิดแบบโยก　[switch baaep yok pid-perd]

tolerance
　the difference between the maximum and minimum permissible values.
許容差（公差）　[kyoyousa (kousa)]
　許容限界の上限と下限の差．
公差带　[gōng chà dài]
toleransi　[toreransi]
공차　[gongcha]

ค่าเผื่อที่ยอมรับได้ [kaa pu-a tii yorm rab dai]
tolerance limit
    the values that are specified as the maximum and minimum permissible limits.
許容限界 [kyoyou genkai]
    特性の許容される上限，及び下限として定められた値．
允许极限 [yǔn xǔ jí xiàn]
batas toleransi [batasu toreransi]
허용 한계 [heoyong hangye]
ค่าเผื่อจำกัด [kaa pu-a cham kad]
tolerance range
    see "tolerance".
許容範囲 [kyoyou han'i]
    "tolerance" 参照．
允许范围 [yǔn xǔ fàn wéi]
rentang toleransi [rentangu toreransi]
허용 범위 [heoyong beomwi]
ขอบเขตค่าเผื่อที่ยอมรับได้ [khorb-khet kaa pu-a tii yorm rab dai]

tool dresser
    a metal shank with tool-stone-grade diamond inset to trim the face of a grinding wheel.
工具ドレッサ [kougu doressa]
    研削ダイヤモンドを埋め込んだ金属軸で，研削といし表面を整えるのに使用する．
工具修整器 [gōng jù xiū zhěng qì]
dresser alat [doresseru arato]
공구 드레서 [gonggu deureseo]
ตัวหัวหุ้มหินเจียร [tu-a hu-a hum hin chiaa]

tooth height of toothed washer
    height of twisted teeth to enhance the locking effect of screws.
歯付座金の歯の高さ [hatuki zagane no ha no takasa]
    ねじのゆるみどめ効果を上げるためにねじった歯の高さ．

齒形弾簧墊圈的齒高 [chǐ xíng tán huáng diàn quān de chǐ gāo]
tinggi gigi ring pipih bergerigi [tingugi gigi ringu pipihu berugerigi]
톱니꼴 와셔의 톱니 높이 [tomnikkol wasyeoui tomni nopi]
ความสูงของแหวนรูปฟัน [kwaam soong khung waaen ruup fan]

toothed lock washer
    spring washer made of thin ring plate with twisted teeth.
歯付座金 [hatuki zagane]
    円環状の薄板にねじれた歯を付けたばね座金．内歯形，外歯形及び内外歯形がある．
齒形弾簧墊圈 [chǐ xíng tán huáng diàn quān]
ring pipih pengunci bergerigi [ringu pipihu pungunsi berugerigi]
톱니꼴 와셔 [tomnikkol wasyeo]
แหวนรูปฟัน [waaen ruup fan]

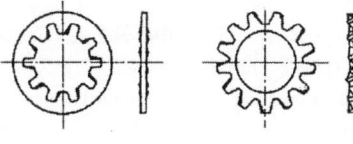

internal tooth type   external tooth type

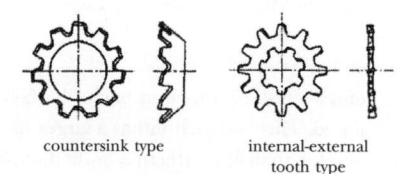

countersink type   internal-external tooth type

Fig T-9 toothed lock washer

top coat
    a painting that is applied on the primer coating.
上塗り [uwanuri]
    下地塗膜の上に上塗り用の塗料を塗る

こと．
外层喷涂　[wài céng pēn tú]
lapis atas　[rapisu atasu]
상도　[sangdo]
สีเคลือบ　[sii klu-ab]

torque
 moment generated around the axis when external force is applied to a helical torsion spring or a torsion bar.
トルク　[toruku]
ねじりコイルばね，トーションバーなどに外力が加えられたときに，軸周りに発生するモーメント．
扭矩　[niǔ jǔ]
momen puntir　[momen puntiru]
토크　[to-keu]
แรงบิด　[rang bid]

torsion bar axis
 a straight line that indicates the center axis of torsion bar.
トーションバー軸線　[toosyon baa zikusen]
トーションバーの中心軸を表す直線．
扭杆轴线　[niǔ gǎn zhóu xiàn]
sumbu batang puntir　[sunbu batangu puntiru]
토션바 축선　[tosyeonba chukseon]
แกนของทอร์ชั่นบาร์　[gaaen khong torsion bar]

torsion bar head
 end section of a torsion bar to be fastened. This end section has a larger diameter than the effective body diameter to have serration or spline groove around the circle, or to have square, or hexagonal cross-section head.
トーションバー端部　[toosyon baa tanbu]
トーションバー端の締結部．一般に有効径より端部径が大きく，セレーション加工又はスプライン加工された

り，四角又は六角に据え込まれたものがある．
扭杆端　[niǔ gǎn duān]
kepala batang puntir　[kepara batangu puntiru]
토션바 끝부　[tosyeonba kkeutbu]
ปลายของทอร์ชั่นบาร์　[plaai khong torsion bar]

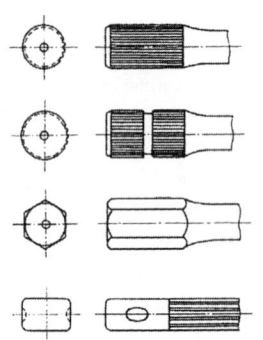

Fig T-10　torsion bar head

torsion bar shank
 see "torsion bar head".
トーションバー把持部　[toosyon baa hazibu]
"torsion bar head" 参照．
扭杆头部　[niǔ gǎn tóu bù]
tankai batang puntir　[tankai batangu puntiru]
토션바 장착부　[tosyeonba jangchakbu]
ก้านของทอร์ชั่นบาร์　[gaan khong torsion bar]

torsion bar spring
 a straight bar with heads on both ends made of spring steel. It provides torsional spring function.
トーションバー　[toosyon baa]
ねじりを利用する棒状のばね．
扭杆弹簧　[niǔ gǎn tán huáng]
pegas batang puntir　[pugasu batangu puntiru]
토션바　[tosyeonba]

สปริงทอร์ชั่นบาร์ [sa-pring torsion bar]

torsion bar spring rate
the moment required to produce unit deflection in a torsion bar.
トーションバーばね定数 [toosyon baa bane zyousuu]
トーションバーに単位のたわみ角を与えるのに必要なモーメント．
扭杆弹簧刚度 [niǔ gǎn tán huáng gāng dù]
derajat perbandingan pegas batang puntir [derajato purubandingan pugasu batangu puntiru]
토션바 스프링 상수 [tosyeonba spring sangsu]
ค่าคงที่สปริงเรทของทอร์ชั่นบาร์ [kaa kong tii sa-pirng rate khong torsion bar]

torsion bar spring with serration ends
a torsion bar with serrations on both ends for applying a torsion moment.
セレーションタイプトーションバー [sereesyon taipu toosyon baa]
ねじりモーメントを加えるための両端末にセレーション加工を施したトーションバー．
锯齿形端部扭杆弹簧 [jù chǐ xíng duān bù niǔ gǎn tán huáng]
pegas batang puntir dengan ujung bergerigi [pugasu batangu puntiru dengan ujungu berugerigi]
세레이션형 토션바 [sereisyeonhyeong tosyeonba]
สปริงทอร์ชั่นบาร์ที่ปลายเจียร์ร่องฟัน [sa-pring torsion bar tii plaai chiaa rong fan]

torsion bar with circular cross section
丸断面トーションバー [marudanmen toosyon baa]
圆形截面扭杆 [yuán xíng jié miàn niǔ gǎn]
batang puntir dengan penambang lingkaran [batangu puntiru dengan punanbangu ringukaran]

원형 단면 토션바 [wonhyeong danmyeon tosyeonba]
สปริงทอร์ชั่นบาร์หน้าตัดรูปวงกลม [sa-pring torsion bar naa tad ruup wong glom]

torsion bar with rectangular cross section
角断面トーションバー [kakudanmen toosyon baa]
矩形截面扭杆 [jǔ xíng jié miàn niǔ gǎn]
batang puntir dengan penampang empat persegi [batangu puntiru dengan punanpangu enpato purosegi]
사각 단면 토션-바 [sagak danmyeon tosyeon-ba]
สปริงทอร์ชั่นบาร์หน้าตัดรูปเหลี่ยม [sa-pring torsion bar naa tad ruup wong liam]

torsion fracture
a fracture caused by applied moment.
ねじり破壊 [neziri hakai]
モーメントが加わることによる破壊．
扭转破坏 [niǔ zhuǎn pò huài]
retakan puntir [retakan puntiru]
비틀림 파괴 [biteullim pagoe]
การแตกจากการบิด [gaan taaek chaak gaan bid]

torsion spring
spring subjected mainly to a twisting moment. In the narrow sense, helical torsion spring.
ねじりばね [neziri bane]
主としてねじりモーメントを受けるばねの総称．狭義には，ねじりコイルばね．
扭转弹簧 [niǔ zhuǎn tán huáng]
pegas puntir [pugasu puntiru]
토션 스프링 [tosyeon spring]
สปริงทอร์ชั่น [sa-pring torsion]

torsion spring coiling machine
forming machine of coils and arms for the helical torsion spring.
トーションマシン [toosyon masin]

torsion spring coiling machine — torsional strain

主として，ねじりコイルばねのコイル部及び腕部を成形する機械．
扭转弹簧成形机　[niǔ zhuǎn tán huáng chéng xíng jī]
mesin penggulung pegas puntir　[mesin pungugurungu pugasu puntiru]
토션 스프링 성형기　[tosyeon spring seonghyeonggi]
เครื่องม้วนสปริงทอร์ชั่น　[kru-ang muan sa-pring torsion]

torsion spring rate
the moment required to produce unit deflection in a torsion spring.
ねじりばね定数　[neziri bane zyousuu]
ねじりばねに単位のたわみ角を与えるのに必要なモーメント．
扭转弹簧刚度　[niǔ zhuǎn tán huáng gāng dù]
derajat perbandingan pegas puntir　[derajato purubandingan pugasu puntiru]
토션 스프링 상수　[tosyeon spring sangsu]
ค่าคงที่สปริงทอร์ชั่น　[kaa kong tii sa-pring torsion]

torsion test of steel wire
a test to study the number of twists, aspects of fracture and twisting, by clamping both ends of a test piece at a specified distance and twisting one end until fracture occurs, while maintaining sufficient tension to prevent deflection. Another test mode is to find whether wire fractures after a specified number of twists.
鋼線のねじり試験　[kousen no neziri siken]
試験片の両端を規定されたつかみ間隔で固くつかみ，たわまない程度に緊張しながらその一方を回転して破断し，そのときのねじり回数，破断の状況，ねじれの状況などを調べる試験．また，規定されたねじり回数までねじったと

き，線が破断しないかどうかを調べることもある．
钢丝扭转试验　[gāng sī niǔ zhuǎn shì yàn]
uji puntir kawat baja　[uji puntiru kawato baja]
강선의 비틀림 시험　[gangseonui biteullim siheom]
การทดสอบการทนต่อแรงบิดในเหล็กเส้น　[gaan tod-sorb gaan ton to-a raaeng bid nai lek-sen]

torsion testing machine
a testing machine to measure the torsion angle and torque by applying torsional load to a test piece.
ねじり試験機　[neziri sikenki]
試験片にねじり荷重を加えねじり角とトルクを測定する試験機．
扭转试验机　[niǔ zhuǎn shì yàn jī]
mesin pengujian puntir　[mesin pungujian puntiru]
비틀림 시험기　[biteullim siheomgi]
เครื่องทดสอบการบิด　[kru-ang tod-sorb gaan bid]

torsional moment
moment generated around the axis when external force is applied to a helical torsion spring or a torsion bar.
ねじりモーメント　[neziri moomento]
ねじりコイルばね，トーションバーなどに外力が加えられたときに，軸周りに発生するモーメント．
扭矩　[niǔ jù]
momen puntir　[momen puntiru]
비틀림 모멘트　[biteullim momenteu]
โมเมนต์ของการบิด　[moment khong gaan bid]

torsional strain
an angular defrection per unit length.
ねじりひずみ　[neziri hizumi]
単位長さ当たりの角変位．

扭转变形　[niǔ zhuǎn biàn xíng]
regangan puntir　[regangan puntiru]
비틀림 변형　[biteullim byeonhyeong]
ความเครียดการบิด　[kwaam kriead gaan bid]
torsional strength
 the maximum torsional stress on an exterior surface calculated from the maximum torsional moment when a material is fractured.
 ねじり強さ　[neziri tuyosa]
 材料をねじりによって，破壊させたとき，その最大ねじりモーメントから計算で求めた外表面の最大ねじり応力．
 扭转强度　[niǔ zhuǎn qiáng dù]
 kuat puntir　[kuato puntiru]
 비틀림 강도　[biteullim gangdo]
 กำลังของการบิด　[gam-lang khong gaan bid]
torsional stress
 shear stress exerted on a cross-section due to torsional action.
 ねじり応力　[neziri ouryoku]
 ねじり作用の結果生じる，横断面にかかるせん断応力．
 扭转应力　[niǔ zhuǎn yīng lì]
 tegangan puntir　[tegangan puntiru]
 비틀림 응력　[biteullim eungnyeok]
 ความเค้นการบิด　[kwaam ken gaan bid]
torsional vibration
 vibration that appears in the form of periodic torsional variation.
 ねじり振動　[neziri sindou]
 ねじりの同期的な変化として現れる振動．
 扭转振动　[niǔ zhuǎn zhèn dòng]
 getaran puntir　[getaran puntiru]
 비틀림 진동　[biteullim jindong]
 การสั่นของการบิด　[gaan san khong gaan bid]
total active length of torsion bar spring
 length of a torsion bar spring to calculate the spring characteristics, considering the effect of the head length.
 トーションバーの有効長　[toosyon baa no yuukoutyou]
 トーションバーのばね特性を計算する基礎となる長さで，つかみ部の影響を考慮したもの．
 扭杆弹簧有效长度　[niǔ gǎn tán huáng yǒu xiào cháng dù]
 panjang aktif total dari pegas batang puntir　[panjangu akutifu totaru dari pugasu batangu puntiru]
 토션바 스프링의 유효 길이　[tosyeonba spring-ui yuhyo giri]
 ระยะใช้งานของทอร์ชั่นบาร์　[ra-ya chai ngaan khong torsion bar]
total coils
 total number of turns between both ends of a coil spring.
 総巻数　[soumakisuu]
 コイルばねのコイルの端から端までの巻数．
 总圈数　[zǒng quān shù]
 jumlah total putaran　[jumurahu totaru putaran]
 총 감김 수　[chong gamgim su]
 ขดรวม　[khod ro-um]
total decarburization
 a phenomenon whereby virtually all carbon is lost from a steel surface when the steel is heated in an atmosphere which is reactive to carbon.
 全脱炭　[zendattan]
 炭素と反応する雰囲気の中で鉄鋼を加熱するとき，表面から炭素がほとんど失われる現象．
 全脱碳　[quán tuō tàn]
 dekarburisasi total　[dekaruburisasi totaru]
 전탈탄　[jeontaltan]
 การขจัดคาร์บอนออกทั้งหมด　[gaan kha-chad carbon ork tang mod]

total deflection

deflection of a spring from no load to the maximum load. In a helical compression spring, the total deflection is the difference between the free length and the solid length.

全たわみ　[zentawami]
無荷重時から最大荷重時までのばねのたわみ．圧縮コイルばねの場合は，自由高さと密着高さとの差をいう．
全变形　[quán biàn xíng]
lendutan total　[rendutan totaru]
전 휨량　[jeon hwimnyang]
การเสียรูปทั้งหมด　[gaan siaa ruup tang mod]

total elongation

the difference between the initial gauge length on a test piece and the elongated gauge length which has subjected to a tensile load in a tensile test.

全伸び　[zennobi]
引張試験において試験片にある引張荷重を加えたとき，その荷重によって伸びた状態における標点間の長さと元の標点距離との差．
全伸长　[quán shēn cháng]
perpanjangan total　[purupanjangan totaru]
전 신율　[jeon sinnyul]
การยืดออกทั้งหมด　[gaan yuud ork tang mod]

total failures

the perfect failures. It means to become the condition where the material does not perform the function at all. It also means the break down and the stop of function.

全破損　[zenhason]
完全に破壊してしまうこと．また，機能を完全に果たさない状態になること．ほかに故障，機能停止などの意も

ある．
全破坏　[quán pò huài]
kegagalan total　[kegagaran totaru]
전 파괴　[jeon pagoe]
ความเสียหายรวม　[kwaam siaa haai ro-um]

total length

the length from one end to the other end of an object.

全長　[zentyou]
物の端から端までの長さ．
全长　[quán cháng]
panjang total　[panjangu totaru]
전장　[jeonjang]
ความยาวรวม　[kwaam ya-o ro-um]

total strain

a difference between an elongated gauge length and an initial gauge length divided by the latter.

全ひずみ　[zenhizumi]
試験片の変形後の標点距離と初期標点距離の差を初期標点距離で割った値．
全变形率　[quán biàn xíng lǜ]
ragangan total　[ragangan totaru]
전 변형　[jeon byeonhyeong]
ความเค้นรวม　[kwaam ken ru-am]

toughness

a property of being tough against impact force.

じん性　[zinsei]
粘り強くて，衝撃破壊を起こしにくい性質．
韧性　[rèn xìng]
kekenyalan　[kekenyaran]
인성　[inseong]
ความเหนียว　[kwaam nieaw]

toughness test

test to examine the distortion or breakdown of the helical torsion spring, torsion bar and spring washer. In the case of spring washer, it is also called a

toughness test.
ねじり試験　[neziri siken]
　ねじりコイルばね，トーションバー，ばね座金類などの変形又は破壊の状況を調べる試験．ばね座金類では，粘り強さ試験ともいう．
扭转试验机　[niǔ zhuǎn shì yàn jī]
uji kekenyalan　[uji kekenyaran]
비틀림 시험　[biteullim siheom]
การทดสอบความเหนียว　[gaan tod-sorb kwaam nieaw]

traceability
　the ability to trace the results of measurement using reasonable standard reference and calibration chain.
トレーサビリティ　[toreesabiritji]
測定結果が適正な標準器と校正連鎖を通じてつながりを持つ状態．
标准依据可循性　[biāo zhǔn yī jù kě xún xìng]
traceabiliti　[toracheabiriti]
트레이서빌리티　[teureiseobilliti]
คุณสมบัติของการสืบย้อนกลับ　[kun-na sombat khong gaan su-sb yorn glab]

trailing leaf for air suspension
　a leaf spring used for air-suspension, connected to the chassis frame before the axle and used mainly as the linkage mechanism.
トレーリングリーフ　[toreeringu riihu]
エアサスペンションに用いられ，主にリンク機構として車軸の前側でシャシフレームとつながれる板ばね．
空气悬挂延簧片　[kōng qì xuán guà yán huáng piàn]
pegas trailing　[pugasu torairingu]
트레이링 판 스프링　[teureiring pan spring]
เหล็กแผ่นเทรลลิ่ง　[lek paaen trailing]

transducer
　a device which converts an input signal into an output signal of different form.
変換器　[henkanki]
入力信号を異なる形式の出力信号に変換する素子．
交換器　[jiāo huàn qì]
pengubah　[pungubahu]
트랜스듀서　[teuranseudyuseo]
ตัวเปลี่ยนสัญญาณ　[to-a pliean san-saan]

transfer load
　load transferred to the end of the plate in the tip-contact method.
伝達荷重　[dentatu kazyuu]
板端法において板の先端に伝達する荷重．
传递载荷　[chuán dì zǎi hé]
beban berpindah　[beban berupindahu]
전달 하중　[jeondal hajung]
โหลดที่ส่งผ่าน　[load tii song paan]

transformation
　the phenomenon whereby a crystalline structure changes to another structure due to rise or fall of temperature. Some transformation does not necessarily involve a change of crystalline structure, such as magnetic transformation.
変態　[hentai]
温度を上昇又は下降させた場合などにある結晶構造から他の結晶構造に変化する現象．磁気変態のように必ずしも結晶構造の変化を伴わないものもある．
状态变化　[zhuàng tài biàn huà]

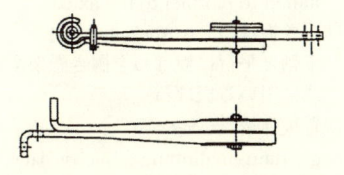

Fig T-11　trailing leaf for air suspension

327

tranformasi [toranforumasi]
변태 [byeontae]
การเปลี่ยนแปลง [gaan plian plaaeng]
transition
遷移 [sen'i]
迁移 [qiān yí]
transisi, peralihan [toransisi, purarihan]
천이 [cheoni]
การส่งผ่าน [gaan song paan]
transition point
  points where the characteristics begin to change or end to change, on the load-deflection diagram of a nonlinear characteristics spring. In the case where the transition area of the characteristics is small enough to be ignored, the transition point may be called the inflection point.
変曲点 [henkyokuten]
非線形特性ばねの荷重-たわみ線図において，特性が変化し始める点又は終了する点．特性の遷移部分が無視できるほど微小な場合は，交会点のことを変曲点と呼ぶこともある．
特性線折点 [tè xìng xiàn shé diǎn]
titik transisi [titiku toransisi]
변곡점 [byeongokjeom]
จุดส่งผ่าน [chud song paan]

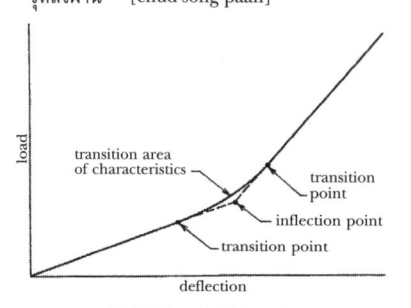

Fig T-12 transition point

transition temperature
  the temperature corresponding to the phenomenon whereby the absorption energy of a material suddenly drops and the fracture surface changes from ductile to brittle in impact tests performed at various temperatures below room temperature.
遷移温度 [sen'i ondo]
ある材料について，常温域から温度を変えて低温域のいろいろな温度で衝撃試験をしたとき，吸収エネルギーが急激に低下し，破面の外観が延性からぜい性に変化する現象に対応する温度．
转变温度 [zhuǎn biàn wēn dù]
suhu transisi [suhu toransisi]
천이 온도 [cheoni ondo]
อุณหภูมิที่ส่งผ่าน [un-na-ha-puum tii song paan]
transverse force
  component force of the load applied to the coil spring in the perpendicular direction to the coil axis.
横方向力 [yokohoukouryoku]
コイルばねに作用する荷重のコイル軸と直交方向との分力．
横向力 [héng xiàng lì]
gaya melintang [gaya merintangu]
횡 방향력 [hoeng banghyangnyeok]
แรงในแนวขวาง [rang nai naaew kwaang]
transverse leaf spring
  a leaf spring used in the suspension system in which the leaf spring is installed to parallel to the axle.
横置き板ばね [yokooki ita bane]
車軸と平行に板ばねを置き懸架する方式に用いる板ばね．
横置板弾簧 [héng zhì bǎn tán huáng]
pegas daun malintang [pugasu daun marintangu]

횡치식 판 스프링 [hoengchisik pan spring]
แหนบแผ่นแนวขวาง [naaep paaen naaew kwaang]

transverse load
    load on a spring applied perpendicular to the direction normally used. In the case of a leaf spring, this means the load applied in the width direction of the leaf.
    横荷重 [yokokazyuu]
    ばねの通常使われる方向とは，直角の方向に加わる荷重．重ね板ばねの場合は，板幅方向に加わる荷重．
    横向载荷 [héng xiàng zǎi hé]
    beban melintang [beban merintangu]
    횡하중 [hoenghajung]
    โหลดในแนวขวาง [load nai naaew kwaang]

transverse stiffness
    stiffness against the load in the perpendicular direction to the normal direction of spring load. In the case of a leaf spring, this means the stiffness in the width direction of the leaf.
    横剛性 [yokogousei]
    ばねの通常使われる方向とは直角の方向からの荷重に対する剛性．重ね板ばねの場合は，板幅方向の剛性．
    横向刚度 [héng xiàng gāng dù]
    kekakuan melintang [kekakuan merintangu]
    가로강성 [garogangseong]
    ความฝืดแนวขวาง [kwaam fuud naaew kwaang]

trapezoidal section
    台形断面 [daikei danmen]
    梯形截面 [tī xíng jié miàn]
    penampang trapesoid [punanpangu torapesoido]
    사다리꼴 단면 [sadarikkol danmyeon]

หน้าตัดรูปสี่เหลี่ยมคางหมู [naa-tad ruup siiliam kaang-muu]

treatment condition
    conditions for heat treatment or chemical treatment.
    処理条件 [syori zyouken]
    熱処理あるいは化学処理のための必要条件．
    处理条件 [chǔ lǐ tiáo jiàn]
    keadaan olahan [keadaan orahan]
    처리 조건 [cheori jogeon]
    เงื่อนไขในการจัดการ [nge-un khai nai gaan chad gaan]

trimming
    a process to remove burrs generated in die forging process from a forged product.
    縁抜き [hutinuki]
    型鍛造で生じたばりを鍛造品から切り離す作業．
    切飞边 [qiē fēi biān]
    penyelesaian [punyeresaian]
    트리밍 [teuriming]
    การตัดขอบ [gaan tad khorb]

trip device
    a device that release a lever or set free a mechanism.
    トリップ装置 [torippu souti]
    レバーやメカニズムを開放する装置．
    断开装置 [duàn kāi zhuāng zhì]
    mesin trip [masin toripu]
    해지 장치 [haeji jangchi]
    เครื่องทริป [kru-ang trip]

triple coil spring
    a spring consisting of three coil springs with different diameters arranged concentrically.
    三重コイルばね [sanzyuu koiru bane]
    大，中，小3個のコイルばねを同心，並列に組み合わせたもの．

triple coil spring — tubular stabilizer bar with ball joint ends

三重螺旋弹簧 [sān chóng luó xuán tán huáng]
pegas ulir tiga lapis [pugasu uriru tiga rapisu]
삼중 코일 스프링 [samjung koil spring]
สปริงขดสามชั้น [sa-pring khod saam shan]

troostite
    a structure formed when martensite is tempered. Troostite is a highly corrosion-resistant structure consisting of minute ferrite and carbide which cannot be distinguished under an optical microscope.
トルースタイト [toruusutaito]
マルテンサイトを焼戻ししたときに生じる組織で，光学顕微鏡で識別できないほど微細なフェライトと炭化物からなる極めて腐食されにくい組織．
屈氏体 [qū shì tǐ]
troostite [toroosutite]
투루스타이트 [turuseutaiteu]
ทรูสไตท์ [troostite]

TTT diagram
    a diagram indicating the start and end lines of transformation by drawing lines connecting the transformation start and end times when eutectoid carbon steel is cooled to and maintained at a number of different temperatures from the austenite region.
T.T.T 線図 [tjii-tjii-tjii senzu]
共析炭素鋼をオーステナイト域から種々の温度に急冷し，その温度に保持して変態の開始時間と終了時間を測定し，これらの点を結んで変態開始線と終了線を表した図．
回火 - 时间 - 温度（TTT）曲线 [huí huǒ — shí jiān — wēn dù (TTT) qū xiàn]
diagram TTT [diaguramu TTT]
티티티 (T.T.T) 선도 [tititi (T.T.T) seondo]
แผนภูมิทีทีที [paaen puum TTT]

tube wall thickness
    a thickness of tube defined by the half of difference between outer and inner diameter.
管肉厚 [kuda nikuatu]
外径と内径の差で定義される管の肉厚．
管壁厚 [guǎn bì hòu]
tebal dinding pipa [tebaru dindingu pipa]
관 두께 [gwan dukke]
ความหนาผนังท่อ [kwaam naa pa-nang to-a]

tubular antiroll bar
    a stabilizer bar made of tubular material.
中空スタビライザ [tyuukuu sutabiraiza]
中空材料を用いたスタビライザ．
中空稳定杆 [zhōng kōng wěn dìng gǎn]
bahan tahan oleng bentuk pipa [bahan tahan orengu bentuku pipa]
중공 스테비라이저 [junggong seutebiraijeo]
แท่งแอนติโรลชนิดท่อ [taaeng antiroll cha-nid to-a]

tubular stabilizer bar
    a stabilizer bar made of tubular material.
中空スタビライザ [tyuukuu sutabiraiza]
中空材料を用いたスタビライザ．
管形稳定杆 [guǎn xíng wěn dìng gǎn]
batang stabilisator bentuk pipa [batangu sutabirisatoru bentuku pipa]
중공 스테비라이저 [junggong seutebiraijeo]
เหล็กกันโคลงชนิดท่อกลวง [lek gan klong cha-id to-a glu-ong]

tubular stabilizer bar with ball joint ends
    a tubular stabilizer bar with ball joint attachment for installation.
ボールジョイント式中空スタビライザ

tubular stabilizer bar with ball joint ends — twist loop

[booru zyointosiki tyuukuu sutabiraiza]
組付用のボールジョイントを持つ中空スタビライザ.
铰接管形稳定杆 [jiǎo jiē guǎn xíng wěn dìng gǎn]
batang stabilisator pipa dengan ujung ball joint [batangu sutabirisatoru pipa dengan ujungu baru jointo]
볼 조인트식 중공 스테비라이저 [bol jointeusik junggong seutebiraijeo]
เหล็กกันโคลงท่อกลวงชนิดปลายข้อต่อลูกบอล [lek gan klong to-a glu-ong cha-id plaai khor toa look ball]

tubular stabilizer bar with screw ends
a tubular stabilizer bar with screw joint attachment for installation.
ねじ式中空スタビライザ [nezisiki tyuukuu sutabiraiza]
組付用のねじ継手を持つ中空スタビライザ.
螺纹连接稳定杆 [luó wén lián jiē wěn dìng gǎn]
batang stabilisator pipa dengan ujung baut [batangu sutabirisatoru pipa dengan ujungu bauto]
나사식 중공 스테비라이저 [nasasik junggong seutebiraijeo]
เหล็กกันโคลงท่อกลวงชนิดปลายสกรู [lek gan klong to-a glu-ong cha-id plaai screw]

tubular torsion bar
a torsion bar formed from tubular material.
中空トーションバー [tyuukuu toosyon baa]
管状の素材を用いて加工されたトーションバー.
管形扭杆 [guǎn xíng niǔ gǎn]
batang puntir bentuk pipa [batangu puntiru bentuku pipa]
중공 토션바 [junggong tosyeonba]

ทอร์ชั่นบาร์ชนิดท่อกลวง [torsion bar cha-nid to-a glu-ong]

tumbler shot peening
a shot peening machine where workpieces are placed inside a rotating barrel-shaped container so that they are equally treated.
回転式ブラスト機 [kaitesiki burasuto ki]
たる状の容器の中に加工物を入れ，加工物が均等に処理できるように容器を回転させショットピーニング加工を行う機械.
回转式喷丸机 [huí zhuǎn shì pēn wán jī]
penyemprotan (peen) berguling [punyenpurotan (peen) beruguringu]
회전식 블라스트기 [hoejeonsik beullaseuteugi]
เครื่องขัดผิวโลหะชนิดหมุน [kru-ang khad piw loo-ha cha nid]

turning tool
a machine tool used to process rotating workpieces with fixed tool.
旋削工具 [sensaku kougu]
工具を固定し，工作物を回転させて使用する工具.
车削工具 [chē xuē gōng jù]
alat pembubut [arato punbubuto]
선삭 공구 [seonsak gonggu]
อุปกรณ์สำหรับการกลึง [up-pa-gorn sam-rab gaan glung]

twist angle
ねじり角 [neziri kaku]
扭转角 [niǔ zhuǎn jiǎo]
sudut pilin [suduto pirin]
비틀림각 [biteullimgak]
มุมบิด [mum bid]

twist loop
a loop which is formed in the process of bending and twisting one coil of an extention coil spring.

ねじりフック [neziri hukku]
コイルのひと巻きをねじり起こして成形した，引張コイルばねのフック．
扭环 [niǔ huán]
lingkar (kait) pilin [ringukaru (kaito) pirin]
비틀림 고리 [biteullim gori]
ตะขอบิด [ta-khor bid]

twist test
    test to examine the distortion or breakdown of the helical torsion spring, torsion bar and spring washer. In the case of spring washer, it is also called a toughness test.
ねじり試験 [neziri siken]
ねじりコイルばね，トーションバー，ばね座金類などの変形又は破壊の状況を調べる試験．ばね座金類では，粘り強さ試験ともいう．
扭转试验 [niǔ zhuǎn shì yàn]
uji pilin [uji pirin]
비틀림 시험 [biteullim siheom]
การทดสอบการบิด [gaan tod-sorb gaan bid]

twisting
    rotating one end of a body with the other end clamped to create angular displacement.
ねじり [neziri]
物体の一端を固定し，他端を回転して相対回転変位を起こすこと．
扭曲 [niǔ qū]
memilin [memirin]
비틀림 [biteullim]
การบิด [gaan bid]

twisting tension
    tension force to the axis direction when the wire is twisted.
ねじり張力 [neziri tyouryoku]
線などをねじった場合に軸方向に働く引張力をいう．

扭转拉力 [niǔ zhuǎn lā lì]
tegangan pilin [tegangan pirin]
비틀림 장력 [biteullim jangnyeok]
แรงดึงบิด [rang dung bid]

two stage progressive tapered leaf spring
    a leaf spring consisting of a tapered main spring and an auxiliary spring to assist the main spring under increased load (helper spring).
親子テーパリーフスプリング [oyako teepa riihu supuringu]
テーパばね板を用い常時荷重を受けるばね（親ばね）と荷重の増加した後に補助的に働くばね（子ばね）とで構成される板ばね．
二阶锥端板簧片 [èr jiē zhuī duān bǎn huáng piàn]
pegas daun bertirus progresif dua tingkat [pugasu daun berutirusu puroguresifu dua tingukato]
주 보조형 테이퍼 판 스프링 [ju bojohyeong teipeo pan spring]
สปริงแผ่นเรียวต่อเนื่อง 2 ช่วง [ta-khor bid]

two-point coiling system
    a coiling system which has two pins directed toward the mandrel in order to provide the material with curvature.
二本ピンコイリングシステム [nihon pin koiringu sisutemu]
コイリングするとき材料に曲率を与えるためのピンが心金に向かって2本ある機構のもの．
二卷簧销系统 [èr juǎn huáng xiāo xì tǒng]
sistim ulir dua titik [sisutin uriru dua titiku]
투핀 코일링 시스템 [tupin koilling siseutem]
ระบบการม้วนชนิดใช้ 2 พิน [ra-bop gaan muan cha-nid chai song pin]

two-stage progressive leaf spring

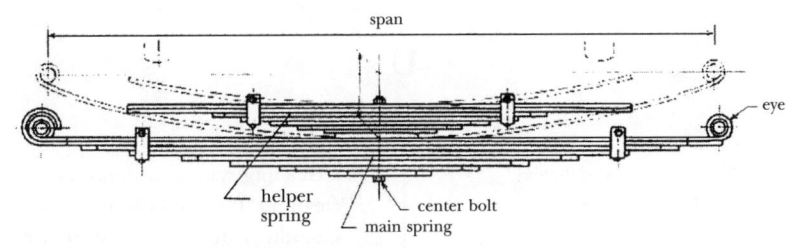

Fig T-13 two-stage progressive leaf spring

two-stage progressive leaf spring
leaf spring comprising a main spring which always supports the load and a helper spring which supports the load supplementary after the load has increased. (see Fig T-13)
親子重ね板ばね [oyako kasaneita bane]
常時荷重を受けるばね（親ばね）と荷重が増加した後に補助的に働くばね（子ばね）とで構成された重ね板ばね。(Fig T-3 参照)
主副重叠弹簧 [zhǔ fù chóng dié tán huáng]
pegas daun dua tahap [pugasu daun dua tahapu]
주. 보조형 비선형 겹판 스프링 [ju.bojohyeong biseonhyeong gyeoppan spring]
สปริงแผ่นต่อเนื่อง 2 ช่วง [sa-pring paaen to-a ne-ung song shu-ong]

U

U section
U 字形断面 [yuuzigata danmen]
U 形断面 [U xíng duàn miàn]
penampang U [punanpangu U]
유 (U) 자형 단면 [you (U) jahyeong danmyeon]
หน้าตัดรูปตัวยู [naa tad ruup tu-a yuu]

U spring
U shaped spring.
U 字形スプリング [yuuzigata supuringu]
U 字形のばね.
U 形弾簧 [U xíng tán huáng]
pegas U [pugasu U]
유 (U) 자형 스프링 [you (U) jahyeong spring]
สปริงรูปตัวยู [sa-pring ruup tu-a yuu]

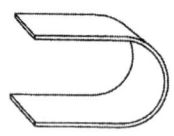

Fig U-1 U spring

ultrasonic cleaning
a cleaning method whereby a work piece is immersed in ultrasonically vibrating cleaning liquid to remove contamination.
超音波洗浄 [tyouonpa senzyou]
洗浄物を超音波振動している洗浄液に浸して汚染物を取り除く洗浄方法.
超声波清洗 [chāo shēng bō qīng xǐ]
pembersih ultrasonik [punberusihu urutorasoniku]
초음파 세정 [choeumpa sejeong]
การทำความสะอาดด้วยอุลตร้าโซนิค [gaan tam kwaam sa-aad du-oi ultrasonic]

ultrasonic cleaning plant
a cleaning system to utilize ultrasonic cleaning. It consists of front processing, ultrasonic cleaning stage, rinsing stage and drying stage.
超音波洗浄装置 [tyouonpa senzyou souti]
超音波洗浄を利用した洗浄装置で，前工程，超音波洗浄工程，リンス工程及び乾燥工程からなる．
超声波清洗装置 [chāo shēng bō qīng xǐ zhuāng zhì]
pabrik pembersih ultrasonik [paburiku punberusihu urutorasoniku]
초음파 세정 장치 [choeumpa sejeong jangchi]
หน่วยทำความสะอาดด้วยอุลตร้าโซนิค [nu-ai tam kwaam sa-aad du-oi ultrasonic]

ultrasonic machining
the removal of material by abrasive bombardment carried by a liquid generated by a ultrasonically (about 20,000 Hz) vibrating tool.
超音波加工 [tyouonpa kakou]
超音波域（約 20,000 Hz）で振動する工具によって液体中でと粒が工作物に衝撃を与える加工方法.
超声波加工 [chāo shēng bō jiā gōng]
bentukan dengan ultrasonik [bentukan dengan urutorasoniku]
초음파 가공 [choeumpa gagong]
กระบวนการอุลตร้าโซนิค [gra bu-an gaan ultrasonic]

ultrasonic test
a nondestructive test for investigating internal defects of a test piece based on acoustic characteristics when sub-

jected to ultrasonic vibrations.
超音波探傷試験　[tyouonpa tansyou siken]
　超音波を試験体中に伝えたとき，試験体が示す音響的性質を利用して，試験体の内部欠陥を調べる非破壊試験．
超声波探伤试验　[chāo shēng bō tàn shāng shì yàn]
uji ultrasonik　[uji urutorasoniku]
초음파 탐상 시험　[choeumpa tamsang siheom]
การทดสอบด้วยคลื่นอุลตร้าโซนิค　[gaan tod sorb du-oi klu-un ultrasonic]

ultrasonic thickness meter
　an instrument for measuring the thickness of an object by applying ultrasonic vibrations and detecting the reflected waves.
超音波厚さ計　[tyouonpa atusakei]
　超音波を被試験材に与え，反射波を検出して厚さを測定する計器．
超声波测厚仪　[chāo shēng bō cè hòu yí]
pengukur ketebalan dengan ultrasonik [pungukuru ketebaran dengan urutorasoniku]
초음파 두께 계기　[choeumpa dukke gyegi]
มิเตอร์วัดความหนาด้วยคลื่นอุลตร้าโซนิค [meter wad kwaam naa du-oi klu-un ultrasonic]

ultrasonic transducer
　a device that converts alternating-current energy above 20 kHz to mechanical vibrations of the same frequency.
超音波振動子　[tyouonpa sindousi]
　20 kHz 以上の交流電流のエネルギーを同じ周波数の機械振動に変換するエネルギー変換器．
超声波振子　[chāo shēng bō zhèn zǐ]
pengubah ultrasonik　[pungubahu urutorasoniku]
초음파 진동자　[choeumpa jindongja]
ตัวถ่ายคลื่นอุลตร้าโซนิค　[to-a taai kluun ultrasonic]

uniaxial stress
　a static of stress in a body under an external load where two of the three principal stresses are zero.
単軸応力　[tanziku ouryoku]
　外力を受けている物体の応力状態で，三つの主応力のうち二つがゼロのときをいう．
軸向应力　[zhóu xiàng yīng lì]
tegangan poros tunggal　[tegangan porosu tungugaru]
단축 응력　[danchuk eungnyeok]
ความเค้นแกนเดียว　[kwaam ken gaaen dieaw]

uniform stress distribution
　an uniformly distributed stress in objects.
等分布応力　[toubunpu ouryoku]
　物体内で一様に分布している応力をいう．
等应力分布　[děng yīng lì fēn bù]
distribusi tegangan merata　[disutoribusi tegangan merata]
등분포 응력　[deungbunpo eungnyeok]
การกระจายความเค้นคงที่　[gaan gra-chaai kwaam ken kong tii]

unscheduled maintenance
　maintenance requirements that had not been pre-planned yet require prompt action.
計画外保全　[keikakugai hozen]
　事前の計画には入っていなかったものの，速やかな対策が必要とされるメンテナンス．
计划外维修　[jì huá wài wéi xiū]
perawatan diluar rencana　[purawatan diruaru rensana]
계획외 보전　[gyehoekoe bojeon]
การบำรุงรักษาอย่างไม่มีกำหนดเวลา　[gaan bam-rung rak-sa yaang mai-mii gam-nod wee-laa]

unsprung weight
ばね下重量　[banesita zyuuryou]
未加載时弾簧上的重量　[wèi jiā zǎi shí tán huáng shàng de zhòng liàng]
berat pegas bawah　[berato pugasu bawahu]
스프링 하 중량　[spring ha jungnyang]
น้ำหนักขณะไม่มีสปริง　[naam-nak kha-na mai-mii sa-pring]

unwinding torque
torque to unwind the power spring.
巻戻トルク　[makimodosi toruku]
ぜんまいを開放する方向に回転させたときに得られるトルク．
回卷扭矩　[huí juǎn niǔ jù]
momen puntir membuka gulungan [momen puntiru menbuka gurungan]
풀림 토-크　[pullim to-keu]
แรงบิดคลายตัว　[rang bid klaai tu-a]

upper control limit
an upper line of the control chart. See "control chart".
上部管理限界　[zyoubu kanri genkai]
管理図の上側限界線．"control chart" 参照．
上控制限　[shàng kòng zhì xiàn]
batas atas kendali　[batasu atasu kendari]
상부 관리 한계　[sangbu gwalli han-gye]
ขอบเขตควบคุมด้านบน　[khob-khed ku-ab kum daan bon]

upset head
an end section of bar whose cross sectional area is increased by upsetting.
据込み頭部　[suekomi toubu]
軸方向に圧縮鍛造したときの圧縮され断面積を増した部分．
镦粗头部　[dūn cū tóu bù]
mengganggu　[mengugangugu]
업셋 헤드　[eopset hedeu]
การขึ้นรูปส่วนหัว　[gaan khun ruup suan ho-a]

upsetting
a forging process in which all or part of a workpiece is compressed axially, to increase the cross-sectional area. Upsetting one end of the workpiece is called heading.
据込み　[suekomi]
材料の軸方向の全体又はある区間を，軸方向に圧縮し，断面積を増す鍛造作業．軸の一端を据え込むときはヘッディングという．
镦粗　[dūn cū]
kepala dilantakan　[kepara dirantakan]
업셋 (축박음)　[eopset (chukbakeum)]
การขึ้นรูป　[gaan khun ruup]

upturned eye
upturned eye of the main leaf.
上卷目玉　[uwamaki medama]
板ばねの目玉の名称で，親板端部を上向きに丸めた目玉．
上卷耳　[shàng juǎn ěr]
gelang keatas　[gerangu keatasu]
윗방향 귀　[witbanghyang gwi]
ม้วนหูด้านบน　[muan huu daan bon]

Fig U-2　upturned eye

V

V block
    a rectangular steel block having a 90° V groove through the center to provide the tight positioning of round workpieces.
    Vブロック　[bui burokku]
    中心線に沿って90°のV溝を持つ鋼製のブロックで，丸い工作物の位置決めを精密に行うのに用いる．
    V型块　[V xíng kuài]
    blok bentuk V　[buroku bentuku V]
    브이 (V) 블록　[beui (V) beullok]
    วี บล็อก　[V block]

vacuum annealing furnace
    a furnace in which the non-oxidizing state is maintained by vacuumizing instead of filling with protective atmosphere. Vacuum may be retained or intended gas may be injected while heating.
    真空焼なまし炉　[sinkuu yakinamasi ro]
    保護雰囲気の代わりに炉内を真空にして酸化しない状態にし，目的のガスを入れたり真空のまま加熱する炉．
    真空退火炉　[zhēn kōng tuì huǒ lú]
    pabrik pelunak vakum　[paburiku pubunaku bakun]
    진공 풀림로　[jingong pullimno]
    เตาอบชนิดสุญญากาศ　[ta-o ob cha-nid soon-yaa-gaad]

vacuum evaporation
    deposition of thin films of metal or other materials on a substrate by evaporation in a vacuum.
    真空蒸着　[sinkuu zyoutyaku]
    真空中の蒸着により，基層の上に金属その他の材料の薄膜を形成すること．
    真空镀膜　[zhēn kōng dù mó]
    evaporasi hampa udara　[evaporasi hanpa udara]
    진공 증착　[jingong jeungchak]
    การระเหยด้วยระบบสุญญากาศ　[gaan ra-heri du-oi ra-bop soon-yaa-gaad]

vacuum heat treatment
    heat treatment whereby workpieces are heated in a vacuum.
    真空熱処理　[sinkuu netusyori]
    真空中で加熱して行う熱処理．
    真空热处理　[zhēn kōng rè chǔ lǐ]
    olah panas hampa udara　[orahu panasu hanpa udara]
    진공 열처리　[jingong yeolcheori]
    การเผาด้วยระบบสุญญากาศ　[gaan pa-o du-oi ra-bop soon-yaa-gaad]

valve spring
    coil spring used for intake and exhaust valves of internal combustion engines.
    弁ばね　[ben bane]
    内燃機関の吸排気弁などに用いるコイルばね．
    阀门弹簧　[fá mén tán huáng]
    pegas katup　[pugasu katupu]
    밸브 스프링　[baelbeu spring]
    สปริงสำหรับปิดเปิดวาล์ว　[sa-pring sam-rap pid perd valve]

valve spring retainer
    a part that holds a valve spring.
    弁ばね受け　[ben bane uke]
    弁ばねを保持させるための部品．
    阀门弹簧座　[fá mén tán huáng zuò]
    penahan pegas katup　[punahan pugasu katupu]

valve spring retainer — vehicle

밸브 스프링 리테이너 [baelbeu spring riteineo]
ตัวกันคลายของสปริงวาล์ว [tu-a gan klaai khong sa-pring valve]

valve spring wire
    a steel wire used for a valve spring of an internal combustion engine. There are a piano wire, a silicone-chrome oil tempered wire, a chrome-vanadium oil tempered wire, a carbon steel oil tempered wire, etc. The round or egg-shaped cross section wires are used.

弁ばね用鋼線 [ben baneyou kousen]
内燃機関のバルブスプリングに用いる鋼線をいい，弁ばね用ピアノ線，シリコンクロム鋼オイルテンパ線，クロムバナジウム鋼オイルテンパ線，炭素鋼オイルテンパ線などがあり，丸や異形断面線が用いられている．

阀门弹簧用钢丝 [fá mén tán huáng yòng gāng sī]
kawat pegas katup [kawato pugasu katupu]
밸브 스프링용 강선 [baelbeu springyong gangseon]
เหล็กเส้นสำหรับผลิตสปริงวาล์ว [lek sen sam-rab pa-lit sa-pring valve]

variable cost
    costs which vary directly with the volume of production. Direct labor and material are examples.

変動費 [hendouhi]
生産量に応じて直接変動する費用．直接労務費と材料費がその例である．

可变成本 [kě biàn chéng běn]
biaya bergerak [biaya berugeraku]
변동비 [byeondongbi]
ค่าใช้จ่ายผันแปร [kha chai-chaai pan-praae]

variable pitch
    uneven coil pitch to make nonlinear characteristics of a helical compression spring.

不等ピッチ [hutou pitti]
圧縮コイルばねのばね特性を非線形とするための均一でないピッチ．

变节距 [biàn jié jù]
jarak antara berubah [jaraku antara berubahu]
부등 피치 [budeung pichi]
ระยะพิทช์ผันแปร [ra-ya pitch pan praae]

variable pitch coil spring
    helical compression spring with uneven pitches.

不等ピッチコイルスプリング [hutou pitti koiru supuringu]
ピッチが均一でない圧縮コイルばね．

变节距弹簧 [biàn jié jù tán huáng]
pegas berjarak antara berubah [pugasu berujaruku antara berubahu]
부등피치 스프링 [budeungpichi spring]
สปริงขดที่มีระยะพิทช์ผันแปร [sa-pring khod tii mii ra-ya pitch pan praae]

Fig V-1 variable pitch coil spring

vehicle
    a self-propelled wheeled machine that transports people or goods on roads.

車両 [syaryou]
道路上で人や物を運搬しつつ，自力走行できる，車両を有する機械．

车辆 [chē liàng]
kendaraan [kendaraan]
차량 [charyang]
ยานพาหนะ [yaan-pa-ha-na]

vehicle spring
    the general term of suspension springs used in railroads or automobiles.
車両用ばね　[syaryouyou bane]
鉄道車両や自動車などに用いる懸架用ばねの総称．
车辆用弹簧　[chē liàng yòng tán huáng]
pegas kendaraan　[pugasu kendaraan]
차량용 스프링　[charyangyong spring]
สปริงสำหรับยานพาหนะ　[sa-pring sam-rab yaan-pa-ha-na]

vernier
    an auxiliary scale which slides along the main scale to permit accurate fractional readings of the least main division of the main scale.
副尺　[hukusyaku]
主尺に沿ってスライドする補助尺で，主尺の最小目盛の分数目盛精度の正確さで読取りを可能にするスケール．
游标　[yóu biāo]
vernier　[berunieru]
부척　[bucheok]
เวอร์เนีย　[vernier]

vernier dial
    a turning dial in which each complete rotation constitutes a fraction of a revolution of the main shaft, permitting fine and accurate adjustment.
バーニヤダイアル　[baaniya daiaru]
1回転が主軸の1回転の部分目盛になるようにした回転目盛で，正確な微調整を可能にするもの．
微调刻度盘　[wēi tiáo kè dù pán]
dial vernier　[diaru berunieru]
버니어 다이얼　[beonieo daieol]
เวอร์เนียไดอัล　[vernier dial]

V-hook over center
    V shaped hook at the end of a helical extension spring.
Vフック　[bui hukku]
引張コイルばねの端部の一種で，V字形のフック．
V形钩环　[V xíng gōu huán]
kait bentuk V　[kaito bentuku V]
브이 (V) 자형 고리　[beui (V) jahyeong gori]
ตะขอรูปตัววีเหนือจุดศูนย์กลาง　[ta-khor ruup tu-a V ne-ua chud-soon-glang]

Fig V-2　V-hook over center

vibrating feeder
    a feeder for bulk materials such as pulverized solids which are transfered by the vibration of a slightly slanted, flat vibrating surface.
振動フィーダ　[sindou fiida]
粉体のようなバルク材に用いられる供給装置で，材料はわずかに傾斜した平らな振動面の振動によって搬送される．
振动送料器　[zhèn dòng sòng liào qì]
pengumpan dengan gataran　[pungunpan dengan gataran]
진동 피더　[jindong pideo]
ตัวป้อนสั่นสะเทือน　[to-a porn san sa-te-un]

vibrating stress
    a time-varying stress.
振動応力　[sindou ouryoku]
時間の関数として変動する応力．
振动应力　[zhèn dòng yīng lì]
tegangan getaran　[tegangan getaran]
진동 응력　[jindong eungnyeok]
ความเค้นสั่นสะเทือน　[kwaam ken san sa-te-un]

vibration absorber
    a device consisting of weight and rubber spring attached to the large am-

vibration absorber — vibrator

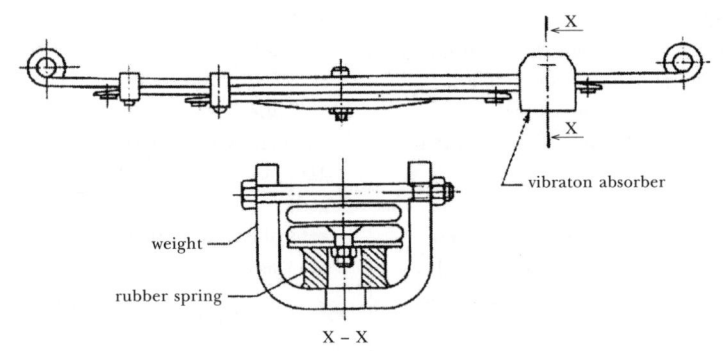

Fig V-3 vibration absorber

plitude section of a leaf spring to suppress resonance. (see Fig V-3)
ダイナミックダンパ　[dainamikku danpa]
重ね板ばねの共振抑制のために，振幅の大きい部分に付けるダンパで，おもり，ゴムばねなどで構成される部品．(Fig V-3 参照)
动态减振器　[dòng tài jiǎn zhèn qì]
peredam getaran　[puredan getaran]
다이내믹 댐퍼　[dainaemik daempeo]
ตัวดูดซับการสั่นสะเทือน　[tu-a dood sap gaan san sa-te-un]

vibration insulating material
a material with a high capacity of absorbing vibration energy through conversion into heat energy.
防振材　[bousinzai]
振動エネルギーを熱エネルギーに変換し，吸収してしまう能力の大きい材料．
减振材料　[jiǎn zhèn cái liào]
bahan peredam getaran　[bahan puredan getaran]
방진재　[bangjinjae]
วัสดุที่ใช้ในการทำฉนวนการสั่นสะเทือน　[wassa-du tii chai nai gaan tam sha-nuan gaan san sa-te-un]

vibration suppression
the prevention of undesirable vibration, either through a damping device or through an active feedback control.
振動制御　[sindou seigyo]
減衰装置あるいはフィードバックコントロールのようなアクティブな手段によって望ましくない振動を防止すること．
减振　[jiǎn zhèn]
kendali getaran　[kendari getaran]
진동 제어　[jindong jeeo]
การควบคุมการสั่นสะเทือน　[gaan ku-ab kum gaan san sa-te-un]

vibration testing machine
a testing machine for the vibration characteristics and durability of springs and other items.
振動試験機　[sindou sikenki]
ばねなどの振動特性，耐久性を調べる試験機．
振动试验机　[zhèn dòng shì yàn jī]
mesin uji getaran　[masin uji getaran]
진동 시험기　[jindong siheomgi]
เครื่องทดสอบการสั่นสะเทือน　[kru-ang todsorb gaan san sa-teun]

vibrator
an equipment that produces vibration impact through electrical or mechanical means.

振動発振機　[sindou hassinki]
　　電気的又は機械的な手段で振動を起こ
　　させ，振動の衝撃を与える装置．
振動器　[zhèn dòng qì]
penggetar　[pungugetaru]
진동 발진기　[jindong baljingi]
ตัวสั่นสะเทือน　　[tu-a san sa-te-un]

vibratory feeder
an equipment that employs vibration to convey the appropriate quantities of a material. Used in part feeders.
振動式搬送装置　[sindousiki hansou souti]
　　振動を利用して物体を適量ずつ送り出
　　す装置．パーツフィーダなどに利用さ
　　れている．
振动式运送装置　[zhèn dòng shì yùn sòng zhuāng zhì]
pengumpan bergetar　[pungunpan berugetaru]
진동식 이송장치　[jindongsik isongjangchi]
ตัวป้อนชนิดสั่นสะเทือน　　[tu-a porn cha-nid san sa-te-un]

vibratory grinding
a method of grinding the surface of a workpiece which is contained in a vibrating container together with an abrasive and water.
振動式研磨　[sindousiki kenma]
　　容器中にといし粒や水などと一緒に品
　　物を入れ，振動によって品物の表面や
　　端面を研磨する方法．
振动式研磨机　[zhèn dòng shì yán mó jī]
penggerindaan bergetar [pungugerindaan berugetaru]
진동식 연마기　[jindongsik yeonmagi]
การเจียร์ด้วยการสั่นสะเทือน　　[gaan chiaa du-oi gaan san sa-te-un]

vibrometer
an instrument designed to measure the amplitude of a vibration.

振動メータ　[sindou meeta]
　　振動の振幅を測定する計器．
振动测试仪　[zhèn dòng cè shì yí]
pengukur getaran　[pungukuru getaran]
진동계　[jindonggye]
อุปกรณ์วัดแอมปริจูดของการสั่นสะเทือน　　[uppa-gorn wad amplitude khong gaan san sa-teun]

Vickers hardness
a hardness value defined by the test load divided by the area of dent in the Vickers hardness test.
ビッカース硬さ　[bikkaasu katasa]
　　ビッカース硬さ試験において，用いた
　　試験荷重を永久くぼみの表面積で除し
　　た値．
维氏硬度　[wéi shì yìng dù]
kekerasan vickers　[kekerasan bikukerusu]
비커스 경도　[bikeoseu gyeongdo]
ความแข็งวิคเกอร์　　[kwaam khaeng Vickers]

viscoelasticity
property of a material which is viscous but which also exhibits certain elastic properties such as the ability to store energy of deformation.
粘弾性　[nendansei]
　　粘性を有するものの変形エネルギーを
　　蓄積する等のある種の弾性体の性質を
　　示す材料特性．
粘弹性　[nián tán xìng]
viskoelastisitas　[bisukoerasutisitasu]
점탄성　[jeomtanseong]
ความหนืดยืดหยุ่น　　[kwaam nu-ud yuud-yun]

viscosity
an index figure which represents the degree of internal friction associated with the flow of fluid.
粘度　[nendo]
　　液体の流れに伴う内部摩擦の程度を表
　　す指標．

粘性　[nián xìng]
kekentalan　[kekentaran]
점성　[jeomseong]
ความหนืด　[kwaam nu-ud]

volute spring

a conical spring formed by winding plate material to allow telescopic spring action.

竹の子ばね　[takenoko bane]
長方形断面の材料の長辺がコイル中心線に平行な円すいコイルばね.
截锥涡卷弹簧　[jié zhuī wō juǎn tán huáng]
pegas volute　[pugasu borute]
볼류트 스프링　[bollyuteu spring]
สปริงก้นหอย　[sa-pring gon hoi]

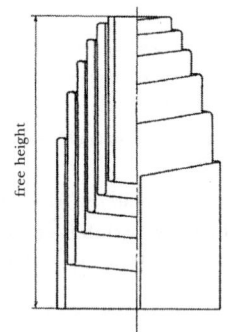

Fig V-4 volute spring

W

Wahl stress corection factor
    typical stress correction factor of the coil spring proposed by Wahl, given by the following formula:
    $K = (4c-1)/(4c-4) + 0.615/c$
    c: spring index D/d
  ワールの応力修正係数　[waaru no ouryoku syuusei keisuu]
    ワールによって提唱された代表的なコイルばねの応力修正係数で，
    $K = (4c-1)/(4c-4) + 0.615/c$
    で表される．
    c：ばね指数 D/d
  瓦耳氏应力修正系数（曲度系数）　[wǎ ěr shì yīng lì xiū zhèng xì shù (qǔ dù xì shù)]
  faktor koreksi tegangan wahl　[fakutoru korekusi tegangan wahuru]
  왈 응력 수정계수　[wal eungnyeok sujeonggyesu]
  สัมประสิทธิ์การปรับความเค้นของวูล　[sam-pra-sit gaan prab kwaam ken khong Wahl]

wall thickness
  肉厚　[nikuatu]
  壁厚　[bì hòu]
  tebal dinding　[tebaru dindingu]
  파이프 두께　[paipeu dukke]
  ความหนาผนัง　[kwaam-naa pa-nang]

warm peening
    shot peening carried out on a spring in a warm temperature. Fatigue strength is improved more than those by ordinary shot peening.
  ホットピーニング　[hotto piiningu]
    ばねに温間で行うショットピーニング．通常のショットピーニングよりも，疲れ強さが向上する．
  热敲击　[rè qiāo jī]
  peen panas　[peen panasu]
  온간 피닝　[ongan pining]
  การพืนนิ่งขณะร้อน　[gaan peening kha-na rorn]

watch spring
    a power spring used for driving watches and a hair spring used for watch speed governor etc.
  時計ばね　[tokei bane]
    駆動用に用いるぜんまいと調速器に用いるひげぜんまいなどのこと．
  钟表弹簧　[zhōng biǎo tán huáng]
  pegas arloji　[pugasu aruroji]
  시계 스프링　[sigye spring]
  สปริงนาฬิกา　[sa-pring na-ri-gaa]

water jet cutting
    a cutting method that uses a jet of pressurized water containing abrasive powder.
  ウォータージェット切断加工　[wootaa zyetto setudan kakou]
    研磨粉の入った水の高圧ジェットで切断する方法．
  喷水切断　[pēn shuǐ qiē duàn]
  pemotongan dengan air jet　[pumotongan dengan airu jeto]
  워터 제트 절단가공　[woteo jeteu jeoldan-gagong]
  การตัดด้วยการพ่นน้ำ　[gaan tad du-oi gaan pon naam]

water quenching
    quenching performed in cold water.
  水焼入れ　[mizu yakiire]
    冷却に水を用いて行う焼入れ．
  水淬火　[shuǐ cuì huǒ]

pengejutan dengan air [pungejutan dengan airu]
물 담금질 [mul damgeumjil]
การชุบแข็งด้วยน้ำ [gaan shub khaaeng du-oi naam]

water soluble
    a property which can dissolve into water.
水溶性 [suiyousei]
    水に溶け込むことができる性質.
水溶性 [shuǐ róng xìng]
larut dalam air [raruto daran airu]
수용성 [suyongseong]
คุณสมบัติการละลายน้ำ [kun-na-som-bat gaan la-laai naam]

waved and toothed lock washer
    it is a kind of toothed spring washer with tooth waved. There are outer tooth, inner tooth, outer and inner tooth, and dish type.
波形歯付座金 [namigata hatuki zagane]
    歯付座金の一種で歯付座金と異なるのは歯部をしわ状（波形の切り起こしをした）にしたばね座金. 外歯, 内歯, 内外歯, 皿型がある.
波形齿垫圈 [bō xíng chǐ diàn quān]
pegas kunci bergigi dan bergelombang [pugasu kunchi berugigi dan berugeronbangu]
파형 이붙이 와셔 [pahyeong ibuchi wasyeo]
แหวนล็อครูปฟันและคลื่น [waaen lock ruup fan]

waved spring washer
    spring washer made by turning wire and formed into a waved shape.
波形ばね座金 [namigata bane zagane]
    線材を巻いて, 波形に成形したばね座金.
波形弹簧垫圈 [bō xíng tán huáng diàn quān]
ring pipih pegas bergelombang [ringu pipihu pugasu berugeronbangu]
파형 스프링 와셔 [pahyeong spring wasyeo]
แหวนสปริงชนิดเป็นลอน [waaen sa-pring cha-nid pen lon]

waved washer
    spring washer made of thin ring plate on which waving is formed.
波形座金 [namigata zagane]
    円環状の薄板に波形を付けたばね座金.
波形垫圈 [bō xíng diàn quān]
ring pipih bergelombang [ringu pipihu berugeronbangu]
파형 와셔 [pahyeong wasyeo]
แหวนกันรั่วชนิดเป็นลอน [waaen gan ru-a cha-nid pen lon]

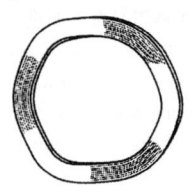

Fig W-1 waved washer

weather resistance
    the ability of a material to withstand the physical and chemical changes over time due to the effects of natural conditions.
耐候性 [taikousei]
    自然条件の影響を受けて時間の経過に伴って起こる材料の物理的及び化学的変化に耐える性質.
耐环境性 [nài huán jìng xìng]
ketahanan terhadap cuaca [ketahanan teruhadapu chuacha]
내후성 [naehuseong]
ความต้านทานสภาพบรรยากาศ [kwaam taan taan sa-paap ban-yaa-gaad]

weathering test
generic term for tests which examine the alteration of surface state after exposing the material or the spring to the light, heat, wind and rain environment. In some cases, this test is carried out being accompanied with the durability test.
耐候試験　[taikou siken]
材料，ばねなどを光，熱，風，雨などの環境下で暴露した場合の表面状況の変化などを評価する試験の総称．場合によっては，耐久性試験と併せて行うこともある．
耐环境试验　[nài huán jìng shì yàn]
uji cuaca　[uji chuacha]
내후 시험　[naehu siheom]
การทดสอบการทนทานต่อบรรยากาศ　[gaan tod-sorb gaan ton taan to-u ban-yaa-gaad]

weatherproof
being able to withstand exposure to weather without damage.
耐候性　[taikousei]
劣化することなく風雨に耐えることのできる性質．
耐环境性　[nài huán jìng xìng]
anti air　[anti airu]
내후성　[naehuseong]
ทนต่อสภาพบรรยากาศ　[ton to-a sa-paap ban-yaa-gaad]

Weibull curve
a curve to show the failure density function of Weibull distribution.
ワイブル曲線　[waiburu kyokusen]
ワイブル分布の故障密度関数を表した線図．
维泊尔曲线　[wéi bó ěr qǔ xiàn]
kurva Weibull　[kuruba weiruburu]
와이블 곡선　[waibeul gokseon]
เส้นโค้งเวย์บูล　[sen-kong Weibull]

Weibull's distribution
the most typical distribution which shows the failure probability about equipment and parts and which is widely used to study the failure problems (fatigue property).
ワイブル分布　[waiburu bunpu]
設備・部品の故障の確率を表す最も代表的な分布で広い範囲にわたって故障の問題（寿命特性）の検討に利用されている．
维泊尔分布　[wéi bó ěr fēn bù]
distribusi weilbull　[disutoribusi weiruburu]
와이블 분포　[waibeul bunpo]
การกระจายเวย์บูล　[gaan gra-chaai Weibull]

wet grinding
the application of coolant to the work and grinding wheel to facilitate the process.
湿式研削　[sissiki kensaku]
研削といしと加工物に冷却液を注いで研削工程を促進させる研削方法．
湿式磨削　[shī shì mó xuē]
pengampelas basah　[punganperasu basahu]
습식 연삭　[seupsik yeonsak]
การเจียร์แบบเปียก　[gaan jia baaep pi-eak]

wheel blade rotor
a rotor with blades curved around the rotation axis to pitch shots utilizing centrifugal force resulting from high-speed rotation.
羽根車　[haneguruma]
高速回転の遠心力を利用してショットを投射するための，回転軸の周囲に曲率を持つ羽根を植え込んだ回転体．
叶轮　[yè lún]
bilah roda　[birahu roda]
임펠러　[impelleo]
ใบพัดปัด　[bai pad pad]

wheel blasting machine
> a shot blasting machine which utilizes a wheel blade rotor.

翼車式ブラスト機　[yokusyasiki burasutoki]
羽根車式のショットブラスト機．
轮式喷砂机　[lún shì pēn shā jī]
mesin penyemprot roda　[mesin punyenpuroto roda]
임펠라식 블라스팅기　[impellasik beullaseutinggi]
เครื่องขัดแบบวงล้อ　[kru-ang khad baaep wong lor]

wide strip
> hot-rolled or cold-rolled strip steel with width of 300 mm and more.

広幅帯鋼　[hirohaba obikou]
300 mm 以上の幅の広い熱間圧延又は冷間圧延された帯状の鋼材．
宽带钢　[kuān dài gāng]
kepingan lebar　[kepingan rebaru]
광폭 강대　[gwangpok gangdae]
เหล็กแผ่นชนิดกว้าง　[lek paaen cha-nid gwaang]

width of wire cross section
> width of a noncircular cross-section of the material.

長径　[tyoukei]
異形断面コイルばねに用いる材料の横断面の幅．
长轴　[cháng zhóu]
lebar penampang kawat　[rebaru punanpangu kawato]

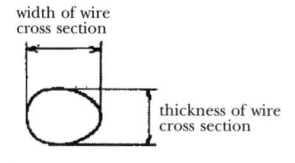

Fig W-2 width of wire cross section

장경　[janggyeong]
ความกว้างของหน้าตัดเหล็กเส้น　[kwaam gwaang khong naa-tad lek sen]

wind up stiffness
> stiffness of a suspension leaf spring against the torsional moment caused by the acceleration of the body around the horizontal axis perpendicular to the moving direction.

ワインドアップ剛性　[waindo appu gousei]
車体の加速度によって走行方向に直角な水平軸の周りに生じるねじりモーメントに抵抗する懸架用板ばねの剛性．
卷曲刚度　[juǎn qū gāng dù]
kekakuan memutar　[kekakuan memutaru]
와인업 강성　[waindeop gangseong]
ความผืดของวิลด์อัพ　[kwaam phu-ud khong wind up]

wind up torque
> torsional moment generated on a leaf spring around the horizontal axis perpendicular to the moving direction caused by the acceleration of the vehicle body.

ワインドアップトルク　[waindo appu toruku]
車体の加速度によって走行方向に直角な水平軸の周りに生じるねじりモーメント．ワインドアップモーメントともいう．
卷曲扭矩　[juǎn qū niǔ jù]
momen puntir memutar　[momen puntiru memutaru]
와인덥 토-크　[waindeop to-keu]
ทอร์คของวิลด์อัพ　[torque khong wind up]

winding barrel
> case retaining a power spring.

ぜんまいケース　[zenmai keesu]
動力用接触形の渦巻ばねを所定の姿に

winding barrel — wire feed speed

保持するケース.
渦巻弾簧　[wō juǎn tán huáng]
tong pelilitan　[tongu puriritan]
태엽 스프링 케이스　[taeyeop spring keiseu]
กล่องสำหรับการพัน　[glong sam-rab gaan pan]

winding torque
　torque needed to wind up the power spring.
　巻上げトルク　[makiage toruku]
　ぜんまいを巻心に巻き付ける方向に回転させるために，必要なトルク.
　卷曲扭矩　[juǎn qū niǔ jǔ]
　moment puntir pelilitan　[momento puntiru puriritan]
　와인딩 토-크　[wainding to-keu]
　แรงทอร์คการพัน　[rang torque gaan pan]

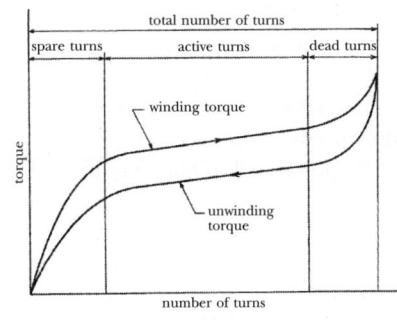

Fig W-3　winding torque

wire cutter
　ワイヤ切断機　[waiya setudanki]
　钢丝切断机构　[gāng sī qiē duàn jī gòu]
　pemotong kawat　[pemotongu kawato]
　와이어 절단기　[waieo jeoldangi]
　เครื่องตัดเส้นลวด　[kru-ang tad sen lu-ad]

wire diameter
　線径　[senkei]
　钢丝直径　[gāng sī zhí jìng]
　diameter kawat　[diameteru kawato]

선경　[seongyeong]
เส้นผ่าศูนย์กลางเส้นลวด　[sen-pa-soon-glaang sen lu-ad]

wire feed
　the feeding of wire during processing.
　線送り　[sen'okuri]
　加工するときに線を供給搬送すること.
　送料　[sòng liào]
　umpan kawat　[umupan kawato]
　선재 이송　[seonjae isong]
　การป้อนเส้นลวด　[gaan porn sen lu-ad]

wire feed length
　the length of wire fed during processing.
　線送り長さ　[sen'okuri nagasa]
　加工するときに線を供給搬送する長さをいう.
　钢丝送进长度　[gāng sī sòng jìn cháng dù]
　panjang umpan kawat　[panjangu unpan kawato]
　선재이송 길이　[seonjaeisong giri]
　ระยะการป้อนเส้นลวด　[ra-ya gaan porn sen lu-ad]

wire feed roller
　a series of roller to feed wire to the coiling point in a coiling machine.
　線送りローラ　[sen'okuri roora]
　コイリング機のコイリングポイントに線を供給する一連のローラ.
　送料滚轮　[sòng liào gǔn lún]
　rol umpan kawat　[roru unpan kawato]
　선재이송 롤러　[seonjaeisong rolleo]
　ความเร็วของล้อส่งป้อนเส้นลวด　[kwaam rew khong lor song porn sen lu-ad]

wire feed speed
　the rate at which wire is fed during processing.
　線送り速度　[sen'okuri sokudo]
　加工部分に供給される素線の供給速

wire feed speed — wire mesh spring

さ．

送料速度　[sòng liào sù dù]
kecepatan umpan kawat　[kechepatan unpan kawato]
선재이송 속도　[seonjaeisong sokdo]
ความเร็วของการป้อนเส้นลวด　[kwaam rew khong gaan porn sen lu-ad]

wire formed hose clip
  a wire clip to grip the hose with a finger tab for squeezing and releasing action.
  ワイヤ式ホースクリップ　[waiyasiki hoosu kurippu]
  つまみをはさんで離すことによりホースをつかむ構造のワイヤクリップ．
  线式管箍　[xiàn shì guǎn gū]
  kawat penjepit selang　[kawato punjepito serangu]
  와이어식 호스 클립　[waieosik hoseu keullip]
  โฮสคลิปชนิดเส้นลวด　[hoseclip cha-nid sen lu-ad]

wire forming
  various shapes of springs made of wire, as called by its material.
  ワイヤフォーミング　[waiya foomingu]
  ばねの材料の種類による名称で，線状の材料を用いた各種形状のばね．
  钢丝成形　[gāng sī chéng xíng]
  pembentukan kawat　[punbentukan kawato]
  와이어 포밍（선세공 스프링의）　[waieo poming (seonsegong springeui)]
  การขึ้นรูปเส้นลวด　[gaan khun roop sen lu-ad]

wire groove
  a groove to ensure that wire is correctly guided. The groove that is provided in the feed rollers or wire guides of coiling machines.
  線溝　[senmizo]
  線が正しく通過できるように付けた溝．コイリングマシンのフィードローラやワイヤガイドに付ける溝のこと．
  滑线槽　[huá xiàn cáo]
  alur kawat　[aruru kawato]
  선재 통과 홈　[seonjae tonggwa hom]
  ร่องของเส้นลวด　[rang khong sen lu-ad]

wire guide
  guiding device or plate for feeding the wire to the forming section correctly in the coiling machine to prevent the buckling.
  ワイヤガイド　[waiya gaido]
  コイリングマシンなどで線を成形部に送り出すときの，座屈防止及び正確な位置への搬送・供給のための案内装置又はプレート．
  钢丝校直装置　[gāng sī xiào zhí zhuāng zhì]
  pemandu kawat　[pumandu kawato]
  와이어 가이드　[waieo gaideu]
  ตัวนำเส้นลวด　[tu-a nam sen lu-ad]

wire harness clamp
  a spring used to hold bundled cable together or to secure bundled cable to other equipment.
  ワイヤハーネスクランプ　[waiya haanesu kuranpu]
  電線の束の離散防止又はその束を機材に固定させるためのばね．
  线束夹子　[xiàn shù jiā zī]
  penjepit kawa harnes　[punjepito kawa harunesu]
  와이어 하네스 크램프　[waieo haneseu keuraempeu]
  ฮาร์เนสแคลมป์จับยึดเส้นลวด　[harness clamp yud sen lu-ad]

wire mesh spring
  steel mesh compressed into a shape of purpose to exert a spring function.

金網ばね　[kanaami bane]
小径の鋼線を編んで作った金網を圧縮成形したばね.
线网弹簧　[xiàn wǎng tán huáng]
pegas jala kawat　[pugasu jara kawato]
망 스프링 (와이어 메쉬 스프링)　[mang spring (waieo meswi spring)]
สปริงลวดตาข่าย　[sa-pring lu-ad taa-khaai]

wire reel
a spool on which wire is wound.
線巻台　[senmakidai]
線を巻き取るための枠.
线材台架　[xiàn cái tái jià]
kumparan kawat　[kunparan kawato]
와이어 릴　[waieo ril]
แท่นม้วนเส้นลวด　[taaen muan sen lu-ad]

wire rod
hot-rolled steel bar wound into a coil. Also called bar-in-coil.
線材　[senzai]
棒状に熱間圧延された鋼でコイル状に巻かれた鋼材. バーインコイルともいう.
线材　[xiàn cái]
batang kawat　[batangu kawato]
선재　[seonjae]
เส้นลวด　[sen lu-ad]

wire spark erosion
electric erosion of a wire induced by electric discharge of an electrode wire for wire discharge process.
放電浸食　[houden sinsyoku]
ワイヤ放電加工などに用いる電極線の放電による線の食食をいう.
放电腐蚀　[fàng diàn fǔ shí]
erosi nyala api kawat　[erosi nyara api kawato]
방전 침식　[bangjeon chimsik]
การกัดกร่อนของเส้นลวดจากการสปาร์คไฟฟ้า
[gaan gad gron khong sen lu-ad chak gaan spark fai-faa]

wire spring
coil spring made of wire.
線ばね　[senbane]
線状の材料を用いたコイルばね.
线弹簧　[xiàn tán huáng]
pegas kawat　[pugasu kawato]
선 스프링　[seon spring]
สปริงเส้นลวด　[sa-pring sen lu-ad]

wire testing
a general name for a test which examines wire's physical or chemical properties.
線材試験　[senzai siken]
素線の物理的又は化学的性質を調べる試験の総称.
线材实验　[xiàn cái shí yàn]
pengujian kawat　[pungujian kawato]
선재 시험　[seonjae siheom]
การทดสอบเส้นลวด　[gaan tod-sorb sen lu-ad]

Wöehler diagram
a graph showing relationship between stress on the vertical axis (linear scale) and number of cycles to failure on the horizontal axis (logarithmic scale). (see Fig S-8)
ベーラー線図　[beeraa senzu]
縦軸に応力（等分目盛），横軸に破壊までの繰返し回数（対数目盛）をとって表した線図．(Fig S-8 参照)
沃尔(S-N) 曲线　[wò ěr (S-N) qǔ xiàn]
diagram Wohler　[diaguran bohureru]
뵐러 선도　[boelleo seondo]
แผนภูมิของวูเลอร์　[paaen puum khong Wohler]

Wöehler test
a test for generating and assessing an S–N diagram by subjecting test pieces manufactured under the same conditions to different repeated stress condi-

tions and determining the number of repeated stress cycles to failure.
ベーラー試験　[beeraa siken]
同じ条件のもとに作られた数個の試験片にそれぞれ異なる大きさの繰返し応力を与え破壊までの応力繰返数を求め，S–N線図を描き，判定するための試験.
沃尔 (S-N) 曲线试验　[wò ěr (S-N) qǔ xiàn shì yàn]
uji Wohler　[uji bohureru]
빌러 시험　[boelleo siheom]
การทดสอบของวูเลอร์　[gaan tod-sorb khong Wohler]

work hardening
the phenomenon whereby the hardness of a metal increases when deformed by processing.
加工硬化　[kakou kouka]
金属が加工変形を受けることによって，硬さが上昇する現象.
加工硬化　[jiā gōng yìng huà]
pengerasan melalui pengerjaan　[pungerasan merarui pungerujaan]
가공 경화　[gagong gyeonghwa]
การทำให้ชิ้นงานแข็งขึ้น　[gaan tam hai shin-ngaan khaeng khun]

workability
an indication of how a material is suitable for processing depending on its intended use.
加工性　[kakousei]
用途に応じて各種の加工に適しているかどうかの程度.
加工性　[jiā gōng xìng]
dapat dikerjakan　[dapato dikerujakan]
가공성　[gagongseong]
ความสามารถในการทำงาน　[kwaam saa-maad nai gaan tam ngaan]

working load
the maximum load that any load-carrying member can withstand in normal use.
常用荷重　[zyouyou kazyuu]
荷重を受ける部品が通常の使用に耐える最大の荷重.
工作负荷　[gōng zuò fù hè]
beban pengerujaan　[beban pungerujaan]
상용 하중　[sangyong hajung]
โหลดที่ใช้งาน　[load tii chai ngaan]

working sequence sheet
a document indicating the work procedures to ensure correct performance of standard procedures.
作業手順書　[sagyou tezyunsyo]
標準作業を正しく運営するため作業の手順を記載した印刷物.
操作指导书　[cāo zuò zhǐ dǎo shū]
petunjuk urutan kerja　[putunjuku urutan keruja]
표준작업 순서 기록지　[pyojunjakeop sunseo girokji]
เอกสารขั้นตอนการทำงาน　[eak-ka-saan khan-torn gaan tam-ngaan]

worm gear spring
a helical compression spring which is inserted into a core using the coil pitch to function as a worm gear.
スプリングウォームギア　[supuringu woomu gia]
圧縮コイルばね形状のばねを軸に挿入し，コイルピッチを利用してウォームギアのウォームの機能を果たすばね.
螺杆弹簧　[luó gǎn tán huáng]
pegas roda gigi worm　[pugasu roda gigi worun]
스프링 웜기어　[spring womgieo]
สปริงเฟืองตัวหนอน　[sa-pring fe-ung tu-a norn]

worm gear type hose clamp
 a hose clamp that is tightened by a worm gear.
 ウォーム式ホースクランプ　[woomusiki hoosu kuranpu]
 ウォームギアで締め付ける構造のホースクランプ．
 螺杆形管夹　[luó gǎn xíng guǎn jiā]
penjepit selang jenis roda gigi worm
[punjepito serangu jenisu roda gigi worun]
 웜식 호스 크램프　[womsik hoseu keuraempeu]
 โฮสคลิปชนิดเฟืองตัวหนอน　[hoseclip cha-nid fe-ung tu-a norn]

wrapping test
 a test in which test wire is tightly wound width a given number of turns around a core of specified diameter to investigate the fracture or other defects.
 巻付試験　[makituke siken]
 試験片を規定の径の心金に規定の回数だけ密接して巻き付け，破断やきずなどの発生状況を調べる試験．
 缠绕试验　[chán rào shì yàn]
uji bungkus　[uji bungukusu]
 감기 시험 (선)　[gamgi siheom (seon)]
 การทดสอบหีบห่อ　[gaan tod-sorb heep ho-a]

X

X-ray diffraction method
    a method to decide the crystal structure and the compound structure by using the X-ray diffraction phenomenon of the crystal lattice.
  X 線回折法　[ekkususen kaisetuhou]
    結晶格子によるX線の回折現象を利用して結晶構造や化合物の構造の決定を行う方法．
  X 线衍射法　[X xiàn yǎn shè fa]
  pengukuran tegangan dengan sinar-X [pungukuran tegangan dengan sinaru X]
  엑스 (X) 선 회절법　[ekseu (X) seon hoejeolbeop]
  การวิเคราะห์แยกโดยเอ็กซ์เรย์　[gaan wi-kra-or yaaek dooi klu-un x-ray]

X-ray stress measuring
    measurement of residual stress using the principle of X-ray diffraction. Typical example for springs is the measurement of compressive residual stress generated by shot peening.
  X 線応力測定　[ekkususen ouryoku sokutei]
    X線回折の原理を用いて行う残留応力の測定．ばねの代表的な例としては，ショットピーニングによる圧縮残留応力の測定があげられる．
  X 射线应力测量法　[X shè xiàn yīng lì cè liáng fǎ]
  difraksi sinar X　[difurakusi sinaru X]
  엑스 (X) 선 응력 측정법　[ekseu (X) seon eungnyeok cheukjeongbeop]
  การวัดค่าความเค้นโดยคลื่นเอ็กซ์เรย์　[gaan wad kaa kwaam ken dooi klu-un X ray]

Y

yellow chromating
    a method of creating a corrosion-resistant film by immersing the workpiece in a bichromate solution after galvanizing in order to prevent zinc from corrosion. The color of the treated surface is yellow.
黄色クロメート処理　[ousyoku kuromeeto syori]
　亜鉛めっき処理後，白さびを防止する目的で重クロム酸塩を主成分とする溶液に品物を浸せきし防せい皮膜を生成させる方法．この処理を施すと黄色の着色が得られる．
铬酸盐光泽处理　[gè suān yán guāng zé chǔ lǐ]
pengkromatan kuning　[pungukuromatan kuningu]
황색 크로메이트 처리　[hwangsaek keuromeiteu cheori]
กระบวนการโครเมตติ้งสีเหลือง　[gra bu-an gaan chromating sii le-ung]

yield
    a phenomenon where strain increases without stress increment during plastic deformation in steel.
降伏　[kouhuku]
　鋼における塑性変形過程で応力が増加することなくひずみが増加する現象．
屈服　[qū fú]
luluh　[ruruhu]
항복　[hangbok]
การล้าตัว　[gaan laa tu-a]

yield elongation
    an elongation observed under yielding condition.
降伏伸び　[kouhuku nobi]
　降伏時に観察される伸び量．
屈服变形　[qū fú biàn xíng]
tegang luluh　[tegangu ruruhu]
항복 신율　[hangbok sinnyul]
การยืดออกของการล้าตัว　[gaan yuud ork khong gaan laa tu-a]

yield point
    the stress at which strain increases without an increase in stress.
降伏点　[kouhukuten]
　応力の増加を伴わずにひずみが増加するときの応力．
屈服点　[qū fú diǎn]
titik luluh　[titiku ruruhu]
항복점　[hangbokjeom]
จุดล้าตัว　[chud laa tu-a]

yield ratio
    the ratio of yield point (normally upper yield point) or yield strength to tensile strength.
降伏比　[kouhukuhi]
　引張強さに対する降伏点（通常上降伏点）又は耐力の割合．
屈强比　[qū qiáng bǐ]
perbandingan hasil　[purubandingan hasiru]
항복비　[hangbokbi]
อัตราส่วนของการล้าตัว　[at-tra su-an khong gaan laa tu-a]

yield strength
    the value obtained by dividing the load at which specified permanent elongation occurs in a tensile test by the original cross-sectional area of the parallel section of tensile test specimen.
耐力　[tairyoku]

引張試験において規定された永久伸びが生じるときの荷重を平行部の原断面積で除した値.

耐力　[nài lì]
kuat luluh　[kuato ruruhu]
내력　[naeryeok]
กำลังล้าตัว　[gam lang laa tu-a]

yield stress
 the stress at which plastic deformation occurs without an increase in load during a tensile test.

降伏応力　[kouhuku ouryoku]
 力の増加が一切ないにもかかわらず，試験中塑性変形が生じるときの応力.

屈服応力　[qū fú yīng lì]
tegangan luluh　[tegangan ruruhu]
항복 응력　[hangbok eungnyeok]
ความเค้นของการล้าตัว　[kwaam ken khong gaan laa to-a]

Young's modulus
 ratio of perpendicular stress generated in a cross-section of a bar to the perpendicular strain. Commonly designated by E. Unit is MPa or N/mm^2.

ヤング率　[yanguritu]
 棒の断面に働く垂直応力と単位長さ当たりの伸び又は縮み（垂直ひずみ）との比．量記号：E，単位記号：MPa又はN/mm^2.

杨氏(弹性)模量　[yáng shì (tán xìng) mó liàng]
modulus young　[modurusu youngu]
세로 탄성 계수　[sero tanseong gyesu]
ค่าสัมประสิทธิ์ของยังก์　[kaa sam-pra-sit khong Young]

Z

Z leaf spring
　　leaf spring shaped in letter "Z".
　　Zリーフ　[zetto riihu]
　　Z字形の板ばね.
　　Z字形板簧　[Z zì xíng bǎn huáng]
　　pegas daun Z　[pugasu daun Z]
　　제트 (Z) 형 판 스프링　[jeteu (Z) hyeong pan spring]
　　สปริงแผ่นแซด　[sa-pring paaen Z]

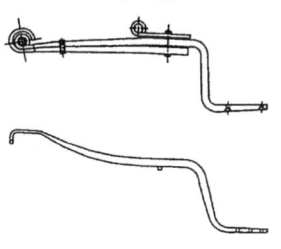

Fig Z-1　Z leaf spring

zero defects
　　a program for improving product quality to the point of perfection.
　　無欠陥　[mukekkan]
　　完璧な水準まで品質を向上させる活動計画.
　　无缺陷（ZD）　[wú quē xiàn (ZD)]
　　tidak ada cacat　[tidaku ada kakato]
　　무결점　[mugyeoljeom]
　　ของเสียเป็นศูนย์　[khorng siaa pen soon]

zigzag spring
　　zigzag-shaped spring.
　　ジグザグばね　[ziguzagu bane]
　　ジグザグ形のばね.
　　蛇 (Z字) 形弹簧　[shé (Z zì) xíng tán huáng]
　　pegas liku-liku　[pugasu riku-riku]
　　지그재그 스프링　[jigeujaegeu spring]
　　สปริงซิกแซก　[sa-pring zig zag]

Fig Z-2　zigzag spring

zigzag spring bending machine
　　a machine or equipment forming a zigzag spring.
　　ジグザグばね成形機　[ziguzagu bane seikeiki]
　　ジグザグばねを成形する機械.
　　蛇 (Z字) 形弹簧成形机　[shé (Z zì) xíng tán huáng chéng xíng jī]
　　mesin penekuk pegas liku-liku　[mesin punekuku pugasu riku-riku]
　　지그재그 스프링 성형기　[jigeujaegeu spring seonghyeonggi]
　　เครื่องดัดสปริงซิกแซก　[kru-ang dad sa-pring zig zag]

zinc chromate
　　a major element of antirust pigment. The zinc chromate film may be created on the surface of zinc plated metal through chromating process.
　　ジンククロメート　[zinku kuromeeto]
　　さび止め顔料の主成分. 亜鉛めっきされた金属表面にクロメート処理を行ってジンククロメート層を形成することもある.
　　镀锌铬法　[dù xīn gè fǎ]
　　proses zinkrolit　[purosesu zinkurorito]
　　징크 크로메이트　[jingkeu keuromeiteu]
　　ซิงค์โครเมต　[zinc chromate]

zinc dust paint
　　protective paint with high zinc dust ratio to be used for steel surface utilizing

zinc dust paint — zinc plating

sacrifice anode action.

亜鉛粉末塗料　[aen hunmatu toryou]
亜鉛粉末を多量に配合した塗料で、犠牲陽極作用によって鉄表面を保護する。

锌粉涂料　[xīn fěn tú liào]

cat debu seng　[chato debu sengu]

아연 분말도료　[ayeon bunmaldoryo]

สีผงสังกะสี　[sii pong]

zinc plating

a process to deposit metallic zinc on a product acting as a cathode when direct current passes through an electrolyte containing zinc or zinc complex ions.

亜鉛めっき　[aen mekki]
亜鉛イオンや亜鉛錯イオンを含む電解質に直流電流を流して陰極側の製品に金属亜鉛を析出させる処理。

镀锌　[dù xīn]

pelapisan seng　[purapisan sengu]

아연 도금　[ayeon dogeum]

การชุบสังกะสี　[gaan shub sang-ga-sii]

索 引
index

日　本　語
Japanese　　359

中　国　語
Chinese　　385

インドネシア語
Indonesian　　411

韓　国　語
Korean　　437

タ　イ　語
Thai　　463

日本語索引

あ

アークハイト　11
アーク溶接　12
アイアニング　172
アイブシュ　115
アイリーフ　115
亜鉛粉末塗料　356
亜鉛めっき　356
亜共析鋼　161
アクチュエータ　4
脚の角度　182
圧延鋼材　255
圧延線　255
圧延方向　88
圧下率　256
圧縮アークコイルばね　11
圧縮応力　58
圧縮角形コイルばね　246
圧縮片絞りばね　21
圧縮コイルばね　150
　——端部　106
圧縮ゴム被覆コイルばね　258
圧縮残留応力　58
圧縮直列組合せコイルばね　266
圧縮強さ　58
圧縮・ねじりコイルばね　57
圧縮ばね　57
　——試験機　57
圧縮引張繰返し応力　57
圧縮引張耐久限度　57
圧縮湾曲コイルばね　74
圧入　31
　——試験　31
　——制御系　234
後処理　232

後熱処理　300
後焼なまし　231
孔あけポンチ　227
穴径　155
穴状浸食　224
穴用 C 形止め輪　172
アナログ計器　8
アニオン電着塗装　9
油焼入れ　214
油焼け　291
編み線　27
荒削り研削　256
粗さ　257
　——測定　257
亜硫酸ガス試験　302
アルカリ試験　6
アルカリ性亜鉛めっき　6
アルキメデスら旋　12
アルミナグリット　8
アルメンストリップ　7
アルメンテスト　7
安全荷重　261
安全装置　261
安全標識　261
安全ベルト用ばね　261
アンチロックブレーキ　10
安定化熱処理　290
案内筒　143

い

E 形止め輪　99
異形コイルばね　172
異形断面ばね　209
板厚標準寸法　270
板形ねじ込みフック　319
板カム　231

359

板ゲージ　270
位置決め　231
一次巻　235
一本ピンコイリング方式　275
移動平均　204
陰極酸化防食処理　38
印刷回路　235
　——基板　235
インターリーフ　170
インターロック　171
インバータ　172

う

ウォータージェット切断加工　343
ウォーム式ホースクランプ　351
受入テスト　2
薄板　318
　——鋼板　271
　——ばね　124
渦巻ばね　283
　——の端末曲げ　315
内側コイルばね　169
内側半径　169
打抜加工性　238
打抜工具　25
打抜力　25, 238
うちのり　169
内ばね　170
　——径　168
内輪　169
上塗り　321
上巻目玉　336
運動エネルギー　176

え

エアブレーキ　5
永久伸び　315
永久ひずみ　225
永久変形　224
英式フック　110

エキスパンダスプリング　114
液体ホーニング　184
エジェクタ　101
S-N 線図　277, 298
X 線応力測定　352
X 線回折法　352
HDD サスペンションスプリング　148
エッチング　112
FRP 板ばね　122
エリゴばね　104
遠隔温度計　311
遠隔測定　311
エングラー度　110
円形コイルばね　45
円形断面ばね　45
エンコーダ　105
円弧カム　44
遠心分離式集じん装置　77
遠心力　41
円すい形圧縮ばね　60
円すい形ばね　60
円すい形ミニブロックスプリング　202
円すいコイルばね　151
塩水浸せき試験　262
円すいといし　1
塩水噴霧試験　262
円すい面　59
延性　95
　——破壊　95
円筒形圧縮コイルばね　77
円筒形ばね　77
円筒形引張コイルばね　77
円筒コイルばね　77
円筒度　84
円盤カム　88
円板クラッチ　88
円盤ばね　124
エンボス　105
塩浴熱処理　261
エンリッチガス　110

お

オイラー定理　112
オイラーの力　112
オイルシール用リングスプリング　214
オイルテンパ線　214
黄色クロメート処理　353
応力　297
　——緩和　247, 299
　——き裂　298
　——クリープ破断時間曲線　301
　——限界　183
　——勾配　298
　——集中係数　297
　——振幅　297
　——対クリープ速度線図　301
　——範囲　299
　——比　299
　——-ひずみ-温度線図　300
　——-ひずみ線図　300
　——腐食割れ　297
　——分布　90
　——誘起マルテンサイト変態　298
応力サイクル　298
　——周波数　298
応力除去　299
　——焼なまし　300
オーステナイト　14
　——系ステンレス鋼　14
　——結晶粒度　14
オーステンパ処理　14
オースホーミング　13
オートメーション　15
オーバーピーニング　218
オープンエンド　215
オープンフラットエンド　215
置割れ　263
屋外暴露試験　216
送り精度　120
送り長さ　121

遅れ破壊　81
押さえ板　234
押さえばね板　15, 244
押込みばね板ナット　239
押曲げ法　233
汚染除去　79
汚染の予防　234
オッシログラフ　216
オペレータ　215
親板　192
親子重ね板ばね　333
親子テーパリーフスプリング　332
温度-時間曲線　312
温度制御　312
温度センサ　312
温度分布　312
温度変換器　312

か

ガータースプリング　137
加圧速度　186
カービングプレス　75
カービングロール　75
カーボン繊維強化ばね　41
カーボンポテンシャル　36
カーリング　74
回帰分析　247
開口フック　215
介在物　165
回転式ブラスト機　331
回転速度　256
回転方向　88
回転曲げ疲労試験　256
回転摩擦係数　49
ガイド長さ　143
ガイドピン　143
火炎焼入れ　124
化学蒸着法　44
化学成分　43
化学ニッケルめっき　44

過共析鋼　161
拡散浸透処理　86
拡散浸透層　86
拡散焼なまし　155
角振動数　9
角線　289
角速度　9
拡大フック　110
角断面トーションバー　323
角断面ブーツバンド　246
角断面ホースクリップ　246
角度測定装置　9
角ばね　245
攪拌器　4
角フック　245
加工硬化　296, 350
加工性　350
加工熱処理　317
化合物層　56
重ね板ばね　178
　　――の滑面端末　181
　　――の高さ　217
　　――用試験機　180
重ねた輪ばね　207
重ね巻き角形断面ねじりばね　95
重ね巻きねじりばね　95
過時効　217
過失系統図　120
荷重　184
　　――限度　185
　　――軸線　185
　　――試験　186
　　――増加　185
　　――-たわみ曲線　186
　　――-伸び曲線　185
荷重測定　186
　　――装置　186
荷重低下　185
　　――率　248
過剰浸炭　303

ガスカーテン　137
ガス窒化　137
硬さ　147
　　――試験　147
片スライドタイプ重ね板ばね　214
硬引線　145
片振応力　238
片振り疲れ限度　118
片持ちばね　35
片持ち梁　35
　　――の長さ　35
カチオン電着塗装　38
活性防食　3
カッティングツール　76
カットワイヤ　75
　　――の大きさ　75
滑面端末　276
渦電流センサ　100
可撓軸　126
可撓性　125
稼働率　243
金網ばね　349
金型　85
　　――用ばね　86
過熱　218
加熱時間　149
加熱試験　149
カバレージ　69
　　――百分率　70
過負荷　218
カム　34
　　――機構　34
　　――形状　34
　　――シャフト　35
　　――従動子　34
　　――制御　34
　　――調整　34
　　――プロフィール　34
下面といし　189
カラーチェック　55

ガラス繊維ばね　　138
ガラスビーズ　　138
渦流探傷試験　　100
環境影響調査　　111
環境影響評価　　110
環境応力き裂　　111
環境工学　　111
環境試験　　111
環境劣化　　111
乾式研削　　94
乾式引抜き　　94
乾湿交互浸せき試験　　94
緩衝器　　271, 278
緩衝材　　75
緩衝用ばね　　2
慣性モーメント　　204
間接工賃　　165
間接測定　　166
間接費　　165
完全焼入れ　　133
完全焼まなし　　133
乾燥炉　　94
官能検査　　265
カンバン　　176
含有量　　62
管理限界　　63
管理図　　63
管理特性　　63

き

機械インピーダンス測定機　　198
機械効率　　198
機械衝撃めっき　　199
機械損失　　199
機械的性質　　199
機械能力指数　　48
機械プレス　　199
規格ばね　　291
規格要求事項　　282
幾何公差　　137

機器用ばね　　11
危険源　　147
危険事象　　147
きしみ音　　290
希釈剤　　87
技術検査　　310
技術仕様書　　311
技術マニュアル　　311
技術メンテナンス　　310
基準線　　247
きず検出器　　125
気体ばね　　137
機能試験　　135
脚長　　182
逆丸フック　　69
CAD　　58
キャビテーション　　39
CAM　　58
キャリパー尺　　33
吸湿性　　161
吸収エネルギー　　2
球状化焼なまし　　283
球状含有物　　138
球状炭化物　　138
求心力　　41
球面座金　　283
急冷時効　　240
供給装置　　121
強磁性体　　121
共振　　251
　——振動数　　251
強制振動　　127
強制破壊　　127
矯正ローラ　　295
共析　　113
　——鋼　　113
極低温工学　　74
局部脱炭　　221
曲率　　74
　——半径　　243

許容応力　6
許容荷重　185
許容限界　321
許容差　320
許容範囲　321
許容品質水準　2
切欠き係数　118
切欠き衝撃試験　212
切欠き衝撃強さ　212
キルド鋼　176
き裂　123
　——成長　70
　——伝ぱ　70
キンク　176
近接検出器　237
金属間化合物　171
金属顕微鏡　200
金属沈殿物　200
金属ばね　200
均熱時間　278

く

空気ばね　5
　——懸架システム　5
空気焼入れ　5
空冷　5
くさび形ピッチツール　229
管肉厚　330
クッション式メッシュスプリング　75
グッドマンの耐久限度線図　139
駆動軸　94
くぼみ　227
　——径　228
　——深さ　228
組合せ二本線ねじりコイルばね　55
組合せばね　55
組立機　13
組立治具　12
組立ライン　12
クラッチスプリング　48

クラッチディスクスプリング　48
クラッチフィンスプリング　48
クラッド　45
クランプ　45
クリアランス　46
クリープ　71
　——限度　71
　——試験　72
　——速度　71
　——強さ　72
　——テンパ　71, 72
　——破断強さ　71
　——ひずみ　72
グリーンクロメート処理　140
クリーンルーム　46
繰返し応力　7
　——振幅　8
繰返し荷重　7
繰返し間欠試験　249
繰返し試験　249
繰返数比　76
繰返し速度　130
繰返し曲げ試験　125
繰出し速度　222
クリップ　46
　——パイプ　47
　——バンド　46
　——ボルト　46
グリップ止め輪　142
グリップバンド　142
クローズドエンド　47
クロスオーバーループ　73
クロスヘッド　73
クロメート処理　44

け

計画外保全　335
蛍光浸透探傷試験　127
形状記憶合金　269
　——ばね　269

日本語索引

ゲージ 136
　——圧 136
　——長さ 136
ゲーナの応力修正係数 139
結晶粗大化 139
結晶粒成長 139
結晶粒度 140
　——分布 140
結晶粒微細化 140
欠点数 212
限界荷重 183
懸架装置 306
懸架ばね 43, 305
懸架用重ね板ばね 305
研削 140
　——圧力 141
　——機 140
　——切断 76
　——端面 143
　——ばね鋼 41
減衰 78
限度見本 27
顕微鏡解析 200
顕微鏡写真 200
研磨加工 2
研磨機 140
研磨材 1
　——吹付 1
研磨ディスク 1
研磨模様 141

こ

子板 18
コイニング 53
コイリング 51
　——工具 52
　——ピン 52
　——ポイント 52
　——方式 52
　——マシン 52

　——ローラ 52
コイル 50
　——外径 216
　——間最小寸法 202
　——間ピッチ 89
　——重心径 196
　——スプリング 152
　——端末 50
　——内径 170
　——の巻方向 88
　——平均径 196
　——平均ピッチ 196
コイルばね 50
　——の応力修正係数 297
　——の展開長さ 84
　——焼入機 241
高温強度 154
高温クリープ特性 153
高温締付試験 153
高温引張試験 314
高温腐食試験 153
硬化 146
交会点 167
光輝焼なまし 28
合金元素 7
合金鋼 6
　——ばね 7
合金成分 6
工具耐用寿命 100
工具ドレッサ 321
硬鋼線 145
　——材 153
交互浸せき腐食試験 7
公差 320
硬質クロムめっき 145
高周波加熱 166
高周波焼入れ 166
公称サイズ 209
孔食 229
剛性 253

365

鋼製グリット　293
合成樹脂ばね　306
鋼製ショット　293
構成刃先　30
鋼製ばね　293
鋼線　293
　　──の逆ねじり試験　252
　　──のキンク試験　177
　　──のねじり試験　324
拘束テンパ　124
高速度鋼　153
鋼帯　293
光沢度　81
恒弾性ばね合金　173
高張力鋼　154
高張力座金　154
工程能力　235
　　──指数　236
工程表　236
降伏　353
　　──応力　354
　　──伸び　353
　　──比　353
降伏点　353
　　──測定試験　83
　　──の引張強さ比　243
高分子材ばね　153
高分子ばね　103
コーティング耐久試験　48
コード巻戻し用ばね　33
コーナー半径　66
コーナープレス　66
互換性　170
誤差限界　112
故障解析　116, 119
故障までの平均時間　197
故障率　116
コッタ　69
　　──継手　69
　　──ピン　69

固定腕長さ　123
子ばね　152
コバルト合金ばね　49
ゴムパッド　258
ゴムばね　258
　　──併用重ね板ばね　181
ゴムマウントタイプ　258
固有振動　206
　　──数　206
固溶化熱処理　280
固溶体　279
コルゲート式メッシュスプリング　68
コレット　55
ころ軸受　255
コンカレント工学　59
こん形状　236
コントロールスプリング　64
Computer Aided Design　58
Computer Aided Manufacturing　58
コンピュータ数値制御　59
コンピュータ制御システム　58
コンピュータ統合生産　58
コンベックス（凸）　64
コンベヤ式加熱炉　64

さ

座　263
サークリップ　44
サージング　305
サーモスタット　317
再結晶温度　245
再結晶焼なまし　245
再現性　249
最小応力　202
最小予圧　202
最大応力　196
最大荷重　196
最大寸法　196
最大トルク　196
最大偏差　196

日本語索引

細粒　122
　　——鋼　122
材料　195
　　——欠陥　195
　　——試験　195
　　——仕様書　195
　　——組織　195
サイレンサ　274
サイレントブシュ　275
作業研究（TMS）　319
作業手順書　350
座屈　29
　　——応力　30
　　——荷重　30
　　——強度　250
　　——強さ　30
　　——長さ　30
擦過腐食　259
作動させる　3
作動長さ　3
さび　259
　　——どめ油　68
サブゼロ処理　302
座巻　106, 109
作用応力　3
皿座ぐり　69
皿ばね　59
　　——座金　60
　　——の凹面　59
　　——の直列積重ね　266
　　——の積み重ね長さ　182
　　——の凸面　64
サルファプリント試験　302
酸洗い　226
　　——検査　227
酸化　219
　　——物層　219
　　——物の巻込み　219
　　——膜　219
三角開先　109

産業廃棄物　166
三重コイルばね　329
酸性亜鉛めっき　3
酸ぜい性　2
酸洗工程　227
酸洗抑制剤　226
残存水素量　249
サンドブラスト　262
散布図　263
サンプルの大きさ　262
残留圧縮応力　249
残留応力　250
　　——分布　250
残留オーステナイト　251
残留磁気　249
残留せん断ひずみ　250
残留引張応力　250

し

仕上げ　123
　　——研磨　123
　　——寸法　123
　　——端面　277
シアン化亜鉛めっき　76
GFRP板ばね　138
C形同心止め輪　33
C形止め輪　33
シーケンス制御　266
シートばね　263
CVD　44
Jリーフ　175
シェービング　269
シェラダイジング　271
磁界強度計　190
四角断面タイプトーションバー　245
時間強度線図　118
自記温度計　317
磁気ばね　191
磁気ひずみ法　191
磁気分離機　191

367

識別　55
磁気変態　191
磁気膜厚計　191
試供体　283
軸　11
　——荷重　16
　——線　16
　——継手　69
　——箱用ばね　16
　——ばね　17
　——用C形止め輪　115
軸方向　16
　——たわみ　15
治具　175
ジグザグばね　355
　——成形機　355
シクネスゲージ　318
次元　87
試験温度　316
試験荷重　315
試験結果　316
試験時間　316
試験条件　315
時限スイッチ　320
試験片　316
試験方法　316
時効　4
　——硬化　4
　——処理　4
自硬性　264
事故発生系統図　114
自在フック式引張コイルばね　265
事象　113
　——記録計　113
システム　306
　——工学　307
下地塗装　235
下巻き目玉　92
実験計画法　83
実行可能性試験　120

実行可能性調査　120
湿式研削　345
湿潤試験　160
質量効果　194
指定荷重　282
支点　133
始動機ばね　292
自動車用変速機　15
自動制御　15
磁粉探傷試験　191
絞りフック　246
絞り丸フック　309
シム　271
締付座金　252
締付試験　45
締付トルク　319
締付けボルト　45
ジャーナル軸受　175
ジャーマンフック　137
シャシ　43
ジャストインタイム　175
車台　15
シャックル　268
　——アイ　268
車両　338
　——用ばね　339
シャルピー衝撃試験　43
自由角度　129
重心　40
集じん装置　96
修正応力　67
周速度　45
自由高さ　130
しゅう動心金　276
自由長さ　130
周波数　130
　——分析　130
自由巻数　212
主応力　234
樹脂系ショット　230

日本語索引

手動装置　144
主ばね　192
寿命試験　182
ショアー硬さ　271
省エネルギー　110
常温硬化　54
常温セッチング　54
衝撃エネルギー　163
衝撃応力　164
衝撃荷重　163
衝撃試験　164
衝撃速度　163
衝撃強さ　164
衝撃波　271
衝撃疲労試験機　163
焼結鉄合金　275
乗降　167
仕様書　282
冗長性　247
上部管理限界　336
錠前ばね　187
常用荷重　350
初応力　168
ショートフック型密着ねじりコイルばね　272
初荷重　168
初期故障　168
初期疲労き裂　164
初張力　168, 234
ショックアブソーバ用ばね　271
ショット　272
　　──形状　274
　　──投射分布　273
　　──の硬さ　273
　　──の強さ　273
　　──密度　273
ショットピーニング　273
　　──投射時間　274
ショットブラスト　272
　　──投射時間　272

初巻きコイル　168
ジョミニー曲線　175
処理条件　329
処理能力　319
真円度　84
心金　51, 52
　　──径　192
　　──式コイリングマシン　193
　　──スライド　193
真空蒸着　337
真空熱処理　337
真空焼なまし炉　337
ジンククロメート　355
シングルリーフ　275
人工時効　12
じん性　326
浸せき工程　163
浸せき塗装　87
伸線加工後の焼入焼戻し鋼線　146
心出し装置　6
浸炭　37
　　──層　38
　　──窒化　36
真直度　295
振動応力　216, 339
振動式研磨　341
振動式搬送装置　341
振動試験機　340
振動数　76
振動制御　340
浸透探傷試験　184
振動の周期　224
振動発振機　341
振動フィーダ　339
振動メータ　341
信頼性　248
浸硫法　303

す

水素ぜい化　160

369

水素量試験　160
水平位置　124
スイベルフック　60, 306
水溶性　344
スイングアクスル式ばね　306
据込み　336
　　──頭部　336
図記号　140
すきまゲージ　121
すきま無しフック　47
すきま腐食　72
スクラップリール　263
スケール　262
　　──落とし　82
スタッド　302
スタビライザ　10, 290
　　──目玉　290
スタビリンク　184
ステップ　294
ステンレス鋼　291
　　──製ショット　291
　　──ばね　291
ストックリール　294
ストリップ　301
ストレインゲージ　296
ストレートスパン　294
ストレートナー　295
ストレスピーニング　300
　　──の初応力　168
ストローク　301
スナップピン　277
スナップリテーナ　277
スパン　281
スプリングウォームギア　350
スプリングバック　284
スプリングピン　288
スペーサ　281
　　──ブシュ　281
すべり摩擦係数　50
スミスの疲れ限度線図　277

スライドパッド　276
スリーブ　276
すりきず　263

せ

静応力　292
ぜい化　105
正規分布　211
成形　128
　　──機　128
　　──性　128
　　──プレス　128
　　──焼入機　22
　　──ロール　128
生産工学　166
静止摩擦　292
　　──係数　50
清浄度　165
制振鋼板　178
ぜい性　29, 272
　　──破壊　29
静的荷重　292
静的試験　292
静電塗装　104
青熱ぜい性　26
静ひずみ　292
精密測定　232
赤外線乾燥装置　167
析出　82, 232
　　──硬化　232
積層断面タイプトーションバー　178
赤熱ぜい性　246
是正処置　67
設計応力　83
設計荷重　82
設計標準　83
切削速度　76
切削抵抗　76
接触荷重　61
接触形ぜんまい　204

セッチング　267
　　——応力　268
　　——荷重　268
　　——高さ　268
接点ばね　61
Zリーフ　355
セメンタイト　39
セメンテーション　39
セラミックばね　41
セル製造方式　39
セレーション形状　267
セレーションタイプトーションバー　323
遷移　328
　　——温度　328
繊維強化プラスチックばね　56
線送り　347
　　——速度　347
　　——長さ　347
　　——ローラ　347
栓形ねじ込みフック　318
線間すきま　280
線径　347
線形特性　183
　　——ばね　183
センサ　265
線材　349
　　——試験　349
線細工ばね　128
旋削工具　331
洗浄処理　254
センタ穴　40
センタカップ　40
センタスペーサ　40
全脱炭　325
センタピン　40
センタボルト　39
全たわみ　326
せん断　269
　　——応力　270

　　——角度　269
　　——荷重　270
　　——強さ　270
　　——ひずみ　270
全長　326
　　——板　133
全伸び　326
全破損　326
線ばね　349
全ひずみ　326
全負荷　134
ぜんまい　232
　　——ケース　19, 346
　　——成形機　283
　　——端部　107
線巻台　349
線溝　348

そ

粗圧延　256
掃引　306
騒音規制　209
相関係数　67
相関分析　67
層状炭化物　178
相対度数　247
総巻数　325
促進寿命テスト　2
測定誤差　198
測定精度　197
測定装置　198
測定値　197
測定範囲　198
測定方法　198
測定用ばね　198
側面フック　214
側面丸フック　134
組織変態　301
塑性域　230
塑性解析　230

371

塑性変形　230
　——層　179
外側ガイド　216
外側コイルばね　217
外半径　115
外輪　217
反り　35
ソルバイト　280

た

ダイアフラムスプリング　85
大気汚染　5
大気湿度　13
耐久限度　109
耐久性　96
　——試験　110
耐クリープ性　71
台形断面　176, 329
耐候試験　345
耐候性　344, 345
耐さび性　260
耐酸性　3
耐湿性　203
対称ばね　306
耐食性　250
ダイナミックダンパ　96, 340
耐熱鋼　149
耐熱性　148
ダイブロック　86
体膨張率　49
ダイマーク　86
耐摩耗性　1
　——被覆　10
耐薬品性　44
ダイヤモンドホイール　85
ダイヤルゲージ　85
耐用寿命　100
代用特性値　8
耐用年数　267
対流炉　64

耐力　353
だ円アークばね　105
だ円コイルばね　104
だ円断面線コイルばね　217
だ円断面線の長・短径　202
だ円ばね　134
倒れ　164
高さ調整　149
竹の子ばね　342
たたき音　244
多段ピーニング　205
脱ガス　80
脱脂　81
ダッシュポット　78
脱水素化　81
脱炭　79
　——深さ　79
縦置き板ばね　187
縦弾性係数　99
縦横比　276
縦割れ　187
ダブルねじりばね　90
ダブルピーニング　91
ダブル引張コイルばね　275
だぼ　207
卵形断面ばね　101
たる形コイルばね　20
たる形テーパコイルばね　20
たる形ばね　20
たわみ　79
　——角　80
　——測定　80
単一皿ばねのたわみ　275
炭化物　35
短径　318
タンジェントテールエンド　308
単軸応力　335
弾性　103
　——域　102
　——係数　203

日本語索引

——限度　102
——座屈　101
——ひずみエネルギー　102
——比率　102
——変形　101
——理論　102
炭素含有量　36
断続試験　171
炭素鋼　36
——ばね　36
弾塑性　103
炭素繊維ばね　36
タンデムアクスル式懸架　308
ダンパ　216
端部のセレーション　266
端末形状　109
端末絞り引張ばね　91
端面　109
——機械加工　190
——研削　106
——摩擦　100
——割れ　100
断面　73
——係数　264
——2次モーメント　137
——の倒れ角　275

ち

力　127
——のモーメント　204
チタン合金ばね　320
窒化　208
——鋼　208
——層　208
——深さ　82
チャート式記録計　43
中間焼なまし　235
中空スタビライザ　330
中空トーションバー　331
鋳鋼グリット　38

鋳鋼ショット　38
中実スタビライザ　279
超音波厚さ計　335
超音波加工　334
超音波振動子　335
超音波洗浄　334
——装置　334
超音波探傷試験　335
長径　346
超硬合金　147
調整用座金　4
調速機　139
長短径比　12
超弾性　238, 303
長方形断面　245
調和分析　147
直接応力　87
直接費　87
直接焼入れ　87
直棒式バースタビライザ　296
直列式コイルドウェーブスプリング　266
直列積重ね　290
直列のばね　289
直角度　225, 289

つ

疲れ限度　117
——線図　118
疲れ寿命　117, 212
疲れ強さ　118
突合せ溶接　32
筒型カム　20
つづみ形圧縮コイルばね　159
釣合いばね　112
つる巻き摩擦ばね　51

て

TMS　319
T.T.T 線図　320, 330

373

Dループ　78
低温処理　189
低温ぜい性　189
低温焼なまし　188
低温炉　189
低荷重硬さ試験　188
定荷重ぜんまい　61
定荷重ばね　61
定期検査　77
定期点検　224
低合金鋼　188
抵抗式連続焼鈍炉　62
抵抗溶接　251
定常クリープ限度　62
ディスクグラインダ　89
ディスクブレーキ　88
低炭素鋼　188
テーパ　308
　　——アイタイプ　310
　　——圧延　309
　　——エンド　309
　　——加工　308
　　——キー　309
　　——コイルばね　51
　　——線圧縮コイルばね　150
　　——線たる形圧縮コイルばね　150
　　——線つづみ形圧縮コイルばね　151
　　——鍛造　308
　　——長さ　310
　　——リーフスプリング　113, 310
　　——ロールベアリング　309
　　——ロールマシン　309
てこ装置　182
デジタル制御　86
鉄–炭素平衡状態図　172
デプスマイクロメータ　82
テフロン皮膜　311
電解研磨　103
展開長さ　84, 114
展開法　83

電気亜鉛めっき　103
電気めっき　104
テンションアニーリング　314
テンションマシン　115
テンションリング　314
展伸度　105
展性　192
伝達荷重　327
電着塗装　104
伝導機構　232
伝熱面熱負荷　113
天然高分子材ばね　206
テンパーカラー　312
テンプレート　313

と

といし　142
　　——交換　141
　　——修整　141
　　——消耗　141
　　——の条件　141
　　——の品質　142
投影検査器　236
等温変態　174
等温焼なまし　173
等価長さ　112
統計的工程管理　292
統計的品質管理　293
銅合金ばね　66
胴締め　29
　　——金具　28
　　——長さ　182
投射エネルギー　25
投射角　163
投射速度　223
投射用羽根車　272
透磁率　225
同心圧縮コイルばね　206
同心度　59
動的応力　98

日本語索引

動的荷重　96
動的測定　97
動的強さ　97
動的特性試験　96
導電率　103
動ばね定数　97
頭部形状　148
等分布応力　335
胴曲り　27
通し送り式研削　319
トーションバー　322
　──軸線　322
　──端部　322
　──の全長　218
　──の有効長　325
　──把持部　322
　──ばね定数　323
トーションマシン　323
通りゲージ　138
特性曲線　42
特性値　43
特性要因図　43
トグルスイッチ　320
時計のぜんまいケース　284
時計ばね　343
時計用ぜんまいばね　47
度数分布　130
止りゲージ　209
止め輪　251, 278
　──の開口部　136
ドライホーニング　94
取付角度　267
トリップ装置　329
トルースタイト　330
トルク　322
トレーサビリティ　327
トレーリングリーフ　327
ドレッシング　93
ドローバーばね　93
ドロップ鍛造機　94

な

内径　169
内燃機関　171
ナイフエッジ　177
内部応力　172
内部硬さ　66
内部組織　66
内部摩擦　171
内面ガイド　169
長さ試験　150
ナックルプレス　177
斜め丸フック　9, 165
波形コイルスプリング　51
波形座金　74, 344
波形歯付座金　344
波形ばね座金　344
軟窒化法　208

に

肉厚　343
二次応力　264
2軸端面研削盤　91
二次巻　253
二次焼入れ　294
二重コイルばね　90
二重フック　91
二重巻きスプリングピン　50
2段行程送り機構　92
二段ピッチコイルばね　95
ニッケル合金ばね　207
ニッケルめっき　207
ニップ　208
　──応力　208
担いばね　21, 37
二番巻　264
二本線ねじりばね　92
二本線引張コイルばね　92
二本ピンコイリングシステム　332
人間機械系　193

375

人間工学　112

ぬ

ヌープ硬さ試験　177

ね

ねじ込みフックプレート　156
ねじ式中空スタビライザ　331
ねじ式中実スタビライザ　279
ねじり　332
　——応力　325
　——回数　213
　——角　331
　——振動　325
　——張力　332
　——強さ　325
　——破壊　323
　——ひずみ　324
　——フック　332
　——モーメント　324
ねじりコイルばね　152
　——端部　107
ねじり試験　327, 332
　——機　324
ねじりばね　323
　——定数　324
熱応力　317
熱可塑性　317
熱間圧延　159
　——帯鋼　159
　——棒鋼　159
熱間加工　159
熱間成形　158
　——重ね板ばね　158
　——コイルばね　158
熱間セッチング　148
熱間鍛造　158
熱間曲げ　157
熱衝撃抵抗　251
熱処理　149
　——したばね用線　145
熱電対　317
熱伝導　148
　——率　316
熱ひずみ　90
熱膨張　316
　——率　50
熱容量　149
燃焼加熱炉　56
燃焼機関　55
粘弾性　341
粘度　341

の

ノッチ効果　211
伸び　105
　——率　105

は

バースタビライザ　19
バー端末形状　269
バーニヤダイアル　339
バーの径　19
バーの長さ　19
パーライト　222
バイアススプリング　24
バイメタル　24
　——温度計　25
バウンドストローク　27
破壊安全率　261
破壊荷重　185
破壊力学　129
　鋼　293
秤ばね　18
はくり試験　223
破砕強さ　73
端受け　106
破損　129
はだ荒れ　257
はだ焼き　37

日本語索引

――硬さ　38
――鋼　37
――層　37
――深さ　37
破断応力　28
破断荷重　28
破断試験　28
破断伸び　28
歯付座金　321
　――の歯の高さ　149, 321
バックラッシュ　18
発せい度　81
バッチ操作　21
バッチ炉　41
パテンチング　222
ばね　284
　――受け具　286
　――荷重　287
　――硬さ　286
　――釘　102
　――クリップ　285
　――ケース　284
　――限界値　285
　――鋼　289
　――座金　187
　――軸線　284
　――試験機　289
　――指数　287
　――下重量　336
　――定数　285
　――選別機　288
　――測定装置　287
　――端面研削盤　286
　――取付けねじ　285
　――取付方式　287
　――長さ　287
　――のアルミニウム被覆　8
　――の入れ子組合せ　206
　――のたわみ　285
　――の端末具　109

――の取付高さ　267
――の取付長さ　267
――の並列重ね合せ　221
――の並列組合せ　220
――の並列配列　220
――の目玉　286
――秤　284
――フック　286
――分離機　288
――用帯鋼　289
――用線　289
――用洋白　207
――用りん青銅　226
――力　286
ばね板　179, 287
　――数　212
　――端部　180
　――中心　179
　――ナット　288
　――の厚さ　181
　――の異方性　9
　――の幅　181
ばね端末　286
　――焼なまし　278
ばね特性　285
　――試験　285
羽根車　345
幅落し　274
破面　129
　――の繊維流れ　121
パラボリック断面　220
パラボリックリーフスプリング　220
ばり　31
ばり取り　78
　――されたばね端末　78
　――装置　79
パレート図　221
バレル研磨　20
板間すきま　136, 215
板間摩擦　131

377

反作用　244
半だ円ばね　265
はんだ付け　278
板端法　320
バンディングプレス　19
反復曲げ試験　252
バンプストローク　31
半丸心金　265
半丸フック　144

ひ

ピアノ線　205, 226
ピーエッチ値　225
ピーク応力　222
ピーニング効果　223
ピーニング時間　223
ピーニングドラム　223
ピーニングノズル　223
PVD　226
引込式心金　252
引抜加工性　93
引抜線　93
引抜速度　93
引抜方向　88
比強度　282
非金属介在物　211
非金属ばね　211
ひげぜんまい　144
微細割れ　144
非時効性　209
非磁性鋼　211
比重量　282
ヒステリシス　161
　　──曲線　161
ヒストグラム　155
ピストンリング　227
ひずみ　89, 295
　　──時効　296
　　──取り焼なまし　295
非接触形ぜんまい　210

非接触測定　210
非線形特性　210
　　──ばね　237
非対称ばね　13
左旋回　181
左巻　181
ビッカース硬さ　341
ピッグテールエンド　227
ピッチ　228
　　──角　228
　　──誤差　229
　　──修正　228
　　──ツール　229
　　──巻きガータスプリング　280
　　──巻きねじりコイルばね　281
引張・圧縮両用コイルばね　90
引張応力　314
引張荷重　314
引張コイルばね　151
　　──端部　107
　　──の長さ　182
引張試験　314
引張強さ　314
引張ばね　114
　　──試験機　315
　　──成形機　315
非鉄金属ばね　210
比透磁率　247
比熱　282
非破壊試験　210
火花試験　282
被覆処理　48
被覆線　70
皮膜厚さ　49
標準化　292
標準値　291
標準偏差　291
表面粗さ　304
表面加工硬化　304
表面硬さ　304

日本語索引

表面欠陥　304
表面硬化処理　304
表面酸化　304
表面仕上げ　304
表面状態　303
表面処理　305
表面脱炭　303
表面張力　305
表面腐食　303
表面保護　304
表面割れ　303
平かいさき　221
平鋼　124
平線　125
平つぶし中空スタビライザ　61
平つぶし中実スタビライザ　60
比例限度　183
ビレット　24
疲労　117
　　──き裂　117
　　──試験　119
　　──寿命　117
　　──破壊　117
　　──破面　117
　　──変形　224
広幅帯鋼　346
品質管理　240
　　──図　240
品質保証　240
ヒンジフック形ねじりコイルばね　154

ふ

ファインブランキング　122
ファスナ　116
　　──ばね　116
ファンタイプ座金　116
フィードクラッチ　120
フィードグリッパ　142
フィードバック制御　121
フィードローラ　120

Vフック　339
Vブロック　337
フィンガ座金　122
フィンガスプリング　122
フーリエ解析　128
フェライト　121
不可逆焼戻しぜい性　173
深絞り　79
負荷装置　186
不活性ガス　166
吹付塗装　284
複合材料　56
　　──ばね　56
複合腐食試験　56
副尺　339
復炭　244
膨れ　25
ブシュ　31
不銹鋼　259
不純物　164
腐食　68
　　──荒れ　68
　　──試験　68
　　──浸透　167
　　──疲れ　68
付属金具　123
縁抜き　329
付着性試験　4
フッキングマシン　157
フック　156
　　──位置　156
　　──内側とコイル間長さ　89
　　──開口　156
　　──形状　156
　　──成形機　287
　　──の高さ　156
　　──の法則　157
プッシャー型炉　238
物理蒸着法　226
不動態　222

379

――防食　222
不等ピッチ　338
　　――圧縮コイルばね　150
　　――コイルスプリング　338
不変荷重　224
フラクトグラフィ　129
ブラシスプリング　29
プラスチックコーティング　230
プラスチックばね　230
ブラスト加工　25
ブランク　25
フランジ　124
ブリネル硬さ　29
ブルーイング　26
ブレインストーム　27
ブレーキ用ばね　28
フレキシブル生産システム　125
プレス　233
　　――クエンチング　233
　　――テンパ　233
フレッチングコロージョン　131
フローチャート　126
　　――記号　126
プログレッシブ重ね板ばね　236
ブロックケージ　136
雰囲気調整炉　237
雰囲気熱処理炉　64
分解修理　218
分散分析　8
粉体塗装　232

へ

平均応力　197
平均修復時間　197
平行度　84, 221
ベイナイト　18
ヘイの耐久限度曲線　144
平面度　125
並列タイプコイルドウェーブスプリング　221

並列多重ばね　205
並列積重ね　290
ベーキング　18
ペーハー値　225
ベーラー試験　350
ベーラー線図　349
へき開破壊　46
へたり　267
ヘヤピンスプリング　144
ベリリウム銅合金　66
ベルト研磨　22
ベルビルばね座金　21
ベルリンアイ　24
ベローズスプリング　21
変換器　327
変曲点　328
変形　80
　　――曲線　80
偏心　99
　　――荷重　99, 186
　　――カム　99
偏析　264
変態　327
ベンチマーキング　22
変動応力疲労限度　119
変動荷重　126
変動費　338
弁ばね　337
　　――受け　337
　　――用鋼線　338

ほ

ポアソン比　231
ポアソン分布　231
保安装置　237
棒鋼　293
防食　68
防振材　340
防せい　259
膨張係数　114

日本語索引

放電浸食　349
ホースクランプ　157
ホース保護用コイル　157
ホーニング　155
ボールジョイント式中空スタビライザ　330
ボールジョイント式中実スタビライザ　279
ボールジョイントタイプ　19
保護塗装　237
保持温度　155
保持時間　155
母集団　231
保証荷重　237
補助ばね　15, 264
補正係数　67
ホットセッチング　159
ホットピーニング　159, 343
ホブ　155
母平均　231
ボルト　26

ま

マーキング　193
マーチャントバー　200
マイクロセンサ　201
マイクロメータ　201
毎分回転数（RPM）　253
前処理　234
マガジンスプリング　190
巻上げトルク　347
巻数　213
巻付試験　351
巻取枠　247
巻方向　88
巻戻し特性曲線　42
巻戻トルク　336
巻戻し摩擦力　132
膜厚　179
マクシブロックスプリング　195

枕ばね　26
マクロ組織試験　190
曲げ応力　23
　　──疲労限度　118
曲げ角度　22
曲げ強度　126
曲げ工具　24
曲げ試験　22
曲げ心金　23
曲げ疲れ強さ　23
曲げ強さ　23
曲げ半径　23
曲げひずみ　23
曲げ方向　216
曲げモーメント　23
曲げ用カム　22
摩擦　131
　　──角　9
　　──クラッチ　131
　　──係数　49
　　──コイルばね　131
　　──抵抗　132
　　──トルク　133
　　──ばね　132
　　──腐食　132
　　──プレス　132
　　──力　131
　　──輪ばね　132
マシンループ　190
マスダンパー　194
マッフル炉　204
摩耗　1
マルエージング鋼　193
　　──ばね　193
丸形ショット　258
丸鋼　257
丸こば　257, 258
丸線　257
丸端面　258
丸断面トーションバー　323

381

マルチスライドプレス　205
マルチリーフスプリング　205
マルテンサイト　194
マルテンパー　194
丸フック　133, 134

み

右巻　253
ミクロ組織検査　201
水焼入れ　343
溝　127
　——付心金　318
　——付きスプリングピン　143
　——付き断面　73, 142
密着位置　279
密着形研削座部　47
密着ストレートフック形ねじりコイルばね
　294
密着高さ　279
密着直線起こし形フックねじりコイルばね
　294
密着長さ　279
密着巻　278
密度　82
密封　263
ミニブロックスプリング　202
耳栓　99
ミリタリラッパ　201
ミルエッジ　201

む

無機亜鉛塗装　170
無機化合物　169
無機材ばね　169
無欠陥　355
無研削コイル端末　106
無人搬送車　14
無電解ニッケルめっき　176

め

メートル系　200
メカトロニクス　199
目立て装置　94
目玉　115
　——付き重ね板ばね　180
　——の中心　286
　——巻き機　115
面取り　42
　——角　24
　——研削盤　42
　——装置　42
　——端面　42

も

モーメント　203
　——曲線　203
　——測定装置　203
モジュール　203
もろさ　129

や

焼入れ　240
　——油　146
　——応力　241
　——温度　242
　——剤　146
　——性　145
　——ひずみ　89
　——変形　241
　——焼戻し　241
　——炉　146
焼なまし　10
　——時間　10
焼ならし　211
焼戻し　313
　——温度　313
　——時間　313
　——ぜい性　311

──マルテンサイト　313
──炉　312
──割れ　313
焼割れ　241
焼け　31
ヤング率　354

ゆ

油圧ばね　160
油圧プレス　160
有機高分子材ばね　215
有限要素法　123
有効スパン　101
有効たわみ　3
有効巻数　212
U字形スプリング　334
U字形断面　334
遊星歯車　229
誘導炉　166
Uフック　114, 187, 243
ゆるみどめ座金　268

よ

溶解　89
溶融亜鉛めっき　158
溶融すずめっき　158
翼車式ブラスト機　346
抑制剤　167
横置き板ばね　328
横置き2連重ね板ばね　92
横荷重　329
横剛性　329
横ずれ　276
　──防止バンド　11
横弾性係数　269
横方向荷重　179
横方向たわみ　178
横方向力　328
予熱　233
　──炉　233

呼び寸法　209
予防保全　234
予巻　281
余裕巻き　78
より線ばね　296
四輪駆動　129

ら

ライナ　184
ラジアル荷重　243
ら旋径　283

り

リーフ長さ　180
離脱荷重　238
リップル　254
リバウンドストローク　244
リバウンド力　244
リブ付き断面　73, 253
リミットスイッチ　183
リムド鋼　254
粒界　139
　──破壊　170
　──腐食　171
　──割れ　170
硫化物　302
流気式炉　5
流体クラッチ　126
流体ばね　127
両腕付きねじりばね　91
両絞り引張コイルばね　283
両スライドタイプ重ね板ばね　26
両スライドタイプテーパリーフスプリング
　27
両端研削　26
両振応力　253
両振曲げ応力疲労限度　119
リラクゼーション曲線　299
リラクゼーション試験　248
リラクゼーション測定　248

リラクゼーション値　248
リレー用ばね　248
臨界冷却速度　72
リングゲージ　254
りん酸塩皮膜処理　225

れ

冷間圧延　54
　──鋼帯　54
冷間加工　54
冷間成形　54
　──圧縮コイルばね　53
　──ばね　53
　──引張コイルばね　53
冷間引抜き　53
　──鋼線　53
冷却液　65
冷却温度　65
冷却材　64
冷却時間　65
冷却装置　65
冷却速度　65
冷却能　65
レール締結用ばね　243
劣化損失　81
レトルト炉　252
連続式ショットピーニング　63
連続パテンチング　62
連続冷却変態曲線　39
連続炉　62

ろ

炉　135
　──冷　65
ろう付け　28
労務費　178
ローラ式矯正装置　255
ロールきず　255
炉床回転式炉　256
炉床振動式炉　63

六角断面タイプトーションバー　152
六角頭部　152
六角棒鋼　152
ロックウエル硬さ　255
ロット　187
　──番号　188
ロッド径　255
露点計　85

わ

ワールの応力修正係数　343
ワイブル曲線　345
ワイブル分布　345
ワイヤガイド　348
ワイヤ式ホースクリップ　348
ワイヤ切断機　347
ワイヤハーネスクランプ　348
ワイヤフォーミング　348
ワインドアップ剛性　346
ワインドアップトルク　346
輪ばね　254
割出板　165
割れ　70
　──検出器　70
　──の深さ　70

中 国 語 索 引

A

阿尔曼试片　7
阿基米德螺旋　12
埃里哥弹簧　104
安全标识　261
安全带弹簧　261
安全负荷　237
安全负荷　261
安全系数　261
安全装置　237
安全装置　261
安装长度　267
安装高度　267
安装角度　267
凹痕形状　236
凹坑　68, 228
凹坑深度　228
凹坑直径　228
奥氏体　14
奥氏体不锈钢　14
奥氏体结晶　14

B

摆锤式冲击试验　43
摆轴弹簧　306
板端法　320
板簧吊耳　24
板间间隙　136
板间摩擦　131
半椭圆弹簧　265
半圆心轴　265
半圆形钩　144
棒钢　293
剥离试验　223
薄板　318

薄钢板　271
保持时间　155
保护膜　237
保护膜厚度　49
保温　155
贝氏（碟形）弹簧垫圈　21
贝氏体　18
比例极限　183
比强度　282
比热　282
比重　282
壁厚　343
边界样品　27
编码器　106
扁钢　124
扁钢丝　125
变换器　172
变节距　338
变节距弹簧　338
变截面单板弹簧　310
变截面钢丝螺旋压缩弹簧　150
变截面钢丝中凹形螺旋压缩弹簧　151
变截面钢丝中凸形螺旋压缩弹簧　150
变形　80, 89, 267
变形点　167
变形曲线　80
标准差　291
标准化　292
标准弹簧　291
标准依据可循性　327
标准值　291
表面处理　305
表面粗糙度　304
表面镀镍　207
表面防护　304
表面腐蚀　303

表面加工硬化　305
表面裂纹　303
表面抛光　304
表面强化效应　223
表面缺陷　304
表面渗碳硬度　38
表面渗碳硬化层　37
表面渗碳硬化层深度　37
表面渗碳硬化处理　37
表面脱碳　303
表面氧化　304
表面硬度　304
表面硬化处理　304
表面张力　305
表面状态　303
并紧位置　279
并列螺旋波弹簧　221
并列组合　290
并列组合弹簧　221
并列组合压缩螺旋弹簧　206
并列组合压缩弹簧　220
并行开发工程　59
玻璃纤维强化弹簧　138
玻璃纤维弹簧　138
玻璃珠　138
波纹弹簧　21
波纹形网格弹簧　68
波形齿垫圈　344
波形垫圈　344
波形弹簧垫圈　344
泊松比　231
泊松分布　231
不等距螺旋压缩弹簧　150
不动态　222
不可逆回火脆性　173
布氏硬度　29
不锈钢　259,291
不锈钢喷丸　291
不锈钢弹簧　291
不压中心(德国)钩环　138

不圆度　84

C

C形挡圈　33
C形同心挡圈　33
材料　195
材料缺陷　195
材料试验　195
材料样本　195
材料组织　195
残留奥氏体　251
残留磁性　249
残余剪切应变　250
残余拉伸应力　250
残余氢含量　249
残余压应力　249
残余应力　250
残余应力分布　250
操作工　215
操作指导书　350
槽　127
测量范围　198
测量方法　198
测量精度　197
测量误差　198
测量用弹簧　198
测量值　197
测量装置　198
侧面圆环　134
插头　274
缠绕试验　351
常化　211
常温立定　54
常温硬化　54
长轴　346
超低温工程　74
超声波测厚仪　335
超声波加工　334
超声波清洗　334
超声波清洗装置　334

超声波探伤试验　335
超声波振子　335
超弹性　238, 303
超硬质合金　147
车辆　338
车辆用弹簧　339
车辆用悬架板弹簧　37
车削工具　331
沉淀　82
衬垫　184
衬套　31
成品尺寸　123
成形　128
成形淬火设备　22
成形垫圈　74
成形滚轮　128
成形机　128
成形性　128
成形压缩弹簧　74
尺寸　87
齿距　228
齿形弹簧垫圈　321
齿形弹簧垫圈的齿高　321
冲裁、(落料)　25
冲裁工具　25
冲裁力　25
冲程量　31
冲击波　271
冲击负荷　163
冲击能　25, 163
冲击疲劳试验机　163
冲击强度　164
冲击试验　164
冲击速度　163
冲击应力　164
冲孔　227
冲孔加工性　238
冲孔力　238
冲压　233
冲压成形　128

冲压加工　25
重叠板弹簧　178, 236
重叠板弹簧的总高度　217
重叠板弹簧试验机　180
重复试验　249
初卷　168, 235
初拉力　168, 234
初期故障　168
初始负荷　168
初始疲劳裂纹　164
初始应力　168
处理能力　319
处理条件　329
触点开关　320
触点弹簧　61
传递载荷　327
传动机构　232
传感器　265
传热面热负荷　113
串联式悬架装置　308
垂直度　225, 290
磁场变态　191
磁场分离机　191
磁场膜厚测试仪　191
磁粉探伤试验　191
磁强计　190
磁性弹簧　191
磁性应变法　191
磁滞(现象)　161
磁滞曲线　162
次级簧片套　264
从动　3
粗糙度　257
粗糙表面　257
粗糙度测量　257
粗切削　256
粗轧　256
促动器　4
促进寿命试验　2
脆化　105

淬火　240
淬火变形　89, 241
淬火冷却剂　146
淬火炉　146
淬火时效　240
淬火温度　242
淬火应力　242
淬火油　146
淬裂　241
脆性　29, 272
脆性失效　29
淬硬性　145

D

D 字形半圆钩环　78
大气湿度　13
大气污染　5
带　301
带槽弹簧销　143
带吊耳孔重叠板弹簧　180
带回转钩环的锥形端　309
带螺纹孔的钩头垫板　156
带式研磨　22
带橡胶软垫重叠板弹簧　181
带销 U 形钩　268
代用特性值　8
带转动螺栓的圆锥端　60
单板弹簧　275
单侧滑动复合板弹簧　214
单卷簧销机构　275
单片碟簧挠度　275
单丝连接双拉簧　275
弹仓弹簧　190
氮化层　208
氮化层深度　82
氮化物　208
弹丸　272
挡圈　252
挡圈开口部　136
刀杆　11

导磁性　225
导电率　103
导轨紧固弹簧　243
倒角　42
倒角角度　24
倒角研磨机　42
倒角装置　42
导热率　316
导向长度　143
导向筒　143
导向销　143
等温变化　174
等温淬火　14
等温退火　174
等效长度　112
等应力板簧　113
等应力分布　335
低负荷硬度试验　188
低合金钢　188
低碳钢　188
低温处理　189, 302
低温脆性　189
低温干燥处理　18
低温炉　189
低温退火　188
底层涂漆　235
底盘　43
点蚀　229
电镀（技术）　104
电镀　103, 104
电弧焊接　12
电解抛光　103
电刷压簧　29
电涡流传感器　100
电阻焊接　251
电阻式连续退火炉　62
吊耳　286
吊耳片　115
吊耳中心　286
碟形弹簧　60

碟形弹簧凹面　59
碟形弹簧的凸面　64
碟形弹簧垫圈　60
碟形弹簧叠加长度　182
顶端淬透性曲线　175
定期检查　77, 224
定位　231
定心装置　6
定形回火　124
动负荷　126
动力阻尼器　96
动能　176
动态测量　97
动态负荷　96
动态减振器　340
动态强度　97
动态弹簧刚度　97
动态应力　98
动特性试验过程　96
镀铬　44
镀锌　356
镀锌铬法　355
端部花键　266
端部退火　278
端部锥形拉伸弹簧　91
端面　109
端面倒角　42
端面滑动的重叠板弹簧　181
端面机械加工　190
端面裂纹　100
端面摩擦　101
端面磨簧机　286
端面磨削　106
端面支承座　106
端圈　106
端圈并紧　47
端圈不并紧　215
端圈不磨　106
端圈磨平不并紧　215
端头形状　109

短钩环扭转弹簧　272
短轴　318
断开装置　329
断口金相学　129
断裂负荷　28
断裂力学　129
断裂强度　46
断裂延伸　28
断裂应力　28
断面倒角　276
截面惯性矩　137
断面积　73
断面纤维组织状态　121
对称板弹簧　306
对焊　32
对合组合碟形弹簧　266
镦粗　336
镦粗头部　336
钝化防蚀　222
多重并列弹簧　205
多道冲压　205
多段表面强化　205
多股螺旋弹簧　296
惰性气体　167

E

E形挡圈　99
额定负荷　282
恩氏度（粘度指数）　110
二次淬火　294
二次卷曲　253
二次喷丸　91
二次退火　231
二次应力　264
二段节距螺旋弹簧　95
二阶锥端板簧片　332
二卷簧销系统　332

F

FRP板弹簧　122

发卡式弹簧 144	非时效性 209
发条 232, 284	非铁金属弹簧 210
阀门弹簧 337	非线性弹簧 237
阀门弹簧用钢丝 338	非线性特性 210
阀门弹簧座 337	沸腾钢 254
法兰盘 124	分布图 263
翻边 74	分度盘 165
反冲力 244	分级淬火 241
反复间歇试验 249	分阶 294
反复弯曲试验 252	分解修理 218
反卷摩擦力 132	粉末镀锌 271
反馈控制 121	粉末涂层 232
反向圆钩环 69	粉碎强度 74
反作用 244	分体钩环拉簧 265
方形钩环 245	蜂窝式压缩弹簧 21
方形截面 245	峰值应力 222
方形截面扭杆 245	腐蚀 68
方形截面线材 289	腐蚀疲劳 68
方形压缩弹簧 246	腐蚀渗透 167
防抱死系统 10	辅助弹簧 15, 152
防尘罩固定用长方形截面板条 246	辅助弹簧片 264
防腐蚀 68	复合板弹簧 205
防腐蚀剂 168	复合材料 56
防滑带 11	复合腐蚀试验 56
防锈 259	负荷 184
防锈油 68	负荷-变形曲线 186
放电腐蚀 349	负荷-拉伸曲线 185
放电试验 282	负荷测定 186
放置开裂 263	负荷测定装置 186
飞边 177	负荷极限 185
非磁性钢 211	负荷力矩 204
非电解镀镍 176	负荷试验 186
非对称板簧 13	负荷损耗 185
非接触测量 210	负荷增大 185
非接触形环状螺旋弹簧 280	负荷轴线 185
非接触形环状扭转弹簧 281	傅利叶分析（变换） 128
非接触形涡卷弹簧 210	覆膜处理 48
非金属介质 211	覆膜耐久试验 48
非金属弹簧 211	副弹簧 18
非破坏性试验 210	附着性试验 4

G

杆长　19
感度试验　265
杆端形状　269
杆径　19, 255
感应电炉　166
干湿交互浸渍试验　94
干式拉伸　94
干式磨削　94
干式研磨　94
干燥烘箱　94
钢　293
钢带　293
钢砂　293
钢丝　293
钢丝成形　348
钢丝的反复扭转试验　252
钢丝的扭结试验　177
钢丝扭转试验　324
钢丝切断机构　347
钢丝切丸　75
钢丝切丸尺寸　75
钢丝送进长度　347
钢丝校直装置　348
钢丝直径　347
钢丸　293
刚性（度）　254
钢制弹簧　293
高度试验　150
高度调整　149
高分子弹簧　153
高分子橡胶弹簧　103
高径比　276
高频淬火　166
高频加热　166
高强度垫圈　154
高强度钢　154
高速钢　153
高温定型试验　153

高温腐蚀试验　153
高温拉伸试验　314
高温强度　154
高温蠕变特性　154
铬酸盐光泽处理　353
公差带　320
公称尺寸　209
公称规格　209
工程表　236
工程能力　235
工程能力指数　236
工具耐久寿命　100
工具修整器　321
工业废弃物　166
工业工程　166
公制　200
工作长度（工作高度）　3
工作负荷　350
共析　113
共析钢　113
共振　251, 305
共振频率　251
沟　127
钩　156
沟槽断面　73, 142
钩环成形机　287
钩环高度　156
钩环机　157
钩环开口　156
钩环内侧与主体之间的尺寸　89
钩环位置　156
钩环形状　157
构成刀刃　31
古德曼疲劳极限图　139
钴合金弹簧　49
鼓形螺旋弹簧　20
固定垫圈　252
固定杆长　123
固溶化热处理　280
固溶体　279

固有振动 206
固有振动频率 206
故障分析 116, 119
故障率 116
故障平均时间 197
故障树 120
挂耳 99
管壁厚 330
管形扭杆 331
管形稳定杆 330
惯性矩 204
光滑端面 277
光洁度 81
光亮退火 28
龟裂 123
滚齿 155
滚动摩擦系数 49
滚动抛光 20
滚动轴承 255
辊式矫直装置 255
过度喷丸 219
过共析钢 161
过量渗碳 303
过热 218
过时效 217
过载 218

H

HDD 悬架弹簧 148
哈夫疲劳极限图 144
含量 62
含碳量 36
焊接 278
合成树脂弹簧 306
合金成分 6
合金钢 6
合金钢弹簧 7
合金元素 7
恒力弹簧 61
恒力涡卷弹簧 61

珩磨 155
恒蠕变极限 62
恒弹性弹簧合金材料 173
横向负载 179
横向刚度 329
横向力 328
横向挠曲 179
横向偏移 276
横向载荷 329
横置板弹簧 328
红外干燥设备 167
后处理 232
弧高 11
弧形螺旋弹簧 11
虎克定律 157
互换性 170
互锁 171
花键形状 267
滑板 276
滑动摩擦系数 50
滑动末端 276
滑动芯轴 276
划痕 263
滑线槽 348
化合物层 56
化学成份 44
化学镀镍 44
化学气相蒸镀法 44
环保镀铬处理 140
环 156
环规 254
环境工程学 111
环境试验 111
环境污染 111
环境应力裂纹 111
环境影响调查 111
环境影响评价 111
环行片弹簧 124
环状弹簧 207, 254
缓冲垫圈 268

中国語索引

缓冲器　75, 271, 278
缓冲式网格弹簧　75
缓冲弹簧　2, 271
簧箍　28
簧圈外径　216
簧圈重心的平均直径　197
回归分析　247
回火　313
回火脆性　311
回火发蓝处理　26
回火裂缝　313
回火炉　312
回火马氏体　313
回火时间　313
回火－时间－温度（TTT）曲线　330
回火温度　313
回火颜色　312
回卷扭矩　336
回弹　284
回弹板　244
回弹冲击　244
回转方向　88
回转钩环　306
回转式喷丸机　331
回转弯曲疲劳试验　256
会诊　27
活塞环　227
活性防腐　3
火焰淬火　124

J

J形片弹簧　175
积层缓冲钢板　178
积层截面扭杆　178
机电化　199
机器用弹簧　11
机械冲击镀锌　199
机械挤压　199
机械损耗　199
机械特性　199

机械效率　198
机械性能参数　48
机械阻抗　198
基准　22
基准线　247
基座　263
集尘器　96
极限冲程　27
极限负荷　183
几何公差　137
继电器弹簧　248
记号　193
计划外维修　335
技术规范　311
技术规范书　282
技术检查　310
技术手册　311
技术维护　311
技术要求　282
计算机辅助设计　58
计算机辅助制造　58
计算机集成制造　59
计算机控制系统　58
加大钩环　110
加工性　350
加工硬化　350
夹箍　46
夹紧　45
夹紧带　46
夹紧箍　142
夹紧管　47
夹紧螺栓　45, 47
夹紧试验　45
夹具　175
夹钳送料　142
加强筋弹簧钢　253
加热时间　149
加热试验　149
夹头　55
加压淬火　233

393

加压回火　233
夹杂物　165
加载速度　187
加载装置　186
剪断角度　269
间断试验　171
间接测量　166
间接成本　165
间接工资　165
间距　196
间隙　18, 46
间隙腐蚀　72
间隙调整套　281
剪断　269
剪切负荷　270
剪切模量　269
剪切强度　270
剪切应变　270
剪切应力　270
剪切应力修正系数　139
碱性电解液镀锌　6
碱性试验　6
减振　340
减振材料　340
减振器　78, 216
交变应力　253
交换器　327
交替浸渍腐蚀试验　7
搅拌器　4
角度测量仪　9
角钢　24
铰接管形稳定杆　331
角速度　9, 45
绞线　27
校正退火　295
校直滚轮　295
校直机　295
接触负荷　61
接触型截锥涡卷弹簧　51
接触形涡卷弹簧　204

接杆形平端空心稳定杆　61
接杆形平端实心稳定杆　60
接管用长方形截面线夹　246
接近检测器　237
接收试验　2
节距　228
节距刀　229
节距误差　229
节距修正　228
截面系数　264
节能　110
截锥涡卷弹簧　342
金属沉淀物　200
金属化合物　171
金属弹簧　200
金相显微镜　200
紧固件弹簧　116
紧固扭矩　319
紧箍弹簧　137
进出　167
进给滚轮　120
进给精度　120
浸硫法　303
浸透探伤试验　184
浸涂　87
浸洗过程　163
经过调质处理的弹簧钢丝　145
精加工　123
晶界腐蚀　171
晶界裂纹　170
晶界破坏　170
晶粒粗大　139
晶粒粗大化（因过热）　139
晶粒度　140
晶粒分布　140
晶粒界　139
晶粒细化　140
精密测量　232
精密冲裁　122
精细弹簧　144

中国語索引

静变形　292
静电镀膜　104
静负荷　292
精磨　123
静摩擦　292
静摩擦系数　50
静态试验　292
径向负荷　243
径向轴承　175
静应力　292
静载荷　224
局部脱碳　221
矩形钢丝弹簧　245
矩形截面扭杆　323
锯齿形端部扭杆弹簧　323
聚四氟乙烯膜　311
卷耳　115
卷耳成形机　115
卷耳套　115
卷簧　51
卷簧工具　53
卷簧滚轮　52
卷簧机　52
卷簧系统　52
卷簧销　52
卷曲刚度　346
卷曲扭矩　346,347
卷绕方向　88
卷绕筒　247
绝缘线　70
均热时间　278
军用包装材料　201

K

卡箍　29
卡箍长度　182
卡箍压力机　19
卡规　33
开机率　243
开口环钩　215

开口铰链　69
开口弹簧挡圈　44
开口销　69
看板　176
抗蠕变性　71
可变成本　338
可靠性　248
可控气热处理炉　237
可塑性　95
可行性试验　120
可行性研究　120
可展性　192
空气淬火　5
空气冷却　5
空气弹簧　5
空气弹簧悬架系统　5
空气悬挂延簧片　327
孔径　155
控制范围　63
控制弹簧　64
控制特性　63
控制图　63
跨度　281
宽带钢　346
扩散层　86
扩散渗透处理　86
扩散退火　155
扩张弹簧　114

L

拉拔后调质钢丝　146
拉拔速度　93
拉拔线　93
拉伸方向　88
拉伸环　315
拉伸加工性　93
拉伸螺旋簧端部　107
拉伸螺旋弹簧　151
拉伸强度　314
拉伸试验　314

拉伸弹簧	114	立定处理时的应力	268
拉伸弹簧成形机	315	力矩	203
拉伸弹簧成形码机	115	力矩测量装置	203
拉伸弹簧钩环内侧长度	182	力矩曲线	203
拉伸弹簧试验机	315	连续进给磨削	319
拉伸应力	314	连续冷却变化曲线	39
拉伸用压缩弹簧	93	连续炉	62
拉伸载荷	314	连续铅浴淬火	62
拉压疲劳极限	57	连续推进式炉	238
拉压双作用螺旋弹簧	90	联轴器	69
拉压循环变应力	57	量具	136
蓝脆	26	量具长	136
劳务费	178	量块	136
肋条断面	73	两端呈圆锥状拉簧	283
冷成形	54	两端滑块接触形片簧	26
冷成形拉伸螺旋弹簧	53	两端滑块接触形锥形片簧	27
冷成形弹簧	53	两端铰接实心稳定杆	279
冷成形压缩螺旋弹簧	53	两端螺纹连接实心稳定杆	280
冷加工	54	两端磨削	26
冷拉拔	53	两端内铰形扭转弹簧	154
冷拉钢丝	53	两端圈并紧并磨平	47
冷却材料	64	料架	294
冷却功率	65	劣化损耗	81
冷却时间	65	裂纹	70
冷却速度	65	裂纹扩展	70
冷却温度	65	裂纹深度	70
冷却液	65	裂纹探测器	70
冷却装置	65	裂纹形成	70
冷压带钢	54	磷化处理	225
冷压延	54	临界冷却速度	73
离合器碟形弹簧	48	流程图	126
离合器弹簧	48	流程图标记	126
离合器翼片弹簧	48	硫化物	302
离散分析	8	流体离合器	126
离心力	41	流体弹簧	127
离心式除尘器	77	硫印试验	302
力	127	六角钢	152
立定处理	268	六角截面扭杆	153
立定处理时的负荷	268	六角头部	152
立定处理时的高度	268	炉	135

中国語索引

炉床振动式炉　63
炉内冷却　65
露点仪　85
露天试验　216
卵形钢丝截面弹簧　101
轮式喷砂机　346
螺杆弹簧　350
螺杆形管夹　351
螺距　89
螺栓　26
螺纹连接稳定杆　331
螺纹心轴　318
螺旋　50
螺旋波形弹簧　51
螺旋端面　50
螺旋角　228
螺旋弹簧　50, 152
螺旋弹簧的展开长度　84
螺旋弹簧销　50
螺旋压缩弹簧　150
螺旋直径　283
落锤式锻造机　94
落料清除装置　263
洛氏硬度　255

M

Maxibloc 弹簧　195
马弗炉　204
马氏体　194
马氏体等温淬火　194
马氏体时效处理钢　193
马氏体时效处理钢弹簧　193
脉动循环疲劳极限　118
脉动应力　238
毛刺　31
密度　82
密封　263
密圈弹簧　278
摩擦　131
摩擦腐蚀　132, 259

摩擦环形弹簧　132
摩擦挤压　132
摩擦角　9
摩擦离合器　131
摩擦力　131
摩擦力矩　133
摩擦螺旋弹簧　131
摩擦弹簧　132
摩擦系数　49
摩擦阻力　133
膜厚　179
磨簧机　140
模具　86
膜具标记　86
模具块　86
膜具用弹簧　86
模量　203
磨料　1
模拟测量仪　8
膜片弹簧　85
磨损　1
磨纹（痕）　141
磨削　140
磨削端面　143
磨削切断　76
磨削压力　141
模压　53
末端　207

N

耐腐蚀试验　68
耐腐蚀性　250
耐环境试验　345
耐环境性　344, 345
耐久寿命　100
耐久性（寿命）　96
耐力　354
耐磨材料　1
耐磨镀层　10
耐磨性　1

397

耐热钢　149
耐热性　148
耐湿性　203
耐酸　3
耐锈　260
耐药性　44
挠度　80
挠度测量　80
挠性　125
挠性轴　126
内部摩擦　171
内部应力　172
内部组织　66
内侧半径　169
内侧面长度　169
内衬片　41
内挡圈　172
内导向面　169
内环　169
内径　169
内片　170
内燃机　271
内弹簧　169, 170
内弹簧径　168
粘弹性　341
粘性　342
镍合金弹簧　207
拧紧工具　116
扭臂杆长度　182
扭臂杆的角度　182
扭齿垫圈齿高　149
扭杆端　322
扭杆全长　218
扭杆弹簧　322
扭杆弹簧刚度　323
扭杆弹簧有效长度　325
扭杆头部　322
扭杆轴线　322
扭环　332
扭结　176

扭矩　322, 324
扭曲　332
扭转变形　325
扭转角　331
扭转拉力　332
扭转破坏　323
扭转强度　325
扭转圈数　213
扭转试验　332
扭转试验机　324, 327
扭转弹簧　152, 323
扭转弹簧成形机　324
扭转弹簧的端部　107
扭转弹簧刚度　324
扭转应力　325
扭转振动　325
浓缩气体　110
努普硬度测试　177

O

欧拉变形理论　112
欧拉力　112

P

pH 值　225
帕累托图　221
盘式离合器　89
抛物线板弹簧　220
抛物线截面　220
喷砂　1
喷射时间　223
喷射速度　223
喷水切断　343
喷涂料　284
喷丸　272, 274
喷丸处理时间　272
喷丸覆盖率　70
喷丸覆盖面积　69
喷丸滚筒　223
喷丸机　262

中国語索引

喷丸密度　273
喷丸强度　273
喷丸时间　274
喷丸试验　7
喷丸叶轮　272
喷丸硬度　273
喷丸嘴　223
膨胀系数　114
批　188
批号　188
批量生产　21
疲劳　117
疲劳(S-N)曲线　298
疲劳变形　224
疲劳极限　109, 117
疲劳裂纹　117
疲劳破坏　117
疲劳破坏面　117
疲劳强度极限图　118
疲劳强度线图　118
疲劳强度　118
疲劳试验　110, 119
疲劳寿命　117, 117, 212
铍青铜合金　66
偏角　80
偏析　264
偏心　99
偏心负荷　99, 186
偏心凸轮　99
偏置钩环　214
偏置弹簧　24
偏置直臂端扭转弹簧　294
片规　270
(板簧)片数　212
片弹簧　125
片形螺纹拧入接头（钩环）　319
片状碳化物　178
频率　76, 130
频率分布　130
频率分析　130

平板凸轮　231
平衡弹簧　18, 112
平均修复时间　197
平均移动　204
平均应力　197
平面度　125
平行度　85, 221
平行切削　221
破坏负荷　185
破坏实验　28
破损　129
破损面　129

Q

起动弹簧　292
起泡　25
汽车变速器　15
气孔　39
气幕　137
气蚀　39
气弹簧　137
气体氮化　137
气体炉　64
气体热处理炉　64
气压计　136
气制动　5
千分表　85
千分尺　201
迁移　328
前处理　234
嵌入式弹簧螺帽　239
嵌入式心轴　252
强迫振动　127
强制破坏　127
翘曲　35
切断工具　76
切飞边　329
切口冲击强度　212
切口冲击试验　212
切口效应　211

399

切向直尾端压缩弹簧　308
切削速度　76
切削阻力　76
琴钢丝　205, 226
氢脆　161
氢脆性　3
清洁度　165
清洁室　46
氢量试验　160
倾斜（偏角）　164
倾斜钩环　165
氰化镀锌　76
球面垫圈　283
球状化退火　283
球状夹杂物　138
球状碳化物　138
驱动轴　94
屈服　353
屈服变形　353
屈服点　353
屈服点测定试验　83
屈服强度　250
屈服应力　354
屈强比　243, 353
屈氏体　330
曲柄连杆式压力机　177
曲率　74
曲率半径　23, 243
曲面成形机（滚或压）　75
曲面滚轧　75
去飞边　274
去毛刺　78
去毛刺装置　79
去氢处理　81
去污　79
去氧化皮　82
去应力后热处理　300
去应力热处理　290
去应力退火　300
全变形　326

全变形率　326
全长　326
全长板　133
全负荷　134
全破坏　326
全伸长　326
圈数　213
全脱碳　325
全中心钩环　133
缺陷数　212

R

燃烧发动机　55
燃烧加热炉　56
热变形　90, 158
热成形钢板弹簧　158
热成形螺旋弹簧　158
热冲击阻力　251
热处理　149
热处理加工　317
热传导　148
热脆性　246
热电偶　317
热定型　159
热定型处理　148
热锻造　158
热加工　159
热浸镀锡　158
热浸镀锌　158
热可塑性　317
热喷丸　159
热膨胀　316
热敲击　343
热容量　149
热弯曲　157
热应力　317
热轧　159
热轧棒料　159
热轧带钢　159
热轧钢棒　200

热轧缘边　201
热胀系数　50
人工时效　12
人机系统　193
人类工程学　112
韧性　326
溶解　89
容许质量水平　2
冗余性　247
柔性制造系统　125
蠕变　71
蠕变定形　71
蠕变回火　72
蠕变极限　71
蠕变强度　72
蠕变失效强度（蠕变失效应力）　71
蠕变试验　72
蠕变速度　71
蠕变应变　72
软氮化法　208
软管保护弹簧　157

S

S-N（疲劳）曲线　277
塞尺　270, 318
塞规　121
三重螺旋弹簧　330
三角形坡口端面　109
扫描　306
砂轮　85, 142
砂轮更换　141
砂轮损耗　141
砂轮修整　141
砂轮修正　93
砂轮质量　142
砂轮状况　141
扇形齿垫圈　116
上卷耳　336
上控制限　336
烧结合金　275

烧伤　31
蛇（Z字）形弹簧　355
蛇（Z字）形弹簧成形机　355
设计标准　83
设计负荷　82
设计应力　83
伸长　105
伸长率　105
深度测微计　82
深铰　79
渗氮钢　208
渗碳　37, 39
渗碳层　38
渗碳钢　37
渗碳钢表面碳含量　36
渗碳体　39
生产线　39
湿润试验　160
湿式磨削　345
失稳　30
失稳高度　30
失稳临界负荷　30
失稳强度　30
失稳应力　30
时间-温度-变形（TTT）曲线　320
蚀刻　112
时效（老化）　4
时效处理　4
时效硬化　4
实心稳定杆　279
十字头　73
史密斯疲劳极限图　277
使凸起（压纹）　105
使用寿命　267
示波器　216
试件　283
事件　113
事件发生系统图　114
事件记录仪　113
试验方法　316

试验负荷	316	送料滚轮	347
实验计划法	83	送料速度	222, 348
试验结果	316	送料装置	121
试验片	316	塑料弹簧	230
试验时间	316	塑性变形	230
试验条件	315	塑性变形层	179
试验温度	316	塑性范围	230
手动装置	145	塑性解析	230
寿命试验	182	塑性破坏	95
数控	59	酸洗	226
树脂弹簧	56	酸洗过程	227
树脂丸粒	230	酸洗检查	227
数字控制	86	酸洗抑制剂	227
双线拉簧	92	酸性电解液镀锌	3
双线扭转弹簧	92	缩式钩环	246
双臂扭簧	91	索氏体	280
双层组合扭转弹簧	55	锁用弹簧	187
双重方形钢丝截面扭转弹簧	95		

T

双重螺旋弹簧	90	台架装置	15
双重扭转弹簧	95	钛合金弹簧	320
双端面磨簧机	91	弹簧	284
双钩环	91	弹簧板	287
双金属	24	弹簧板厚度	181
双金属材料	45	弹簧板间隙	215
双金属温度计	25	弹簧板间隙	208
双体扭转弹簧	90	弹簧板间隙应力	208
双头螺栓	302	弹簧板宽度	181
双行程供料机构	92	弹簧板片	179
双支承重叠板弹簧	92	弹簧板片长度	180
水淬火	343	弹簧板片端部	180
水平位置	124	弹簧板片中心	179
水溶性	344	弹簧盒	19
四轮驱动	129	弹簧测定装置	287
松弛	248	弹簧长度	287
松弛测定法	248	弹簧秤	284
松弛率	248	弹簧带钢	289
松弛试验	248	弹簧挡圈	277, 278
松弛值	248	弹簧的并列组合	220
送料	347	弹簧的铝镀层	8
送料长度	121		

弹簧的螺旋间距　280
弹簧的组合　206
弹簧垫圈　187
弹簧钉　102
弹簧端部　286
弹簧端部的紧固件　109
弹簧端面去毛刺　78
弹簧分离机　288
弹簧分选机　288
弹簧负荷　287
弹簧刚度（常数）　285
弹簧钢　289
弹簧钢片的各向异性　9
弹簧弓　27
弹簧钩环　287
弹簧盒　19, 284
弹簧夹紧螺栓　285
弹簧力　286
弹簧螺母　288
弹簧挠度　285
弹簧内径　170
弹簧试验机　289
弹簧特性　285
弹簧特性试验　285
弹簧线材　289
弹簧销　277, 288
弹簧压板　285
弹簧硬度　286
弹簧用青铜　226
弹簧用银镍合金　207
弹簧中径　196
弹簧轴线　284
弹簧装配系统　287
弹簧座　286
弹塑比　103
弹性　103
弹性比率　102
弹性变形　101
弹性变形能　102
弹性极限　102

弹性理论　102
弹性模量　99, 203
弹性挠度极限　285
弹性区域　102
弹性失稳　101
弹性指数（旋绕比）　287
碳氮共渗　36
碳化物　35
探伤仪　125
碳素钢　36
碳素钢弹簧　36
碳素纤维弹簧　36
碳纤维强化弹簧　41
陶瓷弹簧　41
套管夹　157
特性曲线　42
特性线折点　328
特性要素图　43
特性值　43
梯形截面　176, 329
体积膨胀系数　49
天然高分子弹簧　206
条　301
调速器　139
调整垫　271
调整垫片　281
调整垫圈　4
调质　241
铁磁体　121
铁道车辆用悬架板弹簧　26
铁素体　121
铁－碳状态图（相图）　172
通规　138
通过式喷丸　63
铜焊　28
铜合金弹簧　66
同轴度　59
统计的过程管理　292
统计质量管理　293
头部形状　148

403

投射角度　163
投影仪　236
凸轮　34
凸轮从动件　34
凸轮机构　34
凸轮控制　34
凸轮调整　34
凸轮外形　34
凸轮形状　34
凸轮轴　35
凸起　64
图表记录器　43
涂覆锌粉　170
图号　140
涂塑工艺　230
退火　10
退火时间　10
脱模器（顶出器）　101
脱气　80
脱碳　79
脱碳深度　79
脱脂　81
椭圆截面的长、短径　202
椭圆截面钢丝螺旋弹簧　217
椭圆螺旋弹簧　104
椭圆弹簧　134

U

U 形断面　334
U 形钩　114, 187, 243
U 形钩的孔　268
U 形弹簧　334

V

V 形钩环　339
V 型块　337

W

瓦耳氏应力修正系数（曲度系数）　343
外侧面导向　217

外侧弹簧　217
外层喷涂　322
外轮　217
外圈半径　115
外置偏心圆钩环　190
弯矩　23
弯曲成形心棒　23
弯曲的方向　216
弯曲工具　24
弯曲角度　22
弯曲疲劳强度　23
弯曲强度　23, 126
弯曲试验　22
弯曲凸轮　22
弯曲应变　23
弯曲应力　23
弯曲应力疲劳强度　119
丸粒分布　273
丸粒形状　274
完全淬火　133
完全退火　133
万向节　19
网带式加热炉　64
微观组织试验　190
微裂纹　144
微调刻度盘　339
危险现象　148
危险源　147
微型传感器　201
微振磨损腐蚀　131
维泊尔分布　345
维泊尔曲线　345
维护保养　234
维氏硬度　341
未加载时弹簧上的重量　336
温度传感器　312
温度分布　312
温度记录器　317
温度交换器　312
温度控制　312

温度－时间曲线　312
稳定杆　19, 290
稳定杆连结器　184
稳定器　10
稳定器耳孔　290
涡卷弹簧　283, 347
涡卷弹簧成形机　283
涡卷弹簧的末端弯曲　315
涡卷弹簧端部　107
涡流探伤试验　100
沃尔(S-N)曲线　349
沃尔(S-N)曲线试验　350
无机材料弹簧　169
无机化合物　169
无间隙簧钩　47
无缺陷（ZD）　355
无人运送车　14
无效线匝　78
无心磨削弹簧钢　41
误差极限　112
物理蒸镀法　226

X

X 射线应力测量法　352
X 线衍射法　352
析出　82, 232
析出硬化　232
稀释剂　87
吸湿性　161
吸收能　2
洗涤处理　254
细晶粒　122
细晶粒钢　122
系统　307
系统工程学　307
下侧砂轮　189
下卷耳　92
显微镜图像　200
显微镜图像分析　200
显微组织检查　201

线材　349
线材实验　349
线材台架　349
线成形弹簧　128
线缆收卷弹簧　33
限时开关　320
线式管箍　348
线束夹子　348
线弹簧　349
线网弹簧　349
限位开关　183
纤维强化塑料弹簧　56
线性弹簧　183
线性特性　183
相对导磁率　247
相对频率　247
相关分析　67
相关系数　67
箱式炉　41
橡胶包层压缩弹簧　258
橡胶垫　258
橡胶骨架　258
橡胶弹簧　259
向心力　41
销　69
消除应力　299
肖氏硬度　271
消音轴套　275
楔形键　309
楔形块节距刀　229
谐波分析　147
斜圆钩环　9
卸荷　238
心部硬度　66
锌粉涂料　356
芯轴　52
心轴滑动　193
心轴式卷簧机　193
心轴直径　192
型板　313

形变热处理　13
行程　301
行星齿轮　229
形状记忆合金　269
形状记忆合金弹簧　269
性能试验　135
修边　269
修正处理　67
修正系数　67
修正应力　67
修正装置　94
锈　259
锈蚀程度　81
许用负荷　185
许用应力　6
序列控制　266
悬臂梁　35
悬臂梁长度　35
悬臂弹簧　35
悬架重叠板弹簧　305
悬架弹簧　43, 305
悬架装置　306
旋向　88
旋转炉底炉　256
旋转中心　182
穴状腐蚀　224
循环变负荷　7
循环变应力　7
循环气体炉　5
循环弯曲试验　125
循环弯曲应力下的疲劳强度　119
循环应力幅　8
循环应力疲劳强度　119

Y

压板　234
压薄　172
压并长度　279
压并高度　279
压紧簧板　15

压棱角　66
压力控制系统　234
亚硫酸气体试验　302
压入　31
压入实验　31
压缩残余应力　58
压缩螺旋弹簧端部　106
压缩扭转螺旋弹簧　57
压缩强度　58
压缩弹簧　57
压缩弹簧试验机　57
压缩应力　58
压弯试验法　233
压下率　256
压延方向　88
压中心圆钩环　73
亚共析钢　161
延迟脆性断裂　82
研磨加工　2
研磨盘　1
延伸率　105
盐水浸渍试验　262
盐雾试验　262
盐浴　262
氧化　219
氧化层　219
氧化夹杂物　219
氧化铝磨料　8
氧化膜　219
氧化皮　262
杨氏（弹性）模量　354
样品尺寸　262
叶轮　345
液体珩磨　184
液压　160
液压弹簧　160
一阶应力　234
易碎性　129
异形截面钢丝的长短径比　12
异形截面钢丝螺旋弹簧　209

406

中国語索引

异形螺旋弹簧　172
阴极电镀　38
阴极电镀涂附　9
阴极防腐处理　38
印刷回路　235
印刷回路板　235
应变　296
应变时效　296
应变仪　296
应变硬化　296
应力　297
应力范围　299
应力分布　90
应力幅　297
应力腐蚀裂纹　297
应力极限　183
应力集中系数　118, 297
应力开裂　298
应力喷丸强化　300
应力强化的初始应力　168
应力蠕变断裂时间曲线　301
应力蠕变图　301
应力松弛　299
应力松弛曲线　299
应力梯度　298
应力退火　314
应力修正系数　297
应力循环　298
应力循环周期　298
应力-应变曲线　300
应力-应变-温度曲线　300
应力诱发马氏体变形　298
英式钩环　110
荧光渗透探伤试验　127
硬度　147
硬度试验　147
硬钢丝　145
硬钢丝（冷拉钢丝）　145
硬钢丝盘条　153
硬化　146

硬质镀铬　145
永久变形　224
永久拉伸变形　315
永久应变　225
游标　339
油淬火　214
油淬火钢丝　214
油水分离器　214
油污　291
有机高分子弹簧　215
有限元方法　123
有效长度　101
有效圈数　212
右旋　253
预防污染　234
预卷　281
预热　233
预热炉　233
圆端　257
圆钢　257
圆弧凸轮　45
圆环　134
圆角半径　67
圆截面钢丝弹簧　45
圆盘式磨削机　89
圆盘式制动器　88
圆盘凸轮　88
圆频率　9
圆形端面　258
圆形钢丝切丸　258
圆形截面扭杆　323
圆形金属丝　257
圆形螺纹拧入接头（钩环）　318
圆柱度　84
圆柱螺旋拉伸弹簧　77
圆柱螺旋弹簧　45, 77
圆柱螺旋压缩弹簧　77
圆柱凸轮　20
圆柱形弹簧　77
圆锥螺旋弹簧　151

407

圆锥面　59
圆锥形弹簧　60
圆锥形压缩螺旋弹簧　60
远距离测定　311
远距离温度计　311
允许范围　321
允许极限　321

Z

Z 字形板簧　355
杂质　164
再结晶退火　245
再结晶温度　245
再碳化　244
再现比　76
再现性　249
噪声规范　209
噪音　290
轧制钢材　255
轧制缺陷　255
轧制线材　255
展开长度　84, 114
展开法　83
真空镀膜　337
真空热处理　337
真空退火炉　337
振动测试仪　341
振动频率　130
振动器　341
振动试验机　340
振动式研磨机　341
振动式运送装置　341
振动送料器　339
振动应力　216, 339
振动噪声　244
振动周期　224
镇静钢　176
蒸馏炉　252
正火　211, 222
正时　175

正态分布　211
正应力　87
支撑板簧　21
支点　133
直臂端密圈扭转弹簧　294
直臂端稳定杆　296
直方图　155
直接成本　87
直接淬火　87
直列式波片弹簧　266
直列弹簧　289
直列组合　290
直列组合压缩螺旋弹簧　266
直线度　295
直线跨度　294
止规　209
指形垫圈　122
指针弹簧　122
制动弹簧　28
滞后特性曲线　42
质量保证　240
质量管理　240
质量管理图　240
质量效应　194
质量阻尼　194
中凹形螺旋压簧　160
钟表弹簧　343
钟表用弹簧　47
中间退火　235
中空稳定杆　330
中凸形螺旋弹簧　20
中凸形弹簧　202
中凸形锥形螺旋弹簧　20
中心孔　40
中心螺栓　39
中心密封圈　40
中心销　40
重心　40
轴承座用弹簧　16
轴弹簧　17

中国語索引

轴套　276
轴线　16
轴线呈弧形的椭圆形螺旋弹簧　105
轴向　16
轴向挡圈　115
轴向负荷　16
轴向挠度　16
轴向应力　335
轴用挡圈　142
皱纹　254
猪尾弹簧　227
主副重叠弹簧　333
主簧片　192
主弹簧　192
铸钢喷丸　38
铸钢丸　38
转变温度　328
转速　256
转速（RPM）　253
装配件　123
状态变化　327
锥度　308
锥度加工（锻造）　308
锥度加工（磨削）　308
锥度加工（压延）　309
锥端　309
锥端辊压机　309
锥形长度　310
锥形沉孔　69
锥形吊耳　310
锥形钢丝圆锥形弹簧　202
锥形滚动轴承　309
锥形螺旋弹簧　51
着色检验　55
着色识别　55
自动进退刀离合器　120
自动控制　15
自动湿度,温度调节装置　317
自动装置　15
自硬性　265

自由长度　130
自由高度　130
自由角度　129
自由圈数　212
总圈数　325
总体　231
总体平均　231
纵向裂纹　187
纵置板弹簧　187
组合弹簧　55
组合弹簧簧圈间最小距离　202
阻尼　78
组织变化　301
组装机　13
组装夹具　12
组装线　12
最大尺寸　196
最大负荷　196
最大负荷指示　222
最大扭矩　196
最大偏差　196
最大应力　196
最小应力　202
最小预压力　202
最小最大应力比　299
左旋　181
左旋向　181
作业研究（TMS）　319
座（支承）圈　109
座椅弹簧　263
作用挠度　3
作用应力　3

409

インドネシア語索引

A

aksi koreksi 67
aktuasi 3
aktuator 4
alat bentuk pengulir 52
alat jarak antara 229
alat koreksi gerinda 94
alat pembengkok 24
alat pembubut 331
alat pengatur tengah 6
alat pengering inframerah 167
alat penghilang bram 79
alat pengukur sudut 9
alat pengulir 53
alat penirus 42
alat potong 76
alat untuk bentukan awal 25
aliran serat dari keretakan 122
alur kawat 348
amplitude tegangan 297
amplitude tegangan bolak balik 8
analasisa harmonik 147
analisa frekuensi 130
analisa kerusakan/kesalahan 119
analisa korelasi 67
analisa pengaruh lingkungan 111
analisa plastis 230
analisa regresi 247
analisi fourier 128
analisis grapik mikro 201
analisis kegagalan 116
analisis varian 8
aneal isotermal 174
aneal penuh 133
anneal bebas tegangan 300
anneal melalui proses 235
anneal pada suhu rendah 188
anneal rekristalisasi 245
anneal sampai cemerlang 29
anneal sferoidal 283
anneal tarik 314
anneal untuk meluruskan 295
antar daun 170
antar muka mesin dan orang 193
anti air 345
arah bengkokan 216
arah lilitan 88
arah menarik 88
arah mengulir 88
arah putaran 88
arah rol 88
arah sumbu(aksial) 16
arc height 11
ausform 13
austemper 14
austenit 14
austenit yang ditahan 251
automatis 15

B

bagan batas lelah 118
bagan goodman 139
bagan Haigh 144
bagan ketahanan tegangan 298
bagan kuat lelah 118
bagan Smith 277
bagan tegangan-regang 300
bagan tegangan regang suhu 300
bagian ujung pegas 109
bahan 195
bahan campuran 56

411

bahan feromagnet 121
bahan peredam getaran 340
bahan tahan oleng bentuk pipa 330
bainit 18
baja 293
baja berbutiran halus 122
baja berkekuatan tarik tinggi 154
baja bulat 257
baja dirol 255
baja eutectoid 113
baja hipo eutektoid 161
baja hyper eutectic 161
baja karbon 36
baja karbon rendah 188
baja kecepatan tingg 153
baja kepingan dirol panas 159
baja keras hasil tarikan 145
baja killed 176
baja lembaran 271
baja lembaran berlapis 178
baja maraging 193
baja nitrida 208
baja paduan 6
baja paduan rendah 188
baja pegas 289
baja pegas tergerinda tanpa pusat 41
baja pencegah karat 259
baja rimmed 254
baja tahan karat 291
baja tahan karat austenit 14
baja tahan panas 149
baja tidak magnetis 211
baja yang permukaannya dikeraskan 38
bantalan 75
bantalan karet 258
bantalan luncur 276
bantalan rol 255
bantalan rol tirus 309
batang 19
batang baja 293

batang berpenampang 264
batang dirol panas 159
batang flat bersisi bulat 257
batang kawat 349
batang pemantap 290
batang puntir bentuk pipa 331
batang puntir dengan penambang lingkaran 323
batang puntir dengan penampang empat persegi 323
batang rata 124
batang stabilisator bentuk pipa 330
batang stabilisator padat berujung balljoint 279
batang stabilisator padat berujung baut 280
batang stabilisator pipa dengan ujung ball joint 331
batang stabilisator pipa dengan ujung baut 331
batang stabiliser 19
batang stabiliser lurus 296
batang stabiliser padat 279
batang ukur pedagang 200
batas atas kendali 336
batas beban 185
batas butiran 139
batas defleksi pegas 285
batas kebisingan 209
batas kekenyalan 102
batas kelelahan 117
batas kelelahan pada tegangan getar 118
batas kesalahan 112
batas ketahanan 109
batas perimbangan 183
batas rambat berlanjut 62
batas tegangan 183
batas toleransi 321
bats rambatan 71

batu gerinda 142
baut 26
baut klip 47
baut klip pegas 285
baut penjepit 45
baut pusat 39
baut tak berkepala 302
bead glas 138
bearing jurnal 175
beban 185
beban aksial 16
beban awal 168
beban awal minimum 202
beban batas 183
beban benturan 163
beban berlebih 218
beban berpindah 327
beban berubah-ubah 126
beban bolak balik 7
beban coba 237
beban dinamis 97
beban eksentris 99
beban gandar 16
beban irisan 270
beban jari-jari 243
beban keselamatan 261
beban maksimum 196
beban melengkung 30
beban melintang 329
beban pegas 287
beban penarik 238
beban penentu 268
beban pengerjaan 350
beban penuh 134
beban perusak 28
beban saat pecah 185
beban samping 179
beban statis 292
beban tarik 314
beban tetap 224

beban uji 316
beban yang dirancang 82
beban yang ditentukan 282
began tebaran 263
benchmark 22
benda pengujian 316
benkokan panas 157
bentuk awal yang halus 122
bentuk cekungan 236
bentuk kait 157
bentuk kepala 148
bentuk nok 34
bentuk ujung 109
bentuk ujung batang 269
bentukan awal 25
bentukan dengan ultrasonik 334
bentukan plastis 230
berat jenis 282
berat pegas bawah 336
bercak 78
berlubang karena korosi 229
besi paduan sintered 275
biaya bergerak 338
biaya buruh 178
biaya langsung 87
biaya tidak langsung 165
bilah roda 345
bilet 24
bimetal 24
bingkai kendaraan 15
blok bentuk V 337
blok gage 136
blok ujung 106
bram(dalam permesinan) 31
buang sisi 274
bungkus 276
buruh tidak langsung 165
bush gelang 115
butiran baja 293
butiran baja cor 38

413

butiran halus   122

C

cacat bahan   195
cacat permukaan   304
cairan pendingin   65
cakupan   69
campuran logam dan bukan logam   171
cam bulat   20
cat debu seng   356
celah   123, 208
celah anatara plat   136
celah cincin penyumbat   136
cembung   64
cetakan   86
cilindeer pemandu   143
cincin bantalan   31
cincin dalam   169
cincin luar   217
cincin penahan   252
cincin penahan bentuk E   99
cincin penahan dalam   172
cincin penahan jenis C   33
cincin penahan jenis C yang berpenampang sama   33
cincin pencengkram   142
cincin pengunci   278
cincin pengunci luar   115
cincin tarik   315
cincin torak   227
contoh   283
contoh batas   27
cotter   69
crosshead   73

D

daerah elastis   102
dapat dikerjakan   350
dapat diperkeras   145
daun   179

daun ganda   205
daun gerak balik   244
daun panjang penuh   133
daun pegas   287
daun pegas penekan   15
daun tahap kedua   264
daun tunggal   275
daun utama   192
daur tegangan   298
defleksi menyamping   179
defleksi pegas   285
defleksi sumbu (aksial)   16
dekarburisasi   79, 245
dekarburisasi permukaan   303
dekarburisasi sebagian   221
dekarburisasi total   325
derajat engler   110
derajat kekaratan   81
derajat kilau   81
derajat perbandingan pegas batang puntir   323
derajat perbandingan pegas puntir   324
deteksi sentuhan   237
diagram alur   126
diagram besi-karbon   172
diagram karakteristik   43
diagram kendali mutu   240
diagram pareto   221
diagram TTT   330
diagram Wohler   349
dial vernier   339
diameter batang   255
diameter dalam   169
diameter dalam ulir   170
diameter kawat   347
diameter lubang   155, 228
diameter rata-rata di pusat grapitasi kawat   197
diameter spiral   283
diameter ulir dalam   169

diameter ulir luar 216
diameter ulir rata-rata 196
difraksi sinar X 352
dikeraskan dan sepuh keras setelah
  penarikan 146
dimensi 87
dimensi nominal 209
dimeter poros pemegang 192
disainbantu komputer 58
distorsi 89
distorsi karena kejutan 241
distorsi karena panas 90
distorsi karena pengerasan 89
distribusi butiran peluru 273
distribusi frekuensi 130
distribusi suhu 312
distribusi tegangan 90
distribusi tegangan merata 335
distribusi ukuran butiran 140
distribusi weilbull 345
ditarik dalam keadaan dingin 53
ditumpuk sejajar 290
ditumpuk seri 290
dresser alat 321
dudukan 263

E

efek masa 195
efek peen 223
efiensi mekanis 198
eksentrisitas 99
elastisitas 103
elastoplastik 103
embos 105
encoder 106
endapan 82
endapan logam 200
energi benturan 163
energi kinetik 176
energi regangan elastis 102

energi semprotan 25
energi serapan 2
ergonomik 112
erosi karena pelubangan 224
erosi nyala api kawat 349
etching 112
eutectoid 113
evaporasi hampa udara 337

F

faktor keselamatan pada batas 261
faktor konsentrasi tegangan 297
faktor koreksi 67
faktor koreksi tegangan 297
faktor koreksi tegangan wahl 343
faktor tarik kelelahan 118
ferit 121
flange 124
fleksibel 125
fraktograpi 129
frekuensi 130
frekuensi alami 206
frekuensi daur 76
frekuensi daur tegangan 298
frekuensi getaran 131
frekuensi relatif 247
frekuensi resonansi 251
frekuensi sudut 9
fungsi korelasi 67

G

gage 136
gage pembatas 209
galbani listrik 103
galur 127
garis acuan 247
garis sumbu 16
garis terputus-putus 29
gas mulia 167
gaya 127

gaya bentukan awal 25
gaya gerak balik 244
gaya gesek 131
gaya gesek selama menggulung 132
gaya melintang 328
gaya pegas (beban pegas) 286
gaya pemotong 76
gaya punch 238
gaya sentrifugal 41
gaya sentripetal 41
gelang 115
gelang bantalan sunyi 275
gelang berlin 24
gelang keatas 336
gelang pegas 286
gelang pegas daun 115
gelang pemantap 290
gelang pembuat jarak 281
gelang penyangga 268
gelang terputar kebawah 92
gelombang kejut 271
gerinda pada kedua ujung 26
gerinda tirus 308
gesekan 131
gesekan antar plat 131
gesekan statis 292
gesekan tepi 101
geseran poros pemegang 193
geseran samping 276
getaran alami 206
getaran gaya 127
getaran puntir 325
goresan 263
grafik kontrol 63
grafik perekam 43
grapik mikro 200
grinda cakram 89
grit aluminium 8
gulungan 247
gulungan utama 235

H

hal pendeknya 272
hantaran termal 316
hasil uji 316
hemat energi 110
higroskopik 161
histeresis 161
histogram 155
honing 155
honing kering 94
horning cair 184
Hukum hooke 157

I

indeks kapabilitas proses 235
indeks pegas 287
inklinasi 165
inspeksi teknis 310
instrumen analog 8
intensitas peluru 273
inverter 172
irisan, mengiris 269
ironing 172
isi hidrogen tersisa 249

J

jaminan mutu 240
jarak antar lilitan 280
jarak antar ulir 89
jarak antara 228
jarak antara berubah 338
jarak antara tepi lingkar dalam dan badan pegas 89
jarak minimum (antara ulir) 202
jarak pusat rata-rata antar ulir 196
jari-jari dalam 169
jari-jari lengkung 243
jari-jari luar 115
jari-jari pembengkokan 23

jari-jari pojok 67
jenis lubang tirus 310
jenis sambungan bola 19
jig 175
jig perakitan 12
jpertumbuhan retakan 70
jumlah cacat 212
jumlah daun 213
jumlah daur sampai rusak 212
jumlah pintalan 213
jumlah putaran 213
jumlah total putaran 325
jumlah ulir aktif 212
jumlah ulir bebas 212

K

kait 156
kait bentuk huruf U 114, 187
kait bentuk U 243
kait bentuk V 339
kait bulat 134
kait bulat miring 9
kait bulat terbalik 69
kait datar dengan lubang ulir 156
kait dengan mur 60
kait diperbesar 110
kait german 138
kait pegas 287
kait pemutar 306
kait samping (lingkaran) melereng 165
kait sisi 214
kait tegak lurus 245
kait tengah penuh 133
kalor jenis 282
kamber 35
kanban 176
kandungan 62
kandungan karbon 36
kapabilitas proses 236
kapasitas beban 185

kapasitas panas 149
karakeristik tidak linear 210
karakteristik alernatif 8
karakteristik kontrol 63
karakteristik linier 183
karakteristik pegas 285
karakteristik rambat suhu tinggi 154
karat 259
karbid bersisik 178
karbid bola 138
karbida 35
karet jenis mouunt 258
karet pencegah pergeseran 11
kawat anyaman 27
kawat baja 293
kawat baja ditarik dingin 53
kawat baja pegas 289
kawat baja tinggi karbon (baja keras) 153
kawat berselaput 70
kawat bulat 257
kawat dipotong bulat 258
kawat dirol 255
kawat ditarik keras 145
kawat hasil tarikan 93
kawat musik, kawat piano 205
kawat pegas dikeraskan dan disepuh keras 145
kawat pegas katup 338
kawat penjepit selang 348
kawat persegi 289
kawat piano 226
kawat potongan 75
kawat rata 125
kawat sepuh keras dikeraskan dengan minyak 214
keadaan olahan 329
keadaan permukaan 303
keadaan tetap dalam kelelahan 224
keadaan uji 315

keandalan 249
kebisingan berderak 244
kecepatan beputar 256
kecepatan lingkar 45
kecepatan peen 223
kecepatan pembebanan 187
kecepatan pengulangan 222
kecepatan sudut 9
kecepatan tarikan 93
kecepatan umpan kawat 348
kedalaman dekarburisasi 79
kedalaman lubang 228
kedalaman nitrasi 82
kedalaman pengerasan 37
kedalaman retak 70
kedudukan kait 156
kedudukan rata 124
kegagalan awal 168
kegagalan total 326
kehilangan akibat degradasi 81
kehilangan mekanis 199
kejadian 113
kejadian berbahaya 148
kejut 240
kejut bertahap 294
kejut kempa 233
kejut oli 291
kekakuan melintang 329
kekakuan memutar 346
kekasaran 257
kekasaran permukaan 304
kekekaran 254
kekentalan 342
kekenyalan 326
kekerasan 147
kekerasan brinnel 29
kekerasan inti 66
kekerasan permukaan 304
kekerasan permukaan karburisasi 38
kekerasan Rockwell 255

kekerasan shore 271
kekerasan vickers 341
kekotoran 164
kekuatan pendinginan 65
kekuatan suhu tinggi 154
kekuatan tekan 58
kekusutan 176
kelelahan 117
kelelahan karena korosi 68
kelembaban atmosfer 13
kelengkapan 123
kelonggaran 18, 46
kelunakan 95
kelurusan 295
kemampuan olah 319
kemmmpuan bentuk 128
kempa hubungan buku-jari 177
kempa pengelompokan 19
kemudi empat roda 129
kendali getaran 340
kendali komputer numerik 59
kendali mutu 240
kendali proses secara statistik 293
kendaraan 338
kendaraan otomatis tanpa pengemudi 14
kepala batang puntir 322
kepala bergurat 266
kepala dilantakan 336
kepala hexagon 152
kepinan baja pegas 289
keping baja kanal dingin 54
kepingan almen 7
kepingan baja 293
kepingan lebar 346
kepingan pegas yang anisotrop 9
kerapatan 82
kerapuhan 29, 105, 129
kerapuhan asam 3
kerapuhan biru 26

インドネシア語索引

kerapuhan hidrogen 161
kerapuhan karena sepuh keras 311
kerapuhan pada suhu rendah 189
kerapuhan panas 246
kerapuhan sepuh keras 173
kerataan 125
keretakan 129
kerucut pengampelas 1
kerusakan karena penggerollan 255
kerusakan yang dipaksakan 127
kesalahan jarak antara 229
kesalahan pengukuran 198
kesejajaran 221
kesikuan 225
ketahanan 96
ketahanan terhadap asam 3
ketahanan terhadap cuaca 344
ketahanan terhadap kelelahan 117
ketahanan terhadap zat kimia 44
ketahanan zat kimia 44
ketebalan penampang kawat 318
ketelitian unpan 120
ketepatan pengukuran 197
klip 46
klip cincin 45
klip pegas 285
koefisien gesekan 49
koefisien gesekan luncur 50
koefisien gesekan putar 49
koefisien gesekan statis 50
koefisien gohner 139
koefisien lengkung 74
koefisien muai 114
koefisien muai kubik 49
koefisien pemuaian 50
kolet 55
komposisi kimia 44
komposisi paduan 6
kondisi pasif 222
kondisi roda gerinda 141

konduktifitas listrik 103
konsentisitas 59
kontrol digital 87
kontrol feed-back 121
kontrol otomatis 15
kopling cairan 126
kopling gesekan 131
kopling pipih 89
kopling unpan 120
koreksi jarak antara 228
korosi 68
korosi antar 171
korosi celah 72
korosi karena gesekan 132
korosi karena goresan 131
korosi karena gosokan 259
korosi permukaan 303
kotak pegas 284
kualitas punch 238
kuat belah 46
kuat bengkokan 126
kuat bentur batang tergores 212
kuat benturan 164
kuat dinamis 97
kuat iris 270
kuat jenis 282
kuat lelah 118
kuat lelah pada tegangan lentur 119
kuat lelah pada tegangan lentur terbalik 119
kuat lelah pada teganngan berubah-ubah 119
kuat lelah pemengkokan 23
kuat luluh 354
kuat melengkung 30
kuat pecah karena rambatan 71
kuat pembengkokan 23
kuat puntir 325
kuat rambatan 72
kuat remuk 74

419

kuat tarik 314
kumpaan bahan 294
kumparan kawat 349
kunci dalam 171
kunci tirus 309
kuras otak 27
kurva akhir pegas spriral 315
kurva beban/lenturan 186
kurva CCT 39
kurva histeresis 162
kurva Jominy 175
kurva karakteristik 42
kurva karakteristik selama menggulung 42
kurva laju tegangan creep 301
kurva momen 203
kurva pecah karena tegangan 301
kurva pembentukan 80
kurva pengenduran tegangan 299
kurva perpanjanagn beban (pegas tarik) 185
kurva S-N 277
kurva suhu waktu 312
kurva waktu-suhu-tranformasi 320
kurva Weilbull 345

L

laju benturan 163
laju evaporasi permukaan perpindahan panas 113
laju kerusakan 116
laju mengerol 256
laju pemotong 76
laju pendinginan 65
laju pendinginan kritis 73
laju pengenduran 248
laju perpanjangan 105
laju rambatan 71
langkah 294, 301
langkah gerak balik 244

langkah sampai terantuk 31
lapis atas 322
lapisan anti gesek 10
lapisan campuran 56
lapisan dasar 235
lapisan difusi 86
lapisan nitrida 208
lapisan oksida 219
lapisan pelindung 237
lapisan perubahan plastik 179
lapisan seng anorganik 170
lapisan teflon 311
lapisan tipis oksida 219
lapisan yang mengalami pengerasan 37
larut dalam air 344
larutan rapat 279
las arc 12
las tahanan 251
las ujung 32
lebar daun 181
lebar penampang kawat 346
lecet 25
lembaran tipis 318
lendutan 80
lendutan aktif 3
lendutan total 326
lengkung 29
lengkung penuh pada sisi 134
lenkungan elastis 101
lenturan cakram tunggal 275
lereng tegangan 298
letak rapat 279
lilitan kedua 264
limbah industri 166
lingkar (kait) pilin 332
lingkar tertutup 47
lingkaran D 78
lingkaran ganda 91
lingkaran inggris 110
lingkaran (kait) yang diperkecil 246

インドネシア語索引

lingkaran silang 73
lingkaran terbuka 215
lini merangkai 12
logam paduan super keras 147
lot 188
lubang 228
lubang korosi 68
lubang pusat 40
luluh 353

M

magnet tersisa 249
mampu dibuat ulang 249
mampu saling mengganti 170
mampu tarik 93
mandrel (sumbu) pembengkok 23
mangkuk pusat 40
manual teknis 311
manufaktur bantu komputer 58
manufaktur pola sel 39
manufaktur terintegrasi dengan komputer 59
martemper (kejut bertahap) 194
martensit 194
martensit disepuh keras 313
masukan 165
masukan berbentuk bola 138
mekanika retakan 129
mekanisme nok 34
mekanisme umpan langkah ganda 92
mekatronik 199
melapis dengan celupan 87
melengkung 27, 30
melenting kembali 284
mematenkan 222
membakar 18, 31
membentuk dalam keadaan dingin 54
membilas 254
membuang kerak 82
memilim 332

memperkaya gas 110
mencegah karat 260
mengeringkan hidrogen 81
mengganggu 336
menggulung 74
menghilangkan bram 78
mengukur tanpa menyentuh 210
mengulir 51
menyapu 306
menyerut 269
menyolder 28
mesin gerinda 140
mesin gerinda ujung pegas poros ganda 91
mesin kait 157
mesin pegas tarik 115
mesin pelingkar pegas 287
mesin pembakaran 56
mesin pembakaran internal 171
mesin pembentuk 128
mesin pembentuk gelang 115
mesin penbengkok dan kejut 22
mesin penekuk pegas liku-liku 355
mesin pengejut dan pembentuk 241
mesin penggulung pegas puntir 324
mesin penguji pegas 289
mesin penguji pegas daun 180
mesin pengujian pegas tarik 315
mesin pengujian puntir 324
mesin penirus 42
mesin penyemprot roda 346
mesin perakit 13
mesin rol tirus 309
mesin rup 190
mesin trip 329
mesin uji getaran 340
mesin uji kelelahan benturan 163
mesin uji pegas tekan 57
mesin ulir 52
mesin ulir jenis poros pemegang 193

mesin ulir pegas tarik 315
mesin ulir spiral 283
mesing gerinda ujung pegas 286
metoda benkok tekan 233
metoda magneostriction 191
metoda pengembangan 83
metoda pengukurang 198
metoda sentuh ujung 320
metoda unsur terbatas 123
metode penngujian 316
mikrometer 201
mikrometer pengukur kedalaman 82
mikroskop metalurgi 200
minyak pencegah korosi 68
minyak pengerasan 146
modul 203
modulus E 99
modulus irisan 269
modulus kekenyalan 203
modulus young 354
momen 203
momen gaya 204
momen geometrik 137
momen inersia 204
momen lentur 23
momen puntir 322, 324
momen puntir gesek 133
momen puntir maksimum 196
momen puntir membuka gulungan 336
momen puntir memutar 346
momen puntir pengencang 319
moment puntir pelilitan 347
muai termal 316
mulut kait 156
mur pegas 288
mur pegas tekan 239
mutu roda gerinda 142

N

naik turun 167

neraca pegas 284
nikel perak untuk pegas 207
nilai Cm 48
nilai karakteristik 43
nilai pengenduran 248
nilai ph 225
nilai standar 291
nilai terukur 197
nok 34
nok busur 45
nok cakram 88
nok(cam) pembengkok 22
nok eksentris 99
nok pelat 231
nomor lot 188

O

oksidasi 219
oksidasi permukaan 304
olah kalor 149
olah kalor dalam larutan 280
olah panas bak garam 262
olah panas hampa udara 337
olah panas pelepas tegangan 300
olahan awal 234
olahan dibawah nol 302
olahan pemantap 290
olahan pengampelas 2
olahan permukaan 305
olahan suhu rendah 189
olahan termomekanik 317
operasi rata-rata 243
operator 215
osilograph 216
overhaul 218

P

pabrik pelunak vakum 337
pabrik pembersih ultrasonik 334
paduan brilium tembaga 66

paduan memori bentuk   269
paduan pegas isoelastik   173
paku elastis   102
panjang aktif   3
panjang aktif total dari pegas batang puntir   325
panjang bantang puntir   218
panjang batang   19
panjang bebas   130
panjang dalam   169
panjang daun   180
panjang efektif   101
panjang ekuivalen   112
panjang kait dalam   182
panjang kaki   182
panjang kelengkungan   30
panjang lengan yang tetap   123
panjang pegas   287
panjang pegas yang ditetapkan   267
panjang pemandu   143
panjang pengikat   182
panjang pengukur, pembanding   136
panjang penyangga gantung   35
panjang rapat   279
panjang tambahan   114
panjang tirus   310
panjang total   326
panjang tumpukan pegas cakram   182
panjang umpan   121
panjang umpan kawat   347
panjang yang dikembangkan   84
papan rankaian tercetak   235
pasang   267
pasca anneal   231
pasca perlakuan   232
peen berlebih   219
peen ganda   91
peen panas   159, 343
peen peluru   274
peen tegangan   300

peening ganda   205
pegas   284
pegas arloji   343
pegas bahan anorganik   169
pegas baja   293
pegas baja antikarat   291
pegas baja karbon   36
pegas baja marageing   193
pegas baja paduan   7
pegas bantalan   21
pegas bantu   15, 152
pegas batang puntir   322
pegas batang puntir dengan ujung bergerigi   323
pegas batang tarik   93
pegas bellows   21
pegas bentukan panas   158
pegas berbentuk tong   20
pegas berjarak antara berubah   338
pegas berpenampang tak serupa   209
pegas blok mini   202
pegas blok mini, bentuk kerucut   202
pegas bukan besi   210
pegas bukan logam   211
pegas busur elips   105
pegas cakram kopling   48
pegas cincin   254
pegas cincin datar   124
pegas cincin gesek   132
pegas cincin penyekat minyak   214
pegas cincin tersusun   207
pegas dalam   169
pegas dalam sen   289
pegas dari kawat berhelai helai   296
pegas daun asimetris   13
pegas daun bentukan panas   158
pegas daun berlapis   178
pegas daun bertirus progresif dua tingkat   332
pegas daun celah terbuka   215

pegas daun dengan bantalan karet   181
pegas daun dengan gelang   180
pegas daun dengan ujung geser   181
pegas daun dua tahap   333
pegas daun ganda   205
pegas daun J   175
pegas daun jenis slide   26
pegas daun malintang   328
pegas daun melintang berpenyangga ganda   92
pegas daun memanjang   187
pegas daun mering bersentuh   214
pegas daun parabolik   220
pegas daun pendukung   18
pegas daun pengencang rel   243
pegas daun penumpu   305
pegas daun plastik diperkuat serat   122
pegas daun progresif   236
pegas daun simetris   306
pegas daun tirus   310
pegas daun yg kedua ujunnya bertiris   27
pegas daun Z   355
pegas dengan penguat serat gelas   138
pegas diafragma   85
pegas dibentuk dingin   54
pegas elastomer   103
pegas eligo   104
pegas elips   134
pegas empat persegi panjang   245
pegas fluida   127
pegas gandar berayun   306
pegas gas   137
pegas gaya tetap   61
pegas gelombang terulir   51
pegas gesek   132
pegas gesek vokute terulir   51
pegas heliks   152
pegas heliks ganda   90
pegas hexagon   152

pegas jala kawat   349
pegas jari   122
pegas jaring jenis bantalan   75
pegas jaring jenis corrugate   68
pegas karet   259
pegas katup   337
pegas kawat   349
pegas kawat berpenampang bulat   45
pegas kawat yang dibentuk   128
pegas kendaraan   339
pegas keramik   41
pegas kerucut   60
pegas kerucut heliks   151
pegas kobalt paduan   49
pegas kombinasi   55
pegas kombinasi paralel   220
pegas kombinasi seri   266
pegas komposisi   56
pegas kompresi heliks silinder   77
pegas kopling   48
pegas kopling sekat   48
pegas kotak gandar   16
pegas kotak sediaan   190
pegas kunci bergigi dan bergelombang   344
pegas lengkung kompresi   11
pegas liku-liku   355
pegas logam   200
pegas logam paduan mengingat bentuk   269
pegas lonceng   47
pegas luar   217
pegas magnet   191
pegas matres   86
pegas matres (pegas untuk cetakan)   86
pegas maxiblok   195
pegas miring   24
pegas mortor   204
pegas mortor tanpa kontak   210
pegas organik polimer tinggi   216

インドネシア語索引

pegas pada bantalan　26
pegas paduan nikel　207, 320
pegas paduan tembaga　66
pegas pembawa　37
pegas pemuai　114
pegas penahan　137
pegas penentu　268
pegas penerus　248
pegas pengendali　64
pegas penggulung kabel　33
pegas penghubung　61
pegas pengikat　116
pegas pengimbang　18, 112
pegas penguat gelas　138
pegas pengukur　198
pegas pengunci　187
pegas penjepit rambut　144
pegas penumpu　305
pegas penyangga gantung　35
pegas penyerap　2
pegas peredan kejut　271
pegas perpanjangan heliks　151
pegas pin dengan celah　143
pegas pipih belleville　21
pegas piring kerucut　60
pegas plastis　230
pegas polimer　153
pegas polimer tinggi alami　206
pegas poros　17
pegas progresif　237
pegas punti heliks　152
pegas puntir　323
pegas puntir/ tekan　57
pegas puntir berkaki ganda　91
pegas puntir berkawat ganda　92
pegas puntir berpenampang hexagonal　153
pegas puntir berpenampang kotak　245
pegas puntir berujung kait pendek　272
pegas puntir berujung lurus　294

pegas puntir berujung lurus menepi　294
pegas puntir ganda　90
pegas puntir hinge　154
pegas puntir kombinasi dua kawat　55
pegas puntir penampang berlapis　178
pegas puntir ulir berspasi　281
pegas puntiran rankap　95, 96
pegas rambut　144
pegas rata　125
pegas rem　28
pegas resin sintetis　306
pegas roda gigi worm　350
pegas sabuk keselamatan　261
pegas sasis　43
pegas semi eliptik　265
pegas serat karbon　36
pegas serat komposisi　56
pegas setater　292
pegas silinder　77
pegas spiral　283
pegas standar　291
pegas suspensi HDD　148
pegas tank heliks silinder　77
pegas tarik　114
pegas tarik berkait pipih　319
pegas tarik berkait ulir　318
pegas tarik berkawat ganda　92
pegas tarik berujung spindel　283
pegas tarik kait terpisah　265
pegas tarik ulir bentukan dingin　53
pegas tarikan heliks kerucut ganda　91
pegas tegangan rata　113
pegas tekan　57
pegas tekanan helik, dengan ragam jarak antara　150
pegas tekanan hidrolik　160
pegas tekan beehive　21
pegas tekan heliks　150
pegas tekan heliks, dengan diameter batang

425

tidak konstan 150
pegas tekan heliks, dengan diameter batang
   tidak konstan 151
pegas tekan heliks, dengan diameter batang
   tidak konstan (pegas maxi-blok) 151
pegas tekan heliks jam pasir 158, 160
pegas tekan lengkung 74
pegas tekan penampang kotak 246
pegas tekan tersusun 206
pegas tekan ulir bentukan dingin 53
pegas tempat duduk 263
pegas tenaga 232
pegas tong bertirus 20
pegas trailing 327
pegas U 334
pegas udara 5
pegas ulir 50
pegas ulir aksi ganda (tarik-tekan) 90
pegas ulir bentuk tong 20
pegas ulir berbentuk tak beraturan 173
pegas ulir berjarak garter 280
pegas ulir berlapis karet tekan 258
pegas ulir berpenampang telur 101
pegas ulir dengan bahan tirus 51
pegas ulir dengan penarik ganda 275
pegas ulir elips 104
pegas ulir gaya tetap 61
pegas ulir gelombang pararel 221
pegas ulir gelombang seri 266
pegas ulir gesekan 131
pegas ulir penampang bulat telur 217
pegas ulir pitch ganda 95
pegas ulir silinder 77
pegas ulir tiga lapis 330
pegas untuk peralatan 11
pegas utama 192
pegas volute 342
pegas yang bertindak linier 183
pekerjaan dalam keadaan dingin 54
pelapis 184

pelapisan 9, 45, 48
pelapisan aluminium pada pegas 8
pelapisan dengan difusi 86
pelapisan deposisi uap fisika 226
pelapisan elektropoleik 104
pelapisan karbon nitrida 208
pelapisan (krom) keras 145
pelapisan kromat hijau 140
pelapisan listrik 104
pelapisan nikel 207
pelapisan nikel Kanigen 176
pelapisan nikel secara kimia 44
pelapisan pada permukaan seng tambahan 271
pelapisan plastis 230
pelapisan pospat 225
pelapisan seng 356
pelapisan seng asam 3
pelapisan seng sianida 76
pelapisan serbuk 232
pelat klip 46
pelat tekanan 234
peluru 272
peluru baja 293
peluru baja tahan karat 291
pelurus 295
pemakaian roda gerinda 141
pemanasan awal 233
pemanasan berlebih 218
pemanasan induksi 166
pemandu dalam 169
pemandu kawat 348
pemandu luar 217
pemantapan secara panas 148
pemasukan bukan logam 211
pemasukan oksida 219
pemasukan sulfur 303
pembanding cincin 254
pembanding lolos 138
pembanding regang 296

pembawa kalor 148
pembebasan tegangan 299
pembentukan 80, 128
pembentukan elastis 101
pembentukan kawat 348
pembentukan panas 158
pemberntukan rambatan 71
pembersih ultrasonik 334
pembiruan 26
pembuangan gas 80
pembuat jarak 281
pemegang kunci 277
pemegang pegas 286
pemerikasaan makroskopis 190
pemeriksaan mikroskopis 201
pemeriksaan warna 55
pemesinan ujung 190
pemilah pegas 288
pemisah centrifugal 77
pemisah magnetik 191
pemisah pegas 288
pemisahan 264
pemolesan elektrolitik 103
pemotong gerigi 155
pemotong kawat 347
pemotongan dengan air jet 343
pemudahan dibentuk 192
penahan pegas katup 337
penaksir tebal 318
penampang 73
penampang beralur 73
penampang dasar 176
penampang dengan celah 142
penampang dengan rib 73, 253
penampang empat persegi panjang 245
penampang parabola 220
penampang trapesoid 329
penampang U 334
penandaan, marking 193
penandaan warna 55

penarikan kering 94
pencegahan polusi 234
pencetakan 53
pencucian 93
pendingin 64
pendinginan dalam tungku 65
pendinginan dengan udara 5
penempa jatuhan 94
penempatan 231
penemu cacat 125
penemu retakan 70
penentuan titik mulur 83
pengaduk 4
pengampelas 1
pengampelas basah 345
pengampelas bentuk tong 20
penganeal tahan berkelanjutan 62
pengaratan infiltrasi 167
pengarbonan 37
pengaruh 182
pengaruh goresan 211
pengasaman 226
pengasaran butiran 139
pengatur kecepatan 139
pengaturan nok 34
pengaturan saat panas 159
pengaturan sejajar 220
pengaturan tinggi 149
pengecatan elekrostasis 104
pengecatan semprot 284
pengecekan berkala 224
pengejutan dan sepuh keras 241
pengejutan dengan air 344
pengejutan minyak 214
pengekangan 28
pengencang 116
pengencer 87
pengendalian berurutan 266
pengendalian nok 34
pengendalian suhu 312

pengendapan 232
pengenduran 248
pengenduran tegangan 299
pengerasan 146
pengerasan dalam keadaan dingin 54
pengerasan dengan penuaan 4
pengerasan dengan udara 5
pengerasan diri 265
pengerasan induksi 166
pengerasan langsung 87
pengerasan melalui pengendapan 232
pengerasan melalui pengerjaan 350
pengerasan penuh 133
pengerasan percikan api 124
pengerasan permukaan 37, 304
pengerasan regang 296
pengerasan regang permukaan 305
pengerjaan akhir 123
pengerjaan panas 159
penggerinda ujung 106
penggerindaan akhir 123
penggerindaan bergetar 341
penggerindaan kasar 256
penggerindaan kering 94
penggerindaan umpan terus 319
penggerrindaan 140
penggesekan dalam 171
penggetar 341
penggulung sampah besi 263
penggulungan terbalik 253
penghalang pengasaman 227
penghalusan butiran 140
penghambat 168
penghantaran tenaga 232
penghilangan kontaminasi 79
penghilangan lemak 81
penghomogenan 155
penghubung 69
pengikisan 1
pengikut nok 34

pengimbangan roda gerinda 141
pengkhormatan 44
pengkromatan kuning 353
pengolahan penuaan 4
penguatan, anneal 10
pengubah 327
pengubah suhu 312
pengubah ultrasonik 335
penguji kekerasan beban rendah 188
penguji ketebalan magnetik 191
pengujian arus eddy 100
pengujian bahan 195
pengujian kawat 349
pengujian terputus-putus 171
pengukur bentuk cakra 85
pengukur getaran 341
pengukur ketebalan dengan ultrasonik 335
pengukur medan magnet 190
pengukur perasa 121
pengukur tekanan 136
pengukur tekanan udara 85
pengukuran beban 186
pengukuran dinamis 97
pengukuran kekerasan 257
pengukuran lendutan 80
pengukuran pengenduran 248
pengukuran tak langsung 166
pengukuran tegangan dengan sinar-X 352
pengukuran tekanan secara mekanik 198
pengukurran tepat 232
pengumpan bergetar 341
pengumpan dengan gataran 339
penguraian, pelarutan 89
penilaian pengaruh lingkungan 111
penitridaan 208
penitridan karbon 36
penitritan gas 137

penjalaran retakan 70
penjang material pegas koil 84
penjepit, pemegang 45
penjepit kawa harnes 348
penjepit selang 157
penjepit selang berpenampang kotak 246
penjepit selang jenis roda gigi worm 351
penormalan 211
penuaan 4
penuaan berlebih 217
penuaan buatan 12
penuaan karena lingkungan 111
penuaan karena regang 296
penuaan kejut 240
penurunan beban 185
penyambung batang stabiliser 184
penyambung cotter 69
penyangga, anting anting 268
penyangga gantung 35
penyaring debu 96
penyebaran poison 231
penyekat 263
penyelesaian 329
penyelesaian permukaan 304
penyembur 101
penyembur peen 223
penyemprotan (peen) berguling 331
penyentak 305
penyepuhan deposisi kation 38
penyepuhan mekanis 199
penyepuhan seng basa 6
penyimpangan beban 186
penyimpangan dari bentuk bulat 84
penyimpangan kesejajaran 85
penyimpangan maksimum 196
penyimpangan standar 291
penyisip 271
penympangan dari bentuk silinder 84

penyumbat telinga 99
perawatan diluar rencana 335
perawatan pencegahan 234
perawatan teknis 311
perbandingan daur 76
perbandingan elastis 102
perbandingan hasil 353
perbandingan kerampingan pegas ulir 276
perbandingan panjang-pendek penampang 12
perbandingan tegangan 299
perbandingan titik mulur terhadap kuat tarik 243
percentase cakupan 70
peredam (sisipan ujung) 274
peredam gerak bolak balik 216
peredam getaran 340
peredam masa 194
peredam tumbukan 278
peredamam 78
peredaman dinamis 96
peredan kejut 271
perekam kejadian 113
pergantian roda gerinda 141
perlindungan korosi 68
perlindungan korosi aktif 3
perlindungan korosi katodik 38
perlindungan korosi pasif 222
perlindungan permukaan 304
perlindungan terhadap karat 259
perlit 222
permeabilitas 225
permeabilitas relatif 247
permintaan oksigen secara kimia 44
permintaan spesifikasi 282
permukaan cekung dari pegas cakram 59
permukaan cembung dari pegas 64
permukaan gerinda 143

permukaan kasar 257
permukaan kerucut 59
permukaan retak kelelahan 117
permukaan retakan 129
permukaan ujung 109
perpanjangan 105
perpanjangan pada kerusakan 28
perpanjangan tetap 315
perpanjangan total 326
perpindahan rata-rata 204
perseng 69
persentase perpanjangan 105
pertambahan beban 185
pertumbuhan butiran 139
peruahan bentuk tetap 224
perubahan struktural 301
perunggu pospor untuk pegas 226
petunjuk urutan kerja 350
pin cotter 69
pin pegas 288
pin pegas terulir 50
pin pemandu 143
pin pengulir 52
pin pengunci 277
pin tengah 40
pipa klip 47
piringan pengampelas 1
pita bernampang kotak 246
plastisitas panas 317
plat pengindeks 165
pola 313
polusi udara 5
populasi 231
poros 11
poros fleksibel 126
poros nok 35
poros pemegang luncur 276
poros pemegang setengah bulat 265
poros pemegang yang dapat ditarik 252
poros pemotong ulir 318

poros pengulir 52
potensi karbon 36
potong dengan gerinda 76
potongan sejajar 221
pres 233
pres hidrolik 160
pres pembentuk 128
profil bergurat 267
profil peluru 274
profile nok 34
proses batch 21
proses patenting kontinyu 62
proses pencelupan 163
proses pengasaman 227
proses zinkrolit 355
proyeksi profil 236
psedoelastis 238
punch pelubang 227
pusat daun 179
pusat gelang pegas 286
pusat spacer 41
putaran arah kiri 181
putaran per menit 253

R

ragangan total 326
rambatan 71
rancangan percobaaan 83
rangkaian kontrol tekanan 234
rangkaian tercetak 235
rapat peluru 273
rasio poisson 231
rata-rata populasi 231
reaksi 244
regang 296
regang irisan 270
regang pembengkokan 23
regang risan tersisa 250
regang statis 292
regangan puntir 325

regangan rambatan 72
regangan termal 317
regangan tetap 225
rem angin 5
rem cakram 88
rentang pengendalian 63
rentang pengukuran 198
rentang plastis 230
rentang tegangan 299
rentang toleransi 321
resistensi pengikisan 1
resonansi 251
retak 70
retak antara kristal 170
retak karena kejutan 241
retak karena kelelahan 117
retak karena sepuh keras 313
retak kelelahan 117
retak kelelahan awal 164
retak musim 263
retak permukaan 303
retak rambut 144
retak rapuh tertunda 82
retak tegangan lingkungan 111
retakan antar kristal 170
retakan koreksi tegangan 297
retakan lunak 95
retakan memanjang 187
retakan puntir 323
retakan rapuh 29
retakan tegangan 298
retakan tepi 100
riak 255
ring pemandu 31
ring pipih bergelombang 344
ring pipih bulat 283
ring pipih jari 122
ring pipih kunci jenis kipas 116
ring pipih lenkung 74
ring pipih pegas bergelombang 344

ring pipih pegas kerucut 60
ring pipih penahan 252
ring pipih pengatur 4
ring pipih pengunci bergerigi 321
ring pipih pengunci pegas 187
ring pipih tahan guncangan 268
roda berlian 85
roda gerinda bawah 189
roda gerinda ikatan keramik 41
roda gigi planet 229
roda pengumpan 120
roda penyembur peluru 272
rol dingin 54
rol kasar 256
rol pelengkung 75
rol pengulir 52
rol tirus 309
rol umpan kawat 347
rol untuk meluruskan 295
roll panas 159
roll pembentuk 128
rongga 39
ropemecah jarak antara 229
ruang bersih 46

S

sabuk pencengkram 142
sabuk pengampelas 22
sarana pelindung 237
sarana pembebanan 186
sarana penanganan 145
sarana pengukur beban 186
sarana pengukur momen 203
sarana pengukur pegas 287
sarana pengukuran 198
sarana pengumpan 121
sarana penyelamatan 261
sasis 43
satuan pendingin 65
sebaran biasa 211

sebaran tegangan tersisa   250
sekeras pegas   286
sela   207
selang waktu getaran   224
sembur peluru   273
semburan peluru   272
semburan pengampelas   1
sementasi   39
sementit   39
semprot pasir   262
semprotan   25
sensor   265
sensor arus berlebihan   100
sensor mikro   201
sensor suhu   312
senyawa anorganik   169
sepuh kerah   313
sepuh keras kempa   233
set dingin   54
setengah kait   144
setting   268
shot peening berkelanjutan   63
sifat mekanis   199
simbol diagram alur   126
simbol gambar   140
sistem kendali komputer   58
sistem meter   200
sistem pemasangan pegas   287
sistem pengulir titik tunggal   275
sistem penumpu   306
sistem produksi fleksibel   125
sistem rem anti kunci   10
sistem suspensi udara   5
sistematika kejadian   114
sistematika kelelahan   120
sistim   307
sistim pengulir   52
sistim ulir dua titik   332
skala, sisik kerak, timbangan   262
solder   278

sorbit   280
span   281
span lurus   294
spesifikasi   282
spesifikasi bahan   195
spesifikasi proses   236
spesifikasi teknis   311
spiral archimides   12
stabiliser   10
stabiliser pipa dengan ujung datar   61
stabiliser solid dengan ujung datar   61
standar rancangan   83
standarisasi   292
statistik kendali mutu   293
strip   301
stroke pantulan   27
struktur inti   66
studi fisibilitas   120
studi waktu dan gerak kerja   319
suara berdecit   290
sudut bebas   129
sudut bengkokan   22
sudut benturan   163
sudut gesek   9
sudut irisan   269
sudut jarak antara   228
sudut kaki   182
sudut kemiringan penambang   276
sudut lendutan   80
sudut pemasangan   267
sudut pilin   331
sudut siku   290
sudut tirus   24
suhu disepuh keras   313
suhu kejutan   242
suhu pendinginan   65
suhu rekristalisasi   245
suhu tahan   155
suhu transisi   328
suhu uji   316

sulfida 302
sumber bahaya 147
sumber minor dan mayor penampang bujur telur 202
sumbu batang puntir 322
sumbu beban 185
sumbu pegas 284
super elastisitas 303
superkarburisasi 303
suspensi jenis sumbu tandem 308
susunan bahan 195
susunan pegas 206
susunan sejajar 221
switch batas 183
switch togle 320
switch waktu 320

T

tahan lembab 203
tahan panas 148
tahanan gesek 133
tahanan rambatan 71
tahanan terhadap kejutan termal 251
tahanan terhadap korosi 250
tahanan terhadap tekuk 250
tanda keamanan 261
tanda matres 86
tanda penggerindaan 141
tankai batang puntir 322
tanur sirkulasi udara 5
tarikan awal 168, 234
tarikan dalam 79
tebal daun 181
tebal dinding 343
tebal dinding pipa 330
tebal lapisan 179
tebal lembaran 270
tebal logam lembaran 270
tebal pelapis 49
tegang luluh 353

tegangan 297
tegangan aktif 3
tegangan awal 168
tegangan awal dari tegangan peen 168
tegangan benturan 164
tegangan berdenyut 238
tegangan bolak balik 7, 216
tegangan bolak balik tekan tarik 57
tegangan celah 208
tegangan dalam 172
tegangan dibalik 253
tegangan dinamis 98
tegangan getaran 339
tegangan iris 270
tegangan kedua 264
tegangan kejut 242
tegangan kelengkungan 30
tegangan lansung 87
tegangan lelah tekan tarik 57
tegangan luluh 354
tegangan maksimum 196
tegangan minimum 202
tegangan pembengkokan 23
tegangan permukaan 305
tegangan perusak 28
tegangan pilin 332
tegangan poros tunggal 335
tegangan puncak 222
tegangan puntir 325
tegangan rancangan 83
tegangan rata-rata 197
tegangan sisa tekanan 58
tegangan statis 292
tegangan tarik 314
tegangan tarik tersisa 250
tegangan tekan 58
tegangan tekan tersisa 249
tegangan terkoreksi 67
tegangan tersisa 250
tegangan utama 234

433

tegangan yang dibolehkan 6
tekan heliks kerucut 60
tekanan gerinda 141
tekanan gesek 132
tekanan mekanis 199
tekanan sentuh 61
teknik concurrent 59
teknik industri 166
teknik kriogenik 74
teknik lingkungan 111
teknik sistem 307
telemeter 311
telemetering 311
tembakan dengan plastik 230
tempa inti 66
tempa panas 158
tempa pelengkung 75
temper rambatan 72
tempering pengekang bentuk 124
tenaga euler 112
teori elastis 102
teori euler 113
tepat waktu 175
tepi tirus 42
tepi yang dibulatkan 258
tepi yang dihaluskan 277
termocouple 317
termograp 317
termometer bimetal 25
termostat 318
tes adhesi 4
tes alkali 6
tes almen 7
tetapan pegas 285
tetapan pegas dinamis 97
tidak ada cacat 355
tidak diperlukan lagi 247
tidak menua 209
timah celup panas 158
tinggi bebas 130

tinggi gigi bantalan 149
tinggi gigi ring pipih bergerigi 321
tinggi kait 156
tinggi keseluruhan 217
tinggi penentu 268
tinggi rapat 279
tinggi yang ditetapkan 267
tingkat kualitas yang diterima 2
tingkat pembersihan baja 165
tirai gas 137
tirus 42, 308
titik berat 40
titik luluh 353
titik pengulir 52
titik perubahan kurva 167
titik topang 133
titik transisi 328
toleransi 320
toleransi geometrik 137
tong 19
tong peen 223
tong pegas 284
tong pelilitan 347
traceabiliti 327
tranformasi 328
tranformasi magnetik 191
tranformasi martensit dengan induksi tegangan 298
transformasi isotermal 174
transisi, peralihan 328
transmisi kendaraan 15
troostite 330
tumbuk tirus 308
tumbukan baja cor 38
tumpukan berangkai (dari cangkram tunggal) 266
tumpukan sejajar ganda 205
tungku 135
tungku atmosphere terkendali 64
tungku berguncang berkelanjutan 63

tungku berkelanjutan 62
tungku dengan atmosfer pelindung 237
tungku dengan ban berjalan 64
tungku induksi 166
tungku jenis dorong 238
tungku konveksi 64
tungku pemanasan awal 233
tungku pembakaran 56
tungku pengapian berputar 256
tungku pengerasan 146
tungku pengering 94
tungku retor 252
tungku ruangan 42
tungku sepuh keras 312
tungku suhu rendah 189
tunku muffle 204

U

uir ujung 106
uji beban 186
uji bengkok bolak balik 125
uji bengkok terbalik 252
uji benturan 164
uji bentur batang tergores 212
uji berkala 77
uji berulang 249
uji bungkus 351
uji cetak sulfur 302
uji cuaca 345
uji diluar ruangan 216
uji fisibilitas 120
uji funsional 135
uji jepit 45
uji kandungan hidrogen 160
uji karakter dinamis 96
uji karat celup bolak balik 7
uji kekenyalan 327
uji kekerasan 147
uji kekerasan Knoop 177
uji kekusutan kawat baja 177

uji kelelahan 119
uji kelelahan putar benkok 256
uji kelembaban kabinet 160
uji kepekaan 265
uji ketahanan 110
uji ketinggian 150
uji kontinuitas pelapis 49
uji korosi 68
uji korosi campuran 56
uji korosi gas sulfur 302
uji korosi suhu tinggi 153
uji lingkungan 111
uji partikel magnet 191
uji pemanasan 149
uji pembengkokkan 22
uji penerimaan 2
uji penetrasi cairan 184
uji penetrasi fluoresensi 127
uji pengasaman 227
uji pengencangan suhu tinggi 153
uji pengenduran 248
uji pengupasan lapisan 223
uji percepatan umur 2
uji percik 282
uji pilin 332
uji puntir kawat baja 324
uji puntir terbalik kawat baja 252
uji rambatan 72
uji rendam air garam 262
uji ring pemandu 32
uji rusak 28
uji semprot garam 262
uji serapan dan pengeringan 94
uji sifat pegas 285
uji statis 292
uji tanpa merusak 210
uji tarik 314
uji tarik suhu tinggi 314
uji tumbukan charpy 43
uji ulang 249

435

uji ultrasonik 335
uji usia 182
uji Wohler 350
ujung anneal lunak 278
ujung bentukan 31
ujung daun 180
ujung ekor tangen 308
ujung kuncir 227
ujung lenkungan 109
ujung luncur 276
ujung pegas 286
ujung pegas puntir 107
ujung pegas spiral 107
ujung pegas terhaluskan 79
ujung pegas ulir tarik 107
ujung pegas ulir tekan 106
ujung pegas ulir yang tertutup 47
ujung pisau 177
ujung terbuka 215
ujung terbuka datar 215
ujung terpotong dengan pucuk wajik 109
ujung tirus 309
ujung tirus dengan kait melingkar 309
ujung tumbukan 201
ujung ulir 50
ukuran akhir 123
ukuran butiran 140
ukuran butiran austenit 14
ukuran contoh 262
ukuran kaliper 33
ukuran kawat potongan 75
ukuran maksimum 196
ukuran nominal 209
ulir 50
ulir arah kiri 181
ulir awal 168
ulir cadangan 281
ulir dalam 170
ulir kanan 253

ulir mati 78
ulir padat 278
ulir pelindung selang 157
ulir ujung tertutup dan digerinda 47
ulir ujung tidak digerinda 106
umpan kawat 347
umpan penncengkram 142
umur ketahanan 100
umur ketahanan alat 100
umur layanan 267
unit pelurus rol 255
unsur paduan 7
unsur penggerak 94
usia kelelahan 117

V
vernier 339
viskoelastisitas 341

W
waktu disepuh keras 313
waktu peen 224
waktu peen peluru 274
waktu pemanasan 149
waktu pendinginan 65
waktu penguatan/anneal 10
waktu pengujian 316
waktu rata-rata kerusakan 197
waktu rata-rata perbaikan 197
waktu rendam 278
waktu semburan peluru 272
waktu tahan 155
warna sepuh keras 312
washer kekuatan tarik tinggi 154
wrapper militer 201

Z
zat pengerasan 146
zona karburisasi 38

韓 国 語 索 引

가

가공 경화　296
가공 경화　350
가공 열처리　317
가공성　350
가동율　243
가로강성　329
가속 수명 시험　2
가스 스프링　137
가스 질화　137
가스 커튼　137
가압 속도　187
가열 시간　149
가열시험　149
가터 스프링　137
각 선　289
각단면 브츠 밴드　246
각단면 호스 클램프　246
각도 측정 장치　9
각선　125
각속도　9
각진동수　9
각형 고리　245
각형 단면　245
각형 스프링　246
각형 압축코일 스프링　246
간접 공임　165
간접비　165
간접측정　166
간판　176
간헐 반복 시험　249
감기 시험 (선)　351
감김 방향　88
감김 수　213
감쇠　78

강　293
강대　293, 301
강봉　293
강선　293
강선 역 비틀림 시험　252
강선의 비틀림 시험　324
강선의 킹크 시험　177
강성　254
강자성체　121
강재 스프링　293
강제 그릿　293
강제 쇼트　293
강제 진동　127
강제 파괴　127
개재물　165
거친연삭　256
거칠기　257
건습 반복침적 부식시험　7
건식 연삭　94
건식 인발　94
건식 침적 교차 시험　95
건식 호닝　94
건조로　94
게이지　136
게이지 압력　136
결점수　212
결정 입도　140
결정 입도 분포　140
결정립 미세화　140
결정립 성장　139
결정립 조대화　139
결함 검출기　125
겹판 스프링　178, 205
경강선　145
경강선재　153
경도　147

437

경도 시험　147
경사　165
경사 원형 고리　9, 165
경인선　145
경질 크롬 도금　145
경화　146
경화제　146
계측용 스프링　198
계획외 보전　335
고리　156
고리 개구　156
고리 내측길이　182
고리 높이　156
고리 위치　156
고리 형상　157
고리내측과 코일본체간 거리　89
고리분리형 인장 코일 스프링　266
고무 마운트 형　258
고무 스프링　259
고무 스프링 혼용 겹판 스프링　181
고무 패드　258
고무 피복 압축 코일 스프링　258
고분자 스프링　103
고분자재료 스프링　153
고속도강　153
고온 강도　154
고온 부식 시험　153
고온 인장시험　314
고온 체결 시험　153
고온 크립 특성　154
고용체　279
고용화 열처리　280
고유 진동　206
고유 진동수　206
고장 날때까지의 평균시간　197
고장 해석　116
고장력 와셔　154
고장력강　154
고장율　116
고장의 트리　120

고장해석　119
고정팔 길이　123
고주파 가열　166
고주파 열처리　166
곡률　74
곡률 반경　243
공구 내용수명　100
공구 드레서　321
공급 장치　121
공기 경화　5
공냉　5
공동 현상　39
공상 침식　224
공석　113
공석강　113
공식　229
공정능력　235
공정능력 지수　236
공정표　236
공진　251, 305
공진 진동수　251
공차　320
공칭 응력　87
공칭치수　209
과공석강　161
과부하　218
과시효　217
과열　218
과잉 침탄　303
관 두께　330
관능 검사　265
관리 특성　63
관리 한계　63
관리도　63
관성 모멘트　204
관통 이송 연삭　319
광택도　81
광폭 강대　346
광휘 풀림　29
괴너 응력 수정 계수　139

韓国語索引

교반기　4
교정 롤러　295
교정기　295
구동 축　94
구멍 지름　155
구면 와셔　283
구상 탄화물　138
구상 함유물　138
구상화 풀림　283
구성 날 끝　31
구속 뜨임　124
구심력　41
국부 탈탄　222
굽힘 가공용맨드릴　23
굽힘 각도　22
굽힘 강도　23, 126
굽힘 공구　24
굽힘 모멘트　23
굽힘 반경　23
굽힘 변형률　23
굽힘 시험　22
굽힘 응력　23
굽힘 응력 피로 한도　119
굽힘 피로 강도　23
굽힘방향　216
굽힘용 캠　22
굿맨 내구 한도 선도　139
귀　115
귀감기 기계　115
귀감기 (둘째판 감기 포함)　74
귀마개　99
규격 스프링　291
규격 요구사항　282
균열　70
균열　123
균열 검출기　70
균열 깊이　70
균열 성장　70
균열 시간　278
균열 전파　70

그루브드 스프링 핀　143
그리퍼 이송　142
그린 크로멧 처리　140
그립 링　142
그립 밴드　142
극저온 공학　74
글라스 비-드 (쇼트)　138
금속 산화 개재물　219
금속 스프링　200
금속 침전물　200
금속 현미경　200
금속간 화합물　171
금형　86
금형용 스프링　86
기계 능력 지수　48
기계 프레스　199
기계손실　199
기계적 성질　199
기계적 임피던스 측정기　198
기계충격 도금　199
기계효율　198
기기용 스프링　11
기능 시험　135
기록 온도계　317
기술 시방서　311
기술검사　310
기준선　247
기하학적 공차　137
기호도　140
길이 방향 균열　187
길이 시험　150
깊이 마이크로 미터　82
꼬임 (킹크)　176
꼬임선 스프링　296
끝단부 풀림　278
끝면 기계 가공　190
끝면 다듬질　277
끝부 세레이션　266

439

나

나사박음형 고리　60
나사식 중공 스테비라이저　331
나사형 중실 스테비라이저　280
나사형 후크 프레이트　156
나선경　283
나선형 맨드릴　318
나이프 에지　177
낙하 단조기　94
난형단면 스프링　101
납땜　28, 278
내구성　96
내구성 시험　110
내구한도　109
내구한도 선도　118
내력　354
내륜　169
내마모성　1
내마모성 피막　10
내부 경도　66
내부 길이 (자유길이)　169
내부 마찰　171
내부 응력　172
내부 조직　66
내산성　3
내습성　203
내식강　259
내식성　250, 260
내약품성　44
내연기관　171
내열강　149
내열성　148
내용 연수　267
내용수명　100
내측 가이드　169
내측 멈춤링　172
내측 반경　169
내측 스프링　170
내측 스프링 지름　169

내측 코일 스프링　169
내크리프성　71
내후 시험　345
내후성　344, 345
냉각 속도　65
냉각 시간　65
냉각 온도　65
냉각 장치　65
냉각능　65
냉각액　65
냉각재　64
냉간 가공　54
냉간 성형　54
냉간 성형 스프링　54
냉간 세팅　54
냉간 압연　54
냉간 압연 강대　54
냉간 인발　53
냉간 인발 강선　53
냉간성형 압축 코일 스프링　53
냉간성형 인장 코일 스프링　53
너클 조인트 푸레스　177
노무비　178
노점계　85
노치 효과　211
노치충격 강도　212
노치충격 시험　212
녹　259
높이 조정　149
누름 스프링판　244
누름 스프링판 너트　239
누름 판　234
누름판　15
누-프 경도 시험　177
니켈 도금　207
니켈 합금 스프링　207
닙　208
닙 응력　208

다

다단 피닝　205
다이 마크　86
다이 블록　86
다이내믹 댐퍼　340
다이아몬드 숫돌　85
다이어프램 스프링　85
다이얼 게이지　85
단경　318
단말 형상　109
단매물 스프링　275
단면　73, 109
단면 2 차 모멘트　137
단면 계수　264
단면 균열　100
단면 마찰　101
단면 연삭　106
단면의 기울음 각　276
단속 시험　171
단일 접시 스프링 휨　275
단축 응력　335
담금질　241
담금질 균열　241
담금질 변형　90, 241
담금질 성형 기계　22
담금질 시효　240
담금질 온도　242
담금질 응력　242
담금질 , 뜨임　241
담금질로　146
담금질성　145
담금질유　146
대기 습도　13
대기 오염　5
대류 로　64
대시 포트　78
대체 특성치　8
대칭 스프링　306
더블 인장 코일 스프링　275

도수분포　130
동 스프링 정수　97
동력전달 장치　232
동심 압축 코일 스프링　206
동심도　59
동심스프링의 병열 조합　206
동적 강도　97
동적 응력　98
동적 측정　97
동적 특성 시험　96
동적 하중　97
동합금 스프링　66
되풀림 특성 곡선　42
두께 게이지　318
둘째판 3/4 감김　201
둘째판 감김 (1/4)　264
둥근 끝부 모서리　258
뒤틀림　89
드레싱　93
드레싱 장치　94
등가 길이　112
등분포 응력　335
등온 변태　174
등온 풀림　174
등응력 스프링 (테이퍼 판 스프링)　113
디스크 그라인더　89
디스크 브레이크　88
디지털 제어　87
디핑 도장　87
딥 드로잉　79
뜨임　313
뜨임 균열　313
뜨임 시간　313
뜨임 온도　313
뜨임 취성　311
뜨임로　312

라

라운드 컷트 와이어　258
라이너　184

래틀 노이즈 (따닥따닥 소음)   244
레이디얼 하중   243
레일 클립   243
레토르트 로   252
로   135
로냉   65
로상 진동식로   63
로상 회전로   256
로크웰 경도   255
로트   188
로트 번호   188
롤 홈   255
롤러 베어링   255
롤러식 교정 장치   255
리밋 스위치   183
리바운드 스트로크   244
리코일시 마찰력   132
리플   255
릴   247
릴랙세이션 측정   248
릴레이용 스프링   248
림드 강   254
링 게이지   254
링 스프링   254

마

마그네틱 두께 측정기   191
마루골 단면   73
마루골 스프링강   253
마르에이징강   193
마르에이징강 스프링   193
마르텐사이트   194
마르템퍼   194
마모   1
마모 부식   131, 132
마무리 가공   123
마무리 연마   123
마무리 치수   123
마이크로 센서   201
마이크로미터   201

마찰   131
마찰 계수   49
마찰 링 스프링   132
마찰 부식   259
마찰 스프링   132
마찰 저항   133
마찰 코일 스프링   131
마찰 클러치   131
마찰 토 - 크   133
마찰 프레스   132
마찰각   9
마찰력   131
마찰음   290
마찰형 벌류트 스프링   51
마킹   55, 193
망 스프링 (와이어 메쉬 스프링)   349
맞댄 용접   32
매분 회전수   253
매스 댐퍼   96, 194
매크로 조직 검사   190
맥시블록 스프링   195
맨드릴   52
맨드릴 슬라이드   193
맨드릴 지름   192
맨드릴식 코일링기   193
머리 형상   148
머신 루프   190
머플로   204
멀티 스라이드 프레스   205
멈춤링   252
메인 스프링   192
메커트로닉스   199
메트릭 시스템   200
모듈   203
모따기   42
모따기 단면   42
모따기 연삭반   42
모따기 장치   42
모멘트   203
모멘트 곡선   203

韓國語索引

모멘트 측정 장치　203
모집단　231
모집단 평균　231
모판　192
무결점　355
무기 아연 도장　170
무기재료 스프링　169
무기화합물　169
무연삭 코일끝단　106
무인 운반차　14
무전해 니켈 도금　176
물 담금질　344
물리 증착법 (PVD)　226
미끄럼 끝면　276
미끄럼 마찰 계수　50
미끄럼 심봉　277
미니블록 스프링　202
미립강　122
미립자　122
미세 균열　144
밀 에지　201
밀도　82
밀림　276
밀림방지 클램프　11
밀착 길이　279
밀착 높이　279
밀착 스트레이트 고리형 비틀림코일스프링　294
밀착 위치　279
밀착 직선 수직 고리형 비틀림 코일 스프링　294
밀착감김　278
밀폐 고리　47

바

바깥 반지름　115
바 - 스테비라이저　19
바운드 스트로크　27
바이메탈　24
바이메탈식 온도계　25

바이어스 스프링　24
박리 시험　223
박판　318
박판 강판　271
박판 스프링　125
반발력　244
반복 굽힘시험　125
반복 시험　249
반복 응력　8
반복 응력 진폭　8
반복 하중　7
반복굽힘 시험　252
반복회수 비　76
반원 고리　144
반원 맨드릴　265
반원고리　78
반작용　244
반타원 스프링　265
받침쇠　106
발청도　81
방식　68
방전 침식　349
방진재　340
방청　259
방청유　68
배꼽　207
배럴 연마　20
배치 프로세스　21
백래시　18
밴드의 길이　182
밴딩 지그　28
밴딩 프레스　19
밸브 스프링　337
밸브 스프링 리테이너　338
밸브 스프링용 강선　338
뱃치로　42
버　31
버 제거　78
버 제거 장치　79
버니어 다이얼　339

443

버제거된 스프링끝　79
버클　29
벌림 끝　215
벌림 플랫형 끝　215
벌집형 코일 스프링　21
범프 스트로크　31
베를린 아이　24
베릴륨 동　66
베이나이트　18
베이킹　18
벤치마-크　22
벨로스 스프링　21
벨트 연마　22
벽개 파괴　46
변곡점　167, 328
변동 응력 피로 한도　119
변동 하중　126
변동비　338
변위 측정　80
변태　328
변형　80, 267, 296
변형 곡선　80
변형 시효　296
병렬 다중 접시 스프링　205
병렬 적층　221
병렬형 코일드 웨이브 스프링　221
병열 조합　290
보안 장치　237
보정 계수　67
보조 스프링　15, 152, 264
보증 하중　237
보호 도장　237
복탄　245
복합 부식시험　56
복합 재료　56
복합 재료 스프링　56
볼 조이트 형　19
볼 조인트식 중공 스테비라이저　331
볼 조인트식 중실 스테비라이저　279
볼록면　64

볼록통 테이퍼 코일 스프링　20
볼록통형 스프링　20
볼록통형 코일 스프링　20
볼류트 스프링　342
볼스터 스프링　26
볼트　26
봉 끝단면 형상　269
봉 직경　255
봉의 길이　19
봉의 지름　19
뵐러 선도　349
뵐러 시험　350
부동태　222
부동태 부식방지　222
부등 피치　338
부등 피치 압축 코일 스프링　138
부등피치 스프링　338
부시 (부싱)　31
부시 (부싱) 압입　31
부시 (부싱) 압입 하중 시험　32
부식　68
부식 시험　68
부식 침투　167
부식 피로　68
부식 핏트　68
부척　339
부하 장치　186
분산분석　8
분위기 열처리로　64
분위기 조정로　237
분체 도장　232
분할판　165
분해 수리　218
분화 손실　81
불가역 뜨임 취성　173
불꽃 시험　282
불림　211
불변·하중　224
불순물　164
불활성 가스　167

브러시 스프링　29
브레이드 와이어　27
브레이크 스프링　28
브레인스토밍　27
브리넬 경도　29
브이 (V) 블록　337
브이 (V) 자형 고리　339
블라스팅　25
블랭킹　25
블랭킹 툴　25
블랭킹 휘스　25
블록 게이지　136
블루잉　26
블리스터 (부풀음)　25
비 시효성　209
비 투자율　247
비강도　282
비금속 개재물　211
비금속 스프링　211
비대칭 판 스프링　13
비례한도　183
비선형 특성　210
비선형 특성 스프링　237
비연삭 맞댐 끝　47
비열　282
비자성강　211
비접촉 측정　210
비접촉형 태엽 스프링　210
비중량　282
비철 금속 스프링　210
비커스 경도　341
비틀림　332
비틀림 강도　325
비틀림 고리　332
비틀림 모멘트　324
비틀림 변형　325
비틀림 스프링 끝부분　107
비틀림 시험　327, 332
비틀림 시험기　324
비틀림 응력　325

비틀림 장력　332
비틀림 진동　325
비틀림 코일 스프링　152
비틀림 파괴　323
비틀림 횟수　213
비틀각　331
비파괴 시험　210
빌렛　24

사

사 (四) 륜 구동　129
사각 단면 토션 - 바　323
사각단면 토션바　245
사고발생 계통도　114
사다리꼴 단면　176, 220, 329
사상 기록계　113
사상 (事象)　113
사이드 트림　274
사이렌서　274
사일런트 부시　275
산성 아연 도금　3
산세척　226
산세척 검사　227
산세척 공정　227
산세척 억제제　227
산업폐기물　166
산취성　3
산포도　263
산화　219
산화막　219
산화물 층　219
삼각 끝　109
삼중 코일 스프링　330
상관 계수　67
상관 분석　67
상대 도수　247
상도　322
상부 관리 한계　336
상온 경화　54
상용 하중　350

샌드 블라스트　262
생산공학　166
샤클　268
샤클 아이　268
샬피 충격시험　43
새시　43
새시 스프링　43
서모스탯　318
서브제로 처리　302
석출　82, 232
석출 경화　232
선 스프링　349
선간 틈새　280
선경　347
선삭 공구　331
선세공 스프링　128
선재　349
선재 시험　349
선재 이송　347
선재 통과 홈　348
선재이송 길이　347
선재이송 롤러　347
선재이송 속도　348
선형 특성　183
선형 특성 스프링　183
설계 하중　82
설계응력　83
설계표준　83
설퍼 프린트 시험　302
섬유 강화 프라스틱 스프링　56
성형　128
성형 롤러　128
성형 프레스　128
성형기　128
성형성　128
세라다이징　271
세라믹 스프링　41
세레이션 형상　267
세레이션형 토션바　323
세로 탄성 계수　354

세멘타이트　39
세정처리　254
세팅　268
세팅 높이　268
세팅 응력　268
세팅 하중　268
센서　265
센타레스 연삭 스프링강　41
센터 배꼽　40
센터 볼트　39
센터 스페이서　41
센터 핀　40
셀 제조방식　39
셰이빙　269
소결 철 합금　275
소르바이트　280
소성 변형　230
소성 변형층　179
소성 해석　230
소성역　230
소음 규제　209
쇼아 경도　271
쇼트　272
쇼트 강도　273
쇼트 경도　273
쇼트 밀도　273
쇼트 블라스트　272
쇼트 투사 분포　273
쇼트 피닝　274
쇼트 피닝 시간　274
쇼트 형상　274
쇼트블라스트 임펠라　273
쇼트블라스트 투사 시간　272
수동 장치　145
수명 시험　182
수소 취성　161
수소량 시험　160
수용성　344
수입 시험　2
수정 응력　67

韓国語索引

수지계 쇼트　230
수평 위치　124
순서도　126
순서도 기호　126
숫돌 교환　141
숫돌 소모　141
숫돌 수정　141
숫돌 요건　141
숫돌 품질　142
스냅 리테이너　277
스냅 링　44, 278
스냅 핀　277
스냅링의 개구부　136
스미스 내구 한도 선도　277
스위프　306
스윙 액슬 스프링　306
스케일　262
스케일 제거　82
스크래치　263
스크랩 릴　263
스터드　302
스테비 링크　184
스테비라이저 아이　290
스테빌라이저　10, 290
스테이닝　291
스테인레스강　291
스테인레스강재 쇼트　291
스테인레스강재 스프링　291
스텝　294
스토크 릴 (선대)　294
스트레스 피닝　300
스트레스 피닝 초기응력　168
스트레이트 바 스테비라이저　296
스트레이트 스팬　294
스트레인게이지　296
스트로크　301
스파이럴 스프링　284
스파이럴 스프링 단부 굽힘　315
스파이럴 스프링 케이스　19
스팬　281

스페이서　281
스페이서 부시　281
스프레이 도장　284
스프링　284
스프링 강　289
스프링 경도　286
스프링 고리　287
스프링 고리 성형기　287
스프링 귀　286
스프링 귀 중심　286
스프링 길이　287
스프링 끝　286
스프링 끝부분 부품　109
스프링 너트　288
스프링 백　284
스프링 분리기　288
스프링 상수　285
스프링 선별기　288
스프링 시험기　289
스프링 와셔　187
스프링 웜기어　350
스프링 장착 길이　267
스프링 장착 높이　267
스프링 장착 방식　288
스프링 저울　284
스프링 좌면 연삭기　286
스프링 지수　287
스프링 축선　284
스프링 측정 장치　287
스프링 케이스　284
스프링 클립　285
스프링 클립 볼트　285
스프링 특성　285
스프링 특성 시험　285
스프링 판　179, 287
스프링 판 끝　180
스프링 판 두께　181
스프링 판 이방성　9
스프링 판 중심　180
스프링 핀 (롤 핀)　288

447

스프링 하 중량   336
스프링 하중   287
스프링 한계치   285
스프링 홀더   286
스프링 휨   285
스프링 힘   286
스프링용 강대   289
스프링용 선재   289
스프링용 양백   207
스프링용 인청동   226
스프링의 병렬 배열   220
스프링의 병렬조합   220
스프링의 알루미늄 피복   8
슬라이드 받침쇠   276
슬리브   276
습식 연삭   345
습윤 시험   160
승강   167
시계 스프링   343
시계용 태엽스프링   47
시동기 스프링   292
시료   283
시료 크기   262
시방서   282
시스템   307
시스템 공학   307
시안화 아연 도금   76
시정 조치   67
시퀀스 제어   266
시트   263
시트 스프링   263
시험 결과   316
시험 방법   316
시험 시간   316
시험 온도   316
시험 조건   315
시험 하중   316
시험편   316
시효   4
시효 경화   4

시효처리   4
신뢰성   249
신선 가공후 담금질 . 뜨임   146
신선 재료   93
실   263
실행 가능성 시험   120
실행 가능성 조사   120
실험 계획법   83
심   271
심 맞추기 장치   6
쐐기형 피치 툴   229
씨 (C) 형 동심 스냅 링   33
씨 (C) 형 스냅 링   33

아

아공석강   161
아날로그 계기   8
아니온 전착 도장   9
아랫방향 귀   93
아르키메데스 나선   12
아 - 버   11
아연 도금   356
아연 분말도료   356
아이 리 - 푸   115
아이 부시   115
아이부가 있는 겹판 스프링   180
아이어닝   172
아크 용접   12
아 - 크 하이트   11
아황산 가스 시험   302
안내 길이   143
안내 실린더   143
안내핀   143
안전 벨트용 스프링   261
안전 장치   261
안전 하중   261
안전표지   261
안정화 열처리   290
안지름   169
알루미나 그릿   8

알멘 스트립　7
알멘 테스트　7
알카리 시험　6
압력제어 회로　234
압연 강재　255
압연 방향　88
압연 선재　255
압축 강도　58
압축 스프링　57
압축 스프링 시험기　57
압축 시 (C) 형 코일 스프링　74
압축 아-크 코일 스프링　11
압축 응력　58
압축 인장 내구한도　57
압축 잔류 응력　249
압축 잔류응력　58
압축 직열조합 코일 스프링　266
압축 코일 스프링　150
압축 코일 스프링의 끝부분　106
압축인장 반복응력　57
압축토션 코일 스프링　57
압하율　256
압흔 형상　236
액슬 스프링　17
액슬 박스 스프링　16
액체 호닝　184
액추에이터　4
앤티로크 브레이크 시스템　10
양단 수축형 인장 코일 스프링　283
양단 슬라이드형 겹판 스프링　26
양단 슬라이드형 테이퍼 판 스프링　27
양단 연삭　26
양진 굽힘 응력 피로 한도　119
양진 응력　253
억제제　168
업셋 헤드　336
업셋 (축박음)　336
에너지 효율화　110
에디커렌트 센서　100
에리고 스프링　104

에스엔 (S-N) 곡선　277
에어 브레이크　5
에어 스프링　5
에어 스프링 현가 시스템　6
에치디디 (HDD) 서스펜션 스프링　148
에칭　112
에푸알피 (FRP) 판 스프링　122
엑스 (X) 선 응력 측정법　352
엑스 (X) 선 회절법　352
엠보싱　105
엥글라 도　110
여유 감김 (스파이럴 스프링)　282
여유권수　78
여유성　247
역원 고리　69
연마 모양 (흔적)　141
연마 (연삭) 기　140
연마가공　2
연마재　1
연마재 브라스팅　1
연마판　1
연삭　140
연삭 끝면　143
연삭 맞댐 끝　47
연삭 소손　31
연삭 숫돌　142
연삭 압력　141
연삭 절단　76
연성　95
연성 파괴　95
연소 가열로　56
연소 기관　56
연속 냉각 변태 곡선　39
연속 선재 공급 속도　222
연속 패턴팅　62
연속로　62
연속식 쇼트 피닝　63
연신　105
연신율　105
연질화법　208

449

열 가소성　317
열 전도율　316
열 팽창율　50
열간 가공　159
열간 굽힘　157
열간 단조　158
열간 성형　158
열간 성형 겹판 스프링　158
열간 성형 코일 스프링　158
열간 세팅　148
열간 압연　159
열간 압연 강대　159
열간 압연 봉강　200
열간 압연 환강　159
열변형　90
열용량　149
열응력　317
열전대　317
열전도　148
열처리　149
열처리한 스프링용 선　145
열충격 저항　251
열팽창　316
열풍 순환 로　5
염기성 아연도금　6
염수 분무 시험　262
염수 침적 시험　262
염욕 열처리　262
영구 변형　225
영구 신연　315
영구변형　224
예방 보전　234
예열　233
예열로　233
오른쪽 감기　253
오므린 원형 고리　309
오버피닝　219
오스템퍼링　14
오스테나이트　14
오스테나이트결정입도　14

오스테나이트계 스테인레스강　14
오스포밍　13
오실로그래프　216
오염 예방　234
오염 제거　79
오일 템퍼선　214
오일러 정리　113
오일러의 힘　112
오일씰용 링 스프링　214
오차 한계　112
오픈 후크　215
옥외 폭로 시험　216
온간 피닝　159, 343
온도 변환기　312
온도 분포　312
온도 센서　312
온도 제어　312
온도 - 시간 곡선　312
와류 탐상 시험　100
와이블 곡선　345
와이블 분포　345
와이어 가이드　348
와이어 릴　349
와이어 절단기　347
와이어 포밍 (선세공 스프링의)　348
와이어 하네스 크램프　348
와이어식 호스 클립　348
와인덥 강성　346
와인덥 토 - 크　346
와인딩 토 - 크　347
완전 담금질　133
완전 풀림　133
완충 스프링　271
완충기　271, 278
완충용 스프링　2
왈 응력 수정계수　343
외륜　217
외측 가이드　217
외측 멈춤 링　115
외측 코일 스프링　217

韓国語索引

외팔 보　35
외팔 보 길이　35
왼쪽 감기　181
용융 아연 도금　158
용융 주석 도금　158
용해　89
운동 에너지　176
워터 제트 절단가공　343
원격 온도계　311
원격측정　311
원심력　41
원심분리식 집진장치　77
원주 속도　45
원추 코일 스프링　151
원추면　59
원추형 미니블록 스프링　202
원추형 숫돌　1
원추형 스프링　60
원추형 압축 코일 스프링　60
원추형 인장 스프링　91
원통 코일 스프링　77
원통도　84
원통형 스프링　77
원통형 압축 코일 스프링　77
원통형 인장 코일 스프링　77
원통형 캠　20
원판 캠　88
원판 클러치　89
원핀 코일링 시스템　275
원형 고리　134
원형 단면 선 스프링　45
원형 단면 토션바　323
원형 선재　257
원형 코일 스프링　45
원호 캠　45
웜식 호스 크램프　351
위치 결정　231
위해요소　148
위험원　147
윗방향 귀　336

유 (U) 자형 단면　334
유 (U) 자형 스프링　334
유 (U) 형 고리　114
유 (U) 형 고리　187
유 (U) 형 고리　243
유기 고분자 재 스프링　216
유냉　214
유도로　166
유리 섬유 스프링　138
유리 섬유강화플라스틱 (GFRP) 스프링　138
유성 치차　229
유압 스프링　160
유압 프레스　160
유연성　125
유지 시간　155
유지 온도　155
유체 스프링　127
유체 클러치　126
유한 요소법　123
유화물　302
유효 감김 수　212
유효 변위　3
유효 스팬　101
육각 머리　152
육각단면형 토션바　153
육각봉강　152
음극 산화 방식 처리　38
응력　297
응력 구배　298
응력 균열　298
응력 대 크리프 속도 선도　301
응력 반복수 선도 (S-N 선도)　298
응력 범위　299
응력 부식 균열　297
응력 분포　90
응력 사이클　298
응력 사이클 주파수　298
응력 수정 계수　297
응력 완화　299

451

응력 완화 곡선　299
응력 완화 (이완)　248
응력 유기 마르텐사이트 변태　299
응력 제거　299
응력 제거 풀림　231, 300
응력 진폭　297
응력 집중 계수　297
응력 크리프 파단속도 곡선　301
응력 한계　183
응력-변위 선도　300
응력-변위-온도선도　300
응력비　299
응력제거 열처리　300
응력제거 풀림　295
이 (E) 형 스냅 링　99
이단 피닝　91
이단 피치 코일 스프링　95
이단 행정 이송 기구　92
이동 평균　204
이붙이 와셔의 이의 높이　149
이송 길이　121
이송 정밀도　120
이완 값　248
이완 방지 와셔　268
이완 시험　248
이젝터　101
이중 고리　91
이중 비틀림 스프링　90
이중 암 토션스프링　91
이중 코일 스프링　90
이중감김 각형 단면 비틀림 스프링　96
이중감김 롤 핀　50
이중감김 비틀림 스프링　95
이중선 비틀림 스프링　92
이중선 인장 스프링　92
이중선 조합 비틀림 코일 스프링　55
이차 감기　253
이차 담금질　294
이차 응력　264
이축단면 연삭기　92

이탈 하중　238
이형 단면 선스프링　209
이형 코일 스프링　173
익스팬더 스프링　114
인간공학　112
인간기계 시스템　193
인공 시효　12
인리치 가스　110
인발 가공성　93
인발 방향　88
인발 속도　93
인버터　172
인산염 피막 처리　226
인성　326
인쇄 회로　235
인쇄 회로 기판　235
인입식 맨드릴　252
인장 강도　314
인장 막대 스프링　93
인장 스프링　114
인장 스프링 성형기　315
인장 스프링 시험기　315
인장 스프링 코일링기　115
인장 시험　314
인장 압축 양용 코일 스프링　90
인장 응력　314
인장 잔류 응력　250
인장 코일 스프링　151
인장 코일 스프링의 끝부분　107
인장 풀림　314
인장 하중　314
인코더　106
인터 리프　170
인터로크 장치　171
일정하중 스파이럴 스프링　61
일정하중 스프링　61
일차 감기 (태엽 스프링)　235
임계 냉각 속도　73
임펠라식 블라스팅기　346
임펠러　345

입계   139
입계 부식   171
입계 파괴   170
입계균열   170
잉그리쉬 고리   110

자

자 판   18
자경성   265
자계 강도계   190
자기 변태   191
자기 변형율 법   191
자기 분리기   191
자동 제어   15
자동차용 변속기   15
자동화   15
자리감김   106
자리감김   109
자분 탐상 시험   191
자석 스프링   191
자연 균열   263
자유 각도   129
자유 감김 수   212
자유 길이   130
자유고   130
작동 길이   3
작동시킴   3
작업 연구   319
작용 응력   3
잔류 오스테나이트   251
잔류 응력   250
잔류 응력 분포   250
잔류 전단 변형   250
잔류자기   249
잔존 수소량   249
잠금 스프링   187
장경   346
장고꼴 압축 코일 스프링   160
장단 경 비   12
장착 각도   267

재결정 온도   245
재결정 풀림   245
재료   195
재료 결함   195
재료 시방서   195
재료 시험   195
재료 조직   195
재현성   249
저널 베어링   175
저먼 후크 (측면 원형고리)   138
저스트 인 타임   175
저온 처리   189
저온 취성   189
저온 풀림   188
저온로   189
저울 스프링   18
저탄소강   188
저하중 경도시험   188
저합금 강   188
저항 용접   251
저항식 연속 풀림로   62
적열 취성   246
적외선 건조장치   167
적층단면형 토션바   178
전 변형   326
전 신율   326
전 파괴   326
전 휨량   326
전개법   83
전기 도금   104
전기 아연 도금   103
전단   269
전단 각도   269
전단 강도   270
전단 변형   270
전단 응력   270
전단 하중   270
전달 하중   327
전도율   103
전부하   134

453

전성   192
전신도   105
전열면 열 부하   113
전장   326
전장판 (온길이 판)   133
전착 도장   104
전처리   234
전탈탄   325
전해 연마   103
절단 날   76
절삭 속도   76
절삭 저항   76
점성   342
점탄성   341
접근 검출기   237
접선 꼬리 끝   308
접시 스프링   60
접시 스프링 와셔   60
접시 스프링 적층 길이   182
접시 스프링 직열 적층   266
접시 스프링형 와셔   21
접시스프링 볼록면   64
접시스프링의 오목면   59
접점 스프링   62
접착력 시험   4
접촉 하중   61
접촉형 태엽 스프링   204
정 마찰   292
정규 분포   211
정기 점검   224
정기검사   77
정밀 측정   233
정변형   292
정상 크-립 한도   62
정응력   292
정적 시험   292
정전 도장   104
정지 게이지   209
정지 마찰 계수   50
정하중   292

제이 (J) 형 판 스프링 (에어서스펜션)   175
제진 강판   178
제트 (Z) 형 판 스프링   355
조도 측정   257
조립 라인   12
조립 링 스프링   207
조립기   13
조립지그   12
조미니 커브   175
조속기   139
조압연   256
조임 토-크   319
조정 와셔   4
조종자   215
조직 변태   301
조합 스프링   55
조화 분석   147
종치식 판스프링   187
종탄성계수   99
종횡비 (코일 스프링)   276
좌굴   30
좌굴 강도   30, 250
좌굴 길이   30
좌굴 응력   30
좌굴 하중   30
좌회전   181
주 보조형 테이퍼 판 스프링   332
주 응력   234
주.보조형 비선형 겹판 스프링   333
주강 쇼트   38
주파수 분석   130
중간 풀림   235
중공 스테비라이저   330
중공 토션바   331
중실 스테비라이저   279
중심   40
중심공   40
지그   175
지그재그 스프링   355

454

韓国語索引

지그재그 스프링 성형기　355
지렛대　182
지연 파괴　82
지점　133
지정 하중　282
지지 겹판 스프링　21
지지 스프링　37
직각도　225, 290
직열 스프링　289
직열 적중　290
직열식 코일드 웨이브 스프링　266
직접 담금질　87
직접비　87
진공 열처리　337
진공 증착　337
진공 풀림로　337
진동 발진기　341
진동 시험기　340
진동 응력　216, 339
진동 제어　340
진동 주기　224
진동 주파수　131
진동 피더　339
진동계　341
진동수　76, 130
진동식 연마기　341
진동식 이송장치　341
진동진폭 감쇠 장치　216
진원도　84
진직도　295
질량효과　195
질화　208
질화 깊이　82
질화강　208
질화층　208
집진장치　96
징크 크로메이트　355
짧은고리 밀착 토션 스프링　272

차

차대　15
차량　338
차량용 스프링　339
차축 하중　17
참퍼링 각도　24
챠-트식 기록계　43
처리 조건　329
처리능력　319
천공 펀치　227
천연 고분자재료 스프링　206
천이　328
천이 온도　328
철-탄소 평형 상태도　172
청열 취성　26
청정도　165
청정실　46
체결 볼트　45
체결 시험　45
체결 와셔　252
체적 팽창 계수　49
초경합금　147
초기 고장　168
초기 피로 균열　164
초음파 가공　334
초음파 두께 계기　335
초음파 세정　334
초음파 세정 장치　334
초음파 진동자　335
초음파 탐상 시험　335
초응력　168
초장력　168, 234
초탄성　238
초탄성　303
초하중　168
총 감김 수　325
최대 응력　196
최대 치수　196
최대 토-크　196

455

최대 편차　196
최대 하중　196
최소 예압　202
최소 응력　202
축 방향　16
축 방향 휨　16
축 하중　16
축선　16
축소 고리　246
충격 강도　164
충격 속도　163
충격 시험　164
충격 에너지　163
충격 응력　164
충격 파　271
충격 피로 시험기　163
충격 하중　163
취성　29, 129, 272
취성 파괴　29
취화　105
측면 고리　214
측면 원형 고리　134
측정 정밀도　198
측정방법　198
측정범위　198
측정 오차　198
측정장치　198
측정치　197
층상 탄화물　178
치수　87
침적 공정　163
침탄　37
침탄 질화　36
침탄층　38
침탄층 경도　38
침투 탐상 시험　184
침황처리　303

카

카바레이지　69
카바레이지 백분율　70
카본 퍼텐셜　36
카운터 싱크　69
카티온 전착도장　38
칼라 체크　55
캐드 (CAD)　58
캔틸레버 스프링　35
캘리퍼 게이지　33
캠　34
캠 기구　34
캠 버 (휨)　35
캠 제어　34
캠 조정　34
캠 종동자　34
캠 축　35
캠 푸로필　34
캠 형상　34
캠 (CAM)　58
커브드 와셔　74
커빙 롤　75
커빙 프레스　75
커트 와이어　75
커트 와이어 크기　75
커플링　69
컨네팅로트형 플랫 끝 중공 스태비라이저　61
컨네팅로트형 플랫 끝 중실 스태비라이저　61
컨커런트 공학　59
컴퓨터 수치제어　59
컴퓨터 제어 시스템　58
컴퓨터 통합생산　59
케이블 리와인드 스프링　33
코너 반경　67
코너 프레스　66
코발트 합금 스프링　49
코이닝　53
코일　50
코일 끝　50
코일 바깥지름　216

韓国語索引

코일 스프링　　50, 152
코일 스프링 담금질기　　241
코일 안지름　　170
코일 중심지름　　197
코일 평균 피치　　196
코일 평균지름　　196
코일간 최소 간격　　202
코일간 피치　　89
코일링　　51
코일링 공구　　53
코일링 롤러　　52
코일링 방식　　52
코일링 시작부　　168
코일링 아버　　52
코일링 핀　　52
코일링기　　52
코일스프링의 펼친길이　　84
코일의 감김방향　　88
코터　　69
코터 커플링　　69
코터 핀　　69
코팅　　48
코팅 내구 시험　　49
콘베이어식 가열로　　64
콘트롤 스프링　　64
콜게이트식 메시 스프링　　68
콜릿　　55
쿠션식 메시 스프링　　75
쿠션재　　75
크래드 법　　45
크로메이트처리　　44
크로스 오버형 고리　　73
크로스헤드　　73
크리프　　71
크리프 강도　　72
크리프 뜨임　　71
크리프 뜨임　　72
크리프 변형률　　72
크리프 속도　　71
크리프 시험　　72

크리프 파단 강도　　72
크리프 한도　　71
클램프　　45
클러치 디스크 스프링　　48
클러치 스프링　　48
클러치 휜 스프링　　48
클립　　46
클립 밴드　　46
클립 볼트　　47
클립 파이프　　47
킬드강　　176
타원 단면 선 코일 스프링　　217
타원 단면 선재의 장단지름　　203
타원 스프링　　135
타원 코일 스프링　　104
타원형 아크 스프링　　105
타임 스위치　　320
탄성　　103
탄성 계수　　203
탄성 변형　　102
탄성 스파이크　　102
탄성 영역　　102
탄성 한도　　102
탄성변형 에너지　　102
탄성비율　　102
탄성이론　　102
탄성좌굴　　101
탄소 섬유 강화 스프링　　41
탄소 섬유 스프링　　36
탄소 함유량　　36
탄소강　　36
탄소강 스프링　　36
탄소성　　103
탄창 스프링　　190
탄화물　　35
탈가스　　80
탈수소화　　81
탈지　　81
탈탄　　79
탈탄 깊이　　79

457

태엽　232
태엽 성형기　283
태엽 스프링 끝　107
태엽 스프링 케이스　347
태엽 스프링 케이스(시계용)　284
탠덤 액슬형 현가　308
테이퍼　308
테이퍼 가공(단조)　308
테이퍼 가공(압연)　309
테이퍼 가공(연삭)　308
테이퍼 길이　310
테이퍼 끝　309
테이퍼 롤링 베어링　309
테이퍼 롤링기　309
테이퍼 선재 압축코일 스프링　150
테이퍼 아이 형　310
테이퍼 원단면 미니블럭 코일 스프링　151
테이퍼 원단면 장고꼴 코일 스프링　151
테이퍼 코일 선스프링(미니부록 스프링 등)　51
테이퍼 키　309
테이퍼리프 스프링　310
테크니컬 매뉴얼　311
테크니컬 정비　311
테프론 코팅　311
텐션 링　315
템퍼 칼라　312
템퍼드 마르텐사이트　313
토글 스위치　320
토션 스프링　323
토션 스프링 상수　324
토션 스프링 성형기　324
토션바　322
토션바 끝부　322
토션바 스프링 상수　323
토션바 스프링의 유효 길이　325
토션바 장착부　322
토션바 축선　322
토션바의 전장　218

토-크　322
톱니꼴 와셔　321
톱니꼴 와셔의 톱니 높이　321
통계적 공정관리　293
통계적 품질관리　293
통과 게-지　138
투루스타이트　330
투사 각도　163
투사 속도　223
투사 에너지　25
투영 검사기　236
투자율　225
투핀 코일링 시스템　332
트랜스듀서　327
트레이링 판 스프링　327
트레이서빌리티　327
트리밍　329
특성 곡선　42
특성 요인도　43
특성치　43
틈새　46
틈새 게이지　121
틈새 부식　72
티탄 합금 스프링　320
티티티(T.T.T) 선도　320, 330

파

파괴 안전률　261
파괴 역학　129
파괴 하중　185
파단 시험　28
파단 연신율　28
파단 응력　28
파단 하중　28
파라보릭 판 스프링　220
파레토 도　221
파면　129
파면 섬유 흐름　122
파손　129
파쇄 강도　74

韓国語索引

파스너	116	편진 피로 한도	118
파스너 스프링	116	편진응력	238
파이프 두께	343	펼친 길이	84, 114
파인 블랭킹	122	평강	124
파텐팅	222	평강의 직선도	27
파형 스프링	51	평균 수리 시간	197
파형 스프링 와셔	344	평균응력	197
파형 와셔	344	평면도	125
파형 이붙이 와셔	344	평탄끝	221
판 게이지	270	평행도	85, 221
판 길이	180	평형 스프링	112
판 두께 표준 치수	270	표면 가공 경화	305
판 사이틈	136	표면 결함	304
판 스프링 매수	213	표면 경화	37
판 스프링 판폭	181	표면 경화 처리	304
판 스프링용 시험기	180	표면 경화강	38
판 캠	231	표면 경화깊이	37
판간 간격	215	표면 경화층	37
판간 마찰	131	표면 균열	303
판단법	320	표면 다듬질	304
판스프링 미끄럼 끝면	181	표면 보호	304
판스프링 부품	123	표면 부식	303
판스프링의 높이	217	표면 산화	304
판의 둥근 모서리	257	표면 상태	303
판형 나사 박음 고리	319	표면 장력	305
팔 각도	182	표면 조도	257, 304
팔 길이	182	표면 처리	305
팬형 잠금 와셔	116	표면 탈탄	303
팽창 계수	114	표면경도	304
퍼얼라이트	223	표점 거리	136
펀칭 가공성	238	표준 편차	291
펀칭 휘스	238	표준작업 순서 기록지	350
페라이트	121	표준치	291
페하값	225	표준화	292
편각	80	푸리에 해석	128
편석	264	푸샤형 로	238
편심	99	풀림	10
편심 캠	99	풀림 시간	10
편심 하중	186	풀림 토-크	336
편심하중	99	풀센터 고리	133

459

품질관리   240
품질관리도   240
품질보증   240
프랙토 그래피   129
프레스   233
프레스 굽힘시험법   233
프레스 퀜칭   233
프레스 템파   233
프로그레시브 겹판 스프링   237
프아송 분포   231
프아송비   231
플라스틱 스프링   230
플러그형 나사 박음 고리   318
플라스틱 코팅   230
플랙시블 생산 시스템   125
플랙시블 샤프트   126
플랜지   124
플랫 링 스프링   124
피그테일 끝 (돼지 꼬리 끝)   227
피닝 노즐   223
피닝 드럼   223
피닝 시간   224
피닝 효과   223
피드 롤러   120
피드 클러치   120
피드백 제어   121
피로   117
피로 강도   118
피로 강도 선도   118
피로 균열   117
피로 놋치계수   118
피로 변형   225
피로 수명   117, 212
피로 시험   119
피로 파괴   117
피로 파면   117
피로 한도   117
피막 두께   49, 179
피복선   70
피스톤 링 스프링   227

피아노 선   205
피아노선   226
피치   228
피치 수정   228
피치 오차   229
피치 있는 비틀림 코일 스프링   281
피치 툴   229
피치각   228
피치있는 가터 스프링   280
피크응력   222
피트   228
피트 깊이   228
피트 지름   228
펑거 스프링   122
펑거 와셔   122

하

하면 그라인더   189
하부 도장   235
하중   185
하중 시험   186
하중 저하   185
하중 증가   185
하중 축선   185
하중 측정   186
하중 측정 장치   186
하중 한도   185
하중 - 변위 곡선   186
하중 - 신률 곡선   185
하중저하율   248
한계 하중   183
한도 견본   27
한쪽 끝 스라이드 접촉형 겹판 스프링   214
함유량   62
합금 성분   6
합금강   6
합금강 스프링   7
합금원소   7
합성수지 스프링   306

핫세팅　159
항복　353
항복 신율　353
항복 응력　354
항복비　353
항복점　353
항복점 측정 시험　83
항복점의 인장강도 비　243
항탄성 스프링 합금　173
해지 장치　329
허용 범위　321
허용 응력　6
허용 품질 수준　2
허용 하중　185
허용 한계　321
헤어 스프링 (미소 태엽 스프링)　144
헤어핀 스프링　144
헤이 내구 한도 곡선　144
현가 스프링　305
현가용 겹판 스프링　305
현가장치　306
현미경 사진　200
현미경 조직검사　201
현미경 해석　201
형광 침투 탐상 시험　127
형상 기억 합금　269
형상 기억 합금 스프링　269
형판　313
호닝 가공　156
호브　155
호스 보호용 코일　157
호스 클램프　157
호칭 치수　209
호환성　170
홈　127
홈 단면　73, 142
화염 담금질　124
화학 니켈 도금　44
화학 성분　44
화학 증착법 (CVD)　44

화합물 층　56
확대 고리　110
확산 침투처리　86
확산 침투층　86
확산 풀림　155
확산침투처리　39
환강　257
환경 악화　111
환경 영향 평가　111
환경 응력 균열　111
환경공학　111
환경시험　111
환경영향 조사　111
활성 방식　3
황색 크로메이트 처리　353
회기 분석　247
회전 고리　306
회전 굽힘 피로시험　256
회전 마찰 계수　49
회전 방향　88
회전 속도　256
회전식 블라스트기　331
횡 방향력　328
횡방향 변위　179
횡방향 하중　179
횡치식 2 런 겹판 스프링　92
횡치식 판 스프링　329
횡탄성 계수　269
횡하중　329
후처리　232
후크 법칙　157
후킹 머신　157
휨량　80
흡수 에너지　2
흡습성　161
희석제　87
히스테리시스　161
히스테리시스 곡선　162
히스토그램　155
힌지 고리형 비틀림 코일 스프링　154

461

힘 127
힘의 모멘트 204

タイ語索引

ก

กฎของคาลิเปอร์ 33
กฎของฮุก 157
กฎข้อบังคับเกี่ยวกับเสียง 209
กดขึ้นรูป 128
กรรมวิธีความร้อนเชิงกล 317
กรรมวิธีซัลเฟอร์ไรซิ่ง 303
กระตุ้น 3
กระบวนการเก็บความเรียบร้อยสุดท้าย 123
กระบวนการขัด 2
กระบวนการโครเมตติ้งสีเหลือง 353
กระบวนการชุบแข็งแบบเย็น 54
กระบวนการเดี่ยว 21
กระบวนการทำความสะอาดผิว 25
กระบวนการทำความสะอาดหินเจียรใหม่ 93
กระบวนการทำให้ผิวแข็งในเหล็กคาร์บอนต่ำ 38
กระบวนการทำให้เหล็กบริสุทธิ์ขึ้น 172
กระบวนการที่ใช้ความร้อน 159
กระบวนการนำคาร์บอนกลับคืนสู่ผิวเหล็ก 245
กระบวนการพาเทนติ้งแบบต่อเนื่อง 62
กระบวนการม้วนกลับในการผลิตเพาเวอร์สปริง 253
กระบวนการล้าง 163
กระบวนการล้างผิว 254
กระบวนการหลัง 232
กระบวนการอบ 235
กระบวนการอุลตร้าโซนิค 334
กระบวนการเอาเศษเล็ก ๆ ออกจากผิวไซลินเดอร์โดยการขัด 156
กราฟจุลภาค 200
กราฟแสดงการออสซิเลชั่น 216
กราฟแสดงข้อมูล 263
กราฟแสดงความสัมพันธ์ระหว่างความเค้นและอัตราการเกิดครีป 301
กราฟแสดงโหลดและการยุบตัวของสปริง 42

กริทเหล็กหล่อ 38
กลไกที่เชื่อมเครื่องจักรกับเพลาขับ 232
กลไกป้อนคู่ 92
กล่องจุลทรรศน์สำหรับเช็คโครงสร้างเหล็ก 200
กล่องสปริงนาฬิกา 284
กล่องสำหรับการพัน 347
ก้อนเหล็กก่อนรีดขึ้นรูป 24
ก๊าซเฉื่อย 167
ก๊าซที่มีความเข้มข้นสูง 110
ก้านของทอร์ชั่นบาร์ 322
การกดขึ้นรูป 94
การกระจายของขนาดเม็ดเกรน 140
การกระจายของความเค้นตกค้าง 250
การกระจายของความเครียด 90
การกระจายของพอยซ์ซั่น 231
การกระจายของเม็ดช็อท 273
การกระจายของเม็ดโลหะที่คลอบคลุมผิวโลหะ / การโคเวอร์เรจ 69
การกระจายของอุณหภูมิ 312
การกระจายความเค้นคงที่ 335
การกระจายความถี่ 130
การกระจายปกติ 211
การกระจายเวย์บูล 345
การกลายสภาพด้านแมคคานิค 4
การกัดกร่อน 68
การกัดกร่อนของผิว 303
การกัดกร่อนของเส้นลวดจากการสปาร์คไฟฟ้า 349
การกัดกร่อนจากการเสียดสี 259
การกัดกร่อนที่ผิว 112
การกัดกร่อนเป็นหลุม 229
การกัดกร่อนระหว่างเม็ด 171
การกัดเซาะเป็นรู 224
การกำหนดตำแหน่ง 231
การกำหนดยิลด์พอยท์ 83
การเกิดการครีป 71

463

การเกิดขอบขณะตัด 31
การเกิดความแข็งเอง 265
การเกิดฟองอากาศของก๊าซในของเหลว 39
การเกิดออกซิเดชั่นที่ผิว 304
การแก้ไขพิทช์ 229
การโก่ง/ยุบตัวของสปริง 285
การขจัดก๊าซ 80
การขจัดครีบ 78
การขจัดคาร์บอนออกทั้งหมด 325
การขจัดจาระบี 81
การขจัดผิวสะเก็ด 82
การขจัดไฮโดรเจนออก 81
การขยายตัวของเม็ดเกรน 139
การขยายตัวของรอยแตก 70
การขยายตัวจากความร้อน 316
การขัดเงาด้วยไฟฟ้าและกรด 103
การขัดเงาโดยสายพาน 22
การขัดผิวโดยพ่นเม็ดเหล็ก 272
การขับเคลื่อน 4 ล้อในรถยนต์ 129
การขึ้นรูป 128, 336
การขึ้นรูปด้วยการเจาะรู 25
สปริงขึ้นรูปที่อุณหภูมิห้อง 54
การขึ้นรูปร้อน 158
การขึ้นรูปส่วนหัว 336
การขึ้นรูปเส้นลวด 348
การครีป (ความเครียดเพิ่มขึ้นเมื่อเวลาผ่านไป) 71
การคลายความเค้น 299
การคลายความเค้นโดยให้ความร้อน 300
การคลายตัวของความเค้น 248
การควบคุมกระบวนการโดยใช้สถิติ 293
การควบคุมการฟีดแบ็ค 121
การควบคุมการสั่นสะเทือน 340
การควบคุมคุณภาพ 240
การควบคุมคุณภาพโดยใช้สถิติ 293
การควบคุมแคม 34
การควบคุมซีเควนเชียล 266
การควบคุมแบบตัวเลข 87
การควบคุมแบบอัตโนมัติ 15
การควบคุมอุณหภูมิ 312

การคายความเค้น 299
การเคลื่อนที่ของลำแสงอิเลคตรอน 306
การเคลือบ 48
การเคลือบด้วยเทฟล่อน 311
การเคลือบโดยใช้ไฟฟ้าเป็นตัวนำ 38
การเคลือบโดยทางเคมี (ซีวีดี) 44
การเคลือบทองเหลือง 28
การเคลือบแบบกระจาย 86
การเคลือบแบบจุ่ม 87
การเคลือบป้องกัน 237
การเคลือบลดการเสียดสี 10
การเคลือบสปริงด้วยอลูมิเนียม 8
การเคลือบสังกะสีโดยใช้สารอนินทรีย์ 170
การเคลือบสีด้วยไฟฟ้า 104
การเคลือบสีผุ่น 232
การเคลือบสีพื้น 235
การเคลือบอิออนโดยใช้แผ่นอิเลคโทรด 9
การเคลือบไอน้ำ 226
การโค้งงอขณะยังไม่มีโหลด 27
การเจาะรู 69, 227
การเจียร์ 140
การเจียร์ชนิดป้อนต่อเนื่อง 319
การเจียร์ด้วยการสั่นสะเทือน 341
การเจียร์แบบเปียก 345
การเจียร์แบบแห้ง 94
การเจียร์ปลาย 106
การเจียร์ปลายทั้ง 2 26
การเจียร์สุดท้าย 123
การเจียรหยาบ 256
การเจียร์ให้เรียว 308
การชดเชยหินเจียร์ที่สึก 141
การช็อตด้วยเม็ดพลาสติค 230
การช็อตพีนนิ่ง 274
การชุบกรีนโครเมท 140
การชุบขั้นที่ 2 294
การชุบแข็ง 37, 146
การชุบแข็งความเครียดที่ผิว 305
การชุบแข็งด้วยน้ำ 344
การชุบแข็งด้วยน้ำมัน 214
การชุบแข็งด้วยไนโตรเจน 208

การชุบแข็งโดยทำให้วัตถุเย็นลงอย่างเร็ว 241
การชุบแข็งทันที 87
การชุบแข็งที่ผิว 304
การชุบแข็งที่ผิวเหล็กโดยก๊าซไนโตรเจน 137
การชุบแข็งแบบเหนี่ยวนำกระแสไฟฟ้า 166
การชุบแข็งภายใต้การกดอัด 233
การชุบแข็งภายใต้ศูนย์องศา 302
การชุบแข็งและการอบในกระบวนการรีฮีทดิ้ง 241
การชุบแข็งและอบหลังดึงขึ้นรูป 146
การชุบแข็งสมบูรณ์ 133
การชุบโครเมียม 44
การชุบซิงค์ชนิดด่าง 6
การชุบซิงค์ด้วยไฟฟ้า 103
การชุบด้วยดีบุก / ตะกั่ว 271
การชุบด้วยไฟฟ้า 104
การชุบด้วยวิธีเชิงกล 199
การชุบดีบุก 158
การชุบนิคเกิล 207
การชุบนิคเคิลโดยทางเคมี 44
การชุบนิคเคิลแบบคานิเกน-นิกเคิล 176
การชุบผิวด้วยซิงค์ไซยาไนด์ 76
การชุบพลาสติค 230
การชุบฟอสเฟต 226
การชุบฟอสเฟตด้วยไฟฟ้า 104
การชุบสังกะสี 158, 356
การเชื่อมต่อของแคมแมคคานิซึ่ม 34
การเชื่อมต้านทาน 251
การเชื่อมแบบอาร์ค 12
การเชื่อมปลายให้ติดกัน 32
การเซ็ทตัวของอนุภาคโลหะในของเหลว 200
การเซ็ทถาวร 315
การเซ็ทแบบร้อน 148
การเซ็ทเพื่อเพิ่มความคงทน 268
การดัดงอด้วยความร้อน 157
การดึงขึ้นรูปเย็น 53
การดึงแบบแห้ง 94
การตกตะกอน 232
การตรวจเช็คทางเทคนิค 310
การตรวจเช็คเป็นระยะตามกำหนด 224

การตรวจเช็คร่อยแตกด้วยสี 55
การตรวจสอบโครงสร้างผิวเหล็ก 201
การตรวจสอบรอยร้าวบนผิวโลหะ 129
การตัดขอบ 329
การตัดขอบด้านข้าง 274
การตัดด้วยการพ่นน้ำ 343
การตัดออก 269
การตัดออกด้วยวิธีเจียร์ 76
การต้านทานความชื้น 203
การต้านทานต่อสารเคมี 44
การต้านทานเนื่องจากความฝืด 133
การตีขึ้นรูปขณะร้อน 158
การเตรียมผิว 305
การเตรียมผิวของชิ้นงาน 234
การแตกของผิว 303
การแตกจากการบิด 323
การแตกจากการอบ 313
การแตกจากความเค้นชุบแข็ง 241
การแตกจากแรง 127
การแตกฉีกขาด 123
การแตกด้วยตัวเอง 263
การแตกตามแนวยาว 187
การแตกที่ขอบ 100
การแตกเนื่องจากความเค้นที่สัมพันธ์กับสิ่งแวดล้อม 111
การแตกเนื่องจากความล้า 117
การแตกในผลึกเนื่องจากความร้อน 170
การแตกเปราะ 46
การแตกร้าว 129
การแตกร้าวของเม็ดผลึกเนื่องจากแรงภายนอก 170
การแตกร้าวที่เกิดจากความเค้น 298
การแตกร้าวแบบกัดกร่อนจากความเค้น 297
การแตกหักจากการยืด / ดึง 95
การแต่งปลายด้วยจักรกล 190
การทดสอบกลางแจ้ง 216
การทดสอบการกระแทก 164
การทดสอบการกัดกร่อน 68
การทดสอบการกัดกร่อน, สนิม ในตู้ความชื้นสูง 160

465

การทดสอบการกัดกร่อนโดยก๊าซซัลเฟอร์ไดออกไซด์ 302
การทดสอบการกัดกร่อนโดยพ่นน้ำเกลือ 262
การทดสอบการกัดกร่อนที่อุณหภูมิสูง 153
การทดสอบการกัดกร่อนในสภาวะต่างๆสลับกัน 7
การทดสอบการคงที่ 292
การทดสอบการครีป 72
การทดสอบการโค้งงอ 22
การทดสอบการโค้งซ้ำ 125
การทดสอบการจับยึด 45
การทดสอบการจุดไฟสตาร์ท 282
การทดสอบการแช่น้ำเกลือ 262
การทดสอบการใช้งาน 183
การทดสอบการดัดกลับ 252
การทดสอบการตอบสนอง 265
การทดสอบการแตกด้วยแม่เหล็ก 100
การทดสอบการแตกหัก 28
การทดสอบการทนต่อแรงบิดในเหล็กเส้น 324
การทดสอบการทนทานต่อบรรยากาศ 345
การทดสอบการทำงาน 135
การทดสอบการบิด 332
การทดสอบการบิดกลับ 252
การทดสอบการยึดจับที่อุณหภูมิสูง 153
การทดสอบการยึดติดของผิวเคลือบ 4
การทดสอบการล้า 119
การทดสอบการวัดโหลดที่ลดลงภายใต้ความเครียดคงที่ 248
การทดสอบการสึกกร่อนในภาวะแห้งสลับเปียก 56
การทดสอบการอัดบุช 32
การทดสอบเกี่ยวกับสิ่งแวดล้อม 111
การทดสอบของวูเลอร์ 350
การทดสอบความแข็ง 147
การทดสอบความแข็งแบบนูป 177
การทดสอบความทนทาน 110
การทดสอบความทนทานของการเคลือบสี 49
การทดสอบความเป็นไปได้ 120
การทดสอบความผิดปกติเนื่องจากกระบวนการฮีทติ้ง 149

การทดสอบความล้าหมุนงอ 256
การทดสอบความสูง 150
การทดสอบความเหนียว 327
การทดสอบคุณสมบัติของวัตถุดิบ 195
การทดสอบคุณสมบัติในการเรื่องแสง 127
การทดสอบช็อตพีนนิ่งโดยแถบอัลเมน 7
การทดสอบซัลเฟอร์พรินท์ 302
การทดสอบซ้ำเป็นช่วง ๆ ภายใต้เงื่อนไขคงที่ 249
การทดสอบด้วยการจุ่มและทำให้แห้ง 95
การทดสอบด้วยการสอดของเหลว 184
การทดสอบด้วยคลื่นอุลตร้าโซนิค 335
การทดสอบตามช่วงเวลาที่กำหนด 77
การทดสอบในระดับมหภาค 190
การทดสอบแบบไม่เสียหาย 210
การทดสอบปมในเหล็กเส้น 177
การทดสอบปริมาณไฮโดรเจน 160
การทดสอบเป็นช่วงๆ 171
การทดสอบเพื่อการยอมรับ 2
การทดสอบรอยที่ผิวด้วยกรด 227
การทดสอบแรงกระแทกของแท่งบาก 212
การทดสอบแรงดึง 314
การทดสอบแรงดึงที่อุณหภูมิสูง 314
การทดสอบลักษณะการทำงานแบบไดนามิค 96
การทดสอบลักษณะเฉพาะของสปริง 285
การทดสอบสภาพการแตกโดยการลอกผิวบริเวณเชื่อม 223
การทดสอบเส้นลวด 349
การทดสอบหีบห่อ 351
การทดสอบโหลด 186
การทดสอบอัลคาไลน์ 6
การทดสอบอายุการใช้งานที่อัตราสูงสุด 2
การทนกรด 3
การทนความร้อน 148
การทนทานต่อการครีป 71
การทนทานต่อสนิม 260
การทวนสอบ 249
การทำความสะอาดด้วยอุลตร้าโซนิค 334
การทำคาร์บอนไนไตรด์ 36
การทำงานแบบอัตโนมัติ 15

タイ語索引

การทำช็อทพีนนิ่งหลายครั้ง 205
การทำเซ็ทติ้งขณะร้อน 159
การทำดีพดรออิ้ง 79
การทำไนโตรคาร์บิวไรซิ่ง 208
การทำปฏิกิริยาของออกซิเจนกับธาตุอื่นๆ 219
การทำพีนนิ่งขณะร้อน 159
การทำสเตรสพีนนิ่ง 300
การทำให้เกิดการกลายสภาพ 4
การทำให้แข็งตัวตามระยะเวลา 4
การทำให้คงตัว 290
การทำให้ชิ้นงานแข็งขึ้น 350
การทำให้นูนเป็นลวดลาย 105
การทำให้เป็นโซลิดโซลูชั่น 280
การทำให้เป็นมาตรฐาน 292
การทำให้เย็นโดยใช้ลมเย็น 5
การทำฮีททรีตเมนต์ 149
การทุบขึ้นรูปให้ปลายเรียว 308
การนำความร้อน 148
การนำไฟฟ้า 103
การบดอัด 75
การบวมของชั้นผิวชุบ 25
การบัดกรี 278
การบำรุงรักษาเชิงป้องกัน 234
การบำรุงรักษาทางเทคนิค 311
การบำรุงรักษาอย่างไม่มีกำหนดเวลา 335
การบิด 332
การบิดจากรูปปกติ 89
การบิดจากรูปปกติเนื่องจากการชุบแข็ง 90
การบิดจากรูปปกติเนื่องจากความร้อน 90
การเบคกิ้ง 18
การเบี่ยงเบนจากการขนาน 85
การเบี่ยงเบนจากรูปทรงกระบอก 84
การเบี่ยงเบนจากรูปทรงกลม 84
การเบี่ยงเบนมาตรฐาน 291
การแบนเรียบ 125
การประกันคุณภาพ 240
การประเมินทางสิ่งแวดล้อม 111
การประหยัดพลังงาน 110
การปรับความสูง 149
การปรับแคม 34

การปล่อยให้แข็งตัวเอง 5
การป้องกันการกัดกร่อน 3, 68
การป้องกันการกัดกร่อนทางอ้อม 222
การป้องกันการสึกแบบคาโธดิค 38
การป้องกันผิว 304
การป้องกันมลพิษ 234
การป้องกันสนิม 259
การป้อนเส้นลวด 347
การปั๊มโค้ง 75
การปั๊มอัดด้วยวิธีเชิงกล 199
การเปราะจากการอบ 312
การเปราะแตกจากไฮโดรเจน 161
การเปราะแตกที่อุณหภูมิต่ำ 189
การเปราะแตกอย่างช้าๆ 82
การเปลี่ยนเป็นภาวะมาร์เทนไซต์โดยความเค้น 299
การเปลี่ยนแปลง 328
การเปลี่ยนแปลงของเหล็กชุบแข็งเนื่องจากอุณหภูมิเปลี่ยนไป 240
การเปลี่ยนรูป 80
การเปลี่ยนรูปของโครงสร้าง 302
การเปลี่ยนรูปของวัตถุในกระบวนการแบบเย็น 296
การเปลี่ยนรูปจากการชุบแข็ง 241
การเปลี่ยนรูปถาวรจากการโหลดซ้ำ 225
การเปลี่ยนรูปที่อุณหภูมิเดียว 174
การเปลี่ยนรูปอย่างถาวร 224
การเปลี่ยนรูปอย่างถาวรของสปริง 267
การเปลี่ยนสถานะ 82
การเปลี่ยนหินเจียร์ 141
การแปรรูปสังเคราะห์ 12
การแปรสภาพเกินเวลาที่กำหนด 217
การแปลงสนามแม่เหล็ก 191
การผลิตชนิดจัดอุปกรณ์เป็นสัดส่วนเฉพาะ 39
การผิดรูปจากการคืนตัว 102
การผิดรูปชนิดไม่คืนตัว 230
การเผาด้วยระบบสุญญากาศ 337
การเผาโดยการใช้คลื่นความถี่สูง 166
การพ่นเม็ดโลหะมากเกินกำหนด 219
การพ่นสี 284

467

การพืนนิ่งขณะร้อน 343
การเพรสสลบขอบคม 66
การเพิ่มคาร์บอนพิเศษ 303
การเพิ่มปริมาณคาร์บอน 37
การเพิ่มโหลด 185
การแพร่กระจายของรอยแตก 71
การม้วน 51
การม้วนที่ทำให้ขดที่ติดกัน 279
การม้วนปลายแหนบ 74
การม้วนเริ่มต้นในการผลิตเพาเวอร์สปริง 235
การมาร์คสี 55
การยิงขัดผิว 1
การยิงขัดผิวด้วยเม็ดทราย 262
การยิงเม็ดโลหะ 2 ครั้ง 91
การยึดกัน 39
การยืด 105
การยืดตัวของส่วนที่แตก 28
การยืดหยุ่น 103
การยืดออกของการล้าตัว 353
การยืดออกทั้งหมด 326
การยุบตัวของสปริง 80
การยุบตัวของสปริงแผ่นเดี่ยว 275
การยุบตัวด้านข้าง 179
การยุบตัวที่มีผล 3
การยุบตัวในช่วงยืดหยุ่น 101
การยุบตัวในแนวแกน 16
การยุบตัวลงด้านข้างในสปริงขด 30
การเย็นตัวของชิ้นงานในเตา 65
การแยกออกไป 264
การรมดำ 26
การระดมสมอง 27
การระเหยด้วยระบบสูญญากาศ 337
การร้าวเนื่องจากการใส่โหลดซ้ำ ๆ 117
การรีดขึ้นด้านเดียว 53
การรีดขึ้นรูปปลายเทเปอร์ 309
การรีดโค้ง 75
การรีดที่อุณหภูมิห้อง 54
การรีดร้อน 159
การรีดหยาบ 257
การรีดหุ้ม 45

การเรียงกันเป็นตั้งของสปริงแผ่น 266
การเรียงซ้อนกันของสปริงแผ่น 220
การลบมุม 42
การละลาย 89
การลับแบบแห้ง 94
การลับผิวแบบเปียก 184
การล้างด้วยกรด 226
การล้าตัว 353
การเลื่อนของแกนม้วน 193
การเลื่อนตัวออกด้านข้าง 276
การวัดการคลายตัวของความเค้น 248
การวัดการยุบตัว 80
การวัดความหยาบ 257
การวัดค่าความเค้นโดยคลื่นเอ็กซ์เรย์ 352
การวัดโดยไม่สัมผัส 210
การวัดทางอ้อม 166
การวัดในระยะไกล 311
การวัดแบบไดนามิค 97
การวัดละเอียด 233
การวัดโหลด 186
การวัดอย่างปลอดภัยโดยใช้อุปกรณ์มากกว่า 1 ชนิด 247
การวิเคราะห์การถอดถอน 247
การวิเคราะห์ความถี่ 130
การวิเคราะห์ความผิดปกติ 119
การวิเคราะห์ความล้มเหลวเพื่อหาวิธีการแก้ไข 116
การวิเคราะห์ความสอดคล้อง 147
การวิเคราะห์ค่าที่เปลี่ยนไป 8
การวิเคราะห์คุณสมบัติการเป็นพลาสติก 230
การวิเคราะห์ผลกระทบด้านสิ่งแวดล้อม 111
การวิเคราะห์ผิวโดยกราฟจุลภาค 201
การวิเคราะห์ฟอร์เรียร์ 128
การวิเคราะห์แยกโดยเอ็กซเรย์ 352
การวิเคราะห์หาความสัมพันธ์เชิงปริมาณ 67
การศึกษากลไกของการแตก 129
การศึกษาความเป็นไปได้ 120
การศึกษาด้านเวลาและการเคลื่อนไหว 319
การส่งผ่าน 328
การสั่นของการบิด 325

タイ語索引

การสั่นโดยปราศจากแรงกระทำจากภายนอก 206
การสั่นสะเทือนจากแรงที่มากระทำ 127
การสั่นอย่างรุนแรงของสปริง 251
การสึกกร่อนจากการเสียดสี 131
การสึกกร่อนจากความฝืด 132
การสึกกร่อนจากร่องแตก 72
การสึกจากการเสียดสี 1
การสึกหรอของหินเจียร์ 141
การสูญเสียคาร์บอนที่ผิว 303
การสูญเสียคาร์บอนในโลหะ 79
การสูญเสียคาร์บอนเป็นส่วนๆ 222
การสูญเสียเชิงกล 199
การสูญเสียโหลด 185
การเสี่ยงต่ออันตราย 147
การเสียดทานภายใน 171
การเสียดทานสถิต 292
การเสียรูปทั้งหมด 326
การเสียหายจากการเสื่อมสภาพ 81
การเสื่อมลงในด้านสิ่งแวดล้อม 111
การหน่วง 78
การหมุนซ้าย 181
การโหลดแบบไดนามิค 97
การให้ความร้อนที่อุณหภูมิต่ำ 189
การอบครีบ 72
การอบคลายความเค้น 300
การอบคืนตัว 211
การอบที่อุณหภูมิต่ำ 188
การอบที่อุณหภูมิต่ำเพื่อป้องกันการเกิดอ็อกซิเดชั่น 29
การอบปลาย 278
การอบเพื่อไม่ให้เปลี่ยนรูป 295
การอบเพื่อลดความเค้นตกค้างหลังเชื่อมหรือการเปลี่ยนรูปถาวร 231
การอบฟิกซ์เจอร์ 124
การอบภายใต้การกดอัด 233
การอบแห้งที่อุณหภูมิเดียว 174
การอบให้คงรูป 313
การอบให้ได้ผลึกแบบมาร์เทนไซท์ 194
การอบให้ตกผลึกใหม่ 245

การอบให้เป็นรูปทรงกลม 283
การอบอ่อน 10
การอบอ่อนคลายความตึง 314
การอบอ่อนสมบูรณ์ 133
การออกแบบการทดลอง 83
การอัดบุช 31
การอุ่นให้ความร้อนก่อนปฏิบัติการ 233
การเอาครีบและสเกลออก 20
การเอาสิ่งปนเปื้อนออกไป 79
การเอียง 165
กำลังการโค้งงอ 126
กำลังการชนสูงสุด 74
กำลังของการครีบ 72
กำลังของการบิด 325
กำลังของการล้า 118
กำลังจำเพาะ 282
กำลังต้านการแตกเนื่องจากไดนามิคโหลด 97
กำลังต้านการยุบตัวด้านข้าง 30
กำลังที่ทำให้แตกเนื่องจากการครีป 72
กำลังในการตัด 270
กำลังล้าตัว 354
เกจ 136
เกจบล็อค 136
เกจรูปแหวน 254
เกจวัดความเครียด 296
เกจวัดความหนา 318
เกจสำหรับทำช่องว่างระหว่างชิ้นงาน 121
เกจสำหรับวัดความหนาแผ่น 270
เกจสำหรับวัดช่องว่างระหว่างแผ่น 270
เกลียวแบบอะคิมีเดียน 12
เกียร์ที่ใช้เฟืองเล็กหมุนรอบเฟืองใหญ่ 229
แกนของทอร์ชั่นบาร์ 322
แกนที่รับโหลด 185
แกนม้วน 52
แกนม้วนครึ่งวงกลม 265
แกนม้วนชนิดเคลื่อนที่เข้า-ออก 277
แกนม้วนชนิดมีร่องเกลียว 318
แกนม้วนชนิดหดเข้าออกได้ 252
แกนม้วนแบบโค้ง 23
แกนหลัก, รองของภาพตัดรูปไข่ 203

469

โก เกจ   138

ข

ขด   50
ขดของปลายแผ่นที่ 2   264
ขดที่อยู่ด้านใน   170
ขดปลายปิด   47
ขดรวม   325
ขดแรก   168
ขดสปริงส่วนที่ไม่ได้ใช้งาน   78
ขดสุดท้าย   109
ขดสุดท้ายของสปริงขด   106
ขดสุดท้ายของสปริงขดที่ไม่ได้เจียร์   106
ขนาด   87
ขนาดเกรนของออสเทไนท์   14
ขนาดของกลุ่มตัวอย่าง   262
ขนาดของความเค้น   297
ขนาดของเม็ดเกรน   140
ขนาดที่ใช้ในการระบุทั่ว ๆ ไป   209
ขนาดที่ระบุ   209
ขนาดเม็ดช็อท   75
ขนาดสำเร็จ   123
ขนาดสูงสุด   196
ขนาดโหลดที่รับได้   185
ขบวนการล้างด้วยกรด   227
ขยะอุตสาหกรรม   166
ของเสียเป็นศูนย์   355
ของเหลวที่ทำให้วัตถุเย็นตัว   65
ขอบกัด   201
ขอบเขตของการครีปที่ความเค้นสูงสุด   71
ขอบเขตของความผิดพลาด   112
ขอบเขตควบคุมด้านบน   336
ขอบเขตค่าเผื่อที่ยอมรับได้   321
ขอบที่ลบมุมแล้ว   42
ขอบมน   258
ขอบมนของเหล็กแผ่น   257
ขอบมีด   177
ขอบระหว่างเม็ดเกรน   139
ขอบเรียบ   277
ขั้นตอน   294

ขีดจำกัดของความเค้น   183
ขีดจำกัดของความล้ากดดึง   57
ขีดจำกัดของสัดส่วน   183
เข้าๆออกๆ   167

ค

ครอสเฮด   73
ครีบ   31
คลัชน้ำมัน   126
คลัทช์ชนิดแผ่น   89
คลิปโบลท์สำหรับแหนบสปริง   285
คลิปรัดแหนบ   28, 29, 46
คลิปรัดแหนบสปริง   285
คลื่นกระแทก   271
ความกว้างของหน้าตัดเหล็กเส้น   346
ความกว้างแผ่นแหนบ   181
ความขรุขระของผิว   304
ความแข็ง   147
ความแข็งแกร่งต้านการเปลี่ยนรูป   254
ความแข็งแกร่งที่ทนต่อความร้อนสูงได้   154
ความแข็งของผิวเหล็กที่ชุบแข็ง   38
ความแข็งของเม็ดช็อต   273
ความแข็งของสปริง   286
ความแข็งใจกลางแกน   66
ความแข็งชอร์   271
ความแข็งที่ผิว   304
ความแข็งบริเนล   29
ความแข็งร็อคเวล   255
ความแข็งแรงจากการกระแทก   164
ความแข็งแรงต่อการดึง   314
ความแข็งวิคเกอร์   341
ความคงทน   96
ความเค้น   297
ความเค้นกด   58
ความเค้นกดสูงสุด   58
ความเค้นการบิด   325
ความเค้นแกนเดียว   335
ความเค้นของการยุบตัวด้านข้าง   30
ความเค้นของการล้าตัว   354
ความเค้นคงที่   292

タイ語索引

ความเค้นโค้งงอ 23
ความเค้นจากการกระแทก 164
ความเค้นจากการคำนวณ 67
ความเค้นจากการเซ็ท 268
ความเค้นจากภายนอก 3
ความเค้นจุดแตกหัก 28
ความเค้นณโหลดที่ระดับต่างๆ 8
ความเค้นดึงตกค้าง 250
ความเค้นไดนามิค 98
ความเค้นตกค้าง 250
ความเค้นตกค้างจากการกด 58
ความเค้นตกค้างจากการชุบแข็ง 242
ความเค้นตรง 87
ความเค้นต่ำสุด 202
ความเค้นที่เกิดจากความร้อน 317
ความเค้นที่ผิวขณะโค้งขึ้นรูป 23
ความเค้นทุติยภูมิ 264
ความเค้นนิบ 208
ความเค้นในการตัด 270
ความเค้นปฐมภูมิ 234
ความเค้นภายใน 172
ความเค้นย้อนกลับ 253
ความเค้นรวม 326
ความเค้นเริ่มต้น 168
ความเค้นเริ่มต้นของการทำสเตรสพีนนิ่ง 168
ความเค้นแรงดึง 314
ความเค้นสลับกด-ดึง 57
ความเค้นสลับเป็นจังหวะ 238
ความเค้นสั่นสะเทือน 339
ความเค้นสั่นสะเทือนเป็นช่วง ๆ 216
ความเค้นสูงสุด 222
ความเค้นสูงสุดที่ทำให้หักขณะโค้งขึ้นรูป 23
ความเค้นสูงสุดที่ยอมรับได้ 6
ความเค้นอัดตกค้าง 249
ความเครียดการบิด 325
ความเครียดขณะชุบแข็ง 296
ความเครียดของการโค้งงอ 23
ความเครียดคงที่ 292
ความเครียดครีบ 72
ความเครียดเฉือนตกค้าง 250

ความเครียดตกค้างถาวร 225
ความเครียดในการตัด 270
ความเครียดบิดเบี้ยว 296
ความจุของความเย็นในสารหล่อเย็น 65
ความจุความร้อน 149
ความชื้นในบรรยากาศ 13
ความเชื่อถือได้ 249
ความดันเกจ 136
ความตรง 290, 295
ความต้านทานการกัดกร่อน 250
ความต้านทานการยุบตัว 250
ความต้านทานการสึกจากการเสียดสี 1
ความต้านทานเทอร์มอลช็อค 251
ความต้านทานสภาพบรรยากาศ 344
ความแตกต่างของความลึกของชั้นชุบแข็งเนื่องจากมวลและขนาดที่ต่างกัน 195
ความถี่ 130
ความถี่ของการเกิดรอบความเค้น 298
ความถี่ของการสั่น 131
ความถี่ของการสั่นในสปริงเดี่ยว 206
ความถี่เชิงมุม 9
ความถี่ที่ทำให้เกิดรีโซแนนซ์ 251
ความถี่ที่สอดคล้องกัน 305
ความถี่ในรอบ 76
ความถี่สัมพัทธ์ 247
ความทนทานต่อการล้า 117
ความเปราะ 29, 129, 272
ความเปราะจากกรด 3
ความเปราะจากการรมดำ 26
ความเปราะจากการอบแบบย้อนกลับไม่ได้ 173
ความผิดปกติของพิทช์ 229
ความผิดพลาดจากการวัด 198
ความผิดของวิลด์อัฟ 346
ความผิดที่ขอบ 101
ความผิดแนวขวาง 329
ความผิดในคลัตช์ 131
ความแม่นยำในการป้อน 120
ความแม่นยำในการวัด 198
ความยาวเกจ 136
ความยาวของคาน 35

471

ความยาวของตะขอด้านใน 182
ความยาวของบาร์ 19
ความยาวของแผ่นแหบบ 180
ความยาวของสปริง 287
ความยาวของสปริงขดที่ได้ 84
ความยาวขา 182
ความยาวโซลิด 279
ความยาวตัวนำ 143
ความยาวเทเปอร์ 310
ความยาวเทียบเท่า 112
ความยาวรวม 326
ความยาวรวมของทอร์ชั่นบาร์ 218
ความยาวอิสระในสปริงขดขณะไม่มีโหลด 130
ความยืดหยุ่น 125
ความยืดหยุ่นพิเศษ 303
ความร้อนจำเพาะ 282
ความรุนแรงการช็อต 273
ความเร็วการตัด 76
ความเร็วการหมุน 256
ความเร็วของการกระแทก 163
ความเร็วของการป้อนเส้นลวด 348
ความเร็วของล้อส่งป้อนเส้นลวด 347
ความเร็วเชิงมุม 9
ความเร็วในการพ่น 223
ความเร็วในการโหลด 187
ความเร็วรอบนอก 45
ความเร็วส่ง 222
ความล้มเหลวในระยะแรก 168
ความล้าจากการกัดกร่อน 68
ความล้าในวัตถุ 117
ความลึกของการเคลือบผิวด้วยไนไตรด์ 82
ความลึกของผิวเหล็กที่ชุบแข็ง 37
ความลึกรอยแตก 70
ความลึกหลุม 228
ความสัมพันธ์ระหว่างค่าโหลดและการยุบตัวของสปริง 185
ความสามารถในการชุบแข็ง 145
ความสามารถในการดึง 93
ความสามารถในการทำงาน 350
ความสามารถในการทำงานตามกำหนด 319

ความสามารถในการสับเปลี่ยน 170
ความสูงของพันที่บิด 149
ความสูงของสปริงแผ่นที่ซ้อนกัน 182
ความสูงของแหวนรูปฟัน 321
ความสูงเซ็ท 268
ความสูงโซลิด 279
ความสูงตะขอ 156
ความสูงประกอบ 267
ความสูงแหนบทั้งชุด 217
ความสูงอิสระในสปริงขดขณะไม่มีโหลด 130
ความเสียดทาน 131
ความเสียหายรวม 326
ความหน่วงของมวล 194
ความหนาของการเคลือบ 49
ความหนาของชั้นสี 179
ความหนาของหน้าตัดเส้นลวด 318
ความหนาแน่น 82
ความหนาแน่นของการช็อต 273
ความหนาผนัง 343
ความหนาผนังท่อ 330
ความหนาแผ่นแหบบ 181
ความหนืด 342
ความหนืดยืดหยุ่น 341
ความหยาบ 257
ความเหนียว 192, 326
คัมบัง 176
ค่าการลดลงของโหลดเปรียบเทียบกับโหลดเดิม 248
ค่าของอิงเลอร์ 110
ค่าคงที่สปริง 285
ค่าคงที่สปริงทอร์ชั่น 324
ค่าคงที่สปริงเรทของทอร์ชั่นบาร์ 323
ค่าความเค้นจากการดีไซน์ 83
ค่าความเค้นเฉลี่ย 197
ค่าความเค้นในสปริง 196
ค่าจ้างแรงงาน 178
ค่าจำกัดของการโก่ง/ยุบตัวของสปริง 285
ค่าเฉลี่ยของประชากร 231
ค่าเฉลี่ยเลขคณิตที่ใช้ในการวิเคราะห์เวลา-ซีรี่ย์ 204

ค่าเฉลี่ยเส้นผ่าศูนย์กลางขดวัดโดยแกนกลางแนวดิ่งหน้าตัด 197
ค่าใช้จ่ายทางอ้อม 165
ค่าใช้จ่ายในการผลิต 87
ค่าใช้จ่ายผันแปร 338
ค่าซีเอ็ม 48
ค่าตัวเลขแสดงคุณสมบัติ 43
ค่าที่เบี่ยงเบนไปจากค่าที่ดีไซน์ไว้ 196
ค่าที่วัดได้ 197
คานตรง 35
ค่าเผื่อจำกัด 321
ค่าเผื่อที่ยอมรับได้ 321
ค่าพีเอซ 225
ค่ามาตรฐาน 292
คาร์บอนโพเทนเชียล 36
คาร์ไบด์ 35
คาร์ไบด์ทรงกลมในเหล็ก 138
คาร์ไบด์ละเอียดเป็นชั้น 178
ค่าแรงทางอ้อม 165
ค่าแรงบิดทอร์กสูงสุด 196
ค่าสปริงเรทของแรงไดนามิค 97
ค่าสัมประสิทธิ์ของยังก์ 354
ค่าสัมประสิทธิ์สปริง 287
ค่าโหลดจากการดีไซน์ 82
ค่าโหลดสูงสุด 196
ค่าโหลดสูงสุดบนอุปกรณ์ขัน 237
คุณภาพหินเจียร์ 142
คุณลักษณะที่วัดได้และไม่ได้ 8
คุณสมบัติการเป็นแม่เหล็ก 225
คุณสมบัติการผลิตซ้ำ 249
คุณสมบัติการละลายน้ำ 344
คุณสมบัติของการเจาะ 238
คุณสมบัติของการสืบย้อนกลับ 327
คุณสมบัติความเป็นพลาสติกจากความร้อน 317
คุณสมบัติเชิงกล 199
คุณสมบัติที่ต้องควบคุม 63
คุณสมบัติที่ต่างกันของแผ่นสปริงตามทิศทาง 9
คุณสมบัติที่ต่างกันของสปริงตามปริมาณโหลดที่ใส่ 161
คุณสมบัติในการขึ้นรูป 128

คุณสมบัติในการดูดความชื้น 161
คุณสมบัติในการยืด 95
คุณสมบัติแม่เหล็กสัมพัทธ์ 247
คุณสมบัติยืดหยุ่นเทียม 238
คุณสมบัติเหล็กที่เปราะเนื่องจากกระบวนการแบบร้อน 246
คู่มือด้านทางเทคนิค 311
เครื่องเก็บฝุ่น 96
เครื่องเก็บฝุ่นชนิดใช้แรงหนีศูนย์ 77
เครื่องขัดแบบวงล้อ 346
เครื่องขัดผิวโลหะชนิดหมุน 331
เครื่องขึ้นรูป 128
เครื่องขึ้นรูปและชุบแข็ง 22
เครื่องคลายสปริง 288
เครื่องจักรสำหรับการประกอบ 13
เครื่องเจียร์ 140
เครื่องเจียร์ชนิดเป็นแผ่นขัด 89
เครื่องเจียร์ปลายสปริง 286
เครื่องเจียร์ปลายสปริงแบบ 2 เพลา 92
เครื่องเจียร์ลบมุม 42
เครื่องดัดสปริงซิกแซก 355
เครื่องตรวจวัดสนามแม่เหล็ก 190
เครื่องตรวจหารอยแตก 70
เครื่องตรวจหารอยร้าวด้วยผงแม่เหล็ก 191
เครื่องตัดเส้นลวด 347
เครื่องทดสอบการบิด 324
เครื่องทดสอบการสั่นสะเทือน 340
เครื่องทดสอบความล้าจากการกระแทก 163
เครื่องทดสอบความหนาสนามแม่เหล็ก 191
เครื่องทดสอบแรงกระแทกแบบชาร์ปี้ 43
เครื่องทดสอบสปริงกด 57
เครื่องทดสอบสปริงขดชนิดดึง 315
เครื่องทดสอบแหนบ 180
เครื่องทดสอบโหลดและความทนทานของสปริง 289
เครื่องทริป 329
เครื่องทำตะขอ 157
เครื่องทำสปริงลูป 287
เครื่องบันทึกชนิดแสดงผลเป็นแผนภูมิ 43
เครื่องผสม 4

473

เครื่องเพรส 233
เครื่องเพรสชนิดเลื่อนที่ละหลายแผ่น 205
เครื่องเพรสที่ใช้แรงเสียดทาน 132
เครื่องเพรสแบบคอม้า 177
เครื่องเพรสระบบไฮโดรลิค 160
เครื่องม้วน 52
เครื่องม้วนขดก้นหอย 283
เครื่องม้วนชนิดมีแกนม้วน 193
เครื่องม้วนสปริงขดชนิดดึงและตะขอ 315
เครื่องม้วนสปริงขดที่ชุบแข็งพร้อมกับเพรส 241
เครื่องม้วนสปริงดึง 115
เครื่องม้วนสปริงทอร์ชั่น 324
เครื่องม้วนหูแหนบ 115
เครื่องมือวัด 198
เครื่องยนต์ระบบเผาไหม้ 56
เครื่องยนต์ระบบเผาไหม้ภายใน 171
เครื่องแยกจัดเรียงสปริง 288
เครื่องแยกด้วยแม่เหล็ก 191
เครื่องรีดขึ้นรูปปลายเทเปอร์ 309
เครื่องวัดความแข็งด้วยโหลดต่ำ 188
เครื่องวัดจุดกลั่นตัว 85
เครื่องวัดโหลด 186
แคด 58
แคม 34, 58
แคมชนิดจาน 88
แคมแผ่น 231
แคมแมคคานิซึ่ม 34
แคมเยื้องศูนย์ 99
แคมรูปวงกลม 45
แคมสำหรับขึ้นรูปโค้ง 22
แคลมป์ 45
โครงรถยนต์ 15
โครงรถยนต์ประกอบกับอุปกรณ์รถยนต์ 43
โครงสร้างของวัตถุดิบ 195
โครงสร้างใจกลางแกน 66
โครงสร้างเหล็กระยะเพียร์ไลต์ 223

ง

งานที่ใช้แรงภายนอกทำการเปลี่ยนรูป 102
งานเสียจากการรีด 256

งานเสียบริเวณผิว 304
เงื่อนไขในการจัดการ 329
เงื่อนไขในการทดสอบ 315

จ

จงอย 207
จานต่อเพลา 69
จำนวนขด 213
จำนวนขดที่มีผลต่อค่าสปริงเรต 212
จำนวนขดที่ไม่ทำงาน 212
จำนวนขดในภาวะอิสระ 282
จำนวนของความเค้นสูงสุดที่ทนได้ (ก่อนแตก) 117
จำนวนของเสีย 212
จำนวนครั้งของการบิดจนหักขณะทำการทดสอบการบิด 213
จำนวนแผ่นของแหนบ 213
จำนวนรอบที่ทำให้หัก 212
จิ๊กสำหรับการประกอบ 12
จุดเชื่อมยึด 69
จุดเปลี่ยนการโค้งงอ 167
จุดม้วน 52
จุดล้าตัว 353
จุดส่งผ่าน 328
จุดหมุน 133, 182

ซ

ช่วงการควบคุม 64
ช่วงการเปลี่ยนรูปอย่างถาวร 230
ช่วงของการวัด 198
ช่วงของความเค้น 299
ช่องว่าง 46
ช่องว่างระหว่างแผ่น 136
ช่องว่างระหว่างแผ่นแหนบ 215
ช่องว่างระหว่างแหวนกันหลุด 136
ช็อตพีนนิ่งชนิดต่อเนื่อง 63
ช็อตเหล็กหล่อ 38
ชั้นของการชุบแข็งแบบไนไตรด์ 208
ชั้นของการแพร่กระจาย 86
ชั้นของการสารประกอบ 56

タイ語索引

ชั้นของพลาสติคที่ผิดรูป 179
ชั้นของออกไซด์ 219
ชั้นผิวเหล็กที่ชุบแข็ง 38
ชิ้นงานทดสอบ 316
ชุดของสปริงที่ประกอบแล้ว 206
ชุบแข็งโครเมียม 145
ชุบแข็งจากการขึ้นรูปเย็น 54
ชุบแข็งด้วยน้ำเกลือ 262
ชุบแข็งด้วยเปลวไฟ 124
ชุบซิงค์ชนิดกรด 3
แช็คเคิล 268

ซ

ซัสเพนชั่นชนิดแทนเด็ม 308
ซิงค์โครเมต 355
เซ็ตติ้งที่อุณหภูมิห้อง 54
เซ็ทติ้งโหลด 268
เซ็นเซอร์จับอุณหภูมิ 312
เซ็นเตอร์โบลท์ 40
เซ็นเตอร์พิน 40
เซ็นเตอร์สเปเซอร์ 41
เซอร์คลิป 44
โซลิดโซลูชั่น 279
ไซเลนเซอร์ 274
ไซเลนท์บุช 275

ด

ดัชนีโพรเซสคาพาบิลิตี้ 236
ดายบล็อค 86
ดิสค์เบรค 88
ดีลูป 78
ไดนามิคแดมเปอร์ 96
ไดอัลเกจ 85

ต

ตะขอ 156
ตะขอของสปริง 287
ตะขอครึ่งวงกลม 144
ตะขอชนิดมีร่องเกลียวแบบกด 318
ตะขอชนิดมีร่องเกลียวแบบแผ่น 319

ตะขอด้านข้าง 214
ตะขอด้านที่เอียง 165
ตะขอที่มีขนาดเส้นผ่าศูนย์กลางเล็กกว่าเส้นผ่าศูนย์กลางของขด 246
ตะขอที่มีเส้นผ่าศูนย์กลางนอกใหญ่กว่าขด 110
ตะขอบิด 332
ตะขอแบบเปิดในสปริงดึง 215
ตะขอรูปตัวยู 243
ตะขอรูปตัววีเหนือจุดศูนย์กลาง 339
ตะขอหมุนที่ติดกับสปริงดึง 306
ตะขอเหลี่ยมที่ปลายสปริงดึง 245
ตะปู(ราง)ยึด 102
ตั้งฉาก 225
ตัดขนาน 221
ตัดขอบด้วยเพชร 109
ตัว / สารให้ความเย็น 64
ตัวกระตุ้น 4
ตัวกันคลายของสปริงวาล์ว 338
ตัวกันสะเทือน 271
ตัวควบคุมความเร็วหมุนของเครื่องจักร 139
ตัวฉีด 101
ตัวเชื่อมต่อระหว่างเพลากับเฟรม 278
ตัวเชื่อมยึดเหล็ก 69
ตัวเซ็นเซอร์ 265
ตัวดัดตรงชนิดลูกกลิ้ง 255
ตัวดันเอาแคร์ปออกจากเครื่องเพรส 263
ตัวดูดซับการสั่นสะเทือน 340
ตัวถ่ายคลื่นอุลตร้าโซนิค 335
ตัวนำด้านนอก 217
ตัวนำด้านใน 169
ตัวนำทรงกระบอก 143
ตัวนำเส้นลวด 348
ตัวปรับศูนย์การหมุน 6
ตัวป้อนชนิดสั่นสะเทือน 341
ตัวป้อนสั่นสะเทือน 339
ตัวเปลี่ยนสัญญาณ 327
ตัวยึด 123
ตัวยึดขณะป้อนวัตถุดิบ 142
ตัวยึดสปริง 286
ตัวรัดเข้าด้วยกัน 116

475

ตัวรีดให้ตรง 295
ตัววัดบันทึกอุณหภูมิ 317
ตัววัดอุณหภูมิในระยะไกล 311
ตัวสั่นสะเทือน 341
ตัวห่อหุ้มหินเจียร 321
ตัวอย่างทดสอบ 283
ตัวอย่างแสดงขอบเขต 27
ตาชั่งสปริง 284
ตำแหน่งของมุมระหว่างตะขอที่ปลายทั้งสองของสปริงดึง 156
ตำแหน่งโซลิด 279
ตำแหน่งแนวระนาบ 124
เตาชนิดมีอุปกรณ์ต้านเปลวไฟไม่ให้สัมผัสชิ้นงานโดยตรง 252
เตาชนิดโรตารี่ 256
เตาชุบแข็ง 146
เตาที่ใช้ลมร้อนสูงหมุนเวียนภายใน 5
เตาแบบปิด 204
เตาเผา 135
เตาเผาชนิดต่อเนื่อง 62
เตาเผาชนิดที่ควบคุมบรรยากาศภายในได้ 237
เตาเผาชนิดปรับระดับก๊าซภายใน 64
เตาเผาชนิดผลักชิ้นงานเข้าอย่างต่อเนื่อง 238
เตาเผาชนิดสั่นต่อเนื่อง 63
เตาเผาที่ใช้สายพานลำเลียง 64
เตาเผาแบบเหนี่ยวนำกระแสไฟฟ้า 166
เตาเผาระบบเผาไหม้ 56
เตาเผาสำหรับกระบวนการแบทช์ 42
เตาเผาอุ่นให้ความร้อนต่ำ 233
เตาอบ 312
เตาอบชนิดถ่ายเทความร้อน 64
เตาอบชนิดสูญญากาศ 337
เตาอบต้านทานชนิดต่อเนื่อง 63
เตาอบแห้ง 94
เตาอุณหภูมิต่ำ 189
แตกจากความเปราะ 29

ถ

ถังบรรจุเม็ดเหล็ก 223
แถบ 301

แถบป้องกันการคลายขดของสปริงเกลียว 61
แถบเหล็กแผ่นสปริง 289
แถบอัลเมน 7

ท

ทนต่อสภาพบรรยากาศ 345
ทรูสไตท์ 330
ทฤษฎีการยืดหยุ่น 102
ทฤษฎีของยูเลอร์ 113
ทอร์คของวิลด์อัพ 346
ทอร์ชั่นบาร์ชนิดซ้อนทับขวางกัน 178
ทอร์ชั่นบาร์ชนิดท่อกลวง 331
ทอร์ชั่นบาร์ชนิดหน้าตัดสี่เหลี่ยม 245
ทอร์ชั่นบาร์หน้าตัดรูปหกเหลี่ยม 153
ทำให้แข็งโดยการตกตะกอน 232
ทำให้เม็ดเกรนละเอียดลงมาก 140
ทิศทางการขด 88
ทิศทางการดึง 88
ทิศทางการม้วน 88
ทิศทางการรีด 88
ทิศทางการหมุน 88
ทิศทางของการโค้งงอ 216
ทิศทางแนวเส้นราว 122
ทิศทางสู่ศูนย์กลางของแกน 16
เทศทอยด์ 113
เทเปอร์คีย์ 309
เทอร์โมคอปเปิล 317
เทอร์โมมิเตอร์ชนิดไบเมทัล 25
เทอร์โมสตัด 318
แท่งขัดรูปโดน 1
แท่งลวดเหล็กชนิดคาร์บอนสูง 153
แท่งหกเหลี่ยม 152
แท่งเหล็กกล้า 293
แท่งเหล็กม้วนร้อน 200
แท่งเหล็กรีดร้อน 159
แท่งแอนตี้โรลชนิดท่อ 330
แท่นม้วนเส้นลวด 349
แท่นรอง 263
แท่นสำหรับวางกองเหล็กลวด 294

ธ

ธาตุในโลหะผสม 7

น

นัทเกลียวสปริงแบบกด 239
น้ำมันชุบแข็ง 146
น้ำมันป้องกันการกัดกร่อน 68
น้ำหนักขณะไม่มีสปริง 336
น้ำหนักจำเพาะ 282
แนวแกน 16
แนวแกนของสปริง 284
แนวของการกัดกร่อน 68
แนวของความเค้น 298
โน-โก เกจ 209

บ

บริเวณขดปลายปิด 47
บริเวณที่ยึดหยุ่น 102
บริเวณร่องตัดบนแหนบ 143
บอร์ดวงจรไฟฟ้า 235
บาร์เรล 19
บาร์เรลแคม 20
บาร์เรลสปริง 20
บุช 31
บูทส์แบนด์จากลวดหน้าตัดรูปเหลี่ยม 246
เบนซ์มาร์ค 22
เบรคลม 5
เบ้ากลาง 40
แบนดิ้งเพรส 19
แบริ่งของปลอกสวมเพลา 175
แบริ่งในโรลเลอร์สำหรับขึ้นรูปปลายเทเปอร์ 309
แบริ่งลูกกลิ้ง 255
แบลคแลช 18
โบลท์ 26
โบลท์ยึดคลิป 47
โบลท์ยึดแคลมป์ 45
โบลท์ยึดแหวนรอง 187
ใบพัดปัด 345
ไบไนท์ 18

ไบเมทัล 24

ป

ปฏิกิริยา 244
ปม 176
ประชากร 231
ประสิทธิภาพเชิงกล 198
ปริมาณคาร์บอนในเหล็ก 36
ปริมาณจุ 62
ปริมาณที่รับโหลดได้ 185
ปริมาณไฮโดรเจนตกค้างในเหล็ก 249
ปลอกหุ้มเพลา 276
ปลั๊กอุดหู 99
ปลายกรวยชนิดมีห่วงร้อยโบลท์ 60
ปลายขด 50
ปลายขดชนิดเปิด 215
ปลายขดตรงและสัมผัสกับขดกลม 308
ปลายขดแบบหางหมู 227
ปลายของทอร์ชั่นบาร์ 322
ปลายของแผ่นแหนบ 180
ปลายของสปริงกดรูปก้นหอย 106
ปลายของสปริงดึงรูปก้นหอย 107
ปลายของสปริงพาวเวอร์ 109
ปลายโค้งของสปริงก้นหอย 315
ปลายเทเปอร์ 309
ปลายเทเปอร์ที่มีตะขอกลม 309
ปลายเปิดลักษณะแฟลต 215
ปลายแผ่นแหนบที่ไม่มีหู 276
ปลายสปริง 286
ปลายสปริงก้นหอยทอร์ชั่น 107
ปลายสปริงขด 109
ปลายสปริงที่เจียร์ครีบออกแล้ว 79
ปั้มละเอียด 122
เป็นเนื้อเดียวกัน 155
เป็นรอยเปื้อน 291
เปราะง่าย 105
เปอร์เซ็นต์การโคเวอร์เรจ 70
เปอร์เซ็นต์การยืด 105
ไปยึดคลิป 47

ผ

ผงเหล็กกล้า 293
ผงอลูมิเนียมออกไซด์ 8
ผนึก 263
ผลกระทบจากความเค้นของรู, ร่อง 211
ผลการทดสอบ 316
ผลจากการทำพีนนิ่ง 223
ผลึกแบบมาร์เทนไซท์ 194
ผิวโค้งนูนของสปริงแผ่นรูปกรวย 64
ผิวแตกจากการล้า 117
ผิวที่แตกร้าว 129
ผิวนูนของกรวย 59
ผิวเว้าเข้าของสปริงแผ่น 59
ผิวสเกลของเหล็ก 262
ผิวสำเร็จ 304
ผิวหยาบเป็นรอย 257
ผิวเหล็กชนิดพิเศษไม่เคลื่อนไหว 222
ผิวเหล็กที่ชุบแข็ง 37
แผ่นกั้นกาซในเตาอบ 137
แผ่นกันเลื่อน 11
แผ่นขัด 1
แผ่นชิมปรับขนาดช่อง 271
แผ่นตะขอชนิดมีเกลียว 156
แผ่นบาง 318
แผ่นแบบตัวอักษร 313
แผนผังการควบคุมคุณภาพ 240
แผนผังควบคุม 63
แผ่นฟิล์มออกไซด์เคลือบผิว 219
แผนภาพของเหล็กและคาร์บอน 172
แผนภาพความทนทานต่อความเค้น 298
แผนภูมิกำลังของการล้า 118
แผนภูมิของวูเลอร์ 349
แผนภูมิของสมิธ 277
แผนภูมิความเค้น - ความเครียด 300
แผนภูมิความเค้น - ความเครียด - อุณหภูมิ 301
แผนภูมิความสัมพันธ์ขอบเขตการล้าและความเค้น 118
แผนภูมิทีทีที 330
แผนภูมิพาเรโต 221

แผนภูมิพิกัดความล้ากู๊ดแมน 139
แผนภูมิรูปแบบการเกิดเหตุการณ์ 114
แผนภูมิรูปแบบการผิดปกติ 120
แผนภูมิแสดงค่าที่วัดได้ ณ ความถี่ต่าง ๆ 155
แผนภูมิแสดงเหตุและผล 43
แผนภูมิเฮย์ 144
แผ่นยาง 258
แผ่นรอง 184
แผ่นรับแรงดัน 234
แผ่นสปริง 287
แผ่นสปริงที่เรียงเป็นลำดับตามการยุบตัว 289
แผ่นสปริงยึดนัท 288
แผ่นแสดงดัชนี 165
แผ่นเหล็กกล้า 293
แผ่นเหล็กรับโหลด 276
แผ่นเหล็กรีดเย็น 54
แผ่นแหนบที่มีความยาวมากกว่าความยาวสแปน 133
แผ่นแหบบ 179

พ

พลังงานของการกระแทก 163
พลังงานจลน์ 176
พลังงานที่ใช้ในการทำความสะอาดผิว 25
พลังงานที่ถูกดูดซับไว้ 2
พาเทนติ้ง 222
พิกัดของการล้า, ความเค้นสูงสุด 118
พิกัดของการล้าความเค้น (สูงสุด=ต่ำสุด) 119
พิกัดของการล้าสลับ (ความเค้นสูงสุด-ต่ำสุด) 118
พิกัดของความล้าจากความเค้นโค้งงอ 119
พิกัดของความล้าจากความเค้นโค้งงอที่ค่าความเค้นสูงสุด=ต่ำสุด 119
พิกัดความทนทาน 109
พิกัดความเผื่อทางเรขาคณิต 137
พิทช์ 228
พินที่มีลักษณะเป็นขด 51
พินสำหรับกำหนดตำแหน่งของดาย 143
พื้นผิวเจียร์ของปลายสปริงกด 143
เพลาขับ 94

タイ語索引

เพลาในเครื่องม้วน 11
เพลายึดแคม 35
เพลายึดหยุ่น 126
โพรเซสคาพาบิลิตี้ 235
โพรเซสชาร์ท 236

ฟ

ฟีดคลัชท์ 120
ฟีดโรลเลอร์ 120
เฟรมหมุนสำหรับม้วนลวด 247
แฟกเตอร์ของรอยบากความล้า 118
แฟลตสปริง 125
โฟล์ชาร์ท 126
ไฟฟ้าจักรกล 199

ภ

ภาวะเสี่ยงต่ออันตราย 148

ม

มลภาวะทางอากาศ 5
ม้วนหูด้านบน 336
มาตรการแก้ไข 67
มาตรฐานในการดีไซน์ 83
มาร์คกิ้ง 193
มิเตอร์วัดความหนาด้วยคลื่นอุลตร้าโซนิค 335
มิลลิทารี่ แร็พเพอร์ 201
มุมการยุบตัว 80
มุมของการกระแทก 163
มุมของการเสียดทาน 9
มุมของขา 182
มุมโค้งงอ 22
มุมในการตัด 269
มุมในการประกอบ 267
มุมบาก 24
มุมบิด 331
มุมพิทช์ 228
มุมอิสระ 129
มุมเอียงของลวดหน้าตัด 276
เม็ดเกรนละเอียด 122
เม็ดเกรนหยาบขึ้นจากการโอเว่อร์ฮีท 139

เม็ดช็อต 272
เม็ดช็อตรูปกลม 258
เม็ดช็อท 75
เม็ดช็อทชนิดสเตนเลส 291
แมชชีนลูป 190
แม่พิมพ์ 86
โมดูลัสในการตัด 270
โมดูล 203
โมดูลัสของความยืดหยุ่น 203
โมดูลัสหน้าตัด 264
โมเมนต์ 203
โมเมนต์ของการโค้งงอ 23
โมเมนต์ของการบิด 324
โมเมนต์ของความเฉื่อยรอบแกนหมุน 204
โมเมนต์หน้าตัดสเตป 2 137
โมเมนต์ของแรง 204
ไมโครเซ็นเซอร์ 201
ไมโครมิเตอร์ 201
ไมโครมิเตอร์สำหรับวัดความลึก 82
ไม่แปรสภาพ 209

ย

ยานพาหนะ 338
ยูคลิป 114, 187
เยอรมันฮุก 138
เยื้องศูนย์ 100

ร

รถขนส่งควบคุมโดยคอมพิวเตอร์ 14
ร่อง 127
รอบของการสั่นสะเทือน 224
รอบของความเค้น 298
รอบหมุนต่อนาที 253
รอยขีดข่วน 263
รอยจากแม่พิมพ์ 86
รอยตำหนิของชิ้นงานจากกระบวนการผลิต 195
รอยแตก 70
รอยแตกเริ่มต้นจากการล้า 164
ระดับการเกิดสนิม 81
ระดับการขนาน 221

479

ระดับความมันเงา   81
ระดับความสะอาด   165
ระดับคุณภาพที่ยอมรับได้   2
ระนาบเดียวกัน   59
ระบบ   307
ระบบการควบคุมแรงดัน   234
ระบบการผลิตที่มีความยืดหยุ่นสูง   126
ระบบการผลิตที่รวมโดยคอมพิวเตอร์   59
ระบบการม้วน   52
ระบบการม้วนชนิดใช้ 2 พิน   332
ระบบคน-เครื่องจักร   193
ระบบจัสท์อินไทม์   175
ระบบซัสเพนชั่นในรถยนต์ที่ทำงานโดยลม   6
ระบบที่ควบคุมด้วยคอมพิวเตอร์   58
ระบบที่ใช้คอมพิวเตอร์ควบคุมค่าตัวเลข   59
ระบบเบรคแบบแอนตี้ล็อค   10
ระบบเมตริก   200
ระบบรับแรงสั่นสะเทือนในรถยนต์   306
ระบบสปริงเม้าทิ่ง   288
ระยะ   281
ระยะการกระแทก   31
ระยะการป้อน   121
ระยะการป้อนเส้นลวด   347
ระยะโก่งงอ   182
ระยะของช่องว่างจากปลายตะขอถึงปลายขด   156
ระยะแขนยึด   123
ระยะความสูงส่วนโค้งของแหนบ   35
ระยะจากผิวถึงชั้นที่เสียคาร์บอนในโลหะ   79
ระยะใช้งานของทอร์ชั่นบาร์   325
ระยะด้านใน   169
ระยะดึง   114
ระยะตรง   295
ระยะต่ำสุดระว่างขด   202
ระยะที่ได้   84
ระยะที่มีผล   101
ระยะประกอบ   267
ระยะพิชท์เฉลี่ยระหว่างแกนกลางขด   196
ระยะพิทช์ผันแปร   338
ระยะยุบตัวด้านข้าง   30

ระยะระหว่างขด   89
ระยะระหว่างขอบวงด้านในและตัวสปริง   89
ระยะระหว่างแผ่นแหนบก่อนยึดด้วยเซ็นเตอร์โบลท์   208
ระยะเวลาการเย็นตัว   65
ระยะเวลาคงที่   155
ระยะเวลาในการทดสอบ   316
ระยะเวลาในการทำพื้นนิ่ง   224
ระยะเวลาในการฮีตติ้ง   278
ระยะสโตรค   301
ระยะสโตรคของสปริง   3
ระยะสโตรคขึ้นลง   27
ระยะห่างระหว่างขด   280
รับโหลดเต็มที่   134
รัศมีของการโค้ง   23
รัศมีของส่วนโค้ง   243
รัศมีด้านใน   169
รัศมีนอก   115
รัศมีมุม   67
รางของเส้นลวด   348
รายละเอียดของแคม   34
ริปเปิ้ล   255
รีดขึ้นรูป   128
รูตรงกลางแผ่นแหนบ   40
รูปแบบการม้วนทีละ 1 อัน   275
รูปแบบวงจรไฟฟ้า   235
รูปแบบสำหรับเจียร์ผิว   141
รูปร่างของแคม   34
รูปร่างของปลายแท่ง   269
รูปร่างของปลายสปริงขด   109
รูปร่างตะขอ   157
รูสำหรับใส่พิน   268
เรเดียลโหลด   243
เรียงซ้อนขนานกัน   290
เรียงซ้อนเป็นลำดับตามการยุบตัว   290
แรง   127
แรงกดขณะเจียร์   141
แรงกดจุดสัมผัส   61
แรงกระแทกของแท่งบาก   212
แรงของยูเลอร์   112

タイ語索引

แรงเจาะ 238
แรงเจาะรู 25
แรงดึงตั้งต้นขณะไม่มีโหลด 234
แรงดึงบิด 332
แรงดึงเริ่มต้น 168
แรงดึงออก 238
แรงตัด 76
แรงดึงผิว 305
แรงทอร์คการพัน 347
แรงทอร์คเนื่องจากแรงเสียดทาน 133
แรงในแนวขวาง 328
แรงบิด 322
แรงบิดคลายตัว 336
แรงบิดดึงแน่น 319
แรงเริ่มต้นต่ำสุด 202
แรงสะท้อนกลับ 244
แรงเสียดทาน 132
แรงเสียดทานขณะคลายขด 132
แรงเสียดทานระหว่าง 2 แผ่น 131
แรงหนีศูนย์ 41
แรงหาศูนย์ 41
แรงโหลดของสปริง 286

ล

ลวดกลม 258
ลวดขดสำหรับป้องกันท่อ 157
ลวดขึ้นรูปเย็น 53
ลวดดึงแข็ง 145
ลวดถัก 27
ลวดที่ชุบแล้ว 70
ลวดที่ผ่านการดึง 93
ลวดเปนโน 205
ลวดแผ่นเรียบ 125
ลวดสปริงขึ้นรูป 128
ลวดสปริงชุบแข็งและอบแล้ว 145
ลวดเหล็กดึงแข็ง 145
ลวดเหล็กสำหรับสปริง 289
ลวดเหลี่ยม 289
ลวดอบน้ำมันสำหรับสปริง 214
ลอกออก 269

ล็อต 188
ล็อตนัมเบอร์ 188
ล้อปัดเม็ดเหล็ก 273
ล้อหินเจียร์ด้านล่าง 189
ลักษณะการครีปที่อุณหภูมิสูง 154
ลักษณะการเจียร์ร่องฟัน 267
ลักษณะการช็อต 274
ลักษณะของรอยบุ๋ม 236
ลักษณะของหัวโบลท์ 148
ลักษณะเชิงเส้น 183
ลักษณะปลายแคบเรียวลง 308
ลายแตกขนแมว 144
ลิมิตของความเค้นเมื่อเอาโหลดออก 102
ลิมิตสวิทช์ 183
ลูกกลิ้งสำหรับม้วน 52
ลูกแก้วสำหรับช็อตพีนนิ่ง 138
ลูกคู่ที่ปลายสปริงดึง 91
ลูกชนิดครอสโอเว่อร์ 73
ลูปเต็มวง 69
ลูปปิด 48
ลูปวงเอียง 9
โลหะที่ไม่มีคุณสมบัติเป็นแม่เหล็ก 211
โลหะผสมชนิดสปริงไอโซอิลาสติค 173
โลหะผสมเชฟเมมโมรี่ 269
โลหะผสมทองแดง, นิคเกิล, สังกะสีสำหรับผลิตสปริง 207
โลหะผสมฟอสฟอรัส-ทองเหลืองสำหรับทำสปริง 226
โลหะผสมเหล็กเส้นใยแก้ว 275
โลหะแผ่นบาง 271
ไลน์ประกอบ 12

ว

วัตถุดิบ 195
วัตถุดิบชนิดวัสดุประกอบ 56
วัตถุดิบเหล็กที่มีความสามารถเป็นแม่เหล็กได้สูง 121
วัสดุที่ใช้ในการทำฉนวนการสั่นสะเทือน 340
วิธีการติดปลาย 320
วิธีการทดสอบ 316

481

วิธีการทดสอบการโค้งงอด้วยการกด 233
วิธีการพัฒนาแหนบ 83
วิธีการวัด 198
วิธีการวิเคราะห์โดยการคำนวณ 123
วิธีการสร้างความเครียดและความเค้นด้วยสนามแม่
    เหล็ก 192
วิศวกรรมครีโอเจนิค 74
วิศวกรรมคอนเคอร์เรนท์ 59
วิศวกรรมระบบ 307
วิศวกรรมสิ่งแวดล้อม 111
วิศวกรรมอุตสาหการ 166
วี บล็อก 337
เวลาเฉลี่ยในการใช้งาน 197
เวลาเฉลี่ยในการซ่อม 197
เวลาในการขัดผิว 272
เวลาในการช็อตพีนนิ่ง 274
เวลาในการให้ความร้อน 149
เวลาในการอบ 313
เวลาในการอบอ่อน 10
เวอร์เนีย 339
เวอร์เนียไดอัล 339
เวิร์คเกอร์ 215

ศ
ศูนย์กลางของแผ่นแหนบ 180
ศูนย์กลางหูสปริง 286
ศูนย์ถ่วง 40

ส
สตาบิลิงค์ 184
สตาบิไลเซอร์ 10
สตาร์บิไลเซอร์ 19
สโตรคการสะท้อนกลับ 244
สนิม 259
สนิมซึมลึก 167
สแนปรีเทนเนอร์กันเลื่อน 278
สปริง 284
สปริงกด 57
สปริงกดชนิดขดทรงกระบอก 77
สปริงกดชนิดขึ้นรูปเย็น 53

สปริงกดชนิดติดตะขอที่ปลาย 2 ด้าน 93
สปริงกดชนิดเป็นตาข่าย 75
สปริงกดรูปกรวย 60
สปริงกดรูปไข่ 105
สปริงกดรูปโค้ง 74
สปริงกดรูปร่างบาร์เรล 20
สปริงกดรูปร่างรังผึ้ง 21
สปริงกดรูปอาร์ค 11
สปริงก้นหอย 342
สปริงกลมในแกสเกตริง 315
สปริงกล่องเพลา 16
สปริงกันกระแทก 2
สปริงกันสะเทือน 271
สปริงก๊าซ 137
สปริงการ์เตอร์ชนิดมีระยะระหว่างขด 281
สปริงกึ่งรูปไข่ 265
สปริงเกลียวชนิดไม่สัมผัส 210
สปริงขด 50, 152
สปริงขดก้นหอย 284
สปริงขดก้นหอยเสียดทาน 51
สปริงขดเกลียวชนิดลวดเหลี่ยม 209
สปริงขดขึ้นรูปร้อน 158
สปริงขดความฝืด 131
สปริงขดคู่ 90
สปริงขดชนิดดึงทรงก้นหอย 151, 152
สปริงขดชนิดปลายเรียว 51
สปริงขดชนิดเป็นซีรี่ 266
สปริงขดชนิดม้วนด้านขวา 253
สปริงขดชนิดรูปร่างไม่ปกติ 173
สปริงขดชนิดหมุนซ้าย 181
สปริงขดด้านใน 169
สปริงขดดึงชนิดปลายรูปกรวย 283
สปริงขดดึงชนิดลวดคู่ 55
สปริงขดตัวนอก 217
สปริงขดทำจากลวดขด 296
สปริงขดที่มีพิทช์คู่ 95
สปริงขดที่มีระยะพิทช์ผันแปร 338
สปริงขดประกอบด้วยลวด 2 ชนิด 90
สปริงขดรวมเป็นซีรี่ 266
สปริงขดรับแรงกดชนิดติดยาง 258

สปริงขดรับแรงกดทรงก้นหอย 150
สปริงขดรับแรงกดทรงก้นหอยที่มีปลายเรียว 150
สปริงขดรับแรงกดทรงก้นหอยที่มีระยะพิทช์ต่างๆ 150
สปริงขดรับแรงกดทรงก้นหอยรูปร่างบาร์เรลที่มีปลายเรียว 151
สปริงขดรับแรงกดทรงก้นหอยรูปร่างอาร์วกลาสที่มีปลายเรียว 151
สปริงขดรับแรงกดทรงก้นหอยอาร์วกลาส 160
สปริงขดรับแรงกดรูปเหลี่ยม 246
สปริงขดรับแรงดึงชนิดตะขอแยก 266
สปริงขดรูปกรวย 151
สปริงขดรูปไข่ 104
สปริงขดรูปคลื่น 51
สปริงขดรูปทรงกระบอก 77
สปริงขดรูปร่างบาร์เรลชนิดปลายเรียว 20
สปริงขดรูปวงกลม 45
สปริงขดเรียงอัดซ้อนกัน 220
สปริงขดลวดหน้าตัดรูปเหลี่ยม 246
สปริงขดสามชั้น 330
สปริงขดหน้าตัดรูปไข่ 217
สปริงขดหมุนชนิดปลายติดตะขอ 272
สปริงขดอัดที่มีแกนกลางเดียวกัน 206
สปริงขันเชื่อมท่อ 157
การขึ้นรูปที่อุณหภูมิห้อง 54
สปริงเข็มขัดนิรภัย 261
สปริงคลัชท์ 48
สปริงควบคุม 64
สปริงคานตรง 35
สปริงคาร์บอนไฟเบอร์ 36, 41
สปริงค้ำล้อ 37
สปริงโคบอลท์อัลลอย 49
สปริงจากวัสดุธรรมชาติที่เป็นโพลีเมอร์สูง 206
สปริงจุดสตาร์ท 292
สปริงชนิดขดต่อกัน 69
สปริงชนิดผสม 55, 56
สปริงชนิดลวดกลม 45
สปริงช่วย 15, 152
สปริงใช้ในการปรับสมดุลของโหลดที่ต่างกัน 112
สปริงซัสเพนชั่นสำหรับหัวอ่านฮาร์ดดิสค์ 148
สปริงซิกแซก 355
สปริงเซรามิค 41
สปริงดึง 114
สปริงดึงก้นหอยชนิดขึ้นรูปเย็น 53
สปริงดึงชนิดขดทรงกระบอก 77
สปริงดึงชนิดเส้นลวดเดี่ยว 275
สปริงดึงแบบลวดคู่ 92
สปริงไดอะแฟรม 85
สปริงทอร์ชั่น 323
สปริงทอร์ชั่นชนิดมีระยะระหว่างขด 281
สปริงทอร์ชั่นที่ปลายมีตะขอพับ 154
สปริงทอร์ชั่นบาร์ 323
สปริงทอร์ชั่นบาร์ที่ปลายเจียร่องฟัน 323
สปริงทอร์ชั่นบาร์หน้าตัดรูปวงกลม 323
สปริงทอร์ชั่นบาร์หน้าตัดรูปเหลี่ยม 323
สปริงทอร์ชั่นปลายแขนตรง 294
สปริงทอร์ชั่นปลายแขนเยื้อง 294
สปริงที่ความสัมพันธ์ของการยุบตัวและโหลดไม่เป็นเส้นตรง 237
สปริงที่ความสัมพันธ์ระหว่างโหลดและการยุบตัว 210
สปริงที่ใช้ความฝืดทำให้เกิดการหน่วง 132
สปริงที่ใช้ในการดึงขยาย 114
สปริงที่ทำจากโลหะ 200
สปริงที่ทำจากวัสดุอนินทรีย์ 169
สปริงที่ทำจากวัสดุอินทรีย์ที่เป็นโพลีเมอร์สูง 216
สปริงที่ทำหน้าที่ล็อค 187
สปริงที่เป็นตัววัดเทียบค่า 198
สปริงที่มีตรีสำหรับแผ่นคลัชท์ 48
สปริงที่มีคุณสมบัติรับโหลดเชิงเส้น 183
สปริงที่มีโหลดคงที่ 61
สปริงที่ไม่มีส่วนผสมของโลหะ 211
สปริงที่ไม่มีเหล็กเป็นส่วนผสม 210
สปริงที่วางซ้อนขนานกัน 205
สปริงนาฬิกา 343
สปริงน้ำมัน 127
สปริงบาลานซ์ 18

483

สปริงเบรค 28
สปริงเบาะรอง 263
สปริงแบ็ค 284
สปริงแบบแผ่นชนิดยึดทแยง 92
สปริงแบบแมกซี่บล็อค 195
สปริงแบบแรงบิด 2 ค่า 90
สปริงไบแอส 24
สปริงแปรง 29
สปริงแผ่นกรวย 60
สปริงแผ่นคลัช 48
สปริงแผ่นแซด 355
สปริงแผ่นต่อเนื่อง 2 ช่วง 333
สปริงแผ่นบางรูปแหวน 124
สปริงแผ่นเรียวต่อเนื่อง 2 ช่วง 332
สปริงพลาสติค 230
สปริงพินทรงกระบอก 143
สปริงเพลา 17
สปริงเพาเวอร์ 232
สปริงฟาสเทนเนอร์ 116
สปริงฟิงเกอร์ 122
สปริงเฟืองตัวหนอน 350
สปริงไฟเบอร์ 56
สปริงมอเตอร์ 204
สปริงมาตรฐาน 291
สปริงมินิบล็อค 202
สปริงมินิบล็อครูปกรวย 202
สปริงแม่เหล็ก 191
สปริงยาง 259
สปริงยึดเข็มแทงชนวนปืน 190
สปริงใยแก้ว 138
สปริงใยแก้วเสริมพลาสติก 138
สปริงรัดรางรถไฟ 243
สปริงรับแรงดึงชนิดขดก้นหอยคู่ 91
สปริงรับแรงบิด 2 ทาง 95
สปริงรับแรงบิด 2 ทางชนิดลวดเหลี่ยม 96
สปริงรับแรงบิดขาคู่ 91
สปริงรับแรงบิดลวดคู่ 92
สปริงรับแรงสั่นสะเทือน 305
สปริงรับลมอัด 5
สปริงรูปกรวย 60

สปริงรูปคลื่นขดชนิดวางขนานกัน 221
สปริงรูปตัวยู 334
สปริงรูปทรงกระบอก 77
สปริงรูปร่างที่หนีบผม 144
สปริงรูปวงรี 135
สปริงเรซินสังเคราะห์ 306
สปริงลวดเกลียวรูปไข่ 101
สปริงลวดตาข่าย 349
สปริงลานนาฬิกา 47
สปริงโลหะผสมเชฟเมมโมรี่ 269
สปริงโลหะผสมทองแดง 66
สปริงโลหะผสมไททาเนียม 320
สปริงโลหะผสมแร่นิลและทองแดง 66
สปริงวางเรียงซ้อนขนานกัน 221
สปริงสำหรับการเหวี่ยงของเพลา 306
สปริงสำหรับเครื่องจักร 11
สปริงสำหรับโครงรถยนต์ 43
สปริงสำหรับดึงสายไฟกลับ 33
สปริงสำหรับทำแม่แบบ 86
สปริงสำหรับปิดเปิดวาล์ว 337
สปริงสำหรับยานพาหนะ 339
สปริงสำหรับยึด 137
สปริงสำหรับรับแรงกด / บิด 57
สปริงสำหรับรางรถไฟ 21
สปริงสำหรับรีเลย์ 248
สปริงเส้นผม 144
สปริงเส้นลวด 349
สปริงหน้าสัมผัส 62
สปริงหมอนรองรถไฟ 26
สปริงหีบลม 21
สปริงเหล็กกล้า 293
สปริงเหล็กกันสนิม 291
สปริงเหล็กคาร์บอน 36
สปริงเหล็กผสมนิคเกิล 207
สปริงเหล็กมาร์เรจจิ้ง 193
สปริงเหล็กโลหะผสม 7
สปริงแหนบขึ้นรูปร้อน 158
สปริงแหนบแผ่นชนิดปลายเทเปอร์ 310
สปริงแหนบแผ่นชนิดสมดุล 306
สปริงแหวน 254

タイ語索引

สปริงแหวนที่เรียงซ้อนกัน 207
สปริงแหวนผิวผึด 132
สปริงแหวนลูกสูบ 227
สปริงอีลาสโตเมอร์ 103
สปริงอีลิโก 104
สปริงไฮโดรลิคเพรสเซอร์ 160
สปริงไฮโพลีเมอร์ 153
สเปค 282
สเปคของวัตถุดิบ 195
สเปคด้านเทคนิค 311
สเปเซอร์ 281
สเปเซอร์บุช 281
สภาพผิว 303
สภาพหินเจียร์ 141
สมรรถยะศาสตร์ 112
สลักกันเลื่อน 277
สลักกุญแจสปริง 288
สลักเกลียว 302
สลักชนิดลิ่ม 69
สลักในการม้วน 52
ส่วนโค้งนูน 64
ส่วนตัดรูปสี่เหลี่ยมคางหมู 176
ส่วนประกอบเป็นโลหะผสม 6
ส่วนหัวที่เจียร์รองฟัน 267
สวิตช์ปิด-เปิดแบบโยก 320
สวิตช์เวลา 320
สัญญลักษณ์บนโฟล์ชาร์ท 126
สัญลักษณ์ความปลอดภัย 261
สัญลักษณ์ในแผนภูมิ 140
สัมประสิทธิ์การขยายตัว 114
สัมประสิทธิ์การปรับความเค้นของวูล 343
สัมประสิทธิ์โกเนอร์ 139
สัมประสิทธิ์ของการขยายตัวของความร้อน 50
สัมประสิทธิ์ของการขยายตัวของปริมาตร 49
สัมประสิทธิ์ของความสัมพันธ์ 67
สัมประสิทธิ์ของแรงเสียดทาน 49
สัมประสิทธิ์ของแรงเสียดทานเลื่อน 50
สัมประสิทธิ์ของแรงเสียดทานสถิต 50
สัมประสิทธิ์ของแรงเสียดทานหมุน 49
สายเปียโน 226

สายรัดแหนบ 46, 142
สารขัด 1
สารเติมเพื่อป้องกันสารเคมีอย่างรวดเร็ว 227
สารทำละลาย 87
สารประกอบ 165
สารประกอบของซัลเฟอร์ 302
สารประกอบทรงกลมในเหล็ก 138
สารประกอบโลหะอื่นที่ปนอยู่ในเหล็ก 171
สารประกอบออกไซด์ของเหล็กในเหล็กเชื่อม 219
สารยับยั้ง 168
สารหล่อเย็นที่ใช้ในการชุบแข็ง 146
สารอนินทรีย์ 169
สิ่งเจือปน 164
สิ่งเจือปนที่ไม่ใช่โลหะ 211
สีของแผ่นออกไซด์บนผิวงานเกิดขึ้นหลังจากอบ 312
สีเคลือบ 322
สีผงสังกะสี 356
เส้นกราฟขณะม้วนซ้ำ 42
เส้นโค้งการคายความเค้น 299
เส้นโค้งความเค้น, ครีปและระยะเวลาการแตก 301
เส้นโค้งโจมีนี่ 175
เส้นโค้งซีซีที 39
เส้นโค้งโมเมนต์ 203
เส้นโค้งเวย์บูล 345
เส้นโค้งเวลา, อุณหภูมิ, การแปรรูป 320
เส้นโค้งแสดงการผิดรูป 80
เส้นโค้งแสดงโหลดและการยึดตัวของสปริง 185
เส้นโค้งแสดงโหลดและการยุบตัวของสปริง 186
เส้นโค้งอุณหภูมิ - เวลา 312
เส้นโค้งเอส-เอน 277
เส้นโค้งฮีสเทอร์เรซีส 162
เส้นผ่าศูนย์กลางของแกนม้วน 192
เส้นผ่าศูนย์กลางของขดกันหอย 283
เส้นผ่าศูนย์กลางของขดด้านใน 169
เส้นผ่าศูนย์กลางของแท่ง 255
เส้นผ่าศูนย์กลางของบาร์ 19
เส้นผ่าศูนย์กลางเฉลี่ยของขด 196

485

เส้นผ่าศูนย์กลางด้านในของขด 170
เส้นผ่าศูนย์กลางนอกของขด 216
เส้นผ่าศูนย์กลางภายใน 169
เส้นผ่าศูนย์กลางรู 155
เส้นผ่าศูนย์กลางเส้นลวด 347
เส้นผ่าศูนย์กลางหลุม 228
เส้นลวด 349
เส้นลวดเหล็กกล้า 293
เส้นอ้างอิง 247
เสียงกระทบกัน 244
เสียงดังเอี๊ยดๆ 290

ห

หน่วยทำความสะอาดด้วยอุลตร้าโซนิค 334
หน้าตัดของวัสดุ 73
หน้าตัดของวัสดุแบบแกน 73
หน้าตัดของวัสดุแบบมีร่อง 73
หน้าตัดที่เป็นร่องและขอบยกขึ้น 253
หน้าตัดแนวยาวของแหนบ 245
หน้าตัดพาราโบลิคในแหนบ 220
หน้าตัดรูปตัวยู 334
หน้าตัดรูปสี่เหลี่ยมคางหมู 329
หลุมบนผิวแผ่นเหล็ก 228
ห่วงกลมด้านข้าง 134
ห่วงกลมด้านปลาย 134
ห่วงที่อยู่ตรงกลางขด 133
ห้องคลีนรูม 46
หัวข้อกำหนดเฉพาะ 282
หัวพ่นเม็ดเหล็ก 223
หัวหกเหลี่ยม 152
หินขัด 142
หินเจียร์ชนิดมีผงเพชร 85
หีบบรรจุสปริง 284
หูของสตาบิไลเซอร์ที่ไม่ใส่บุช 115
หูสตาบิไลเซอร์ชนิดลูกบอล 19
หูสตาบิไลเซอร์ชนิดใส่บุชยาง 258
หูสปริง 286
หูเหล็กกันโคลง 290
หูแหนบ 115
หูแหนบชนิดม้วนลง 93

หูแหนบเบอร์ลิน 24
เหตุการณ์ 113
เหล็กกล้า 293
เหล็กกล้าคาร์บอนต่ำ 188
เหล็กกล้าทนความร้อน 149
เหล็กกล้ารอบสูง 153
เหล็กกล้าไฮเปอร์ยูเต็กตอยด์ 161
เหล็กกล้าไฮโปยูเต็กตอยด์ 161
เหล็กกันโคลง 290
เหล็กกันโคลงชนิดท่อกลวง 330
เหล็กกันโคลงท่อกลวงชนิดปลายข้อต่อลูกบอล 331
เหล็กกันโคลงท่อกลวงชนิดปลายสกรู 331
เหล็กกันโคลงแบบกลวงปลายแบนชนิดแท่งเชื่อมต่อ 61
เหล็กกันโคลงแบบตรง 296
เหล็กกันโคลงแบบตัน 279
เหล็กกันโคลงแบบตันชนิดปลายเป็นเกลียว 280
เหล็กกันโคลงแบบตันชนิดปลายเป็นบอลจอยท์ 279
เหล็กกันโคลงแบบตันปลายแบนชนิดแท่งเชื่อมต่อ 61
เหล็กกันสนิม 291
เหล็กแกร่ง 147
เหล็กคาร์บอน 36
เหล็กโครงสร้างซอร์ไบท์ 280
เหล็กโครงสร้างซีเมนไทท์ 39
เหล็กโครงสร้างเฟอร์ไรท์ 121
เหล็กโครงสร้างมาร์เทนไซท์ที่ผ่านการอบ 313
เหล็กช็อต 293
เหล็กใช้สำหรับชุบแข็งด้วยในโตรเจน 208
เหล็กแถบรีดร้อน 159
เหล็กบริสุทธิ์จากการถลุง 176
เหล็กแผ่นชนิดกว้าง 346
เหล็กแผ่นเทรสลิ่ง 327
เหล็กแผ่นบางที่ใช้รองรับการสั่นสะเทือน 178
เหล็กแผ่นยาว 124
เหล็กแผ่นรีด 255
เหล็กมาราจิ้ง 193
เหล็กเม็ดเกรนละเอียด 122

タイ語索引

เหล็กยูเทคทอยด์ 113	แหวนกั้นด้านใน 172
เหล็กริม 254	แหวนกั้นแบบแกนร่วมชนิด ซี 73
เหล็กรีดแผ่นชนิดหน้าตัดกลม 257	แหวนกันรั่วชนิดเป็นลอน 344
เหล็กแรงดึงสูง 154	แหวนกันเลื่อน 278
เหล็กโลหะผสม 7	แหวนด้านนอก 217
เหล็กโลหะผสมต่ำ 188	แหวนตัวใน 169
เหล็กโลหะผสมทนสนิม 259	แหวนปรับแต่ง 4
เหล็กสเตนเลสออสเทไนท์ 14	แหวนรองชนิดโค้ง 75
เหล็กสปริง 289	แหวนรองทรงกลม 283
เหล็กสปริงที่เจียร์ผิวนอกเพื่อลดขนาด 41	แหวนรองฟิงเกอร์ 122
เหล็กเส้นรีด 255	แหวนรองยึดโบลท์ 252
เหล็กเส้นสำหรับผลิตสปริงวาล์ว 338	แหวนรัด 142
แหนบชนิด FRP 122	แหวนรูปฟัน 321
แหนบชนิด เจ 175	แหวนแรงดึงสูง 154
แหนบชนิดที่ปลายเลื่อนได้ 181	แหวนล็อคชนิดฟันรูปพัด 116
แหนบชนิดม้วนหูข้างเดียว 215	แหวนล็อครูปฟันและคลื่น 344
แหนบชนิดมียางกันสะเทือน 181	แหวนสปริงกันน้ำมันรั่ว 214
แหนบชนิดมีหู 180	แหวนสปริงชนิดเป็นลอน 344
แหนบชนิดไม่สมมาตร 13	แหวนสปริงเบลล์วิลล์ 21
แหนบชนิดหลายแผ่นความยาวต่าง ๆ กัน 205	แหวนสปริงป้องกันการหลวม 268
แหนบชนิดหูเทเปอร์ 310	แหวนสปริงรูปกรวย 60
แหนบช่วย 15, 18, 244, 264	โหลด 185
แหนบที่มีความเค้นเท่ากัน 113	โหลดเกินกำหนด 218
แหนบแผ่นกลาง 170	โหลดของการกระแทก 163
แหนบแผ่นชนิดโพรเกรสซีฟ 237	โหลดของการดึง 314
แหนบแผ่นเดี่ยว 275	โหลดของเพลา 17
แหนบแผ่นแนวขวาง 329	โหลดของสปริง 287
แหนบแผ่นในการรับโหลด 192	โหลดคงที่ 224, 292
แหนบแผ่นบางชนิดซ้อนกันหลายชั้น 178	โหลดจำกัด 183
แหนบแผ่นรับแรงสั่นสะเทือน 305	โหลดด้านข้าง 179
แหนบสปริงชนิดปลาย 2 ข้างเลื่อนได้ 26	โหลดที่กำหนด 283
แหนบสปริงปลายเรียวชนิดเลื่อนได้ทั้ง 2 ข้าง 27	โหลดที่ใช้งาน 350
	โหลดที่ใช้ในการทดสอบ 316
แหนบหลัก 192	โหลดที่ทำให้เกิดการยุบตัวด้านข้าง 30
แหนบชนิดพาราโบลิค 220	โหลดที่ระดับต่างๆ 7
แหบบแผ่นแนวยาว 187	โหลดที่ส่งผ่าน 327
แหวนกั้น 252	โหลดในการตัด 270
แหวนกั้นชนิด ซี 33	โหลดในแนวแกนของขด 16
แหวนกั้นชนิด อี 99	โหลดในแนวขวาง 329
แหวนกั้นด้านนอก 115	โหลดในระดับที่ปลอดภัย 261

487

โหลดเป็นจังหวะ 126
โหลดเบื้องศูนย์ 99, 186
โหลดเริ่มต้น 168
โหลดสุดท้ายขณะหัก 185
โหลดสูงสุดที่รับได้ 28
ไหม้ 31

อ

องค์ประกอบการเกิดความเค้น 297
องค์ประกอบการแก้ไขความเค้น 297
องค์ประกอบทางเคมี 44
องค์ประกอบในการแก้ไข 67
ออสเทนไนท์ตกค้าง 251
ออสเทไนท์ 14
ออสเทมเพอริ่ง 14
ออสฟอร์มมิ่ง 13
อัตราการครีป 71
อัตราการคลายตัวของความเค้น 248
อัตราการดึง 93
อัตราการนำความร้อน 316
อัตราการผลิตสูงสุด 243
อัตราการยืด 105
อัตราการยืดหยุ่น 102
อัตราการเย็นตัว 65
อัตราการเย็นตัววิกฤติ 73
อัตราการระเหยของผิวร้อน 113
อัตราการรีด (ความหนาก่อนและหลังการรีด) 256
อัตราการหมุน 76
อัตราความเครียดคงที่ขณะทำครีบเทสต์ 62
อัตราความปลอดภัย 261
อัตราความล้มเหลว 116
อัตราโค้ง 74
อัตราส่วนของการล้าตัว 353
อัตราส่วนของความเค้น 299
อัตราส่วนของความสูงอิสระและเส้นผ่าศูนย์กลางเฉลี่ยในสปริงขด 276
อัตราส่วนของจุดล้าต่อกำลังการดึง 243
อัตราส่วนของพอยซ์ซัน 231
อัตราส่วนความกว้างและความหนาของหน้าตัดเส้น

วด 12
อายุการใช้งาน 100, 267
อายุการใช้งานของอุปกรณ์ 100
อาร์คไฮท์ 11
อำนาจแม่เหล็กตกค้างในวัตถุ 249
อิงลิชฮุก 110
อี โมดูลัส 99
อีเลคโตพลาสติกซิตี้ 103
อุณหภูมิการเย็นตัว 65
อุณหภูมิขณะชุบแข็ง 242
อุณหภูมิคงที่ 155
อุณหภูมิตกผลึกใหม่ 245
อุณหภูมิที่ใช้ในการทดสอบ 316
อุณหภูมิที่ส่งผ่าน 328
อุณหภูมิในการอบ 313
อุณหภูมิสูงเกินกำหนด 218
อุปกรณ์ 175
อุปกรณ์การเจาะรู 25
อุปกรณ์การทำความเย็น 66
อุปกรณ์การม้วน 53
อุปกรณ์จัดครีบ 79
อุปกรณ์ควบคุมด้วยมือ 145
อุปกรณ์ชนิดอนาล็อก 8
อุปกรณ์ตรวจจับการเข้าใกล้ 237
อุปกรณ์ตรวจจับรอยบนวัตถุดิบ 125
อุปกรณ์ดัด 76
อุปกรณ์ดัดชนิดหมุน 155
อุปกรณ์แต่งหินเจียร์ 94
อุปกรณ์ทำพิทช์ 229
อุปกรณ์ทำให้ตรง 295
อุปกรณ์ทำให้แห้งโดยรังสีอินฟราเรด 167
อุปกรณ์ที่ใช้ในการลบมุม 42
อุปกรณ์ที่ใช้ในการโหลด 106
อุปกรณ์ที่ใช้ประคองซีทของสปริงขด 109
อุปกรณ์นิรภัย 261
อุปกรณ์ในการทำให้โค้งงอ 24
อุปกรณ์ในการโหลด 186
อุปกรณ์บันทึกข้อมูล 114
อุปกรณ์ปรับความเร็วสำหรับรถอัตโนมัติ 15
อุปกรณ์ป้องกัน 237

อุปกรณ์เปลี่ยนกระแสตรงเป็นกระแสสลับ   172
อุปกรณ์เปลี่ยนอุณหภูมิ   312
อุปกรณ์พิทช์ชนิดลิ่ม   229
อุปกรณ์ลดการสั่น   78, 216
อุปกรณ์ล็อคแบบสลับ   171
อุปกรณ์วัดค่าของสปริง   287
อุปกรณ์วัดที่แสดงค่าพารามิเตอร์ของการสั่น
   199
อุปกรณ์วัดมุม   9
อุปกรณ์วัดโมเมนต์   203
อุปกรณ์วัดลักษณะ , ขนาดของวัตถุ   236
อุปกรณ์วัดแอมปริจูดของการสั่นสะเทือน   341
อุปกรณ์สำหรับการกลึง   331
อุปกรณ์สำหรับติดหินเจียร์เข้ากับเพลา   124
อุปกรณ์สำหรับป้อนชิ้นงาน   121
อุปกรณ์สำหรับยึดชิ้นงานในเครื่องกลึง   55
เอกสารขั้นตอนการทำงาน   350
เอ็ดดี้เคอเรนท์เซนเซอร์   100
เอ็นโค้ดเดอร์   106
แอมปริจูดของความเค้นณโหลดที่ระดับต่างๆ   8
โอเว่อร์ฮอลล์   218

ฮ

ฮาร์นเนสแคลมป์จับยึดเส้นลวด   348
โฮสคลิปจากลวดหน้าตัดรูปเหลี่ยม   246
โฮสคลิปชนิดเฟืองตัวหนอน   351
โฮสคลิปชนิดเส้นลวด   348

参 考 文 献

1) JIS 工業用語大辞典 第 5 版，(財) 日本規格協会
   Glossary of Technical Terms in Japanese Industrial Standards 5th edition
2) 日・中・韓・英 ばね用語辞典，1996，(社) 日本ばね工業会
   Technical Dictionary of Springs Jpanese-Chinese-Korean-English
3) 英和・和英 工業技術用語集，日外アソシエーツ
   Engineering Terms 50000
4) 文部省 学術用語集 機械工学編（増訂版），丸善
5) 自動車用語和英辞典，(社) 自動車技術会
   Japanese-English Dictionary for Automobiles
6) Encyclopedia of Spring Design, Spring Manufacturers Institute
7) Handbook of Spring Design, Spring Manufacturers Institute
8) Dictionary of Engineering Second Edition, McGRAW-HILL
9) Kenneth G. Budinski, Michael K. Budinski: Sixth Edition Engineering Materials Properties and Selection, Prentice Hall

6 カ国語 ばね用語事典

定価：本体 5,600 円（税別）

2004 年 9 月 10 日　第 1 版第 1 刷発行
2015 年 9 月 1 日　　　第 2 刷発行

編　　著　社団法人 日本ばね工業会

発 行 者　揖斐　敏夫

発 行 所　一般財団法人 日本規格協会
　　　　　〒108-0073　東京都港区三田 3 丁目 13-12 三田 MT ビル
　　　　　　　　　　　http://www.jsa.or.jp/
　　　　　　　　　　　振替　00160-2-195146

印 刷 所　株式会社 平文社
制　　作　有限会社 カイ編集舎

権利者との
協定により
検印省略

© Japan Spring Manufacturers Association, 2004　Printed in Japan
ISBN978-4-542-20309-9

● 当会発行図書、海外規格のお求めは、下記をご利用ください．
　営業サービスチーム：(03) 4231-8550
　書店販売：(03) 4231-8553　注文 FAX：(03) 4231-8665
　JSA Web Store：http://www.webstore.jsa.or.jp/